SECOND EDITION

ELECTROMAGNETICS OF TIME VARYING COMPLEX MEDIA

Frequency and Polarization Transformer

SECOND EDITION

ELECTROMAGNETICS OF TIME VARYING COMPLEX MEDIA

Frequency and Polarization Transformer

Dikshitulu K. Kalluri

CRC Press
Taylor & Francis Group
Boca Raton London New York

CRC Press is an imprint of the
Taylor & Francis Group, an **informa** business

CRC Press
Taylor & Francis Group
6000 Broken Sound Parkway NW, Suite 300
Boca Raton, FL 33487-2742

First issued in paperback 2019

© 2010 by Taylor and Francis Group, LLC
CRC Press is an imprint of Taylor & Francis Group, an Informa business

No claim to original U.S. Government works

ISBN-13: 978-1-4398-1706-3 (hbk)
ISBN-13: 978-1-138-37424-9 (pbk)

Library of Congress Cataloging-in-Publication Data

Kalluri, Dikshitulu K.
   Electromagnetics of time varying complex media : frequency and polarization transformer / Dikshitulu K. Kalluri. -- 2nd ed.
       p. cm.
   Includes bibliographical references and index.
   ISBN 978-1-4398-1706-3 (hardcover : alk. paper)
   1. Electromagnetism. 2. Plasma (Ionized gases) 3. Electromagnetic waves. I. Title.

QC760.K36 2010
537--dc22   2010006624

Visit the Taylor & Francis Web site at
http://www.taylorandfrancis.com

and the CRC Press Web site at
http://www.crcpress.com

This book is dedicated

to my wife Kamala

and to my children Srinath, Sridhar, and Radha

Contents

Part I Theory: Electromagnetic Wave Transformation in a Time-Varying Magnetoplasma Medium

Part II   Numerical Simulation: FDTD for Time-Varying Medium

Part III Application: Frequency and Polarization Transformer—Switched Medium in a Cavity

Foreword to the First Edition

I am pleased to write a Foreword to this book by Professor D. K. Kalluri, for he has been a friend since I first became aware of his work in 1992. In that year, Professor C. S. Joshi of the University of California Los Angeles and coworkers were undertaking a series of experiments, based on ideas developed by a number of us theorists, to study the frequency shift of radiation by means of moving ionization fronts in plasmas. This work was closely related to that which Professor Kalluri had been doing for many years, and so we arranged for him to spend the summer of 1992 in Berkeley. In this way a fruitful collaboration, and a good friendship, was initiated. The transformation of an electromagnetic wave by an inhomogeneous medium has been discussed in many books. The Doppler effect, which is a frequency change due to a moving boundary, is also a standard topic in many books. However, the transformation of the frequency of an electromagnetic wave by a general time-varying medium is rarely discussed. The frequency change in a non-moving medium is contrary to the usual experience of wave change. Because of his background in electrical engineering and electromagnetics, Professor Kalluri could weave, in a unique way, this topic of current research interest into a book on Electromagnetics of Complex Media varying in space and time. By using simple ideal models, he has focused on the effect of the time-varying parameters in conjunction with one or more additional kinds of complexities in the properties of the medium, and thus made the subject more accessible. The reader can be assured that Professor Kalluri is highly competent to write this book, for he has contributed original research papers on the frequency shifting of electromagnetic radiation using a transient magnetized plasma. He has expertise, and many publications, dealing with moving media, the use of Laplace transforms to study the effect of boundaries on transient solution, the generation of a downshifted wave whose frequency may be controlled by the strength of a static magnetic field (transformation of a whistler wave), and the turning off of the external magnetic field (which showed that an original whistler wave is converted into a wiggler magnetic field). In short, the book that follows is written by a highly competent author who treats important subjects. I hope the reader will enjoy it as much as I have.

Andrew M. Sessler
Berkeley, California

Preface

A simple electromagnetic medium is modeled by three electromagnetic parameters, permittivity (ε), permeability (μ), and conductivity (σ), all of which are positive scalar constants. A complex medium is one in which at least one of these parameters is not a positive scalar constant.

A number of excellent graduate-level books on the electromagnetics of inhomogeneous (space-varying) anisotropic (electromagnetic properties depend on direction of the fields) and dispersive complex media are available. Examples of such books include *Radio Wave Propagation in Ionosphere* by K. G. Budden, published by Cambridge University Press in 1966 and *Theory of Reflection* by John Lekhner, published by Martinus Nijhoff Publishers in 1987. Motivation for development of the theory, numerical simulation, and experimental work includes application in designing systems using (a) shortwave propagation in the ionosphere and (b) optics of anisotropic crystals. In all these studies the medium is assumed to have time-invariant but space-varying electromagnetic properties. The dominant effect of a system involving a time-invariant medium is conservation of the frequency of the signal, and the spatial variation of the medium induces a change in the wave number. The properties of the medium do change with time in these systems but very slowly compared to the period of the propagating wave, and the effects are of secondary importance. For example, the electron density profile in the ionosphere changes from night to day and hence the permittivity of the medium changes in time but the signal frequency is hardly affected by such a change. Hence the time-varying aspect of the medium is generally ignored.

With the availability of powerful ultrafast pulse sources, it is now possible to ionize a medium in a short time and thus change the permittivity of the medium in a time that is short compared to the period of the propagating wave. An idealization of this experimental situation is called flash ionization (sudden-switching) of the medium. The temporal electron density profile is approximated as a step profile with zero rise time. Across such a time discontinuity, the wave number k is conserved but induces a change in the frequency. This property can be used to create a frequency transformer. One can visualize a black box whose input frequency f_i is changed to an output frequency f_o by the time-varying medium.

Part I of this edition, Theory: Electromagnetic Wave Transformation in a Time-Varying Magnetoplasma Medium, contains a slightly enhanced version of the entire first edition of the book. The time-varying medium, under consideration, continues to be mostly the magnetoplasma because, at this time, it is the only complex medium whose electromagnetic properties can be changed significantly by ionization or deionization of the plasma medium or by switching on or off the background magnetic field. The relative permittivity ε_p can be changed by orders of magnitude. All the misprints in the first edition have been corrected.

Part II, Numerical Simulation: FDTD for Time-Varying Medium, concentrates on the development of an important tool for numerical simulation of electromagnetic wave interaction based on the finite-difference time-domain (FDTD) technique. It is based on expressing the constitutive relation between the current density **J** and the electric field **E** in the time domain as an auxiliary differential equation. While Maxwell's equations are discretized, as usual, by using the central difference approximations, the auxiliary differential equation is solved exactly over a time step by assuming constant values of the plasma parameters (ω_p, ω_b, ν) at the center of the time step. The algorithm, called "exponential time-stepping" in the literature, is therefore valid for large values of the plasma parameters, and the limitation of the usual technique based on central difference approximation of the derivative of **J**, which severely restricts the values of $\omega_b \Delta t$ and $\nu \Delta t$, is mitigated. Location of **J** at the center of Yee's cube yields the leap-frog step-by-step algorithm. Appendix K, which is a reprint of the paper published in 1998, discusses this aspect in detail. The author believes that the modified FDTD used in this book is appropriate for the time-varying magnetoplasma medium, as evidenced by the successful simulations given in Part II as well as Part III.

The first two parts assume unbounded medium or simple boundaries with free space. For the practical case of a finite rise time of the ionization process, the frequency-transformed traveling waves can escape the time-varying medium through the boundary, before the full interaction with the time-varying medium takes place. The experimental difficulty of the generation of a large volume of uniform plasma in a short time can be mitigated if the plasma is bounded in a cavity. In a cavity, the switching angle ϕ_0 is another controlling parameter. The input parameters, including the parameters describing the polarization of the input wave and the system parameters of the switched magnetoplasma medium influence the frequencies and the phases of the output waves. Thus a switched magnetoplasma in a cavity acts like a generic frequency and polarization transformer.

Part III, Application: Frequency and Polarization Transformer—Switched Medium in a Cavity, discusses this application. The author believes that this is an important application emerging from the theories developed in Part I. In particular, it is shown that one can use this technique to transform from 2.45 GHz, an inexpensive source in a consumer product (microwave oven), to 300 GHz terahertz radiation with enhanced electric field. Appendix Q, which is a reprint of a recently published paper, discusses this aspect. The frequency transformer that accomplishes this transformation has a whistler wave as the input wave and the time-varying medium in the cavity is created by switching off the ionization source, thus converting the magnetoplasma medium in the cavity to free space.

The author is grateful to Professor Alexeff for volunteering to write a brief account of the experimental work done so far on this topic by his and other groups. The author would achieve his objective if this book helps to stimulate additional experimental work by many, including himself. When the third edition is published Part IV will grow, indicating the progress in experimental work. Experiments based on switching in a cavity can be done more easily and can bring more researchers with moderate laboratory equipment into the area. The author is planning such experiments based on a significantly improved and easier technique in development.

This edition starts with an overview of the material covered in the book on the frequency transformation effect of the time-varying magnetoplasma medium. It contains 14 chapters and 17 appendices. The appendices (except Appendix A) are reprints (with minor changes) of papers published by the author and his doctoral students. This collection at one place should help a new researcher of the area to have comprehensive access to the subject. Chapters 11 and 12 deal with the concepts and serve the purpose of giving an overall picture before digging into the deeper exposure in Appendices I through Q.

Preface to the First Edition

After careful thought I have developed two graduate courses in electromagnetics, each with a different focus. One of them assumes a simple electromagnetic medium, that is a medium described by scalar permittivity (ε), permeability (μ), and conductivity (σ). The geometrical effect due to the shape and size of the boundaries is the main focus. There are many excellent textbooks, for example, *Advanced Electromagnetics* by Constantine A. Balanis, published by John Wiley & Sons in 1989, to serve the needs of this course.

Recent advances in material science suggest that materials can be synthesized with any desired electromagnetic property. The optimum properties for a given application have to be understood and sought after. To facilitate developing such appreciation in our graduate students, I have developed another course called the "Electromagnetics of Complex Media." The focus here is to bring out the major effects due to each kind of complexity in the medium properties. The medium is considered complex if any of the electromagnetic parameters ε, μ, and σ are not scalar constants. If the parameters are functions of signal frequency, we have temporal dispersion. If the parameters are tensors, we have anisotropy. If the parameters are functions of position, we have inhomogeneity. A plasma column in the presence of a static magnetic field is at once dispersive, anisotropic, and inhomogeneous. For this reason, I have chosen plasma as the basic medium to illustrate some aspects of the transformation of an electromagnetic wave by a complex medium.

An additional aspect of medium complexity that is of current research interest arises out of the time-varying parameters of the medium. Powerful lasers that produce ultrashort pulses for ionizing gases into plasmas can permit fast changes in the dielectric constant of a medium. An idealization of this process is called the sudden creation of the plasma or the sudden switching of the medium. More practical processes require a model of a time-varying plasma with an *arbitrary rise time*.

The early chapters use a mathematical model that usually has one kind of complexity. The medium is often assumed to be unbounded in space or has a simple plane boundary. The field variables and the parameters are often assumed to vary in one spatial coordinate. This eliminates the use of heavy mathematics and permits the focus to be on the effect. The last chapter, however, has a section on the use of the finite-difference time-domain method for the numerical simulation of three-dimensional problems.

The main effect of switching a medium is to shift the frequency of the source wave. The frequency change is contrary to the usual experience of wave change we come across. The exception is the Doppler effect, which is a frequency change due to a moving boundary. The moving boundary is a particular case of a time-varying medium.

The primary title is to indicate that the book will serve the needs of students who study electromagnetics as a base for a number of disciplines that use "complex materials."

Examples are electro-optics, plasma science and engineering, microwave engineering, and solid-state devices. The aspects of "Electromagnetic Wave Transformation by a Complex Medium" that are emphasized in the book are the following:

1. Dispersive medium

2. Tunneling of power through a plasma slab by evanescent waves

3. Characteristic waves in an anisotropic medium

4. Transient medium and frequency shifting

5. Green's function for unlike anisotropic media

6. Perturbation technique for unlike anisotropic media

7. Adiabatic analysis for modified source wave

All the above topics use one-dimensional models.

The following topics are covered briefly: (1) chiral media; (2) surface waves; and (3) periodic media. The topics that are not covered include (1) nonlinear media; (2) parametric instabilities; and (3) random media. I hope to include these topics in the future in an expanded version of the book to serve a two-semester course or to give a choice of topics for a one-semester course.

Problems are added at the end of the book for the benefit of those who would like to use the book as a textbook. The background needed is a one-semester undergraduate electromagnetics course that includes a discussion of plane waves in a simple medium. With this background, a senior under-graduate student or a first-year graduate student can easily follow the book. The solution manual for the problems is available.

The secondary title of the book emphasizes the viewpoint of frequency change and is intended to draw the attention of new researchers who wish to have a quick primer into the theory of using magnetoplasmas for coherent generation of tunable radiation. I hope the book will stimulate experimental and additional theoretical and numerical work on the remarkable effects that can be obtained by the temporal and spatial modification of magnetoplasma parameters. A large part of the book contains research published by a number of people, including the author of this book, in recent issues of several research journals. Particular attention is drawn to the reprints given in Appendices B through H. The book also contains a number of unpublished results.

The RMKS system of units is used throughout the book. The harmonic time and space variations are denoted by $\exp(j\omega t)$ and $\exp(-jkz)$, respectively.

Acknowledgments

The acknowledgments for this edition should be read along with the acknowledgments of the first edition since I continue to interact with most of those in that list. However, in the last few years there are two doctoral graduates who worked on time-varying media and shared with me the excitement of exploring this area of research. They are Dr. Monzurul Eshan and Dr. Ahmad Khalifeh. My present doctoral students Sebahattin Eker and Jinming Chen are deeply involved in this research. They are coauthors of some of the publications that are included in the appendices. Their help is thankfully acknowledged.

I am grateful to Professor Igor Alexeff of the University of Tennessee for writing Chapter 13 on experimental work.

A very special thanks to Jinming, who in his last semester of doctoral work undertook the task of critically reviewing the manuscript and used his considerable computer skills to format the text and figures to the specified standard.

Nora Konopka and Catherine Giacari at CRC encouraged me to submit a proposal for the second edition instead of publishing a separate research monograph on the Frequency and Polarization Transformer. The second edition has thus become a comprehensive work on Time-Varying Media. They along with other members of the team at CRC, Ashley Gasque and Kari Budyk, connected with bringing this book into its final shape, deserve my special thanks.

Acknowledgments to the First Edition

I am proud of my present and past doctoral students who have shared with me the trials and tribulations of exploring a new area of research. Among them, I would particularly like to mention Dr. V. R. Goteti, Dr. T. T. Huang, and Mr Joo Hwa Lee. A very special thanks to Joo Hwa, who in his last semester of doctoral work undertook the task of critically reviewing the manuscript and used his considerable computer skills to format the text and figures to the specified standard. I am grateful to the University of Massachusetts Lowell for granting the sabbatical during the spring of 1996 for the purpose of writing this book. The support of my research by Air Force Office of Scientific Research and Air Force Laboratories during 1996–1997 is gratefully acknowledged. I am particularly thankful to Dr. K. M. Groves for acting as my Focal Point at the Laboratory and for contributing to the research. The encouragement of my friends and collaborators, Professor Andrew Sessler of the University of California, and Professor Igor Alexeff of the University of Tennessee helped me a great deal in doing my research and writing this book.

My peers with whom I had the opportunity to discuss research of mutual interest include Professors A. Baños, Jr., S. A. Bowhill, J. M. Dawson, M. A. Fiddy, O. Ishihara, C. S. Joshi, T. C. Katsoules, H. H. Kuehl, S. P. Kuo, M. C. Lee, W. B. Mori, E. J. Powers, Jr., T. C. K. Rao, B. Reinisch, G. Sales, B. V. Stanic, N. S. Stepanov, D. Wunsch, and B. J. Wurtele and Drs. V. W. Byszewski, S. J. Gitomer, P. Muggli, R. L. Savage, Jr., and S. C. Wilks. The anonymous reviewers of our papers also belong to this group. I wish to recognize this group as having played an important role in providing us with the motivation to continue our research.

My special thanks go to Dr. Robert Stern at CRC for waiting for the manuscript and, once submitted, processing it with great speed.

Finally, I am most thankful to my wife Kamala for assisting me with many aspects of writing the book and to my children Srinath, Sridhar, and Radha for standing by me, encouraging me, and giving up their share of my time for the sake of research.

Author

Dikshitulu K. Kalluri, PhD, is professor of Electrical and Computer Engineering at the University of Massachusetts, Lowell. Born in Chodavaram, India, he received his B.E. degree in electrical engineering from Andhra University, India; a DII Sc degree in high-voltage engineering from the Indian Institute of Science in Bangalore, India; earned a master's degree in electrical engineering from the University of Wisconsin, Madison, and his doctorate in electrical engineering from the University of Kansas, Lawrence.

Dr. Kalluri began his career at the Birla Institute, Ranchi, India, advancing to the rank of professor, heading the Electrical Engineering Department, then serving as (dean) assistant director of the Institute. He has collaborated with research groups at the Lawrence Berkeley Laboratory, the University of California Los Angeles, the University of Southern California, and the University of Tennessee, and has worked for several summers as a faculty research associate at Air Force Laboratories. Since 1984, he has been with the University of Massachusetts Lowell, where he is coordinator of the doctoral program and co-director of the Center for Electromagnetic Materials and Optical Systems (CEMOS). As part of the center, he recently established the Electromagnetics and Complex Media Research Laboratory.

Dr. Kalluri, a fellow of the Institute of Electronic and Telecommunication Engineers and a member of Eta Kappa Nu and Sigma Xi, has published many technical articles and reviews.

Overview*

0.1 Introduction

This book deals with time-varying *complex media* and their applications. Some of these complexities are nonhomogeneity, dispersion, bi-isotropy, anisotropy, and combinations thereof. Medium complexity is expressed through constitutive relations. A comprehensive and rigorous presentation of constitutive relations for various kinds of complex media is given in Ref. [1].

An additional aspect of medium complexity that is of current research interest [2] arises out of the time-varying parameters of the medium. Transformation of the frequency of an electromagnetic wave (EMW) by a general time-varying medium is rarely discussed in books on electromagnetics, even though the Doppler effect (a frequency change due to a moving medium) is a standard topic in many books. The moving-medium problem is a particular case of a time-varying medium.

The frequency change in a nonmoving medium is contrary to usual experiences. The book deals with frequency shifts of several orders of magnitude that can be achieved by adding the complexity of time-varying parameters on top of the complexity of an anisotropic medium—in particular, the magnetoplasma medium. A time-varying magnetoplasma medium can act as a frequency transformer with a large frequency transformation ratio of the frequency of the output wave to the input wave (frequency upshifting) or a very small frequency transformation ratio (frequency downshifting). This remarkable effect is discussed in Part I (Theory: Electromagnetic Wave Transformation in a Time-Varying Magnetoplasma Medium) of the book by using simple ideal models for the geometry of the problem as well as constitutive relations of the magnetoplasma medium. The subject thus becomes more accessible and the focus is on the effect rather than on the achievement of high accuracy in the results. Accurate results can be obtained by using the finite-difference time-domain (FDTD) method for the numerical simulation of three-dimensional problems [3], which is discussed in Part II (Numerical Simulation: FDTD for Time-Varying Medium) of this book.

---

* © SPIE Sections 0.1 through 0.7 are based on a chapter written by Kalluri, titled 'Frequency-shifts induced by a time-varying magnetoplasma medium' pp. 245–266, in the book *Introduction to Complex Mediums for Optics and Electromagnetics*, edited by Weighlhofer, W. S. and Lakhtakia, A., SPIE, Bellington, WA, USA, 2003. With permission.

The chapter is organized as follows. Section 0.2 discusses the effect of a temporal discontinuity as opposed to a spatial discontinuity in the properties of the medium, and provides a simple explanation for the frequency shift caused by a temporal discontinuity. Section 0.3 discusses constitutive relations for a time-varying plasma medium. The sudden switching of an unbounded plasma medium is considered in Section 0.4. Section 0.5 deals with the more realistic problem of switching a plasma slab, and Section 0.6 discusses applications of frequency-shifting research that are under development. Section 0.7 explains the theory of wave propagation in a time-varying magnetoplasma medium and the possibility of effecting a big change in the relative permittivity of the medium by changing the ionization level or the background magnetic field. Such a big change in the relative permittivity leads to a large frequency shift. Section 0.8 provides an overview of Part III (Application: Frequency and Polarization Transformer—Switched Medium in a Cavity) and Section 0.9 provides an overview of Part IV (Experiments).

(a)

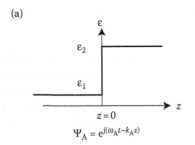

Spatial discontinuity in a time-
invariant medium

$$\Psi_I = \Psi_R = \Psi_T \mid_{z=0} \text{ for all } t$$
$$\therefore \ \omega_A = \omega_I = \omega_R = \omega_T$$

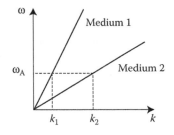

(b)

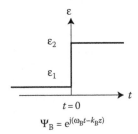

Temporal discontinuity in a space-
invariant medium

$$\Psi_B = \Psi_N \mid_{t=0} \text{ for all } z$$
$$\therefore \ k_B = k_1 = k_2$$

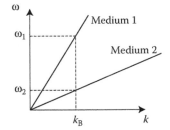

FIGURE 0.1 Comparison of the effects of temporal and spatial discontinuities.

0.2 Frequency Change due to a Temporal Discontinuity in Medium Properties

Let us consider normal incidence on a spatial discontinuity in the dielectric properties of a medium, of a plane wave propagating in the z-direction. The spatial step profile of the permittivity ε is shown at the top of Figure 0.1a.

The permittivity suddenly changes from ε_1 to ε_2 at $z = 0$. Let us also assume that the permittivity profile is time invariant. The phase factors of the incident, reflected, and transmitted waves are expressed as $\psi_A = e^{j(\omega_A t - k_A z)}$, where $A = I$ for the incident wave, $A = R$ for the reflected wave, and $A = T$ for the transmitted wave. The boundary condition of the continuity of the tangential component of the electric field at $z = 0$ *for all t* requires

$$\omega_I = \omega_R = \omega_T = \omega_A. \tag{0.1}$$

This result can be stated as follows: the frequency ω is conserved across a spatial discontinuity in the properties of the electromagnetic medium. As the wave crosses from one medium to the other in space, the wave number k changes as dictated by the change in the phase velocity, not considering absorption here. The bottom part of Figure 0.1a illustrates this aspect graphically. The slopes of the two straight lines in the ω–k diagram are the phase velocities in the two media. Conservation of ω is implemented by drawing a horizontal line, which intersects the two straight lines. The k values of the intersection points give the wave numbers in the two media.

A dual problem can be created by considering a temporal discontinuity in the properties of the medium. Let an unbounded medium (in space) undergo a sudden change in its permittivity at $t = 0$. The continuity of the electric field at $t = 0$ now requires that the phase factors of the wave existing before the discontinuity occurs, called a source wave, must match with phase factors, Ψ_N, of the newly created waves in the altered or switched medium, when $t = 0$ is substituted in the phase factors. This must be true for all values of z. Thus comes the requirement that k is conserved across a temporal discontinuity in a spatially invariant medium. Conservation of k is implemented by drawing a vertical line in the ω–k diagram, as shown in the bottom part of Figure 0.1b. The ω values of the intersection points give the frequencies of the newly created waves [4–7].

0.3 Time-Varying Plasma Medium

Any plasma is a mixture of charged particles and neutral particles. The mixture is characterized by two independent parameters for each of the particle species. These parameters are particle density N and temperature T. There is a vast amount of literature on plasmas. A few references of direct interest to the reader of this chapter are provided here [8–10]. These deal with modeling a magnetoplasma as an electromagnetic medium. The models are adequate in exploring some of the applications where the medium can be considered to have time-invariant electromagnetic parameters.

There are some applications in which thermal effects are unimportant. Such a plasma is called a *cold plasma*. A *Lorentz plasma* [8] is a further simplification of the medium. It is assumed that the electrons interact with each other in a Lorentz plasma only through collective space-charge forces and that the heavy positive ions and neutral particles are at rest. The positive ions serve as a background that ensures the overall charge-neutrality of the mixture. In this chapter, the Lorentz plasma is the dominant model used to explore the major effects of a nonperiodically time-varying electron density profile $N(t)$.

The constitutive relations for this simple model viewed as a dielectric medium are given by

$$\mathbf{D} = \varepsilon_0 \varepsilon_p \mathbf{E}, \tag{0.2}$$

where

$$\varepsilon_p = 1 - \frac{\omega_p^2}{\omega^2} \tag{0.3}$$

and

$$\omega_p^2 = \frac{q^2 N}{m \varepsilon_0}. \tag{0.4}$$

In these equations, q and m are absolute values of the charge and mass of the electron, respectively, and ω_p^2 is the square of plasma frequency proportional to electron density.

A sketch of ε_p versus ω is given in Figure 0.2. The relative permittivity ε_p is real-positive-valued only if the signal frequency is larger than the plasma frequency. Hence, ω_p is a cutoff frequency for the isotropic plasma. Above the cutoff, a Lorentz plasma behaves like a dispersive dielectric with the relative permittivity lying between 0 and 1.

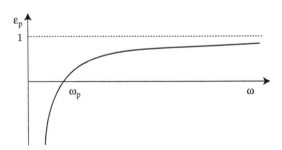

FIGURE 0.2   Relative permittivity ε_p versus angular frequency ω for a plasma medium.

The relative permittivity of the medium can be changed by changing the electron density N, which in turn can be accomplished by changing the ionization level. The sudden change in the permittivity shown in Figure 0.1b is an idealization of a rapid ionization. Quantitatively, the sudden-change approximation can be used if the period of the source wave is much larger than the rise time of the temporal profile of the electron density. A step change in the profile is referred to in the literature as sudden creation [2,5] or flash ionization [2].

Experimental realization of a small rise time is not easy. A large region of space has to be ionized uniformly at a given time. Joshi et al. [11], Kuo [12], Kuo and Ren [13], as well as Rader and Alexeff [14] developed ingenious experimental techniques to achieve these conditions and demonstrated the principle of frequency shifting using an isotropic plasma (see Part IV of this book). One of the earliest pieces of experimental evidence of frequency shifting quoted in the literature is a seminal paper by Yablonovitch [15]. Savage et al. [16] used the ionization front to upshift the frequencies. Ionization front is a moving boundary between unionized medium and plasma [17]. Such a front can be created by a source of ionizing radiation pulse, say, a strong laser pulse. As the pulse travels in a neutral gas, it converts it into a plasma, thus creating a moving boundary between the plasma and the unionized medium. However, the ionization-front problem is somewhat different from the moving-plasma problem. In the front problem, the boundary alone is moving and the plasma is not moving with the boundary.

The constitutive relation (Equation 0.2), based on the dielectric model of a plasma, does not explicitly involve the current density $\mathbf{J}$ in the plasma. The constitutive relations that involve the plasma current density $\mathbf{J}$ are given by

$$\mathbf{D} = \varepsilon_0 \mathbf{E}, \tag{0.5}$$

$$\mathbf{J} = -q N \mathbf{v}. \tag{0.6}$$

The velocity $\mathbf{v}$ of the electrons is given by the force equation

$$m \frac{dv}{dt} = -q\mathbf{E}. \tag{0.7}$$

In Equation 0.7, the magnetic force due to the wave's magnetic field $\mathbf{H}$ is neglected since it is much smaller [8] than the force due to the wave's electric field. The magnetic force term $(-q\mathbf{v} \times \mathbf{H})$ is nonlinear. Stanic [18] studied this problem as a weakly nonlinear system.

Since ion motion is neglected, Equation 0.6 does not contain ion current. Such an approximation is called radio approximation [9]. It is used in the study of radiowave propagation in the ionosphere. Low-frequency wave propagation studies take into account the ion motion [9].

Substituting Equations 0.5 through 0.7 into the Ampere–Maxwell equation

$$\nabla \times \mathbf{H} = \mathbf{J} + \frac{\partial \mathbf{D}}{\partial t}, \tag{0.8}$$

we obtain

$$\nabla \times \mathbf{H} = j\omega\varepsilon_0\varepsilon_p(\omega)\mathbf{E}, \tag{0.9}$$

where $\varepsilon_p(\omega)$ is given by Equation 0.3 and an $\exp(j\omega t)$ time dependence has been assumed. For an arbitrary temporal profile of the electron density $N(t)$, Equation 0.6 is not valid [2,19,20]. The electron density $N(t)$ increases because of the new electrons born at different times. The newly born electrons start with zero velocity and are subsequently accelerated by the fields. Thus, all the electrons do not have the same velocity at a given time during the creation of the plasma. Therefore,

$$\mathbf{J}(t) \neq -qN(t)\mathbf{v}(t), \tag{0.10}$$

but

$$\Delta\mathbf{J}(t) = -q\Delta N_i\mathbf{v}_i(t), \tag{0.11}$$

instead. Here ΔN_i is the electron density added at t_i, and $\mathbf{v}_i(t)$ is the velocity at time t of these ΔN_i electrons born at time t_i. Thus, $\mathbf{J}(t)$ is given by the integral of Equation 0.11 and not by Equation 0.10. The integral of Equation 0.11, when differentiated with respect to t, gives the constitutive relation between $\mathbf{J}$ and $\mathbf{E}$ as follows:

$$\frac{d\mathbf{J}}{dt} = \varepsilon_0\omega_p^2(\mathbf{r}, t)\mathbf{E}(\mathbf{r}, t). \tag{0.12}$$

Equations 0.8 and 0.12 and the Faraday equation

$$\nabla \times \mathbf{E} = -\mu_0\frac{\partial\mathbf{H}}{\partial t} \tag{0.13}$$

are needed to describe the electromagnetics of isotropic plasmas.

Propagation of an EMW traveling in the z-direction with $\mathbf{E} = \hat{x}E$ and $\mathbf{H} = \hat{y}H$ can be studied by assuming that the components of the field variables have harmonic space variation, that is,

$$F(z, t) = f(t)e^{-jkz}. \tag{0.14}$$

Substituting Equation 0.14 into Equations 0.8, 0.12, and 0.13, we obtain the wave equations

$$\frac{d^2E}{dt^2} + \left[k^2c^2 + \omega_p^2(t)\right]E = 0 \tag{0.15}$$

and

$$\frac{d^3H}{dt^3} + \left[k^2c^2 + \omega_p^2(t)\right]\frac{dH}{dt} = 0 \tag{0.16}$$

for E and H.

0.4 Sudden Creation of an Unbounded Plasma Medium

The geometry of the problem is shown in Figure 0.3. A plane wave of frequency ω_0 is propagating in free space in the z-direction. Suddenly at $t = 0$, an unbounded plasma medium of plasma frequency ω_p is created. Thus arises a temporal discontinuity in the properties of the medium. The solution of Equation 0.16, when ω_p is a constant, can be obtained as

$$H(t) = \sum_{m=1}^{3} H_m \exp(j\omega_m t), \tag{0.17}$$

where

$$\omega_m[\omega_m^2 - (k^2 c^2 + \omega_p^2)] = 0. \tag{0.18}$$

The ω–k diagram [2] for the problem under discussion is obtained by graphing (Equation 0.18). Figure 0.4 shows the ω–k diagram, where the top and bottom branches are due to the factor in the square brackets equated to zero, and the horizontal line is due to the factor $\omega = 0$. The line $k = $ constant is a vertical line that intersects the ω–k diagram at the three points marked as 1,

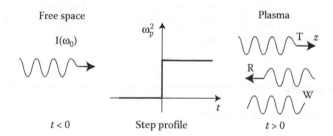

FIGURE 0.3 Suddenly created unbounded plasma medium.

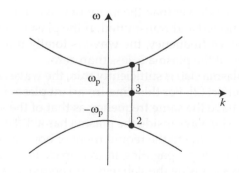

FIGURE 0.4 ω–k diagram and wiggler magnetic field.

2, and 3. The third mode is the *wiggler* mode [2,5,21]. Its real-valued fields are

$$\mathbf{E}_3(x,y,z,t) = 0, \tag{0.19}$$

$$\mathbf{H}_3(x,y,z,t) = \hat{y}H_0 \frac{\omega_p^2}{\omega_0^2 + \omega_p^2} \cos(kz), \tag{0.20}$$

$$\mathbf{J}_3(x,y,z,t) = \hat{x}H_0 k \frac{\omega_p^2}{\omega_0^2 + \omega_p^2} \sin(kz). \tag{0.21}$$

It is of zero frequency but varies in space. Such wiggler fields are used in a free electron laser (FEL) to generate coherent radiation [22]. Its electric field is zero but it has a magnetic field due to the plasma current $\mathbf{J}_3$. In the presence of a static magnetic field in the z-direction, the third mode becomes a traveling wave with a downshifted frequency. This aspect is discussed in Section 7.5.

Modes 1 and 2 have frequencies given by

$$\omega_{1,2} = \pm\sqrt{\left(\omega_p^2 + k^2 c^2\right)}, \tag{0.22}$$

where ω_2 has a negative value. Since the harmonic variations in space and time are expressed in the phase factor $\exp(\omega t - kz)$, a negative value for ω gives rise to a wave propagating in the negative z-direction. It is a backward propagating wave or for convenience can be referred to as a wave reflected by the discontinuity. Modes 1 and 2 have higher frequencies than the source wave. These are upshifted waves.

0.5   Switched Plasma Slab

The interaction of an EMW with a plasma slab is experimentally more realizable than with an unbounded plasma medium. When an incident wave enters a preexisting plasma slab, the wave must experience a spatial discontinuity. If the plasma frequency is lower than the incident wave frequency, the incident wave is partially reflected and transmitted. If the plasma frequency is higher than the incident wave frequency, the wave is totally reflected because the relative permittivity of the plasma is less than zero.

However, if the plasma slab is sufficiently thin, the wave can be transmitted due to a tunneling effect [2]. For this time-invariant plasma, the reflected and transmitted waves have the same frequency as that of the source wave; they are called A waves. The wave inside the plasma has a different wave number but the same frequency due to the requirement of boundary conditions.

When a source wave is propagating in free space and suddenly a plasma slab is created, the wave inside the slab region experiences a temporal discontinuity in the properties of the medium. Hence the switching action generates new waves whose frequencies are upshifted and then the waves propagate out of the slab; they are called B waves. The phenomenon is illustrated in Figure 0.5. In Figure 0.5a, the source wave of frequency ω_0 is propagating in free space. At $t = 0$, a slab of the plasma frequency ω_p is created. The A waves in Figure 0.5b have the same frequency as that of the source wave. The B waves are newly created waves due to the sudden switching of the plasma slab and have upshifted frequencies $\omega_1 = \sqrt{\left(\omega_0^2 + \omega_p^2\right)} = -\omega_2$. The negative value for the frequency of the second B wave shows that it is a backward prop-agating wave. These waves, however, have the same wave number as that of the source wave as long as they remain in the slab region. As the B waves come out of the slab, they encounter a spatial discontinuity and therefore the

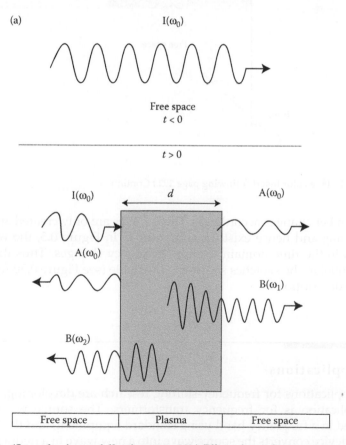

FIGURE 0.5 **(See color insert following page 202.)** Effect of switching an isotropic plasma slab. A waves have the same frequency as the incident wave frequency (ω_0), and B waves have upshifted frequency $\omega_1 = \sqrt{\left(\omega_0^2 + \omega_p^2\right)} = -\omega_2$. (a) The waves are sketched in the time domain to show the frequency changes. (b) Shows a more accurate description.

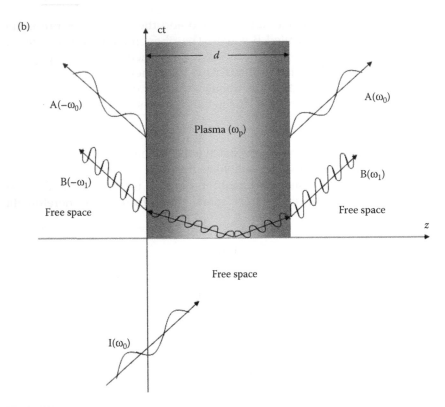

FIGURE 0.5 **(See color insert following page 202.)** Continued.

wave number changes accordingly. The B waves are only created at the time of switching and hence exist for a finite time. In Figure 0.5, the waves are sketched in the time domain to show frequency changes. Thus the arrowhead symbol in the sketches shows the *t* variable (see Figure 0.5b for a more accurate description).

0.6 Applications

Many applications for frequency-shifting research are developing. An obvious application is for frequency transformers. The source wave can be generated in a frequency band using standard equipment, and the switched plasma device converts the source wave into a new wave in a frequency band not easily accessible by other methods. The frequency-shifting mechanism can be applied for plasma cloaking of satellites and aircrafts, and for producing short-chirped-pulses as ultra-wide-band signals [23]. Application to photonics has been dealt with in detail by Nerukh et al. [24].

0.7 Time-Varying Magnetoplasma Medium

A plasma medium in the presence of a static magnetic field behaves like an anisotropic dielectric [1,2]. Therefore, the theory of EMW propagation in this medium is similar to the theory of light waves in crystals. Of course, in addition, account has to be taken of the highly dispersive nature of the plasma medium.

A cold magnetoplasma is described by two parameters: the electron density N and the quasistatic magnetic field. The first parameter is usually given in terms of the plasma frequency ω_p. The strength and the direction of the quasistatic magnetic field have a significant effect on the dielectric properties of the plasma. The parameter that is proportional to the static magnetic field is the electron gyrofrequency ω_b, defined in Section 7.1. The cutoff frequency of the magnetoplasma is influenced by ω_p as well as ω_b. An additional important aspect of the dielectric properties of the magnetoplasma medium is the existence of a resonant frequency. At the resonant frequency, the relative permittivity ε_p goes to infinity. As an example, for longitudinal propagation defined in Section 7.2, resonance occurs when the signal frequency ω is equal to the electron gyrofrequency ω_b. For the frequency band $0 < \omega < \omega_b$, $\varepsilon_p > 1$ and can have very high values for certain combinations of ω, ω_p, and ω_b; for instance, when $f_p = \omega_p/2\pi = 10^{14}\,\text{Hz}$, $f_b = \omega_b/2\pi = 10^{10}\,\text{Hz}$, and $f = 10\,\text{Hz}$, $\varepsilon_p = 9 \times 10^{16}$ [9]. A big change in ε_p can thus be obtained by collapsing the electron density, thus converting the magnetoplasma medium into free space. A big change in ε_p can also be obtained by collapsing the background quasistatic magnetic field, thus converting the magnetoplasma medium into an isotropic plasma medium. These aspects are discussed in the remaining parts of this section.

0.7.1 Basic Field Equations

The electric field $\mathbf{E}(\mathbf{r}, t)$ and the magnetic field $\mathbf{H}(\mathbf{r}, t)$ satisfy the Maxwell–Curl equations:

$$\nabla \times \mathbf{E} = -\mu_0 \frac{\partial \mathbf{H}}{\partial t}, \tag{0.23}$$

$$\nabla \times \mathbf{H} = \varepsilon_0 \frac{\partial \mathbf{E}}{\partial t} + \mathbf{J}. \tag{0.24}$$

In the presence of a quasistatic magnetic field $\mathbf{B}_0$, the constitutive relation for the current density is given by

$$\frac{d\mathbf{J}}{dt} = \varepsilon_0 \omega_p^2(\mathbf{r}, t)\mathbf{E} - \mathbf{J} \times \boldsymbol{\omega}_b(\mathbf{r}, t), \tag{0.25}$$

where

$$\omega_b = \frac{q\mathbf{B}_0}{m} = \omega_b \hat{\mathbf{B}}_0. \tag{0.26}$$

Therein, $\hat{\mathbf{B}}_0$ is a unit vector in the direction of the quasistatic magnetic field, ω_b is the gyrofrequency, and

$$\omega_p^2(\mathbf{r}, t) = \frac{q^2 N(\mathbf{r}, t)}{m\varepsilon_0}. \tag{0.27}$$

0.7.2 Characteristic Waves

Next, the solution for a plane wave propagating in the z-direction in a homogeneous, time-invariant unbounded magnetoplasma medium can be obtained by assuming

$$f(z, t) = \exp\left[j(\omega t - kz)\right], \tag{0.28}$$

$$\omega_p^2(z, t) = \omega_p^2, \tag{0.29}$$

$$\omega_b(z, t) = \omega_b, \tag{0.30}$$

where f stands for the components of the field variables $\mathbf{E}$, $\mathbf{H}$, or $\mathbf{J}$.

The well-established *magnetoionic theory* [8–10] is concerned with the study of plane wave propagation of an arbitrarily polarized plane wave in a cold anisotropic plasma, where the direction of phase propagation of the plane wave is at an arbitrary angle to the direction of the static magnetic field. As the plane wave travels in such a medium, the polarization state continuously changes. However, there are specific normal modes of propagation in which the state of polarization is unaltered. Plane waves with left (L wave) or right (R wave) circular polarization are the normal modes in the case of wave propagation along the quasistatic magnetic field. Such propagation is labeled as *longitudinal propagation*. The ordinary wave (O wave) and the extraordinary wave (X wave) are the normal modes for *transverse propagation*, where the direction of propagation is perpendicular to the static magnetic field. In this chapter, propagation of the R wave in a time-varying plasma is discussed. An analysis of the propagation of other characteristic waves can be found elsewhere [2].

0.7.3 R-Wave Propagation

The relative permittivity for R-wave propagation is [2,8]

$$\varepsilon_{pR} = 1 - \frac{\omega_p^2}{\omega(\omega - \omega_b)} = \frac{(\omega + \omega_{c1})(\omega - \omega_{c2})}{\omega(\omega - \omega_b)}, \tag{0.31}$$

where ω_{c1} and ω_{c2} are the cutoff frequencies given by

$$\omega_{c1,c2} = \mp\frac{\omega_b}{2} + \sqrt{\left(\frac{\omega_b}{2}\right)^2 + \omega_p^2}. \tag{0.32}$$

Equation 0.31 is obtained by eliminating $\mathbf{J}$ from Equation 0.24 with the help of Equation 0.25 and recasting it in the form of Equation 0.9.

The dispersion relation is obtained from

$$k_R^2 c^2 = \omega^2 \varepsilon_{pR} = \frac{\omega(\omega + \omega_{c1})(\omega - \omega_{c2})}{(\omega - \omega_b)}, \tag{0.33}$$

where k_R is the wave number for the R wave and c is the speed of light in free space. When expanded, this equation becomes

$$\omega^3 - \omega_b\omega^2 - (k_R^2 c^2 + \omega_p^2)\omega + k_R^2 c^2\omega_b = 0. \tag{0.34}$$

Figure 0.6 shows a graph of ε_p versus ω while Figure 0.7 gives the ω–k diagram for R-wave propagation. The R wave is a characteristic wave of longitudinal propagation. For this wave, the medium behaves like an isotropic plasma except that ε_p is influenced by the strength of the quasistatic magnetic field. Particular attention is drawn to the specific feature, visible in Figure 0.6, showing $\varepsilon_p > 1$ for the R wave in the frequency band $0 < \omega < \omega_b$. This mode of propagation is called *whistler mode* in the literature on ionospheric physics and *helicon mode* in the literature on solid-state plasmas. Sections 7.6 and 7.7 deal with the transformation of the whistler wave by a transient magnetoplasma medium and the consequences of such a transformation. An isotropic plasma medium does not support a whistler wave.

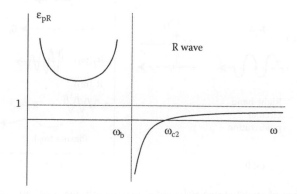

FIGURE 0.6 Relative permittivity for R-wave propagation.

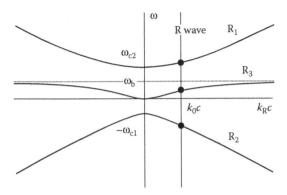

FIGURE 0.7 ω–k diagram for R-wave propagation.

0.7.4 Sudden Creation

In this section, the problem of sudden creation of the plasma in the presence of a static magnetic field in the z-direction is analyzed. The geometry of the problem is given in Figure 0.8. The source wave is assumed to be an R wave. The sudden creation is equivalent to creating a temporal discontinuity in the dielectric properties of the medium. In such a case, the wave number k_0 is conserved across the temporal discontinuity. For a given k_0 of the source wave, we draw a vertical line that intersects the branches in the ω–k diagram in Figure 0.7 at three points. The frequencies of these waves are different from the source frequency. The medium switching, in this case, creates three R waves labeled as R_1, R_2, and R_3. Whereas R_1 and R_3 are transmitted waves, R_2 is a reflected wave.

A physical interpretation of the waves can be given in the following way. The electric and magnetic fields of the incident wave and the quasistatic magnetic

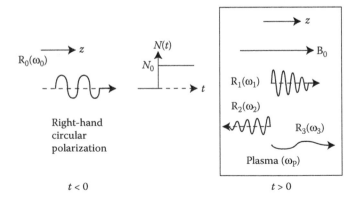

FIGURE 0.8 Effect of switching an unbounded magnetoplasma medium. Sketches of the B waves generated in the plasma are given for R-incidence.

field accelerate the electrons in the newly created magnetoplasma, which in turn radiate new waves. The frequencies of the new waves and their fields can be obtained by adding contributions from the many electrons whose positions and motions are correlated by the collective effects supported by the magnetoplasma medium. Such a detailed calculation of the radiated fields seems to be quite involved. A simple, but less accurate, description of the plasma effect is obtained by modeling the magnetoplasma medium as a dielectric medium whose refractive index is computed through magnetoionic theory [9]. The frequencies of the new waves are constrained by the requirements that the wave number k_0 is conserved over the temporal discontinuity and the refractive index n is the one that is applicable to the type of wave propagation in the magnetoplasma. This gives a conservation law [25] $k_0 c = \omega_0 = n(\omega)$, from which ω can be determined. Solution of the associated electromagnetic initial value problem gives the electric and magnetic fields of the new waves.

0.7.5 Frequency-Shifting Characteristics of Various R Waves

The shift ratio and the efficiency of the frequency-shifting operation can be controlled by the parameters ω_p and ω_b. The results are presented by normalizing all frequency variables with respect to the source wave frequency ω_0. This normalization is achieved by taking $\omega_0 = 1$ in numerical work.

For R waves, the curves of $\omega-\omega_p$ and $\omega-\omega_b$ are sketched in Figure 0.9a and b, respectively. In Figure 0.10, results are presented for the R_1 wave: values on the vertical axis give the frequency-shift ratio since the frequency variables are normalized with respect to ω_0. This is an upshifted wave and the shift ratio increases with ω_p as well as ω_b. From Figure 0.10 it appears that, by

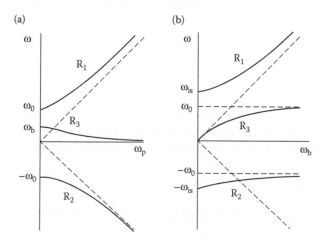

FIGURE 0.9 Frequency shifting of R waves. Sketches of (a) ω versus ω_p, (b) ω versus ω_b; $\omega_{is} = \left(\omega_0^2 + \omega_p^2\right)^{1/2}$.

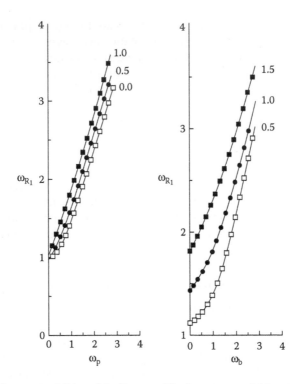

FIGURE 0.10   Frequency shifting of the R_1 wave. The frequency variables are normalized with respect to the source wave frequency by taking $\omega_0 = 1$. Shown is the frequency-shift ratio versus (a) ω_p and (b) ω_b. The numbers on the curves are (a) ω_b and (b) ω_p.

a suitable choice of ω_p and ω_b, one can obtain any desired large frequency-shift. However, the wave generated can have weak fields associated with it and the power density S_1 can be low. This point is illustrated in Table 0.1 by considering two sets of values for the parameters (ω_p, ω_b). For the set (0.5, 0.5), the shift ratio is 1.2 but the power density ratio S_1/S_0 is 0.57; however, for the set (2.0, 2.0), the shift ratio is 3.33 but the power density ratio is only 0.07. Similar remarks apply to other waves.

The R_2 wave is a reflected wave. It is an upshifted wave and the shift ratio increases with ω_p but decreases with ω_b [2]. The R_3 wave in Figure 0.11 is a transmitted wave that is downshifted. The shift ratio decreases with ω_p

TABLE 0.1

R_1 Wave Shift Ratio and Power Density for Two Sets of (ω_p, ω_b)

ω_0	ω_p	ω_b	E_1/E_0	H_1/H_0	S_1/S_0	ω_{R1}/ω_0
1	0.5	0.5	0.83	0.69	0.57	1.20
1	2.0	2.0	0.39	0.18	0.07	3.33

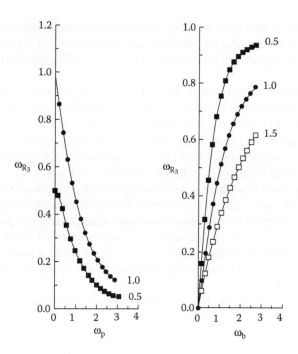

FIGURE 0.11 Frequency shifting of the R$_3$ wave, $\omega_0 = 1$. Shown is the frequency-shift ratio versus (a) ω_p and (b) ω_b. The numbers on the curves are (a) ω_b and (b) ω_p. Shown is the frequency-shift ratio versus (a) ω_p and (b) ω_b. The numbers on the curves are (a) ω_b and (b) ω_p.

and increases with ω_b. When $\omega_b = 0$, ω_{R3} becomes zero. The electric field E_3 becomes zero, and the magnetic field degenerates to the wiggler magnetic field [26]. This result is in conformity with the result for the isotropic case discussed in Section 0.4.

0.7.6 Frequency Upshifting with Power Intensification

Suppose that a source wave is present in an unbounded magnetoplasma medium. The parameters are such that the refractive index is greater than one. A wave propagating in the *whistler mode* is an example of such a wave. By collapsing the ionization, which can be achieved by removing the source of ionization, it is possible to obtain a frequency-upshifted wave with power intensification. The two limiting cases of (i) sudden collapse and (ii) slow decay are considered in this section. The starting point is the wave equation for the magnetic field of the R wave in a time-varying magnetoplasma medium obtained from Equations 0.23 through 0.25 as follows:

$$\frac{d^3H}{dt^3} - j\omega_b(t)\frac{d^2H}{dt^2} + \left[k_R^2 c^2 + \omega_p^2(t)\right]\frac{dH}{dt} - j\omega_b(t)k_R^2 c^2 H = 0. \qquad (0.35)$$

The sudden collapse case is solved as an initial value problem and the slow decay case is solved using an adiabatic approximation [2]. In either case, the final upshifted frequency is the source frequency multiplied by the refractive index.

When the source frequency is much less than the electron gyrofrequency, the refractive index of the whistler wave n_w is quite large, and the electric field is intensified by a factor of $n_w/2$ for the case of sudden collapse and by a factor of $n_w/\sqrt{2}$ for the case of slow decay. The corresponding intensification factors for the power density are $n_w/4$ and $n_w/2$, respectively. A physical explanation of the results, based on energy balance, is as follows. The energy in the whistler mode is predominantly the magnetic energy due to the plasma current. After the plasma collapses, the plasma current collapses and the magnetic energy due to plasma current is converted into wave electric and magnetic energies, giving rise to frequency-upshifted waves with enhanced electric fields and power density. Recent work by Bakov et al. [27] on energy relations in a general time-varying magnetoplasma can be applied to the study of frequency shifts of the other modes.

Figure 0.12 illustrates the results for an exponential decay profile:

$$\omega_p^2(t) = \omega_{p0}^2 e^{-bt}. \tag{0.36}$$

All variables are normalized with reference to the source wave quantities. The source wave frequency $\omega_0 = 1$ so that ω_1 is the frequency-upshift ratio. The parameters ω_b and ω_{p0} are assigned the values 100 and 150, respectively. A value of 0.01 is assigned to b, which therefore describes a slow decay, since $100/2\pi$ cycles of the source wave are accommodated in one time constant. The independent variable, time, is normalized with respect to the period of the source wave. Figure 0.12 also contains, for comparison, horizontal lines that are the results for sudden collapse. It is clear that frequency upshifting with power intensification occurs.

0.7.7 Generation of a Controllable Helical Wiggler Magnetic Field

A plasma in the presence of a quasistatic magnetic field supports a whistler wave. When the static magnetic field is switched off, the energy of the whistler wave is converted into the energy of a helical wiggler magnetic field [2]. The analysis is based on the solution of Equation 0.35 in which ω_p^2 is assumed to be constant but ω_b is a function of time.

Figure 0.13 illustrates the results for an exponential decay profile

$$\omega_b(t) = \omega_{b0} e^{-bt}. \tag{0.37}$$

All variables are normalized with frequency with reference to the source wave quantities. The parameters ω_{b0} and ω_p are assigned the values 100 and

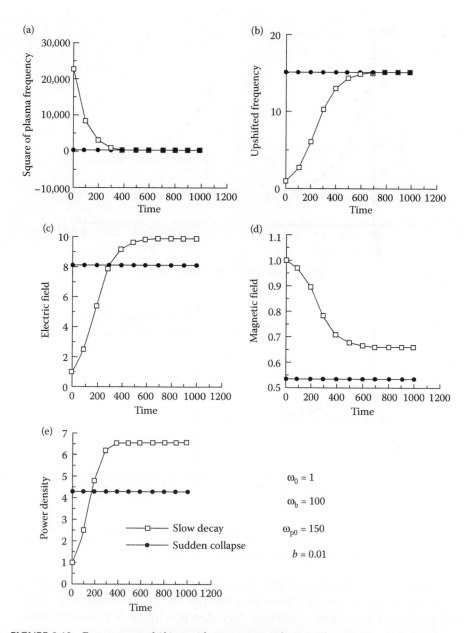

FIGURE 0.12 Frequency upshifting with power intensification of a whistler wave. Values on the vertical axes are normalized with respect to source wave quantities. The source frequency ω_0 is taken as 1. The horizontal timescale is normalized with respect to the period of the source wave.

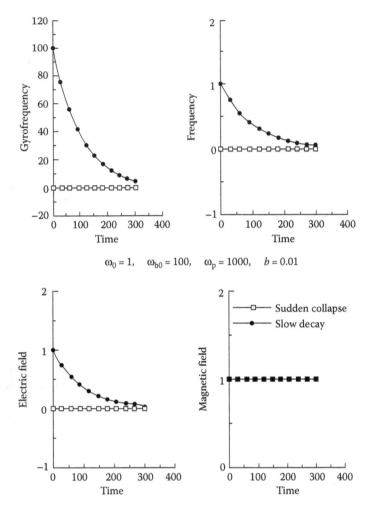

FIGURE 0.13 Conversion of a whistler wave into a helical magnetic wiggler field. Values on the vertical axes are normalized with respect to source wave quantities; $\omega_0 = 1$. The horizontal timescale is normalized with respect to the period of the source wave.

1000, respectively. A value of 0.01 is assigned to b. Figure 0.13 also shows, for comparision, horizontal lines that are the results for sudden collapse. Irrespective of the rate of collapse of B_0, a strong wiggler field is generated.

0.8   Overview of Part III of this Book

In Parts I and II, the theory and the numerical simulation of the interaction of an EMW with a time-varying magnetoplasma are discussed. An unbounded or a simple plane boundary in free space is assumed.

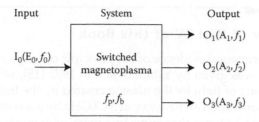

FIGURE 0.14 Frequency transformer concept.

A frequency transformer concept is developed as depicted in Figure 0.14.

The system parameters f_b and f_p influence the output frequencies and amplitudes.

A very large or a very small frequency-shift ratio is obtained when the source wave is a whistler wave and the magnetoplasma is switched off.

In these cases, unless the plasma is created very fast, the newly created traveling waves leave the boundaries before the interaction with the time-varying medium is completed.

However, if the time-varying medium is created in a cavity, the interaction continues since the waves are bounded by the cavity walls.

Figure 0.15 shows the frequency transformer concept when the switching of the medium takes place in a cavity. The polarization of the source wave determined by the parameters δ, $\tan \gamma$, and the switching angle (ϕ_0) are the additional parameters of the input wave that can control the polarization of the output waves. Thus the system box in Figure 0.15 is a "frequency and polarization transformer."

In this part, the "frequency and polarization transformer" concept is discussed using the theory for instantaneous change of the medium and FDTD for the more realistic space and time profiles of the system parameters. A most useful application of this principle is discussed at length in Appendix Q. It is shown that it is possible to transform a 2.45 GHz source radiation (source in the microwave oven consumer product) to 300 GHz radiation, with signification enhanced strength of the electric field.

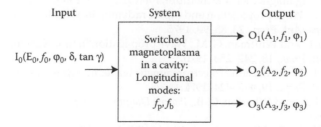

FIGURE 0.15 Frequency and polarization transformer concept.

0.9   Overview of Part IV of this Book

The earliest experimental evidence of the principle of frequency shifting by a dynamic plasma was given by Yablonovitch in 1973 [15], who observed the spectral broadening of light by the plasma created by the light pulse itself.

Joshi et al. [11] upshifted an RF wave at 33.3 GHz by 5% with 10% efficiency. They also observed 2.3 times the source frequency with much lower efficiency.

Chapter 13 is written by Professor Alexeff, whose group is one of three that conducted "Proof of the Principle" experiments [11–16] when the plasma medium is isotropic. Kuo et al. did the only such experiment involving magnetoplasmas [28].

0.10   Conclusion

It is hoped that the theory, FDTD simulation techniques, and the results presented in this book will result in design and performance of the experiments, which in turn will lead to practical devices. One of them being pursued by the author is the frequency transformer to transform 2.45 GHz to 300 GHz.

References

1. Weiglhofer, W. S., Constitutive characterization of simple and complex mediums, In: W. S. Weiglhofer and A. Lakhtakia (Eds), *Introduction to Complex Mediums for Optics and Electromagnetics*, SPIE, Bellingham, WA, 2003.
2. Kalluri, D. K., *Electromagnetics of Complex Media*, CRC Press LLC, Boca Raton, FL, 1999.
3. Lee, J. H. and Kalluri, D. K., Three dimensional FDTD simulation of electromagnetic wave transformation in a dynamic inhomogeneous magnetized plasma, *IEEE Trans. Ant. Prop.*, 47, 1146–1151, 1999.
4. Auld, B. A., Collins, J. H., and Zapp, H. R., Signal processing in a nonperiodically time-varying magnetoelastic medium, *Proc. IEEE*, 56, 258–272, 1968.
5. Jiang, C. L., Wave propagation and dipole radiation in a suddenly created plasma, *IEEE Trans. Ant. Prop.*, 23, 83–90, 1975.
6. Felsen, B. L. and Whitman, G. M., Wave propagation in time-varying media, *IEEE Trans. Ant. Prop.*, 18, 242–253, 1970.
7. Fante, R. L., Transmission of electromagnetic waves into time-varying media, *IEEE Trans. Ant. Prop.*, 19, 417–424, 1971.
8. Heald, M. A. and Wharton, C. B., *Plasma Diagnostics with Microwaves*, Wiley, New York, 1965.
9. Booker, H. G., *Cold Plasma Waves*, Kluwer, Higham, MA, 1984.
10. Tannenbaum, B. S., *Plasma Physics*, McGraw-Hill, New York, 1967.

11. Joshi, C. J., Clayton, C. E., Marsh, K., Hopkins, D. B., Sessler, A., and Whittum, D., Demonstration of the frequency upshifting of microwave radiation by rapid plasma creation, *IEEE Trans. Plasma Sci.*, 18, 814–818, 1990.

12. Kuo, S. P., Frequency up-conversion of microwave pulse in a rapidly growing plasma, *Phys. Rev. Lett.*, 65, 1000–1003, 1990.

13. Kuo, S. P. and Ren, A., Experimental study of wave propagation through a rapidly created plasma, *IEEE Trans. Plasma Sci.*, 21, 53–56, 1993.

14. Rader, M., Dyer, F., Matas, A., and Alexeff, I., Plasma-induced frequency shifts in microwave beams, In: *Conf. Rec. Abstracts, IEEE Int. Conf. Plasma Sci.*, Oakland, CA, p. 171, 1990.

15. Yablonovitch, E., Spectral broadening in the light transmitted through a rapidly growing plasma, *Phys. Rev. Lett.*, 31, 877–879, 1973.

16. Savage, Jr., R. L., Joshi, C. J., and Mori, W. B., Frequency up-conversion of electromagnetic radiation upon transmission into an ionization front, *Phys. Rev. Lett.*, 68, 946–949, 1992.

17. Lampe, M. and Walker, J. H., Interaction of electromagnetic waves with a moving ionization front, *Phys. Fluids*, 21, 42–54, 1978.

18. Stanic, B., Drljaca, P., and Boskoic, B., Electron plasma waves generation in suddenly created isotropic plasma, *IEEE Trans. Plasma Sci.*, 26, 1514–1519, 1998.

19. Stepanov, N. S., Dielectric constant of unsteady plasma, *Sov. Radiophys. Quan. Electr.*, 19, 683–689, 1976.

20. Banos, Jr., A., Mori, W. B., and Dawson, J. M., Computation of the electric and magnetic fields induced in a plasma created by ionization lasting a finite interval of time, *IEEE Trans. Plasma Sci.*, 21, 57–69, 1993.

21. Wilks, S. C., Dawson, J. M., and Mori, W. B., Frequency up-conversion of electromagnetic radiation with use of an overdense plasma, *Phys. Rev. Lett.*, 61, 337–340, 1988.

22. Granastein, V. L. and Alexeff, I., *High-Power Microwave Sources*, Artech House, Boston, MA, 1987.

23. Mori, W. B., (Ed), Special issue on generation of coherent radiation using plasmas, *IEEE Trans. Plasma Sci.*, 21(2), 1993.

24. Nerukh, A. G., Scherbatko, L. V., and Marciniak, M., *Electromagnetics of Modulated Media with Applications to Photonics*, IEEE/LEOS Poland chapter, Warsaw, 2001; printed by National Institute of Telecommunications, Department of Transmission and Fiber Technology.

25. Kalluri, D. K., Effect of switching a magnetoplasma medium on a traveling wave: Conservation law for frequencies of newly created waves, In: *Conf. Rec. Abstracts, IEEE Int. Conf. Plasma Sci.*, Oakland, CA, p. 129, 1990.

26. Kalluri, D. K., Effect of switching a magnetoplasma medium on a travelling wave: Longitudinal propagation, *IEEE Trans. Ant. Prop.*, 37, 1638–1642, 1989.

27. Bakunov, M. I. and Grachev, I. S., Energy relations for electromagnetic waves in a time-varying magnetoplasma medium, *Proc. SPIE*, 4467, 78–86, 1991.

28. Kuo, S. P., Bivolaru, D., Orlick, L., Alexeff, I., and Kalluri, D. K., A transmission line filled with fast switched periodic plasma as a wideband frequency transformer, *IEEE Trans. Plasma Sci.*, 29, 365–370, 2001.

Part I

Theory: Electromagnetic Wave Transformation in a Time-Varying Magnetoplasma Medium

1

Isotropic Plasma: Dispersive Medium

1.1 Introduction

Plasma is a mixture of charged particles and neutral particles. The mixture is *quasineutral*. Plasma is characterized by two independent parameters for each of the particle species. These are particle density N and temperature T. Plasma physics deals with such mixtures. There exists a vast amount of literature on this topic. A few references [1–3] of direct interest to the reader of this book are given at the end of this chapter. They deal with modeling a magnetized plasma as an electromagnetic medium. The models are adequate in exploring some of the applications where the medium can be considered to have time-invariant electromagnetic parameters.

There are some applications where the thermal effects are unimportant. Such a plasma is called a *cold plasma*. A *Lorentz plasma* [1] is a further simplification of the medium. In this model, it is assumed that electrons interact with each other only through collective space charge forces and that heavy positive ions and neutral particles are at rest. Positive ions serve as a background that ensures overall charge neutrality of the mixture. In this book, the Lorentz plasma will be the dominant model used to explore the major effects of a nonperiodically time-varying electron density profile $N(t)$. Departure from the model will be made only when it is necessary to bring in other relevant effects. Refs. [1,2] show the approximations made to arrive at the Lorentz plasma model.

1.2 Basic Field Equations for a Cold Isotropic Plasma

The electric field $\mathbf{E}(\mathbf{r}, t)$, the magnetic field $\mathbf{H}(\mathbf{r}, t)$, and the velocity field of the electrons $\mathbf{v}(\mathbf{r}, t)$ in the isotropic Lorentz plasma satisfy the following equations:

$$\nabla \times \mathbf{E} = -\mu_0 \frac{\partial \mathbf{H}}{\partial t}, \tag{1.1}$$

$$\nabla \times \mathbf{H} = \varepsilon_0 \frac{\partial \mathbf{E}}{\partial t} + \mathbf{J}, \tag{1.2}$$

3

$$m\frac{d\mathbf{v}}{dt} - q\mathbf{E,} \tag{1.3}$$

where $\mathbf{J}$ is the free electron current density in the plasma, q is the absolute value of the charge of an electron, and m is the mass of an electron. The relation between the current density and the electric field in the plasma will depend on the ionization process [4–7] that creates the plasma. Let us assume that the electron density profile $N(t)$ in the plasma is known and that the created electrons have zero velocity at the instant of their birth. From Equation 1.3, the velocity at t of the electrons born at t_i is given by

$$\mathbf{v}_i(\mathbf{r}, t) = -\frac{q}{m}\int_{t_i}^{t} \mathbf{E}(\mathbf{r}, \tau)\, d\tau. \tag{1.4}$$

The change in current density at t due to the electrons born at t_i can be computed as follows:

$$\Delta\mathbf{J}(\mathbf{r}, t) = -q\Delta N_i\mathbf{v}_i(\mathbf{r}, t). \tag{1.5}$$

Here ΔN_i is the electron density added at t_i and is given by

$$\Delta N_i = \left[\frac{\partial N}{\partial t}\right]_{t=t_i}\Delta t_i. \tag{1.6}$$

Therefore, the current density is given by (see Appendix A)

$$\mathbf{J}(\mathbf{r}, t) = \frac{q^2}{m}\int_{0}^{t}\frac{\partial N(\mathbf{r}, \tau)}{\partial \tau}\, d\tau\int_{\tau}^{t}\mathbf{E}(\mathbf{r}, \alpha)\, d\alpha + \mathbf{J}(\mathbf{r}, 0). \tag{1.7}$$

The expression for $\mathbf{J}$ can be simplified (see Appendix A) as follows:

$$\mathbf{J}(\mathbf{r}, t) = \varepsilon_0\int_{0}^{t}\omega_p^2(\mathbf{r}, \tau)\mathbf{E}(\mathbf{r}, \tau)\, d\tau + \mathbf{J}(\mathbf{r}, 0). \tag{1.8}$$

Here ω_p^2 is the square of the plasma frequency proportional to electron density N and is given by

$$\omega_p^2(\mathbf{r}, \tau)\frac{q^2N(r, t)}{m\varepsilon_0}. \tag{1.9}$$

See Ref. [1] for a physical explanation of the term "plasma frequency." Differentiating Equation 1.8, we obtain a differential equation for $\mathbf{J}$:

$$\frac{d\mathbf{J}}{dt} = \varepsilon_0\omega_p^2(\mathbf{r}, t)\, \mathbf{E}(\mathbf{r}, t). \tag{1.10}$$

Thus, the equations in a time-varying and space-varying isotropic plasma are Equations 1.1, 1.2, and 1.10. In this description, we use the variable **J** instead of **v**. Taking the curl of Equation 1.1 and eliminating **H** and **J** by using Equations 1.2 and 1.10, the wave equation for **E** can be derived:

$$\nabla^2 \mathbf{E} - \nabla(\nabla \cdot \mathbf{E}) - \frac{1}{c^2}\frac{\partial^2 \mathbf{E}}{\partial t^2} - \frac{1}{c^2}\omega_p^2(\mathbf{r}, t)\,\mathbf{E} = 0. \tag{1.11}$$

Similar efforts will lead to a wave equation for the magnetic field:

$$\nabla^2 \dot{\mathbf{H}} - \frac{1}{c^2}\frac{\partial^2 \dot{\mathbf{H}}}{\partial t^2} - \frac{1}{c^2}\omega_p^2(\mathbf{r}, t)\mathbf{H} + \varepsilon_0 \nabla \omega_p^2(\mathbf{r}, t) \times \mathbf{E} = 0, \tag{1.12}$$

where

$$\dot{\mathbf{H}} = \frac{\partial \mathbf{H}}{\partial t}. \tag{1.13}$$

In deriving Equation 1.12, the equation

$$\nabla \cdot \mathbf{H} = 0 \tag{1.14}$$

is used. The last term in Equation 1.12 contains **E**; however, if ω_p^2 varies only with t, its gradient is zero and Equation 1.12 becomes

$$\nabla^2 \dot{\mathbf{H}} - \frac{1}{c^2}\frac{\partial^2 \dot{\mathbf{H}}}{\partial t^2} - \frac{1}{c^2}\omega_p^2(\mathbf{r}, t)\dot{\mathbf{H}} = 0. \tag{1.15}$$

1.3 One-Dimensional Equations

Let us consider the particular case where (a) the variables are functions of only one spatial coordinate, say the z coordinate; (b) the electric field is linearly polarized in the x-direction; and (c) the variables are denoted by

$$\mathbf{E} = \hat{x}E(z, t), \tag{1.16}$$

$$\mathbf{H} = \hat{y}H(z, t), \tag{1.17}$$

$$\mathbf{J} = \hat{x}J(z, t), \tag{1.18}$$

$$\omega_p^2 = \omega_p^2(z, t). \tag{1.19}$$

The basic equations for E, H, and J take the following simple form:

$$\frac{\partial E}{\partial z} = -\mu_0 \frac{\partial H}{\partial t}, \tag{1.20}$$

$$-\frac{\partial H}{\partial z} = \varepsilon_0 \frac{\partial E}{\partial t} + J, \tag{1.21}$$

$$\frac{dJ}{dt} = \varepsilon_0 \omega_p^2(z,t)E, \tag{1.22}$$

$$\frac{\partial^2 E}{\partial z^2} = \frac{1}{c^2}\frac{\partial^2 E}{\partial t^2} - \frac{1}{c^2}\omega_p^2(z,t)E = 0, \tag{1.23}$$

$$\frac{\partial^2 \dot{H}}{\partial z^2} - \frac{1}{c^2}\frac{\partial^2 \dot{H}}{\partial t^2} - \frac{1}{c^2}\omega_p^2(z,t)\dot{H} + \varepsilon_0 \frac{\partial}{\partial z}\omega_p^2(z,t)E = 0. \tag{1.24}$$

If ω_p^2 varies only with t, Equation 1.24 reduces to

$$\frac{\partial^2 \dot{H}}{\partial z^2} - \frac{1}{c^2}\frac{\partial^2 \dot{H}}{\partial t^2} - \frac{1}{c^2}\omega_p^2(z,t)\dot{H} = 0. \tag{1.25}$$

Let us next consider that the electric field is linearly polarized in the y-direction, that is,

$$\mathbf{E} = \hat{y}E(z,t), \tag{1.26}$$

$$\mathbf{H} = -\hat{x}H(z,t), \tag{1.27}$$

$$\mathbf{J} = \hat{y}J(z,t). \tag{1.28}$$

The basic equations for E, H, and J are given by Equations 1.20 through 1.25. It immediately follows that the basic equations for circular polarization, where

$$\mathbf{E} = \left(\hat{x} \mp j\hat{y}\right)E(z,t), \tag{1.29}$$

$$\mathbf{H} = \left(\pm j\hat{x} + \hat{y}\right)H(z,t), \tag{1.30}$$

and

$$\mathbf{J} = \left(\hat{x} \mp j\hat{y}\right)J(z,t), \tag{1.31}$$

are once again given by Equations 1.20 through 1.25. In the above equations, the upper sign is for right circular polarization and the lower sign is for left circular polarization.

There is one more one-dimensional solution that has some physical significance. Let

$$\mathbf{E} = \hat{z} E(z, t), \tag{1.32}$$

$$\mathbf{H} = 0, \tag{1.33}$$

$$\mathbf{J} = \hat{z} J(z, t). \tag{1.34}$$

From Equation 1.32, it follows that

$$\nabla \times \mathbf{E} = 0. \tag{1.35}$$

From Equations 1.1, 1.2, and 1.10,

$$\varepsilon_0 \frac{\partial E}{\partial t} = -J, \tag{1.36}$$

$$m \frac{d\mathbf{v}}{dt} = -q\mathbf{E}, \tag{1.37}$$

$$\frac{dJ}{dt} = \varepsilon_0 \omega_p^2 (z, t) E. \tag{1.38}$$

Also,

$$\nabla \cdot \mathbf{E} = \frac{\partial E}{\partial z}. \tag{1.39}$$

From Equation 1.11,

$$\frac{\partial^2 E}{\partial t^2} + \omega_p^2 (z, t) E = 0. \tag{1.40}$$

Equation 1.40 can also be obtained from Equations 1.36 and 1.38. Equation 1.40 can easily be solved if ω_p^2 is not a function of time:

$$E = E_0(z) \cos \left[\omega_p(z) t \right], \tag{1.41}$$

$$\mathbf{E} = \hat{z} E. \tag{1.42}$$

Equation 1.42 represents longitudinal oscillation of the electrostatic field at the plasma frequency. In a warm plasma, this oscillation will be converted into an electrostatic wave, which is also called the *Langmuir wave* or the electron plasma wave [1]. For a warm plasma, Equation 1.3 needs modification due to the pressure-gradient forces that are caused by the thermal effects. In metals at optical frequencies, *Fermi velocity* plays the role of a thermal velocity (see Ref. [8]).

1.4 Profile Approximations for Simple Solutions

Subsequent chapters discuss solutions based on some simple approximations for the ω_p^2 profile as it transitions from one medium to the other. We assume asymptotic values for it deep inside each medium and a nonperiodic change of the parameter from medium 1 to medium 2 over a length scale L (range length) and a timescale T_r (rise time). Figure 1.1 and Equations 1.43 through 1.47 describe the profiles:

$$\omega_p^2(z,t) = \omega_{p1}^2 + (\omega_{p2}^2 - \omega_{p1}^2)f_1(z)f_2(t), \tag{1.43}$$

$$f_1(z) = 0, \quad -\infty < z < -L_1, \tag{1.44}$$

$$f_1(z) = 1, \quad L_2 < z < \infty, \tag{1.45}$$

$$f_2(t) = 0, \quad -\infty < t < -T_{r1}, \tag{1.46}$$

$$f_2(t) = 1, \quad T_{r2} < t < \infty. \tag{1.47}$$

In the above equations, ω_{p1}^2 is the plasma frequency of medium 1 as $z \to -\infty$ and $t \to -\infty$ and ω_{p2}^2 is the plasma frequency of medium 2 as $z \to \infty$ and $t \to \infty$. Functions $f_1(z)$ and $f_2(t)$, sketched in Figure 1.1, describe the transition of plasma frequency from its asymptotic values.

If the range length $L_1 + L_2 = L$ is much less than the significant length of the problem, say the wavelength λ of the source wave in medium 1, then the profile can be approximated as a spatial step profile and the problem can be solved as a boundary value problem using appropriate boundary conditions. If $L \gg \lambda$, then adiabatic analysis based on approximation techniques such as Wentzel–Kramers–Brillouin (WKB) can be used. If L is comparable to λ, a perturbation technique [9] based on Green's function for a sharp boundary of unlike media can be used. Lekner [9] gives an excellent account of the application of this technique to the calculation of reflection coefficient for a spatial dielectric profile. The inhomogeneous media problem (spatial profile) was investigated extensively in view of its applications in optics [9] and ionospheric physics [10].

If the rise time $T_{r1} + T_{r2} = T_r$ is much less than the significant period of the problem, say period t_0 of the source wave in medium 1, then the profile can be approximated as a temporal step profile and the problem can be solved as an initial value problem using appropriate initial conditions. If $T_r \gg t_0$, then adiabatic analysis based on approximation techniques such as WKB can be used. If the rise time T_r is comparable to t_0, then a perturbation technique based on Green's function for the switched medium can be used. These are some of the techniques explored in subsequent chapters.

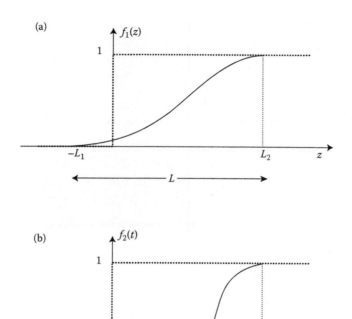

FIGURE 1.1 (a) Spatial profile. (b) Temporal profile.

Another simple profile of great interest is the ionization front profile (Figure 1.2) defined by

$$\omega_p^2(z,t) = \omega_{p0}^2 \left[u(z + v_F t)\right]. \tag{1.48}$$

In the above equation, u is a unit step function,

$$u(\varphi) = \begin{cases} 0, & \varphi < 0, \\ 1, & \varphi \geq 0, \end{cases} \tag{1.49}$$

and v_F is the velocity of the ionization front. Such a profile can be created by a source of ionizing radiation pulse, say a strong laser pulse. As the pulse travels in a neutral gas, it converts it into a plasma, thus creating a moving boundary between the plasma and the unionized gas. Section 10.3 deals with such a problem. In passing, it can be mentioned that the ionization front

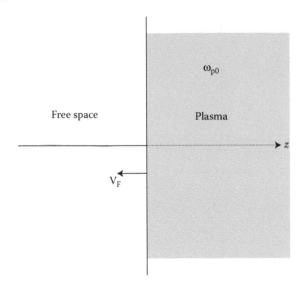

FIGURE 1.2   Ionization front.

problem is somewhat different from the moving plasma problem. In the front problem, the boundary alone is moving and the plasma is not moving with the boundary. The moving plasma problem has been dealt with by a number of authors [11–15], and is not discussed in this book.

1.5 Dispersive Media

The dielectric constant ε_r of a dielectric is frequency independent over wide bands of frequencies, and in these bands the dielectric is nondispersive. In this book, the term "pure dielectric" or "simply dielectric" is used to describe a nondispersive dielectric with a dielectric constant greater than or equal to one.

Plasma can be modeled as a dielectric. However, its dielectric constant depends on the frequency of the signal and is given by

$$\varepsilon_p = \left(1 - \frac{\omega_p^2}{\omega^2}\right). \tag{1.50}$$

This dependence makes it highly dispersive and leads to a qualitative difference in the properties of waves scattered by a temporal change in the electron density of the plasma medium as compared with such scattering by a pure dielectric.

In the presence of a static magnetic field, the dielectric constant not only is a function of frequency, but also becomes a tensor and the medium has to be modeled as a dispersive anisotropic medium. The properties of waves scattered by a temporal change in the electron density of the plasma medium in the presence of a static magnetic field involve many more modes [16]. Subsequent chapters deal with these topics.

References

1. Heald, M. A. and Wharton, C. B., *Plasma Diagnostics with Microwaves*, Wiley, New York, 1965.
2. Booker, H. G., *Cold Plasma Waves*, Kluwer, Hingham, MA, 1984.
3. Tannenbaum, B. S., *Plasma Physics*, McGraw-Hill, New York, 1967.
4. Baños, Jr., A., Mori, W. B., and Dawson, J. M., Computation of the electric and magnetic fields induced in a plasma created by ionization lasting a finite interval of time, *IEEE Trans. Plasma Sci.*, 21, 57, 1993.
5. Kalluri, D. K., Goteti, V. R., and Sessler, A. M., WKB solution for wave propagation in a time-varying magnetoplasma medium: Longitudinal propagation, *IEEE Trans. Plasma Sci.*, 21, 70, 1993.
6. Lampe, M. and Ott, E., Interaction of electromagnetic waves with a moving ionization front, *Phys. Fluids*, 21, 42, 1978.
7. Stepanov, N. S., Dielectric constant of unsteady plasma, *Soviet Radiophys. Quant. Electron*, 19, 683, 1976.
8. Forstmann, F. and Gerhardts, R. R., *Metal Optics Near the Plasma Frequency*, Springer, New York, 1986.
9. Lekner, J., *Theory of Reflection*, Kluwer, Boston, 1987.
10. Budden, K. G., *Radio Waves in the Ionosphere*, Cambridge University Press, Cambridge, 1961.
11. Chawla, B. R. and Unz, H., *Electromagnetic Waves in Moving Magneto-plasmas*, The University Press of Kansas, Lawrence, 1971.
12. Kalluri, D. K. and Shrivastava, R. K., Radiation pressure due to plane electromagnetic waves obliquely incident on moving media, *J. Appl. Phys.*, 49, 3584, 1978.
13. Kalluri, D. K. and Shrivastava, R. K., On total reflection of electromagnetic waves from moving plasmas, *J. Appl. Phys.*, 49, 6169, 1978.
14. Kalluri, D. K. and Shrivastava, R. K., Brewster angle for a plasma medium moving at relativistic speed, *J. Appl. Phys.*, 46, 1408, 1975.
15. Kalluri, D. K. and Shrivastava, R. K., Electromagnetic wave interaction with moving bounded plasmas, *J. App. Phys.*, 44, 4518, 1973.
16. Kalluri, D. K., Frequency-shifting using magnetoplasma medium, *IEEE Trans. Plasma Sci.*, 21, 77, 1993.

2

Space-Varying Time-Invariant Isotropic Medium

2.1 Basic Equations

Let us assume that the electron density varies only in space (inhomogeneous isotropic plasma). From Equation 1.9,

$$\omega_p^2 = \omega_p^2(\mathbf{r}).\tag{2.1}$$

One can then construct basic solutions by assuming that the field variables vary harmonically in time:

$$\mathbf{F}(\mathbf{r}, t) = \mathbf{F}(\mathbf{r})\exp(j\omega t).\tag{2.2}$$

In the above equation, $\mathbf{F}$ stands for any of $\mathbf{E}$, $\mathbf{H}$, or $\mathbf{J}$. Equations 1.1, 1.2, and 1.10 then reduce to

$$\nabla \times \mathbf{E} = -j\omega\mu_0\mathbf{H},\tag{2.3}$$

$$\nabla \times \mathbf{H} = j\omega\varepsilon_0\mathbf{E} + \mathbf{J},\tag{2.4}$$

$$j\omega\mathbf{J} = \varepsilon_0\omega_p^2(\mathbf{r})\mathbf{E}.\tag{2.5}$$

Combining Equations 2.4 and 2.5, we can write

$$\nabla \times \mathbf{H} = j\omega\varepsilon_0\varepsilon_p(\mathbf{r}, \omega)\mathbf{E},\tag{2.6}$$

where ε_p is the dielectric constant of isotropic plasma and is given by

$$\varepsilon_p(\mathbf{r}, \omega) = 1 - \frac{\omega_p^2(\mathbf{r})}{\omega^2}.\tag{2.7}$$

The wave equations 1.11 and 1.12 reduce to

$$\nabla^2 \mathbf{E} + \frac{\omega^2}{c^2}\varepsilon_p(\mathbf{r})\mathbf{E} = \nabla(\nabla \cdot \mathbf{E}), \tag{2.8}$$

$$\nabla^2 \mathbf{H} + \frac{\omega^2}{c^2}\varepsilon_p(\mathbf{r})\mathbf{H} = -\frac{\varepsilon_o}{j\omega}\nabla\omega_p^2(\mathbf{r}) \times \mathbf{E}$$

$$= -j\omega\varepsilon_0 \nabla\varepsilon_p(\mathbf{r}) \times \mathbf{E}. \tag{2.9}$$

The one-dimensional equations 1.20 through 1.22 take the form

$$\frac{\partial E}{\partial z} = -j\omega\mu_0 H, \tag{2.10}$$

$$-\frac{\partial H}{\partial z} = j\omega\varepsilon_0 E + J, \tag{2.11}$$

$$j\omega J = \varepsilon_0 \omega_p^2(z)E. \tag{2.12}$$

Combining Equations 2.11 and 2.12 or from Equation 2.6,

$$-\frac{\partial H}{\partial z} = j\omega\varepsilon_0 \varepsilon_p(z, \omega)E, \tag{2.13}$$

where

$$\varepsilon_p(z, \omega) = 1 - \frac{\omega_p^2(z)}{\omega^2}. \tag{2.14}$$

Figure 2.1 shows the variation of ε_p with ω. ε_p is negative for $\omega < \omega_p$, 0 for $\omega = \omega_p$, and positive but less than 1 for $\omega > \omega_p$. The one-dimensional wave equations 1.23 and 1.24, in this case, reduce to

$$\frac{d^2 E}{dz^2} + \frac{\omega^2}{c^2}\varepsilon_p(z, \omega)E = 0, \tag{2.15}$$

$$\frac{d^2 H}{dz^2} + \frac{\omega^2}{c^2}\varepsilon_p(z, \omega)\, H = \frac{1}{\varepsilon_p(z, \omega)}\frac{d\varepsilon_p(z, \omega)}{dz}\frac{dH}{dz}. \tag{2.16}$$

Equation 2.15 can easily be solved if we further assume that the dielectric is homogeneous (ε_p is not a function of z). Such a solution is given by

$$E = E^+(z) + E^-(z), \tag{2.17}$$

$$E = E_0^+ \exp(-jkz) + E_0^- \exp(+jkz), \tag{2.18}$$

where

$$k = \frac{\omega}{c}\sqrt{\varepsilon_p} = k_0\sqrt{\varepsilon_p} = k_0 n. \tag{2.19}$$

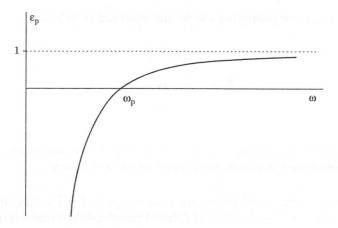

FIGURE 2.1 Dielectric constant versus frequency for a plasma medium.

Here n is the refractive index and k_0 is the free space wave number. The first term on the right-hand side of Equation 2.17 represents a traveling wave in the positive z-direction (positive-going wave) of angular frequency ω and wave number k and the second term represents a similar but negative-going wave. The z-direction is the direction of phase propagation, and the velocity of phase propagation v_p (phase velocity) of either of the waves is given by

$$v_p = \frac{\omega}{k}. \qquad (2.20)$$

From Equation 2.10,

$$H^+(z) = \frac{1}{(-j\omega\mu_0)}(-jk)E_0^+ \exp(-jkz) = \frac{1}{\eta}E^+(z), \qquad (2.21)$$

$$H^-(z) = -\frac{1}{\eta}E^-(z), \qquad (2.22)$$

where

$$\eta = \sqrt{\frac{\mu_0}{\varepsilon_0\varepsilon_p}} = \frac{\eta_0}{n} = \frac{120\pi}{n}\,(\text{ohms}). \qquad (2.23)$$

Here η is the intrinsic impedance of the medium. The waves described above, which are one-dimensional solutions of the wave equation in an unbounded isotropic homogeneous medium, are called *uniform plane waves* in the sense that the phase and the amplitude of the waves are constant in a plane. Such one-dimensional solutions in a coordinate-free description are called transverse electric and magnetic waves (TEMs), and for an arbitrarily directed wave, their properties can be summarized as follows:

$$\hat{\mathbf{E}} \times \hat{\mathbf{H}} = \hat{k}, \tag{2.24}$$

$$\hat{\mathbf{E}} \cdot \hat{k} = 0, \quad \hat{\mathbf{H}} \cdot \hat{k} = 0, \tag{2.25}$$

$$E = \eta H. \tag{2.26}$$

In the above equations, $\hat{k}$ is a unit vector in the direction of phase propagation. Stated in words, the properties are as follows:

1. The unit electric field vector, the unit magnetic field vector, and the unit vector in the direction of (phase) propagation form a mutually orthogonal system.
2. There is no component of the electric or the magnetic field vector in the direction of propagation.
3. The ratio of the electric field amplitude to the magnetic field amplitude is given by the intrinsic impedance of the medium.

Figure 2.2 shows the variation of dielectric constant ε_p with frequency for a typical real dielectric. For such a medium, in a broad frequency band, ε_p can be treated as not varying with ω and denoted by the dielectric constant ε_r. We will next consider the step profile approximation for a dielectric profile. In each of the media 1 and 2 the dielectric can be treated as homogeneous. The solution for this problem is well known but is included here to provide a comparison with the solution for a temporal step profile discussed in Chapter 3.

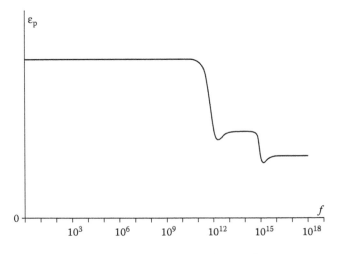

FIGURE 2.2   Sketch of the dielectric constant of a typical material.

2.2 Dielectric–Dielectric Spatial Boundary

The geometry of the problem is shown in Figure 2.3. Let the incident wave in medium 1, also called the source wave, have the fields

$$\mathbf{E}_i(x, y, z, t) = \hat{x} E_0 \exp[j(\omega_i t - k_i z)], \tag{2.27}$$

$$\mathbf{H}_i(x, y, z, t) = \hat{y} \frac{E_0}{\eta_1} \exp[j(\omega_i t - k_i z)], \tag{2.28}$$

where ω_i is the frequency of the incident wave and

$$k_i = \omega_i \sqrt{\mu_0 \varepsilon_1} = \frac{\omega_i}{v_{p1}}. \tag{2.29}$$

The fields of the reflected wave are given by

$$\mathbf{E}_r(x, y, z, t) = \hat{x} E_r \exp[j(\omega_r t + k_r z)], \tag{2.30}$$

$$\mathbf{H}_r(x, y, z, t) = -\hat{y} \frac{E_r}{\eta_1} \exp[j(\omega_r t + k_r z)], \tag{2.31}$$

where ω_r is the frequency of the reflected wave and

$$k_r = \omega_r \sqrt{\mu_0 \varepsilon_1}. \tag{2.32}$$

The fields of the transmitted wave are given by

$$\mathbf{E}_t(x, y, z, t) = \hat{x} E_t \exp[j(\omega_t t - k_t z)], \tag{2.33}$$

$$\mathbf{H}_t(x, y, z, t) = \hat{y} \frac{E_t}{\eta_2} \exp[j(\omega_t t - k_t z)], \tag{2.34}$$

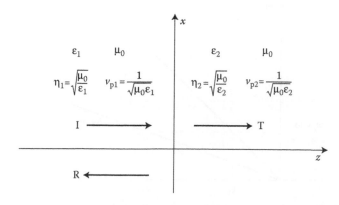

FIGURE 2.3 Dielectric–dielectric spatial boundary problem.

where ω_t is the frequency of the transmitted wave and

$$k_t = \omega_t \sqrt{\mu_0 \varepsilon_2} = \frac{\omega_t}{v_{p2}}. \tag{2.35}$$

The boundary condition of the continuity of the tangential component of the electric field at the interface $z = 0$ can be represented as

$$\mathbf{E}(x,y,0^-,t) \times \hat{z} = \mathbf{E}(x,y,0^+,t) \times \hat{z}, \tag{2.36}$$

$$\left[\mathbf{E}_i(x,y,0^-,t) + \mathbf{E}_r(x,y,0^-,t)\right] \times \hat{z} = \left[\mathbf{E}_t(x,y,0^+,t)\right] \times \hat{z}. \tag{2.37}$$

The above equations must be true for all x, y, and t. Thus we have

$$E_0 \exp[j\omega_i t] + E_r \exp[j\omega_r t] = E_t \exp[j\omega_t t]. \tag{2.38}$$

Since Equation 2.38 must be satisfied for all t, the coefficients of t in the exponents of Equation 2.38 must match:

$$\omega_i = \omega_r = \omega_t = \omega. \tag{2.39}$$

The above result can be stated as follows: the frequency ω is conserved across a spatial discontinuity in the properties of an electromagnetic medium. As the wave crosses from one medium to the other in space, the wave number k changes as dictated by the change in phase velocity (Figure 2.4).

From Equations 2.39 and 2.38, we obtain

$$E_0 + E_r = E_t. \tag{2.40}$$

The second independent boundary condition of continuity of the tangential magnetic field component is written as (Equation 2.41)

$$\left[\mathbf{H}_i(x,y,0^-,t) + \mathbf{H}_r(x,y,0^-,t)\right] \times \hat{z} = \left[\mathbf{H}_t(x,y,0^+,t)\right] \times \hat{z}. \tag{2.41}$$

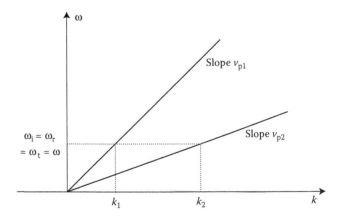

FIGURE 2.4   Conservation of frequency across a spatial boundary.

The reflection coefficient $R_A = E_r/E_0$ and the transmission coefficient $T_A = E_t/E_0$ are determined using Equations 2.40 and 2.41. From Equation 2.41, we have

$$\frac{E_0 - E_r}{\eta_1} = \frac{E_t}{\eta_2}. \tag{2.42}$$

The results are

$$R_A = \frac{\eta_2 - \eta_1}{\eta_2 + \eta_1} = \frac{n_1 - n_2}{n_1 + n_2}, \tag{2.43}$$

$$T_A = \frac{2\eta_2}{\eta_2 + \eta_1} = \frac{2n_1}{n_1 + n_2}. \tag{2.44}$$

The significance of subscript A in the above equation is to distinguish from the reflection and transmission coefficients due to a temporal discontinuity in the properties of the medium. This aspect is discussed in Chapter 3.

We next show that the time-averaged power density of the source wave is equal to the sum of the time-averaged power density of the reflected wave and the transmitted wave:

$$\left|\frac{1}{2}\text{Re}\left[\mathbf{E_i} \times \mathbf{H_i^*} \cdot \hat{z}\right]\right| = \left|\frac{1}{2}\text{Re}\left[\mathbf{E_r} \times \mathbf{H_r^*} \cdot \hat{z}\right]\right| + \left|\frac{1}{2}\text{Re}\left[\mathbf{E_t} \times \mathbf{H_t^*} \cdot \hat{z}\right]\right|. \tag{2.45}$$

The left-hand side is

$$\text{LHS} = \frac{1}{2}E_0 H_0 = \frac{1}{2}\frac{E_0^2}{\eta_1}, \tag{2.46}$$

whereas the right-hand side is

$$\text{RHS} = \frac{1}{2}E_0^2\left[\frac{|R_A|^2}{\eta_1} + \frac{|T_A|^2}{\eta_2}\right] = \frac{1}{2}\frac{E_0^2}{\eta_1}\left[\frac{(\eta_2 - \eta_1)^2 + 4\eta_1\eta_2}{(\eta_2 + \eta_1)^2}\right] = \frac{1}{2}\frac{E_0^2}{\eta_1}. \tag{2.47}$$

2.3 Reflection by a Plasma Half-Space

Let an incident wave of frequency ω traveling in free space ($z < 0$) be incident normally on the plasma half-space ($z > 0$) of plasma frequency ω_p. The intrinsic impedance of the plasma medium is given by

$$\eta_p = \frac{\eta_0}{n_p} = \frac{120\pi\omega}{\sqrt{\omega^2 - \omega_p^2}}. \tag{2.48}$$

From Equation 2.43, the reflection coefficient R_A is given by

$$R_A = \frac{\eta_p - \eta_0}{\eta_p + \eta_0} = \frac{1 - n_p}{1 + n_p} = \frac{\Omega - \sqrt{\Omega^2 - 1}}{\Omega + \sqrt{\Omega^2 - 1}}, \qquad (2.49)$$

where

$$\Omega = \frac{\omega}{\omega_p} \qquad (2.50)$$

is the source frequency normalized with respect to the plasma frequency. The power reflection coefficient $\rho = |R_A|^2$ and the power transmission coefficient τ $(= 1 - \rho)$ versus Ω are shown in Figure 2.5. In the frequency band $0 < \omega < \omega_p$, the characteristic impedance η_p is imaginary. The electric field $\mathbf{E}_P$ and the magnetic field $\mathbf{H}_P$ in the plasma are in *time quadrature*; the real part of $(\mathbf{E}_P \times \mathbf{H}_P^*)$ is zero. The wave in the plasma is an *evanescent wave* [1] and carries no real power. The source wave is totally reflected and $\rho = 1$. In the frequency band $\omega_p < \omega < \infty$, the plasma behaves as a dielectric and the source wave is partially transmitted and partially reflected. The time-averaged power density of the incident wave is equal to the sum of the time-averaged power densities of the reflected and transmitted waves.

Metals at optical frequencies are modeled as plasmas with plasma frequency ω_p of the order of 10^{16}. Refining the plasma model by including the collision frequency will lead to three frequency domains of conducting, cutoff, and dielectric phenomena [2]. A more comprehensive account of modeling metals as plasmas is given in Refs. [3,4]. The associated phenomena of attenuated

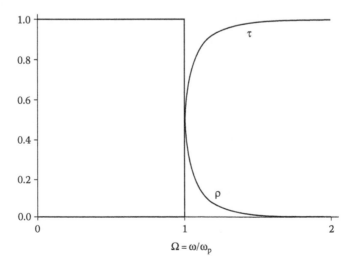

FIGURE 2.5 Sketch of the power reflection coefficient (ρ) and the power transmission coefficient (τ) of a plasma half-space.

total reflection (ATR), surface plasmons, and other interesting topics are based on modeling metals as plasmas. A brief account of surface waves is given in Section 2.5.2.

2.4 Reflection by a Plasma Slab [5]

In this section, we will consider oblique incidence to add to the variety in problem formulation. The geometry of the problem is shown in Figure 2.6. Let x–z be the plane of incidence and $\hat{k}$ be the unit vector along the direction of propagation. Let the magnetic field $\mathbf{H}^I$ be along y and the electric field $\mathbf{E}^I$ lie entirely in the plane of incidence. Such a wave is described in the literature by various names: (1) p wave; (2) TM wave; and (3) parallel polarized wave. Since the boundaries are at $z = 0$ and $z = d$, it is necessary to use x, y, and z coordinates in formulating the problem and express $\hat{k}$ in terms of x and z coordinates. The problem therefore appears to be two dimensional.

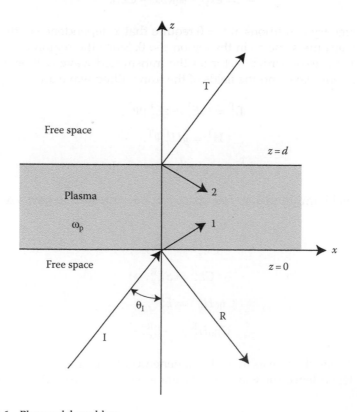

FIGURE 2.6 Plasma slab problem.

Equations 2.51 through 2.56 describe the incident wave:

$$\mathbf{H}^I = \hat{y}H_y^I \exp[-jk_0\hat{k} \cdot \mathbf{r}], \tag{2.51}$$

$$\hat{k} = \hat{x}S + \hat{z}C, \tag{2.52}$$

$$S = \sin\theta_I, \qquad C = \cos\theta_I, \tag{2.53}$$

$$\mathbf{E}^I = (\hat{y} \times \hat{k})E^I \exp[-jk_0\hat{k} \cdot \mathbf{r}] = (\hat{x}E_x^I + \hat{z}E_z^I)\Psi^I, \tag{2.54}$$

$$E_x^I = CE^I, \quad E_z^I = -SE^I, \quad E^I = \eta_0 H_y^I, \tag{2.55}$$

$$\Psi^I = \exp[-jk_0(Sx + Cz)]. \tag{2.56}$$

The reflected wave is written as

$$\mathbf{E}^R = (\hat{x}E_x^R + \hat{z}E_z^R)\Psi^R, \tag{2.57}$$

$$\mathbf{H}^R = \hat{y}H_y^R\Psi^R, \tag{2.58}$$

$$\Psi^R = \exp[-jk_0(Sx - Cz)]. \tag{2.59}$$

The boundary conditions at $z = 0$ require that x dependence in the region $z > 0$ remains the same as in the region $z < 0$. Since the region $z > d$ is also free space, the exponential factor for the transmitted wave will be the same as the incident wave and the fields of the transmitted wave are

$$\mathbf{E}^T = (\hat{x}E_x^T + \hat{z}E_z^T)\Psi^T, \tag{2.60}$$

$$\mathbf{H}^T = \hat{y}H_y^T\Psi^T, \tag{2.61}$$

$$\Psi^T = \exp[-jk_0(Sx + Cz)]. \tag{2.62}$$

All the field amplitudes in free space can be expressed in terms of E_x^I, E_x^R, or E_x^T:

$$CE_z^{I,T} = -SE_x^{I,T}, \tag{2.63}$$

$$CE_z^R = SE_x^R, \tag{2.64}$$

$$C\eta_0 H_y^{I,T} = E_x^{I,T}, \tag{2.65}$$

$$C\eta_0 H_y^R = -E_x^R. \tag{2.66}$$

We next consider waves in the homogeneous plasma region $0 < z < d$. The x and z dependence for waves in plasma are derived from the exponential factor Ψ^P:

$$\Psi^P = \exp[-jk_0(Sx + qz)], \tag{2.67}$$

where q has to be determined. The wave number in the plasma is given by $k_p = k_0\sqrt{\varepsilon_p}$. Thus we have

$$q^2 + S^2 = \varepsilon_p = 1 - \frac{\omega_p^2}{\omega^2}, \tag{2.68}$$

$$q_{1,2} = \pm\sqrt{C^2 - \frac{\omega_p^2}{\omega^2}} = \pm\sqrt{\frac{C^2\Omega^2 - 1}{\Omega^2}}. \tag{2.69}$$

The two values for q indicate the excitation of two waves in the plasma: the first wave is a positive wave and the second wave is a negative wave. For $\Omega < 1/C$, q is imaginary and the two waves in the plasma are evanescent. The next section deals with this frequency band.

The fields in the plasma can be written as

$$\mathbf{E}^P = \sum_{m=1}^{2} (\hat{x}E_{xm}^P + \hat{z}E_{zm}^P)\Psi_m^P, \tag{2.70}$$

$$\mathbf{H}^P = \sum_{m=1}^{2} \hat{y}H_{ym}^P \Psi_m^P, \tag{2.71}$$

$$\Psi_m^P = \exp[-jk_0(Sx + q_m z)]. \tag{2.72}$$

All the field amplitudes in the plasma can be expressed in terms of E_{x1}^P and E_{x2}^P:

$$\eta_0 H_{ym}^P = \frac{\varepsilon_p}{q_m} E_{xm}^P = \eta_{ym} E_{xm}^P, \tag{2.73}$$

$$E_{zm}^P = -\frac{S}{q_m} E_{xm}^P. \tag{2.74}$$

The above relations are obtained from Equations 2.3 and 2.6 by noting that $\partial/\partial x = -jk_0 S$ and $\partial/\partial z = -jk_0 q$; they can also be written from inspection. Assuming that E_x^I is known, the unknowns reduce to four: E_{x1}^P, E_{x2}^P, E_x^R, and E_x^T. They can be determined from the four boundary conditions of continuity of the tangential components E_x and H_y at $z = 0$ and $z = d$. In matrix form

$$\begin{bmatrix} 1 & 1 & -1 & 0 \\ C\eta_{y1} & C\eta_{y2} & 1 & 0 \\ \lambda_1 & \lambda_2 & 0 & -1 \\ \lambda_1 C\eta_{y1} & \lambda_2 C\eta_{y2} & 0 & -1 \end{bmatrix} \begin{bmatrix} E_{x1}^P \\ E_{x2}^P \\ E_x^R \\ E_x^T \end{bmatrix} = \begin{bmatrix} 1 \\ 1 \\ 0 \\ 0 \end{bmatrix} E_x^I, \tag{2.75}$$

where

$$\lambda_1 = \exp\left[jk_0(C - q_1)d\right],$$
$$\lambda_2 = \exp\left[jk_0(C - q_2)d\right]. \tag{2.76}$$

Solving Equation 2.75,

$$E_{x1}^P = 2\lambda_2(1 - C\eta_{y2})\frac{E_x^I}{\Delta}, \tag{2.77}$$

$$E_{x2}^P = -2\lambda_1(1 - C\eta_{y1})\frac{E_x^I}{\Delta}, \tag{2.78}$$

$$E_x^R = (1 - C\eta_{y1})(1 - C\eta_{y2})(\lambda_2 - \lambda_1)\frac{E_x^I}{\Delta}, \tag{2.79}$$

$$E_x^T = 2\lambda_1\lambda_2 C(\eta_{y1} - \eta_{y2})\frac{E_x^I}{\Delta}, \tag{2.80}$$

where

$$\Delta = \lambda_2(1 + C\eta_{y1})(1 - C\eta_{y2}) - \lambda_1(1 - C\eta_{y1})(1 + C\eta_{y2}). \tag{2.81}$$

The power reflection coefficient $\rho = \left|E_x^R/E_x^I\right|^2$ is given by

$$\rho = \frac{1}{1 + \left(\left((2C\epsilon_p q_1)/(C^2\epsilon_p^2 - q_1^2)\right)\cosec(2\pi\Omega q_1 d_p)\right)^2}, \tag{2.82}$$

where

$$d_p = \frac{d}{\lambda_p}, \tag{2.83}$$

$$\lambda_p = \frac{2\pi c}{\omega_p}. \tag{2.84}$$

In the above equations, λ_p is the free space wavelength corresponding to the plasma frequency and is used to normalize the slab width. Substituting for q_1, in the range of real q_1

$$\rho = \frac{1}{1 + B\cosec^2 A}, \quad \frac{1}{C} < \Omega < \infty, \tag{2.85}$$

and in the range of imaginary q_1

$$\rho = \frac{1}{1 - B\cosech^2|A|}, \quad 0 < \Omega < \frac{1}{C}, \tag{2.86}$$

where A and B are given by

$$A = 2\pi\Omega q_1 d_p = 2\pi\sqrt{C^2\Omega^2 - 1}\,d_p, \tag{2.87}$$

$$B = \frac{4C^2\varepsilon_p^2 q_1^2}{\left(C^2\varepsilon_p^2 - q_1^2\right)^2} = \frac{4C^2\Omega^2(\Omega^2 - 1)^2(C^2\Omega^2 - 1)}{(2C^2\Omega^2 - \Omega^2 - C^2)^2}. \tag{2.88}$$

The power transmission coefficient $\tau = (1 - \rho)$ and is given by

$$\tau = \frac{|B|}{|B| + \sin^2 |A|}, \quad \frac{1}{C} < \Omega < \infty, \tag{2.89}$$

$$\tau = \frac{|B|}{|B| + \sinh^2 |A|}, \quad 0 < \Omega < \frac{1}{C}. \tag{2.90}$$

From Equation 2.85, $\rho = 0$ when $\sin A = 0$ or

$$A = n\pi, \quad n = 0, 1, 2, \ldots. \tag{2.91}$$

There is one more value of Ω for which $\rho = 0$. From Equation 2.88, when $C\varepsilon_p = q_1$, $B = \infty$, and from Equation 2.89, $\tau = 1$ and $\rho = 0$. This point corresponds to a frequency Ω_B given by

$$\Omega_B^2 = \frac{C^2}{2C^2 - 1}. \tag{2.92}$$

It is easily shown that Ω_B exists only for the p wave and is greater than $1/C$. In fact this point corresponds to the *Brewster angle* [4]. Figure 2.7 shows Ω_B versus $\cos\theta_B$, where θ_B is the Brewster angle.

Thus, in the frequency band $\Omega > 1/C$, the variation of ρ with slab width is oscillatory. Stratton [6] discussed these oscillations for a dielectric slab and associated them with the interference of internally reflected waves in the slab. In the case of the plasma slab, the variation of ρ with the source frequency is also oscillatory, since in this range the plasma behaves like a dispersive dielectric. The maxima of ρ are less than or equal to 1 and occur for Ω satisfying the transcendental equation

$$\tan A = fA, \tag{2.93}$$

where

$$f = \frac{C^2\Omega^2(\Omega^2 - 1)(C^2 + \Omega^2 - 2C^2\Omega^2)}{(2C^2\Omega^4 - 2C^2\Omega^2 + C^2 - \Omega^2)(1 + \Omega^2 - 2C^2\Omega^2)}. \tag{2.94}$$

Figure 2.8 shows ρ versus Ω for a normalized slab width $d_p = 1.0$ and $C = 0.5$ ($\theta_I = 60°$). The inset shows the variation of q with Ω. The oscillations can be clearly seen in the real range of q. Heald and Wharton [2] show similar curves

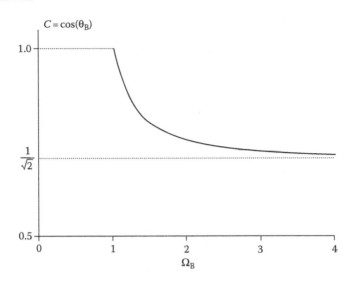

FIGURE 2.7　Brewster angle for a plasma medium.

for normal incidence on a lossy plasma slab using parameters normalized with reference to the source wave quantities.

In Figure 2.8, $\rho \approx 1$ in the imaginary range of q, showing total reflection of the source wave. However, it will be shown that there can be considerable *tunneling of power* for sufficiently thin plasma slabs.

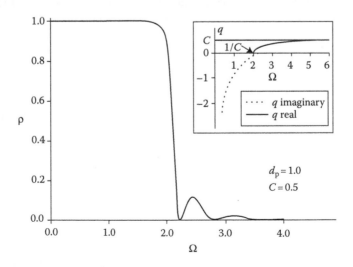

FIGURE 2.8　ρ versus Ω for an isotropic plasma slab (parallel polarization).

2.4.1 Tunneling of Power through a Plasma Slab [5,7]

For the frequency band $\Omega < 1/C$, the characteristic roots are imaginary and the waves excited in the plasma are evanescent. The incident wave is completely reflected by the semi-infinite plasma. However, in the case of a plasma slab, some power gets transmitted through it even in this frequency band. It can be shown that this tunneling effect is due to the interaction of the electric field of the positive wave with the magnetic field of the negative wave and vice versa.

The above statement is supported by examining the power flow through Poynting vector calculation. From Equations 2.77 through 2.81, one can obtain

$$E_x^T = 2C\varepsilon_p q_1 \exp\left[jk_0 Cd\right] \frac{E_x^I}{\Delta}, \tag{2.95}$$

$$E_{x1}^P = (q_1 + C\varepsilon_p)q_1 \exp\left[jk_0 q_1 d\right] \frac{E_x^I}{\Delta}, \tag{2.96}$$

$$E_{x2}^P = -(-q_1 + C\varepsilon_p)q_1 \exp\left[-jk_0 q_1 d\right] \frac{E_x^I}{\Delta}, \tag{2.97}$$

where

$$\Delta = 2q_1 C\varepsilon_p \cos(k_0 q_1 d) + j(q_1^2 + C^2\varepsilon_p^2)\sin(k_0 q_1 d). \tag{2.98}$$

The magnetic field components H_{y1}^P, H_{y2}^P, and H_y^T can be obtained from Equations 2.73 and 2.65.

The power crossing any plane in the plasma parallel to the interface can be found by calculating the z-component of the Poynting vector associated with the plasma waves. This is given by

$$S_z^P = \frac{1}{2}\mathrm{Re}\left[\left(E_{x1}^P \Psi_1^P + E_{x2}^P \Psi_2^P\right)\left(H_{y1}^{P*}\Psi_1^{P*} + H_{y2}^{P*}\Psi_2^{P*}\right)\right], \tag{2.99}$$

which on expansion gives

$$S_z^P = \frac{1}{2}\mathrm{Re}\left[E_{x1}^P \Psi_1^P H_{y1}^{P*}\Psi_1^{P*}\right] + \frac{1}{2}\mathrm{Re}\left[E_{x2}^P \Psi_2^P H_{y1}^{P*}\Psi_1^{P*}\right]$$
$$+ \frac{1}{2}\mathrm{Re}\left[E_{x1}^P \Psi_1^P H_{y2}^{P*}\Psi_2^{P*}\right] + \frac{1}{2}\mathrm{Re}\left[E_{x2}^P \Psi_2^P H_{y2}^{P*}\Psi_2^{P*}\right], \tag{2.100}$$

where Ψ_1^P and Ψ_2^P are given by Equation 2.72. The contribution from each of the four terms on the right-hand side of Equation 2.100 is now discussed for the two cases of q_1 real $\Omega > 1/C$ and q_1 imaginary $\Omega < 1/C$. It is to be noted that the second and third terms give contributions due to the cross interaction of fields in the positive and negative waves.

Case 1: q_1 real
It is easy to see that since the second and third terms are equal but opposite in sign, there is no contribution to the net power flow from the cross interaction. The sum of the first and fourth terms is equal to $(1/2)$ Re $[(E_x^T \Psi^T) (H_y^{T*} \Psi^{T*})]$, where Ψ^T is given by Equation 2.62. Thus, the power crossing any plane parallel to the slab interface gives exactly the power that emerges into free space at the other boundary of the slab.

Case 2: q_1 imaginary
It can be shown that each of the first and fourth terms is zero. This is because the corresponding electric and magnetic fields are in time quadrature. Furthermore, the second term is equal to the third term and each is equal to $(1/4)$ Re $[(E_x^T \Psi^T) (H_y^{T*} \Psi^{T*})]$. Thus, the power flow in the tunneling frequency band $(\Omega < 1/C)$ comes entirely from the cross interaction.

At $\Omega = 1$, $\varepsilon_p = 0$ and from Equation 2.88, $B = 0$ but $A \neq 0$. From Equation 2.89, $\tau = 0$. This point exists only for the p wave.
At $\Omega = 1/C$, $A=0$ and $B = 0$. By evaluating the limit, we obtain the power transmission coefficient τ_c at this point:

$$\tau_c = \frac{1}{1 + S^4 \pi^2 d_p^2}, \quad \Omega = \frac{1}{C}. \tag{2.101}$$

Numerical results of τ versus Ω are presented in Figures 2.9 and 2.10 for the tunneling frequency band. In Figure 2.9, the angle of incidence is taken to be 60° and the curves are given for various d_p. Between $\Omega = 0$ and 1 there is a value of $\Omega = \Omega_{max}$ where the transmission is maximum. The maximum value of the transmitted power decreases as d_p increases.

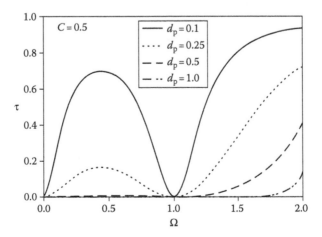

FIGURE 2.9 Transmitted power for an isotropic plasma slab (parallel polarization) in the tunneling range $(\Omega < 1/C)$ for various d_p.

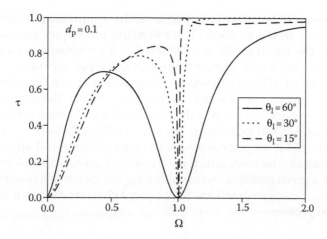

FIGURE 2.10 Transmitted power for an isotropic plasma slab (parallel polarization) in the tunneling range ($\Omega < 1/C$) for various angles of incidence.

In Figure 2.10, τ versus Ω is shown with the angle of incidence as the parameter. It is seen that the point of maximum transmission moves to the right as the angle of incidence decreases and merges with $\Omega = 1$ for $\theta_I = 0°$ (normal incidence). The maximum value of the transmitted power decreases as θ_I increases. This point is given by the solution of the transcendental equation

$$\tanh |A| = f\,|A|, \tag{2.102}$$

where f and A are given in Equations 2.94 and 2.87, respectively. It is suggested that by measuring Ω_{max} experimentally one can determine either the plasma frequency or the slab width, knowing the angle of incidence.

2.5 Inhomogeneous Slab Problem

Let us next consider the case where ε_p is a function of z in the region $0 < z < d$. To simplify and focus on the effect of the inhomogeneity of the properties of the medium, let us revert back to the normal incidence. The differential equation for the electric field is given by Equation 2.15. This equation cannot be solved exactly for a general $\varepsilon_p(z)$ profile.

However, there are a small number of profiles for which exact solutions can be obtained in terms of *special functions*. As an example, the problem of the *linear profile* can be solved in terms of *Airy functions*. An account of such solutions for a dielectric profile $\varepsilon_r(z)$ is given in Refs. [8,9] and for a plasma profile $\varepsilon_p(z, \omega)$ in Refs. [2,10]. Abramowitz and Stegun [11] give detailed information on special functions.

A vast amount of literature is available on the approximate solution for an inhomogeneous media problem. The following techniques are emphasized in this book. The actual profile is considered as a perturbation of a profile for which the exact solution is known. The effect of the perturbation is calculated through the use of a Green's function. The problem of a *fast profile* with finite range-length can be so handled by using a step profile as the reference. At the other end of the approximation scale, the *slow profile* problem can be handled by adiabatic and WKBJ techniques. See Refs. [1,8] for dielectric profile examples and Refs. [2,10] for plasma profile examples. Numerical approximation techniques can also be used, which are extremely useful in obtaining specific numbers for a given problem and in validating the theoretical results obtained from the physical approximations. Subsequent chapters discuss these aspects for temporal profiles.

2.5.1   Periodic Layers of the Plasma

Next, we consider a particular case of an inhomogeneous plasma medium which is periodic. Let the dielectric function

$$\varepsilon_p(z, \omega) = \varepsilon_p(z + mL, \omega), \tag{2.103}$$

where m is an integer and L is the spatial period. If the domain of m is all integers, $-\infty < m < \infty$, we have an unbounded periodic medium. The solution of Equation 2.15 for the infinite periodic structure problem [1,9,12] will be such that the electric field $E(z)$ differs from the electric field at $E(z + L)$ by a constant:

$$E(z + L) = CE(z). \tag{2.104}$$

The complex constant C can be written as

$$C = \exp(-j\beta L), \tag{2.105}$$

where β is the complex propagation constant of the periodic medium. Thus one can write

$$E(z) = E_\beta(z) \exp(-j\beta z). \tag{2.106}$$

From Equation 2.106,

$$E(z + L) = E_\beta(z + L) \exp(-j\beta(z + L)). \tag{2.107}$$

Also from Equations 2.104 and 2.106,

$$E(z + L) = E(z) \exp(-j\beta L) = E_\beta(z) \exp(-j\beta z) \exp(-j\beta L)$$
$$= E_\beta(z) \exp(-j\beta(z + L)). \tag{2.108}$$

From Equations 2.107 and 2.108, it follows that $E_\beta(z)$ is periodic:

$$E_\beta(z + L) = E_\beta(z). \tag{2.109}$$

Equation 2.106, where $E_\beta(z)$ is periodic, is called the Bloch wave condition. Such a periodic function can be expanded in a Fourier series:

$$E_\beta(z) = \sum_{m=-\infty}^{\infty} A_m \exp\left(\frac{-j2m\pi z}{L}\right), \tag{2.110}$$

and $E(z)$ can be written as

$$E(z) = \sum_{m=-\infty}^{\infty} A_m \exp\left[-j\left(\beta + \frac{2m\pi}{L}\right)z\right] = \sum_{m=-\infty}^{\infty} A_m \exp(-j\beta_m z), \tag{2.111}$$

where

$$\beta_m = \beta + \frac{2m\pi}{L}. \tag{2.112}$$

Taking into account both positive-going and negative-going waves, $E(z)$ can be expressed as [1]

$$E(z) = \sum_{m=-\infty}^{\infty} A_m \exp(-j\beta_m z) + \sum_{m=-\infty}^{\infty} B_m \exp(+j\beta_m z). \tag{2.113}$$

Let us next consider wave propagation in a periodic layered medium with plasma layers alternating with free space. The geometry of the problem is shown in Figure 2.11. Adopting the notation of [12], a unit cell consists of free space from $-l < z < l$ and a plasma layer of plasma frequency ω_{p0} from $l < z < l + d$. The thickness of the unit cell is $L = 2l + d$. The electric field $E(z)$ in the two layers of the unit cell can be written as

$$E(z) = A \exp\left(-j\frac{\omega}{c}z\right) + B \exp\left(-j\frac{\omega}{c}z\right), \quad -l \leq z \leq l, \tag{2.114}$$

$$E(z) = C \exp\left[-jn\frac{\omega}{c}(z-l)\right] + D \exp\left[jn\frac{\omega}{c}(z-l)\right], \quad l \leq z \leq l + d, \tag{2.115}$$

where n is the refractive index of the plasma medium:

$$n = \sqrt{\varepsilon_P} = \sqrt{1 - \frac{\omega_{p0}^2}{\omega^2}}. \tag{2.116}$$

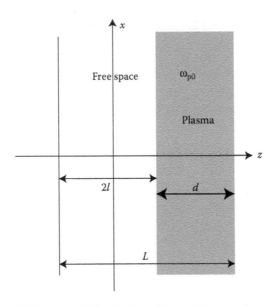

FIGURE 2.11 Unit cell of unbounded periodic media consisting of a layer of plasma.

The continuity of tangential electric and magnetic fields at the boundary translates into the continuity of E and $\partial E/\partial z$ at the interfaces:

$$E(l^-) = E(l^+), \tag{2.117}$$

$$\frac{\partial E}{\partial z}(l^-) = \frac{\partial E}{\partial z}(l^+), \tag{2.118}$$

$$E(l + d^-) = E(l + d^+), \tag{2.119}$$

$$\frac{\partial E}{\partial z}(l + d^-) = \frac{\partial E}{\partial z}(l + d^+). \tag{2.120}$$

From Equations 2.117 and 2.118,

$$A \exp\left(-j\frac{\omega}{c}l\right) + B \exp\left(j\frac{\omega}{c}l\right) = C + D, \tag{2.121}$$

$$A \exp\left(-j\frac{\omega}{c}l\right) - B \exp\left(j\frac{\omega}{c}l\right) = n(C - D). \tag{2.122}$$

Since $l + d^+ = -l^+ + L$, from Equation 2.104

$$E(l + d^+) = E(-l^+ + L) = \exp(-j\beta L)E(-l^+). \tag{2.123}$$

From Equations 2.119, 2.123, 2.114, and 2.115,

$$C \exp\left(-jn\frac{\omega}{c}d\right) + D \exp\left(jn\frac{\omega}{c}d\right) = \exp(-j\beta L)[A \exp\left(-j\frac{\omega}{c}l\right) + B \exp\left(j\frac{\omega}{c}l\right). \tag{2.124}$$

From Equations 2.120, 2.123, 2.114, and 2.115,

$$- n\frac{\omega}{c} C \exp\left(-jn\frac{\omega}{c}d\right) + n\frac{\omega}{c} D \exp\left(jn\frac{\omega}{c}d\right)$$

$$= \exp(-j\beta L)\left[-j\frac{\omega}{c} A \exp\left(-\frac{j\omega l}{c}\right) + j\frac{\omega}{c} B \exp\left(j\frac{\omega}{c}l\right)\right]. \qquad (2.125)$$

Equations 2.121, 2.122, 2.124, and 2.125 can be arranged as a matrix:

$$\begin{bmatrix} e^{-j\omega l/c} & e^{+j\omega l/c} & -1 & -1 \\ e^{-j\omega l/c} & -e^{j\omega l/c} & -n & n \\ e^{-j\omega l/c} & e^{+j\omega l/c} & -e^{j(\beta L-n\omega d/c)} & -e^{j(\beta L+n\omega d/c)} \\ e^{-j\omega l/c} & -e^{+j\omega l/c} & -ne^{j(\beta L-n\omega d/c)} & ne^{j(\beta L+n\omega d/c)} \end{bmatrix} \begin{bmatrix} A \\ B \\ C \\ D \end{bmatrix} = 0. \qquad (2.126)$$

A nonzero solution for the fields can be obtained by equating the determinant of the square matrix in Equation 2.126 to zero. This leads to the following dispersion relation:

$$\cos \beta L = \cos \frac{n\omega d}{c} \cos \frac{2\omega l}{c} - \frac{1}{2}\left[n + \frac{1}{n}\right]\sin \frac{n\omega d}{c}\sin \frac{2\omega l}{c}. \qquad (2.127)$$

The above equation can be studied by using reference values β_r and $\omega_r = \beta_r c = (2\pi c/\lambda_r)$. Thus Equation 2.127, in the normalized form, can be written as

$$\cos\left(\frac{\beta}{\beta_r}\frac{2\pi L}{\lambda_r}\right) = \cos\left(\frac{n\omega}{\omega_r}\frac{2\pi d}{\lambda_r}\right)\cos\left(\frac{2\omega}{\omega_r}\frac{2\pi l}{\lambda_r}\right)$$

$$- \frac{1}{2}\left(n + \frac{1}{n}\right)\sin\left(\frac{n\omega}{\omega_r}\frac{2\pi d}{\lambda_r}\right)\sin\left(\frac{2\omega}{\omega_r}\frac{2\pi l}{\lambda_r}\right). \qquad (2.128)$$

Figure 2.12 shows the graph ω/ω_r versus β/β_r for the following values of the parameters: $L = 0.6\lambda_r$, $d = 0.2\lambda$, and $\omega_{p0} = 1.2\omega_r$. We note from Figure 2.12 that the wave is evanescent in the frequency band $0 < \omega/\omega_r < 0.611$. This stop band is due to the plasma medium in the layers. If the layers are dielectric, this stop band will not be present. We have again a stop band in the frequency domain $0.877 < \omega/\omega_r < 1.225$. This stop band is due to the periodicity of the medium. Periodic dielectric layers do exhibit such a forbidden band. This principle is used in optics to construct *Bragg reflectors* with extremely large reflectance [9]. Figure 2.13 shows the ω–β diagram for dielectric layers with the refractive index $n = 1.8$. The first stop band in this case is given by $0.537 < \omega/\omega_r < 0.779$.

2.5.2 Surface Waves

Let us backtrack a little bit and consider the oblique incidence of a p wave on a plasma half-space. In this case only the outgoing wave will be excited in the

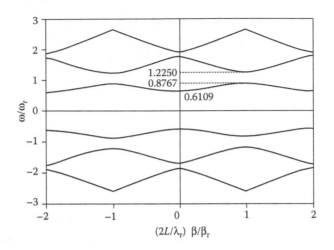

FIGURE 2.12 Dispersion relation of periodic plasma medium for $d = 0.2\lambda_r$, $L = 0.6\lambda_r$.

plasma half-space and Equation 2.75 becomes

$$\begin{bmatrix} 1 & -1 \\ C\eta_{y1} & 1 \end{bmatrix} \begin{bmatrix} E_{x1}^P \\ E_x^R \end{bmatrix} = \begin{bmatrix} 1 \\ 1 \end{bmatrix} E_x^I. \qquad (2.129)$$

The solution of Equation 2.129 gives fields of the reflected wave and the wave in the plasma half-space in terms of fields of the incident wave. If E_x^I is zero (no incident wave), we expect $E_{x1}^P = E_x^R = 0$. This is true in general, with one exception. The exception occurs when the determinant of the square

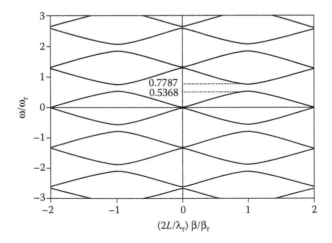

FIGURE 2.13 Dispersion relation of periodic dielectric layers for $d = 0.2\lambda_r$, $L = 0.6\lambda_r$. The refractive index of the dielectric layer is 1.8.

matrix on the left-hand side of Equation 2.129 is zero. In such a case, E^P_{x1} and E^R_x can be nonzero even if E^I_x is zero, indicating the possibility of the existence of fields in free space even if there is no incident field. For the exceptional solution, the reflection coefficient is infinity, that is, $E^R_x \neq 0$, but $E^I_x = 0$. The solution is an eigenvalue solution and gives the dispersion relation for surface plasmons. From Equations 2.129 and 2.73, the dispersion relation is obtained as

$$1 + C\eta_{y1} = 1 + C\frac{\varepsilon_p}{q_1} = 0. \tag{2.130}$$

Noting that

$$k_0 q_1 = \sqrt{k_0^2 \varepsilon_p - k_x^2} \tag{2.131}$$

and

$$k_0^2 C^2 = (k_0^2 - k_x^2), \tag{2.132}$$

where

$$k_0 = \frac{\omega}{c}, \tag{2.133}$$

$$k_x = k_0 S, \tag{2.134}$$

the dispersion relation can be written as

$$k_x^2 = \frac{\omega^2}{c^2} \frac{\varepsilon_p}{1 + \varepsilon_p}. \tag{2.135}$$

The exponential factors for $z < 0$ and $z > 0$ can be written as

$$\psi^{P1} = \exp[-jk_x x - jk_0 q_1 z], \quad z > 0, \tag{2.136}$$

$$\psi^R = \exp[-jk_x x + jk_0 C z], \quad z < 0. \tag{2.137}$$

It can be shown that k_x will be real and both

$$\alpha_1 = jk_0 q_1 \tag{2.138}$$

and

$$\alpha_2 = jk_0 C = j\sqrt{k_0^2 - k_x^2} \tag{2.139}$$

will be real and positive if

$$\varepsilon_p < -1. \tag{2.140}$$

When Equation 2.140 is satisfied,

$$\psi^{P1} = \exp(-\alpha_1 z)\exp(-jk_x x), \quad z > 0, \tag{2.141}$$

$$\psi^{R} = \exp(\alpha_2 z)\exp(-jk_x x), \quad z < 0, \tag{2.142}$$

where α_1 and α_2 are positive real quantities. Equations 2.141 and 2.142 show that the waves, while propagating along the surface, attenuate in the direction normal to the surface. For this reason, the wave is called a surface wave. In a plasma when $\omega < \omega_p/\sqrt{2}$, ε_p will be less than -1. Thus an interface between free space and a plasma medium can support a surface wave.

Defining the refractive index of the surface mode as n, that is,

$$n = \frac{c}{\omega}k_x, \tag{2.143}$$

Equation 2.135 can be written as

$$n^2 = \frac{\Omega^2 - 1}{2(\Omega^2 - 1/2)}, \quad \Omega < \frac{1}{\sqrt{2}}. \tag{2.144}$$

For $0 < \omega < \omega_p/\sqrt{2}$, n is real and the surface wave propagates. In this interval of ω, the dielectric constant $\varepsilon_p < -1$. The surface wave mode is referred to, in the literature, as Fano mode [13] and also as a nonradiative surface plasmon. In the interval $1 < \Omega < \infty$, n^2 is positive and less than 0.5. However, the α values obtained from Equations 2.138 and 2.139 are imaginary. The wave is not bound to the interface and the mode, called Brewster mode, in this case is radiative and referred to as a radiative surface plasmon. Figure 2.14 sketches $\varepsilon_p(\omega)$ and $n^2(\omega)$. Another way of presenting the information is through the Ω–K diagram, where Ω and K are normalized frequency and wave number, respectively, that is,

$$\Omega = \frac{\omega}{\omega_p}, \tag{2.145}$$

$$K = \frac{ck_x}{\omega_p} = n\Omega. \tag{2.146}$$

From Equations 2.143 and 2.144, we obtain the relation

$$K^2 = n^2\Omega^2 = \frac{\Omega^2(\Omega^2 - 1)}{2\Omega^2 - 1}. \tag{2.147}$$

Figure 2.15 shows the Ω–K diagram.

The Fano mode is a surface wave called the surface plasmon and exists for $\omega < \omega_p/\sqrt{2}$. The surface mode does not have a low-frequency cutoff. It

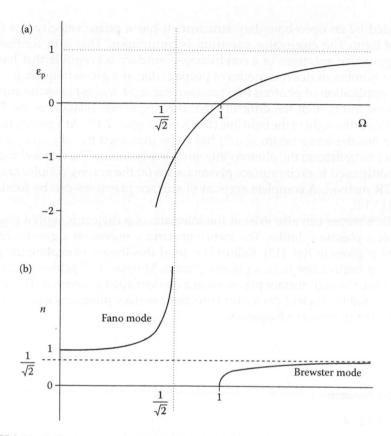

FIGURE 2.14 Refractive index for surface plasmon modes.

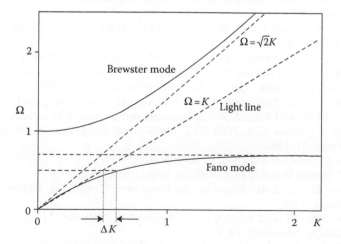

FIGURE 2.15 Ω–K diagram for surface plasmon modes.

is guided by an open-boundary structure. It has a phase velocity less than that of light. The eigenvalue spectrum is continuous. This is in contrast to the eigenvalue spectrum of a conducting-boundary waveguide that has an infinite number of discrete modes of propagation at a given frequency.

The application of photons (electromagnetic light waves) to excite surface plasmons meets with the difficulty that the dispersion relation of the Fano mode lies to the right of the light line $\Omega = K$ (see Figure 2.15). At a given photon energy $\hbar\omega$, the wave vector $\hbar(\omega/c)$ has to be increased by $\Delta k_x = (\omega_p/c)\Delta K$ in order to transform the photons into surface plasmons. The two techniques commonly used to excite surface plasmons are (a) the grating coupler and (b) the ATR method. A complete account of surface plasmons can be found in Refs. [13,14].

Surface waves can also exist at the interface of a dielectric with a plasma layer or a plasma cylinder. The theory of surface waves on a gas-discharge plasma is given in Ref. [15]. Kalluri [16] used this theory to explore the possibility of backscatter from a plasma plume. Moissan [17] achieved plasma generation through surface plasmons in a device called a *surfatron*. The surfatron is a highly efficient device for launching surface plasmons that produce plasma at a microwave frequency.

References

1. Ishimaru, A., *Electromagnetic Wave Propagation, Radiation, and Scattering*, Prentice-Hall, Englewood Cliffs, NJ, 1991.
2. Heald, M. A. and Wharton, C. B., *Plasma Diagnostics with Microwaves*, Wiley, New York, 1965.
3. Forstmann, F. and Gerhardts, R. R., *Metal Optics Near the Plasma Frequency*, Springer, New York, 1986.
4. Boardman, A. D., *Electromagnetic Surface Modes*, Wiley, New York, 1982.
5. Kalluri, D. K. and Prasad, R. C., Thin film reflection properties of isotropic and uniaxial plasma slabs, *Appl. Sci. Res.* (Netherlands), 27, 415, 1973.
6. Stratton, J. A., *Electromagnetic Theory*, McGraw-Hill, New York, 1941.
7. Kalluri, D. K. and Prasad, R. C., Transmission of power by evanescent waves through a plasma slab, *1980 IEEE International Conference on Plasma Science*, Madison, p. 11, 1980.
8. Lekner, J., *Theory of Reflection*, Kluwer, Boston, 1987.
9. Yeh, P., *Optical Waves in Layered Media*, Wiley, New York, 1988.
10. Budden, K. G., *Radio Waves in the Ionosphere*, Cambridge University Press, Cambridge, 1961.
11. Abramowitz, M. and Stegun, I. A., *Handbook of Mathematical Functions*, Dover Publications, New York, 1965.
12. Kuo, S. P. and Faith, J., Interaction of an electromagnetic wave with a rapidly created spatially periodic plasma, *Phys. Rev. E*, 56, 1, 1997.

13. Boardman, A. D., Hydrodynamic theory of plasmon–polaritons on plane surfaces, In: A. D. Boardman (Ed.), *Electromagnetic Surface Modes*, Chapter 1, Wiley, New York, 1982.
14. Raether, H., *Surface Plasmons*, Springer, New York, 1988.
15. Shivarova, A. and Zhelyazkov, I., Surface waves in gas-discharge plasmas, In: A. D. Boardman (Ed.), *Electromagnetic Surface Modes*, Chapter 12, Wiley, New York, 1982.
16. Kalluri, D. K., Backscattering from a plasma plume due to excitation of surface waves, *Final Report Summer Faculty Research Program*, Air Force Office of Scientific Research, 1994.
17. Moissan, M., Beaudry, C., and Leprince, P., A new HF device for the production of long plasma column at a high electron density, *Phys. Lett.*, 50A, 125, 1974.

3

Time-Varying and Space-Invariant Isotropic Plasma Medium

3.1 Basic Equations

Let us assume that the electron density varies only in time:

$$\omega_p^2 = \omega_p^2(t). \tag{3.1}$$

One can then construct the basic solutions by assuming that the field variables vary harmonically in space:

$$\mathbf{F}(\mathbf{r}, t) = \mathbf{F}(t) \exp(-j\mathbf{k} \cdot \mathbf{r}). \tag{3.2}$$

Equations 1.1, 1.2, and 1.10 then reduce to

$$\mu_0 \frac{\partial \mathbf{H}}{\partial t} = j\mathbf{k} \times \mathbf{E}, \tag{3.3}$$

$$\varepsilon_0 \frac{\partial \mathbf{E}}{\partial t} = -j\mathbf{k} \times \mathbf{H} - \mathbf{J}, \tag{3.4}$$

$$\frac{d\mathbf{J}}{dt} = \varepsilon_0 \omega_p^2(t) \mathbf{E}. \tag{3.5}$$

From Equations 3.4 and 3.5 or from Equation 1.11, we obtain the wave equation for $\mathbf{E}$:

$$\frac{d^2 \mathbf{E}}{dt^2} + \left[k^2 c^2 + \omega_p^2(t) \right] \mathbf{E} - \mathbf{k}(\mathbf{k} \cdot \mathbf{E}) = 0. \tag{3.6}$$

From Equation 1.15,

$$\frac{d^2 \dot{\mathbf{H}}}{dt^2} + \left[k^2 c^2 + \omega_p^2(t) \right] \dot{\mathbf{H}} = 0. \tag{3.7}$$

The one-dimensional Equations 1.20 through 1.25 reduce to

$$\mu_0 \frac{\partial H}{\partial t} = jkE, \tag{3.8}$$

$$\varepsilon_0 \frac{\partial E}{\partial t} = -jkH - J, \tag{3.9}$$

$$\frac{dJ}{dt} = \varepsilon_0 \omega_p^2(t) E, \tag{3.10}$$

$$\frac{d^2 E}{dt^2} + \left[k^2 c^2 + \omega_p^2(t) \right] E = 0, \tag{3.11}$$

$$\frac{d^2 \dot{H}}{dt^2} + \left[k^2 c^2 + \omega_p^2(t) \right] \dot{H} = 0. \tag{3.12}$$

If we consider H rather than $\dot{H}$ as the dependent variable, Equation 3.12 becomes

$$\frac{d^3 H}{dt^3} + \left[k^2 c^2 + \omega_p^2(t) \right] \frac{dH}{dt} = 0. \tag{3.13}$$

The solution of Equations 3.11 and 3.13, when ω_p^2 is a constant, can easily be obtained as

$$E(t) = \sum_{m=1}^{3} E_m \exp(j\omega_m t), \tag{3.14}$$

$$H(t) = \sum_{m=1}^{3} H_m \exp(j\omega_m t), \tag{3.15}$$

where

$$\omega_{1,2} = \pm\sqrt{k^2 c^2 + \omega_p^2}, \tag{3.16}$$

$$\omega_3 = 0. \tag{3.17}$$

Given E_0 and H_0, the initial values of the electric and magnetic fields E_m and H_m can be determined by solving the initial value problem. A detailed discussion of such a solution is given in the next section by considering a canonical problem of the sudden-switching of an unbounded homogeneous plasma medium.

3.2 Reflection by a Suddenly Created Unbounded Plasma Medium

The geometry of the problem is shown in Figure 3.1. A plane wave of frequency ω_0 is propagating in free space in the z-direction. Suddenly at $t = 0$, a plasma

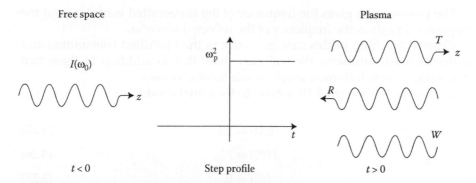

FIGURE 3.1 Suddenly created unbounded plasma medium.

medium is created. Thus a temporal discontinuity in the dielectric properties of the medium is created. Let the exponential wave function of the source wave, reflected wave, and the transmitted wave be given by

$$\Psi^I = \exp\left[j(\omega_0 t - k_0 z)\right], \tag{3.18}$$

$$\Psi^R = \exp\left[j(-\omega_r t - k_r z)\right], \tag{3.19}$$

$$\Psi^T = \exp\left[j(\omega_t t - k_t z)\right]. \tag{3.20}$$

In implementing the initial conditions of the continuity of the electric and magnetic fields at $t = 0$ for all z, we need to have the same coefficient of z in Equations 3.18 through 3.20 giving the condition

$$k_0 = k_r = k_t = k. \tag{3.21}$$

In the case of temporal discontinuity, the wave number is conserved. The free space wave number is given by

$$k_0 = \frac{\omega_0}{c}, \tag{3.22}$$

and the wave number in the plasma is given by

$$k_p = \frac{\omega}{c}\sqrt{\varepsilon_p} = \sqrt{\omega^2 - \omega_p^2}, \tag{3.23}$$

where ω is the frequency of the waves in the plasma. From Equations 3.21 through 3.23, we have the equation $\omega_0^2 = \omega^2 - \omega_p^2$ leading to

$$\omega = \pm\sqrt{\omega_0^2 + \omega_p^2}. \tag{3.24}$$

The positive sign gives the frequency of the transmitted wave* ω_t and the negative sign gives the frequency of the reflected wave* ω_r.

Plasma switching in this case gives rise to the upshifted transmitted and reflected wave solutions. We will next show that in addition to these two traveling wave solutions, a *wiggler* mode solution exists.

Equations 3.8 through 3.10, subject to the initial conditions

$$E(0) = E_0, \tag{3.25}$$

$$H(0) = H_0, \tag{3.26}$$

$$J(0) = 0, \tag{3.27}$$

describe an initial value problem that can be solved easily by several techniques. We choose to use the Laplace transform technique [1,2] to lay the foundation for the solution of the more difficult transient problem of a switched plasma half-space discussed in the next chapter. Defining the Laplace transform of a function $f(t)$,

$$\pounds\{f(t)\} = F(s) = \int_0^\infty f(t)\exp(-st)\,dt, \tag{3.28}$$

and noting that

$$\pounds\left\{\frac{df}{dt}\right\} = sF(s) - f(0), \tag{3.29}$$

we can convert Equations 3.8 through 3.10 into a matrix algebraic equation with the initial conditions appearing on the right-hand side as the excitation vector:

$$\begin{bmatrix} \mu_0 s & -jk & 0 \\ -jk & \varepsilon_0 s & 1 \\ 0 & -\varepsilon_0\omega_p^2 & s \end{bmatrix} \begin{bmatrix} H(s) \\ E(s) \\ J(s) \end{bmatrix} = \begin{bmatrix} \mu_0 H_0 \\ \varepsilon_0 E_0 \\ 0 \end{bmatrix}. \tag{3.30}$$

The time domain solution can then be obtained by computing the state transition matrix [1] or more simply by solving for the s domain field variables and taking the Laplace inverse of each of the variables. The s-domain variables

---

* In the literature on the subject, several alternative names were used for the waves generated by the switching action. Alternative names for the transmitted wave are: (1) right-going wave; (2) positive-going wave; and (3) forward-propagating wave. Alternative names for the reflected wave are: (1) left-going wave; (2) negative-going wave; and (3) backward-propagating wave.

are given by

$$H(s) = H_0 \frac{s^2 + \omega_p^2 + jkcs}{s(s^2 + \omega_p^2 + k^2c^2)}, \tag{3.31}$$

$$E(s) = E_0 \frac{s + jkc}{s^2 + \omega_p^2 + k^2c^2}, \tag{3.32}$$

$$J(s) = \frac{H_0}{c} \frac{\omega_p^2(s + jkc)}{s(s^2 + \omega_p^2 + k^2c^2)}. \tag{3.33}$$

It can be noted that there are two poles in the expression for $E(s)$, whereas $H(s)$ and $J(s)$ have an additional pole at the origin. The time domain solution obtained by computing the residues at the poles can be written as the sum of three modes given in Equations 3.14, 3.15, and 3.34:

$$J(t) = \sum_{m=1}^{3} J_m \exp(j\omega_m t). \tag{3.34}$$

The frequency and the fields of each mode are listed below:

Mode 1 ($m = 1$):

$$\omega_1 = \sqrt{k^2c^2 + \omega_p^2(t)} = \sqrt{\omega_0^2 + \omega_p^2}, \tag{3.35}$$

$$\frac{E_1}{E_0} = \frac{\omega_1 + \omega_0}{2\omega_1}, \tag{3.36}$$

$$\frac{H_1}{H_0} = \frac{\omega_0}{\omega_1} \frac{E_1}{E_0}, \tag{3.37}$$

$$J_1 = -j\varepsilon_0 \frac{\omega_p^2}{\omega_1} E_1. \tag{3.38}$$

Mode 2 ($m = 2$):

$$\omega_2 = -\sqrt{\omega_0^2 + \omega_p^2} = -\omega_1, \tag{3.39}$$

$$\frac{E_2}{E_0} = \frac{\omega_2 + \omega_0}{2\omega_2} = \frac{\omega_1 - \omega_0}{2\omega_1}, \tag{3.40}$$

$$\frac{H_2}{H_0} = \frac{\omega_0}{\omega_2} \frac{E_2}{E_0} = -\frac{\omega_0}{\omega_1} \frac{E_2}{E_0}, \tag{3.41}$$

$$J_2 = -j\varepsilon_0 \frac{\omega_p^2}{\omega_2} E_2. \tag{3.42}$$

Mode 3 ($m = 3$):

$$\omega_3 = 0, \tag{3.43}$$

$$\frac{E_3}{E_0} = 0, \tag{3.44}$$

$$\frac{H_3}{H_0} = \frac{\omega_p^2}{\omega_0^2 + \omega_p^2}, \tag{3.45}$$

$$J_3 = j\frac{\omega_0}{c}H_3, \tag{3.46}$$

$$v_3 = -\frac{1}{Nq}J_3. \tag{3.47}$$

Modes 1 and 2 are transverse electromagnetic waves (EMWs) whose electric and magnetic fields are related by the intrinsic impedance of the plasma medium at the frequency ω_m (ω_m is an algebraic quantity including sign):

$$\frac{E_1}{H_1} = \frac{E_0}{H_0}\frac{\omega_1}{\omega_0} = \frac{\eta_0}{\sqrt{1 - \omega_p^2/\omega_1^2}} = \frac{\eta_0}{\sqrt{\varepsilon_p(\omega_1)}} = \eta_{p1}, \tag{3.48}$$

$$\frac{E_2}{H_2} = \eta_0\frac{\omega_2}{\omega_0} = \eta_{p2}. \tag{3.49}$$

Mode 3 is the wiggler mode discussed in the next section [3–5]. In the presence of a static magnetic field in the z-direction, the third mode becomes a traveling wave with a downshifted frequency. See Chapter 7 for a thorough discussion of this aspect.

3.3 ω–k Diagram and the Wiggler Magnetic Field

The ω–k diagram for the problem under discussion can be obtained from Equation 3.13 by assuming the time variation of the fields as $\exp(j\omega t)$. Under this assumption, the differentiation in time domain is equivalent to the multiplication by $(j\omega)$ in the frequency domain:

$$(j\omega)^3 + \left[k^2c^2 + \omega_p^2\right](j\omega) = 0,$$
$$\omega\left[\omega^2 - (k^2c^2 + \omega_p^2)\right] = 0. \tag{3.50}$$

Figure 3.2 shows the ω–k diagram where the top and bottom branches are due to the factor in the square brackets equal to zero, and the horizontal line

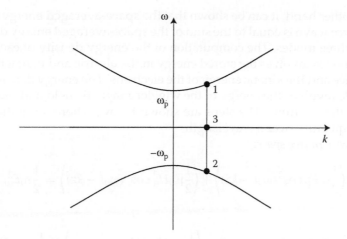

FIGURE 3.2 ω–k diagram and the wiggler magnetic field.

is due to the factor $\omega = 0$. The line $k = $ constant is a vertical line and intersects the ω–k diagram at the three points shown as 1, 2, and 3. The third mode ($\omega = 0$) is the wiggler mode. Its real fields are

$$\mathbf{E}_3(x, y, z, t) = 0, \tag{3.51}$$

$$\mathbf{H}_3(x, y, z, t) = \hat{y} H_0 \frac{\omega_p^2}{\omega_0^2 + \omega_p^2} \cos(kz), \tag{3.52}$$

$$\mathbf{J}_3(x, y, z, t) = -\hat{x} H_0 k \frac{\omega_p^2}{\omega_0^2 + \omega_p^2} \sin(kz), \tag{3.53}$$

$$\mathbf{v}_3(x, y, z, t) = -\frac{1}{Nq} \mathbf{J}_3(x, y, z, t). \tag{3.54}$$

3.4 Power and Energy Considerations

In this section, we show that unlike the case of spatial discontinuity, for a temporal discontinuity, the real power density of the source wave is not equal to the sum of real power densities of the three modes [6]. Let $S_m = E_m H_m$. From the expressions for the fields of the various modes given above,

$$\left| \frac{S_1}{S_0} \right| + \left| \frac{S_2}{S_0} \right| + \left| \frac{S_3}{S_0} \right| = \frac{1}{2} \frac{\omega_0}{\omega_1} \left[1 + \frac{\omega_0^2}{\omega_1^2} \right] \neq 1. \tag{3.55}$$

Note that the power density of the third mode is zero.

On the other hand, it can be shown that the space-averaged energy density of the source wave is equal to the sum of the space-averaged energy densities of all the three modes. The computation of the energy density of each of the first two modes involves the stored energy in the electric and magnetic fields in free space and the kinetic energy of the electrons. The energy density of the third mode involves the energy of the wiggler magnetic field and the kinetic energy of the electrons. The steps are shown below, where $\langle w \rangle$ indicates an averaged quantity over a wavelength.

Source wave in free space:

$$\langle w_0 \rangle = \left\langle \frac{1}{2}\varepsilon_0 E_0^2 \cos^2(\omega_0 t - kz) \right\rangle + \left\langle \frac{1}{2}\mu_0 H_0^2 \cos^2(\omega_0 t - kz) \right\rangle = \frac{1}{2}\varepsilon_0 E_0^2. \quad (3.56)$$

Mode 1:

$$\langle w_1 \rangle = \left\langle \frac{1}{2}\varepsilon_0 E_1^2 \cos^2(\omega_1 t - kz) \right\rangle + \left\langle \frac{1}{2}\mu_0 H_1^2 \cos^2(\omega_1 t - kz) \right\rangle + \frac{1}{2}Nmv_1^2. \quad (3.57)$$

From Equation 3.38 and the relation $J_1 = -qNv_1$, the expression for the instantaneous velocity field v_1 can be obtained as

$$v_1(x, y, z, t) = -\frac{qE_1}{m\omega_1} \sin(\omega_1 t - kz). \quad (3.58)$$

Substituting Equation 3.58 into Equation 3.57 and simplifying, we obtain

$$\frac{\langle w_1 \rangle}{\langle w_0 \rangle} = \left(\frac{E_1}{E_0}\right)^2 = \frac{1}{4}\left[1 + \frac{\omega_0}{\omega_1}\right]^2, \quad (3.59)$$

Mode 2:
Similarly,

$$\frac{\langle w_2 \rangle}{\langle w_0 \rangle} = \left(\frac{E_2}{E_0}\right)^2 = \frac{1}{4}\left[1 - \frac{\omega_0}{\omega_1}\right]^2. \quad (3.60)$$

Mode 3:
From Equations 3.51 through 3.52 or Equations 3.43 through 3.47, we have

$$w_3 = \frac{1}{2}\mu_0 [H_3 \cos(kz)]^2 + \frac{1}{2}mN\left[\frac{k}{Nq}H_3 \sin(kz)\right]^2 \quad (3.61)$$

and

$$\frac{\langle w_3 \rangle}{\langle w_0 \rangle} = \frac{1}{2}\left[\frac{\omega_p^2}{\omega_0^2 + \omega_p^2}\right]^2\left[1 + \frac{\omega_0^2}{\omega_p^2}\right] = \frac{1}{2}\frac{\omega_p^2}{\omega_0^2 + \omega_p^2}. \quad (3.62)$$

From Equations 3.59, 3.60, and 3.62, we have

$$\frac{\langle w_1 \rangle}{\langle w_0 \rangle} + \frac{\langle w_2 \rangle}{\langle w_0 \rangle} + \frac{\langle w_3 \rangle}{\langle w_0 \rangle} = 1. \quad (3.63)$$

3.5 Perturbation from Step Profile*

A step profile is a useful approximation for a fast profile with a small rise time. In this section, we will consider a perturbation technique to compute the correction terms for a fast profile. An important step in this technique is the construction of Green's function $G(t, \tau)$ for the switched plasma medium, which is considered in the next section.

Figure 3.3 shows the geometry of the problem. The equation for the electric field in the ideal sudden-switching case (step profile) is given by

$$\frac{d^2 E_0(t)}{dt^2} + \tilde{\omega}^2(t)E_0(t) = 0, \tag{3.64}$$

where $\tilde{\omega}^2(t) = k^2 c^2 + \tilde{\omega}_p^2(t)$, $\tilde{\omega}_p^2(t)$ is a step function and k is the conserved wave number [1].

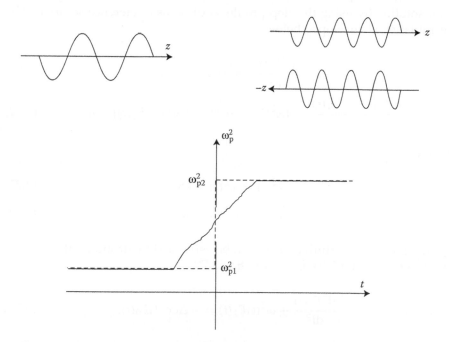

FIGURE 3.3 Perturbation from step profile. Geometry of the problem. (From Huang, T., et al., *IEEE Trans. Plasma Sci.*, 26(1), 1998. With permission.)

* © 1998 IEEE. Sections 3.5 through 3.10 and Figures 3.3 through 3.10 are reprinted from *IEEE Trans. Plasma Sci.*. Huang, T. T., et al., *IEEE Trans. Plasma Sci.*, 26(1), xxx–xxx, 1998. With permission.

The solution of Equation 3.64 is already known [3]:

$$E_0(t) = e^{j\omega_1 t}, \quad t < 0, \tag{3.65}$$

$$E_0(t) = T_0 e^{j\omega_2 t} + R_0 e^{-j\omega_2 t}, \quad t > 0, \tag{3.66}$$

where

$$T_0 = \frac{\omega_2 + \omega_1}{2\omega_2}, \tag{3.67}$$

$$R_0 = \frac{\omega_2 - \omega_1}{2\omega_2}, \tag{3.68}$$

$$\omega_1^2 = k^2 c^2 + \omega_{p1}^2, \tag{3.69}$$

$$\omega_2^2 = k^2 c^2 + \omega_{p2}^2. \tag{3.70}$$

For a general profile (Figure 3.3) of $\omega_p^2(t)$, we seek to have a perturbation solution by using the step profile solution as a reference solution. The formulation is developed below:

$$\frac{d^2 E(t)}{dt^2} + \omega^2(t) E(t) = 0, \tag{3.71}$$

$$\frac{d^2 [E_0(t) + E_1(t)]}{dt^2} + [\tilde{\omega}^2(t) + \Delta\omega^2(t)][E_0(t) + E_1(t)] = 0, \tag{3.72}$$

where

$$\omega^2(t) = \omega_p^2(t) + k^2 c^2, \tag{3.73}$$

$$\Delta\omega^2(t) = \omega^2(t) - \tilde{\omega}^2(t). \tag{3.74}$$

The first-order perturbation E_1 can be estimated by dropping the second-order term, $\Delta\omega^2(t) E_1(t)$, from Equation 3.72:

$$\frac{d^2 E_1(t)}{dt^2} + \tilde{\omega}^2(t) E_1(t) = -\Delta\omega^2(t) E_0(t). \tag{3.75}$$

Treating the right-hand side of Equation 3.75 as the driving function $F(t)$, we can write the solution of E_1:

$$E_1(t) = G(t, \tau) * F(\tau) = \int_{-\infty}^{+\infty} G(t, \tau)[-\Delta\omega^2(\tau) E_0(\tau)] \, d\tau, \tag{3.76}$$

where $G(t, \tau)$ is Green's function, which satisfies the second-order differential equation

$$\ddot{G}(t,\tau) + \tilde{\omega}^2(t)G(t,\tau) = \delta(t - \tau). \tag{3.77}$$

The approximate one-iteration solution for the original time-varying differential equation takes the form

$$E(t) \approx E_0(t) + E_1(t) = E_0(t) + \int_{-\infty}^{\infty} G(t,\tau)[-\Delta\omega^2(\tau)E_0(\tau)]\, d\tau. \tag{3.78}$$

More iterations can be made until the desired accuracy is reached. The general N-iterations solution is

$$E(t) \cong E_0(t) + \sum_{n=1}^{N} E_n(t), \tag{3.79}$$

where

$$E_n(t) = \int_{-\infty}^{\infty} G(t,\tau)[-\Delta\omega^2(\tau)E_{n-1}(\tau)]\, d\tau. \tag{3.80}$$

3.6 Causal Green's Function for Temporally-Unlike Plasma Media [7–9]

From the causality requirement,

$$G(t, \tau) = 0, \quad t < \tau. \tag{3.81}$$

For $t > \tau$, $G(t, \tau)$ will be determined from the impulse source conditions [10], that is,

$$G(\tau^+, \tau) = G(\tau^-, \tau), \tag{3.82}$$

$$\frac{\partial G(\tau^+, \tau)}{\partial t} - \frac{\partial G(\tau^-, \tau)}{\partial t} = 1. \tag{3.83}$$

From Equation 3.81, $G(\tau^-, \tau) = 0$ and $\partial G(\tau^-, \tau)/\partial t = 0$. Thus, Equations 3.82 and 3.83 reduce to

$$G(\tau^+, \tau) = 0, \tag{3.84}$$

$$\frac{\partial G(\tau^+, \tau)}{\partial t} = 1. \tag{3.85}$$

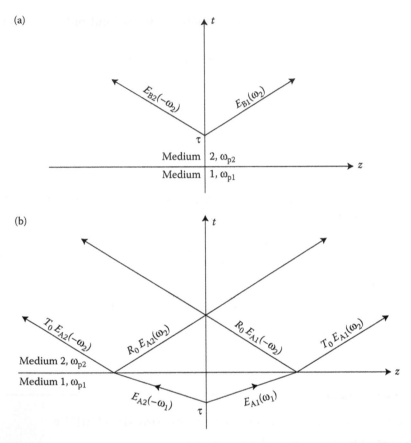

FIGURE 3.4 Geometrical interpretation of the Green's function for temporally-unlike plasma media. (From Huang, T., et al., *IEEE Trans. Plasma Sci.*, 26(1), 1998. With permission.)

A geometrical interpretation of the steps involved in constructing Green's function is given in Figure 3.4, where the horizontal axis is the spatial coordinate z along which the waves are propagating. The horizontal axis ($t = 0$) is also the temporal boundary between the two unlike plasma media.

In Figure 3.4a, we consider the case $\tau > 0$. The impulse source is located in the second medium. From Equation 3.77, Green's function can be written as

$$G(t, \tau) = E_{B1}(\tau)e^{j\omega_2 t} + E_{B2}(\tau)e^{-j\omega_2 t}, \quad t > \tau > 0. \tag{3.86}$$

E_{B1} and E_{B2} are determined from Equations 3.84 and 3.85:

$$E_{B1} = \frac{1}{2j\omega_2}e^{-j\omega_2 \tau}, \tag{3.87}$$

$$E_{B2} = -\frac{1}{2j\omega_2}e^{+j\omega_2 \tau}. \tag{3.88}$$

In Figure 3.4b, we consider the case $\tau < 0$. The impulse source is located in the first medium. Thus $G(t, \tau)$ in the interval $\tau < t < 0$ can be written as

$$G(t, \tau) = E_{A1}(\tau)e^{j\omega_1 t} + E_{A2}(\tau)e^{-j\omega_1 t}, \quad \tau < t < 0. \tag{3.89}$$

E_{A1} and E_{A2} are once again determined by Equations 3.84 and 3.85:

$$E_{A1} = \frac{1}{2j\omega_1}e^{-j\omega_1 \tau}, \tag{3.90}$$

$$E_{A2} = -\frac{1}{2j\omega_1}e^{+j\omega_1 \tau}. \tag{3.91}$$

Referring to Figure 3.4b, we note that the two waves launched by the impulse source at $t = \tau$ travel along the positive z-axis (transmitted wave) and along the negative z-axis (reflected wave). They reach the temporal interface at $t = 0$ where the medium undergoes a temporal change. Each of these waves gives rise to two waves at the upshifted frequency ω_2. The amplitude of the four waves in the interval $0 < t < \infty$ can be obtained by taking into account the amplitude of each of the incident waves (incident on the temporal interface) and the scattering coefficients. The geometrical interpretation of the amplitude computation of the four waves is shown in Figure 3.4b. Green's function for this interval $t > 0$ and $\tau < 0$ is given by

$$G(t, \tau) = E_{A1}(\tau)\left[T_0 e^{j\omega_2 t} + R_0 e^{-j\omega_2 t}\right] + E_{A2}(\tau)\left[R_0 e^{j\omega_2 t} + T_0 e^{-j\omega_2 t}\right],$$

$$\tau < 0 < t. \tag{3.92}$$

Explicit expressions for $G(t, \tau)$ valid for various regions of the (t, τ) plane are given in Figure 3.5. The difference between this causal Green's function and the Green's function for spatially unlike dielectric media given in Refs. [11,12] can be noted.

3.7 Transmission and Reflection Coefficients for a General Profile [7,8]

The approximate amplitude of the electric field after one iteration ($N = 1$) can be obtained from Equation 3.76:

$$E(t) \approx E_0(t) + E_1(t)$$

$$= E_0(t) - \int_{-\infty}^{+\infty} d\tau \, G(t, \tau)\Delta\omega^2(\tau)[E_0(\tau)]. \tag{3.93}$$

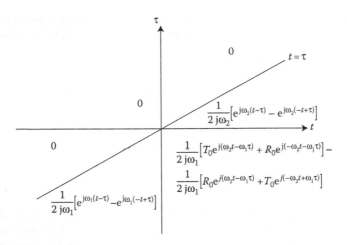

FIGURE 3.5 Causal Green's function for temporally-unlike plasma media. (From Huang, T., et al., *IEEE Trans. Plasma Sci.*, 26(1), 1998. With permission.)

For $t > 0$,

$$E(t) \approx E_0(t) - \int_0^t d\tau \frac{[e^{j\omega_2(t-\tau)} - e^{j\omega_2(-t+\tau)}]}{2j\omega_2} \Delta\omega^2(\tau)[T_0 e^{j\omega_2\tau} + R_0 e^{-j\omega_2\tau}]$$

$$- \int_{-\infty}^0 d\tau \frac{\begin{bmatrix} +T_0 e^{j(\omega_2 t-\omega_1\tau)} + R_0 e^{j(-\omega_2 t-\omega_1\tau)} \\ -R_0 e^{j(\omega_2 t+\omega_1\tau)} - T_0 e^{j(-\omega_2 t+\omega_1\tau)} \end{bmatrix}}{2j\omega_1} \Delta\omega^2(\tau)[e^{j\omega_1\tau}]. \qquad (3.94)$$

In the asymptotic limit of $t \to \infty$, $E_1(t)$ can be written as

$$E_1(t) = R_1 e^{-j\omega_2 t} + T_1 e^{+j\omega_2 t}, \qquad (3.95)$$

where R_1 and T_1 are the first-order correction terms of the reflection coefficient and the transmission coefficient, respectively.

$$R_1 = -\int_0^{+\infty} d\tau \Delta\omega^2(\tau) \frac{-(R_0 + T_0 e^{j^2\omega_2\tau})}{2j\omega_2} - \int_{-\infty}^0 d\tau \Delta\omega^2(\tau) \frac{(R_0 - T_0 e^{j^2\omega_1\tau})}{2j\omega_1},$$

$$\qquad (3.96)$$

$$T_1 = -\int_0^{+\infty} d\tau \Delta\omega^2(\tau) \frac{(T_0 + R_0 e^{-j^2\omega_2\tau})}{2j\omega_2} - \int_{-\infty}^0 d\tau \Delta\omega^2(\tau) \frac{(T_0 - R_0 e^{j^2\omega_1\tau})}{2j\omega_1}.$$

$$\qquad (3.97)$$

Higher-order correction terms ($N > 1$) can be obtained by using more iterations.

3.8 Transmission and Reflection Coefficients for a Linear Profile [7,8]

As a particular example, we use Equations 3.96 and 3.97 to compute R_1 and T_1 for a profile of $\omega_p^2(t)$ rising linearly from 0 (at $t = T_r/2$) to ω_{p2}^2 (at $t = T_r/2$) with a slope of ω_{p2}^2/T_r. Here T_r is the rise time of the profile. The function $\Delta\omega^2(\tau)$ takes the form of $\omega_{p2}^2(\tau/T_r - 1/2)$ for $0 < \tau < T_r/2$ and $\omega_{p2}^2(\tau/T_r + 1/2)$ for $-T_r/2 < \tau < 0$. It is zero on the rest of the real line.

$$
R_1 = -\int_0^{T_r/2} d\tau\,\omega_{p2}^2\left(\frac{\tau}{T_r} - \frac{1}{2}\right)\frac{-\left(R_0 + T_0 e^{j2\omega_2\tau}\right)}{2j\omega_2} - \int_{-T_r/2}^0 d\tau\,\omega_{p2}^2\left(\frac{\tau}{T_r} + \frac{1}{2}\right)\frac{\left(R_0 - T_0 e^{j2\omega_1\tau}\right)}{2j\omega_1}
$$

$$
= j\omega_{p2}^2\left[-\frac{T_0\left(e^{j\omega_2 T_r} - 1 - j\omega_2 T_r\right)}{8\omega_2^3 T_r} + \frac{R_0 T_r}{16\omega_2} + \frac{T_0\left(e^{-j\omega_1 T_r} - 1 + jT_r\omega_1\right)}{8\omega_1^3 T_r} + \frac{R_0 T_r}{16\omega_1}\right],
$$

$$(3.98)$$

$$
T_1 = -\int_0^{T_r/2} d\tau\,\omega_{p2}^2\left(\frac{\tau}{T_r} - \frac{1}{2}\right)\frac{\left(T_0 + R_0 e^{-j2\omega_2\tau}\right)}{2j\omega_2} - \int_{-T_r/2}^0 d\tau\,\omega_{p2}^2\left(\frac{\tau}{T_r} + \frac{1}{2}\right)\frac{\left(T_0 - R_0 e^{j2\omega_1\tau}\right)}{2j\omega_1}
$$

$$
= j\omega_{p2}^2\left[+\frac{R_0\left(e^{-j\omega_2 T_r} - 1 - j\omega_2 T_r\right)}{8\omega_2^3 T_r} - \frac{T_0 T_r}{16\omega_2} + \frac{R_0\left(e^{-j\omega_1 T_r} - 1 + jT_r\omega_1\right)}{8\omega_1^3 T_r} + \frac{T_0 T_r}{16\omega_1}\right].
$$

$$(3.99)$$

The solution for other profiles can be obtained similarly.

However, to perform the integrations in Equations 3.96 and 3.97 for an arbitrary profile, it can be necessary to expand the exponentials in Equations 3.96 and 3.97 in a power series and then do the integration. For a given T_r it is possible to choose the relative positioning of $\omega_p^2(t)$ and $\tilde{\omega}_p^2(t)$ such that

$$
\int_{-\infty}^{+\infty} \Delta\omega^2(\tau)\,d\tau = 0. \tag{3.100}
$$

Such a choice will improve the accuracy of the power reflection coefficient if we choose to approximate the exponential terms in Equations 3.98 and 3.99 by keeping only one term (the dominant term). For the linear profile under consideration, Equation 3.100 is satisfied, and keeping only the dominant term, we obtain

$$
R \approx R_0 + R_1 = R_0 - \frac{\omega_{p2}^2 T_r^2}{24} T_0, \tag{3.101}
$$

$$
T \approx T_0 + T_1 = T_0 - \frac{\omega_{p2}^2 T_r^2}{24} R_0. \tag{3.102}
$$

3.9 Validation of the Perturbation Solution by Comparing with the Exact Solution [7,8]

To illustrate the validity of the method, the perturbation solution using Equations 3.98 and 3.99 is compared with the exact solution of a linear profile. For a linear profile, Equation 3.71 can be arranged in the form of the Airy differential equation by introducing a new variable ξ. The basic steps are shown below:

$$\xi = \omega^2(t) = k^2 c^2 + \omega_{p1}^2 + \frac{\omega_{p2}^2 - \omega_{p1}^2}{T_r} t, \quad 0 < t < T_r, \tag{3.103}$$

$$\frac{d\xi}{dt} = \frac{\omega_{p2}^2 - \omega_{p1}^2}{T_r} = \beta, \tag{3.104}$$

$$\frac{d^2 E}{dt^2} + \omega^2(t) E = 0, \tag{3.105}$$

$$\frac{d^2 E}{d\xi^2} \beta^2 + \xi E = 0, \tag{3.106}$$

$$\frac{d^2 E}{d\xi^2} + \frac{\xi}{\beta^2} E = 0. \tag{3.107}$$

The solution of Equation 3.107 is given by

$$E(t) = \begin{cases} e^{j\omega_1 t}, & t < 0, \\ c_1 Ai\left[-\beta^{-2/3}\xi\right] + c_2 Bi\left[-\beta^{-2/3}\xi\right], & 0 < t < T_r, \\ T e^{j\omega_2 t} + R e^{-j\omega_2 t}, & t > T_r, \end{cases} \tag{3.108}$$

where Ai and Bi are the Airy functions [13].

The four coefficients c_1, c_2, T, and R in Equation 3.108 can be determined by using the continuity conditions on E and $\dot{E}$ at the two temporal interfaces $t = 0$ and $t = T_r$. Thus, the scattering coefficients R and T are obtained but not given here to save space. It can be noted that the power reflection coefficient computed from the exact solution depends on T_r but is independent of the location of the starting point of the linear profile.

Figures 3.6 and 3.7 compare the perturbation solution (broken line) based on Equations 3.98 and 3.99 with the exact solution (solid line) for the power reflection coefficient $[\rho = |R|^2 (\omega_1/\omega_2)]$ and power transmission coefficient $[\tau = |T|^2 (\omega_1/\omega_2)]$ for a linear profile. The results are presented in a normalized form by taking $f_1 = \omega_1/2\pi = 1.0$. The other parameters are $f_{p1} = \omega_{p1}/2\pi = 0$ and $f_{p2} = \omega_{p2}/2\pi = 1.2$. The perturbation solution tracks the exact solution closely up to $T_r = 0.25$. (At this point, the power reflection coefficient is $1/4$ of the value for sudden-switching.) More iterations can

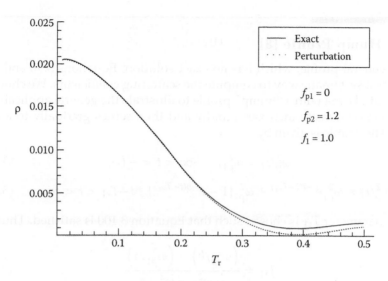

FIGURE 3.6 Power reflection coefficient (ρ) versus rise time (T_r) for a linear profile. (From Huang, T., et al., *IEEE Trans. Plasma Sci.*, 26(1), 1998. With permission.)

be taken by increasing N in Equation 3.80 to extend the range of validity of the perturbation solution. It can be noted that both ρ and τ decrease as T_r increases. A reduction in ρ does not result in an increase in τ. At a temporal discontinuity, $\rho + \tau \neq 1$ [6,14].

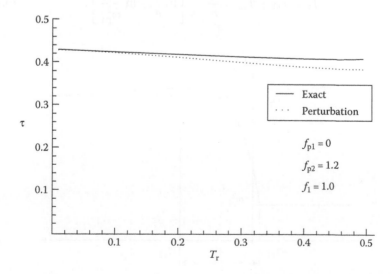

FIGURE 3.7 Power transmission coefficient (τ) versus rise time (T_r) for a linear profile. (From Huang, T., et al., *IEEE Trans. Plasma Sci.*, 26(1), 1998. With permission.)

3.10   Hump Profile [8]

For a general profile, which has no exact solution, Equations 3.96 and 3.97 provide a systematic way to compute the scattering coefficients. We choose a transient plasma with a "hump" profile to illustrate the general applicability. The electron density increases rapidly and then settles gradually to a new level. The profile is given by

$$\omega_p^2(t) = \omega_{p1}^2, \quad -\infty < t < -T_{r1},$$
(3.109)

$$\omega_p^2(t) = \omega_{p1}^2 e^{-a(t+T_{r1})} + \omega_{p2}^2[1 - e^{-b(t+T_{r1})}], \quad -T_{r1} < t < \infty.$$
(3.110)

The parameter T_{r1} is chosen such that Equation 3.100 is satisfied. Thus we obtain

$$T_{r1} = \frac{\left(\omega_{p2}^2/b\right) - \left(\omega_{p1}^2/a\right)}{\omega_{p2}^2 - \omega_{p1}^2}.$$
(3.111)

Figure 3.8 shows the profile. The profile has an extremum at $t = T_{r2}$, where

$$T_{r2} = \frac{1}{b-a}\left[\ln\frac{b}{a} + \ln\frac{\omega_{p2}^2}{\omega_{p1}^2}\right] - T_{r1}.$$
(3.112)

The rise time T_r is defined as

$$T_r = T_{r1} + T_{r2} = \frac{1}{b-a}\left[\ln\frac{b}{a} + \ln\frac{\omega_{p2}^2}{\omega_{p1}^2}\right].$$
(3.113)

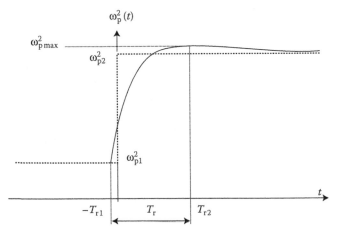

FIGURE 3.8   Sketch of a hump profile. (From Huang, T., et al., *IEEE Trans. Plasma Sci.*, 26(1), 1998. With permission.)

In the above, we restrict the two parameters a and b by the inequality

$$1 < \frac{b}{a} < \frac{\omega_{p2}^2}{\omega_{p1}^2}, \tag{3.114}$$

which ensures that T_{r1} is positive and the extremum is a maximum.

Equation 3.71 has no obvious exact solution for the "hump" profile. However, with the perturbation approach of Equations 3.96 and 3.97, we can solve for the transmission and reflection coefficients explicitly. For the hump profile,

$$\Delta\omega^2(t) = \begin{cases} 0, & -\infty < t < -T_{r1}, \\ \omega_{p1}^2 e^{-a(t+T_{r1})} + \omega_{p2}^2[1 - e^{-b(t+T_{r1})}] - \omega_{p1}^2, & -T_{r1} < t < 0, \\ \omega_{p1}^2 e^{-a(t+T_{r1})} + \omega_{p2}^2[1 - e^{-b(t+T_{r1})}] - \omega_{p2}^2, & 0 < t < \infty, \end{cases} \tag{3.115}$$

and the transmission and reflection coefficients, based on one iteration, are

$$R \approx R_0 - \int_0^{+\infty} d\tau \left[\omega_{p1}^2 e^{-a(\tau+T_{r1})} - \omega_{p2}^2 e^{-b(\tau+T_{r1})} \right] \frac{(R_0 + T_0 e^{2j\omega_2\tau})}{-2j\omega_2}$$

$$- \int_{-T_{r1}}^0 d\tau \left[\omega_{p1}^2 \left(e^{-a(\tau+T_{r1})} - 1 \right) - \omega_{p2}^2 \left(e^{-b(\tau+T_{r1})} - 1 \right) \right] \frac{(R_0 - T_0 e^{2j\omega_1\tau})}{2j\omega_1}$$

$$= R_0 + \frac{R_0}{2j\omega_2} \left(\frac{\omega_{p1}^2 e^{-aT_{r1}}}{a} - \frac{\omega_{p2}^2 e^{-bT_{r1}}}{b} \right) + \frac{T_0}{2j\omega_2} \left(\frac{\omega_{p1}^2 e^{-aT_{r1}}}{a - 2j\omega_2} - \frac{\omega_{p2}^2 e^{-bT_{r1}}}{b - 2j\omega_2} \right)$$

$$+ \frac{T_0}{2j\omega_1} \left((\omega_{p2}^2 - \omega_{p1}^2) \frac{1 - e^{-2j\omega_1 T_{r1}}}{2j\omega_1} - \omega_{p1}^2 \frac{e^{-aT_{r1}} - e^{-2j\omega_1 T_{r1}}}{a - 2j\omega_1} \right.$$

$$+ \omega_{p2}^2 \frac{e^{-bT_{r1}} - e^{-2j\omega_1 T_{r1}}}{b - 2j\omega_1} \left. \right) - \frac{R_0}{2j\omega_1} \left(\frac{\omega_{p1}^2}{a} \left(1 - e^{-aT_{r1}} - aT_{r1} \right) \right.$$

$$- \frac{\omega_{p1}^2}{b} \left(1 - e^{-bT_{r1}} - bT_{r1} \right) \left. \right), \tag{3.116}$$

$$T \approx T_0 - \int_0^{+\infty} d\tau \left[\omega_{p1}^2 e^{-a(\tau+T_{r1})} - \omega_{p2}^2 e^{-b(\tau+T_{r1})} \right] \frac{(T_0 + R_0 e^{-2j\omega_2\tau})}{2j\omega_2}$$

$$- \int_{-T_{r1}}^0 d\tau \left[\omega_{p1}^2 \left(e^{-a(\tau+T_{r1})} - 1 \right) - \omega_{p2}^2 \left(e^{-b(\tau+T_{r1})} - 1 \right) \right] \frac{(T_0 - R_0 e^{2j\omega_1\tau})}{2j\omega_1}$$

$$= T_0 - \frac{T_0}{2j\omega_2}\left(\frac{\omega_{p1}^2 e^{-aT_{r1}}}{a} - \frac{\omega_{p2}^2 e^{-bT_{r1}}}{b}\right) - \frac{R_0}{2j\omega_2}\left(\frac{\omega_{p1}^2 e^{-aT_{r1}}}{a+2j\omega_2} - \frac{\omega_{p2}^2 e^{-bT_{r1}}}{b+2j\omega_2}\right)$$

$$+ \frac{R_0}{2j\omega_1}\left((\omega_{p2}^2 - \omega_{p1}^2)\frac{1 - e^{-2j\omega_1 T_{r1}}}{2j\omega_1} - \omega_{p1}^2\frac{e^{-aT_{r1}} - e^{-2j\omega_1 T_{r1}}}{a - 2j\omega_1}\right.$$

$$+ \left.\omega_{p2}^2\frac{e^{-bT_{r1}} - e^{-2j\omega_1 T_{r1}}}{b - 2j\omega_1}\right) - \frac{T_0}{2j\omega_1}\left(\frac{\omega_{p1}^2}{a}\left(1 - e^{-aT_{r1}} - aT_{r1}\right)\right.$$

$$- \left.\frac{\omega_{p1}^2}{b}\left(1 - e^{-bT_{r1}} - bT_{r1}\right)\right), \tag{3.117}$$

Using Equations 3.116 and 3.117, the power reflection and transmission coefficients are computed and presented in Figures 3.9 and 3.10. The parameters are $f_{p1} = 0.6, f_{p2} = 1.2, f_1 = 1.0$, and $b/a = 2.5$. The value of T_r is changed from 0^+ to 0.5, by varying the individual values of a and b. The choice of a constant value for the ratio b/a results in the same $\omega_{p\,\text{max}}^2$ for various profiles with different T_r values. The unexpected slight increase of ρ at the start of the curve is worth noting. When T_r is 0^+, the effect of the hump is not felt and the scattering coefficients are those of the reference profile. As T_r increases, the effect of the hump is a rapid change of ω_p from ω_{p1} to $\omega_{p\,\text{max}}$ and the power scattering coefficients are close to the values expected for sudden-switching from ω_{p1} to $\omega_{p\,\text{max}}$ rather than ω_{p1} to ω_{p2}. As T_r increases further, the rapid change is no longer close to the sudden-switching approximation. R_1 and T_1 have significant values and result in the decrease of the power scattering coefficients.

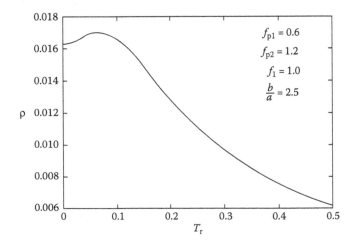

FIGURE 3.9 Power reflection coefficient (ρ) versus rise time ($T_r = T_{r1} + T_{r2}$) for a hump profile. (From Huang, T., et al., *IEEE Trans. Plasma Sci.*, 26(1), 1998. With permission.)

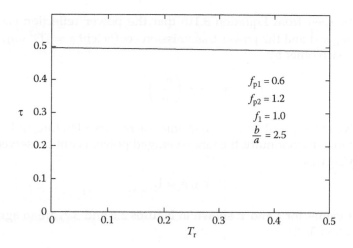

FIGURE 3.10 Power transmission coefficient (τ) versus rise time (T_r) for a hump profile. (From Huang, T., et al., *IEEE Trans. Plasma Sci.*, 26(1), 1998. With permission.)

By adjusting the four parameters of the hump profile, many practical transient plasma profiles can be studied.

3.11 Comparison Identities

If R_0 and T_0 are the scattering coefficients for the reference profile and R and T are the scattering coefficients for the actual profile, it can be shown that [15]

$$R^2 - T^2 = -\frac{\omega_1}{\omega_2} \tag{3.118}$$

and

$$RT_0 - R_0T = \frac{F(T_{r1}, T_{r2})}{2j\omega_2}, \tag{3.119}$$

where

$$F(T_{r1}, T_{r2}) = \int_{-T_{r1}}^{T_{r2}} \Delta\omega^2(t) E(t) E_0(t)\, dt, \tag{3.120}$$

and $\Delta\omega^2$ is given by Equation 3.74.

Equations 3.118 and 3.119 are the "comparison identities" analogous to such identities given in Ref. [11] for the space-varying case. Equation 3.118 can be called a conservation law. It does not alter for different profiles as long as they are identical for $t \le T_{r1}$ and $t \le T_{r2}$.

It can be seen from Equation 3.118 that the power reflection coefficient $\rho = |R|^2 \, (\omega_1/\omega_2)$ and the power transmission coefficient $\tau = |T|^2 \, (\omega_1/\omega_2)$ are related to each other by

$$\tau - \rho = \left(\frac{\omega_1}{\omega_2}\right)^2, \tag{3.121}$$

irrespective of the profile and the rise time. As remarked before, unlike in the case of spatial discontinuity, the time-averaged power is not conserved in the time-varying case:

$$\tau + \rho \neq 1. \tag{3.122}$$

The exact values for ρ and τ, shown in Figures 3.6 and 3.7, are in agreement with Equation 3.121.

References

1. Derusso, P. M., Roy, R. J., and Close, C. H., *State Variables for Engineers*, Wiley, New York, 1965.
2. Aseltine, J. A., *Transform Methods in Linear System Analysis*, McGraw-Hill, New York, 1958.
3. Jiang, C. L., Wave propagation and dipole radiation in a suddenly created plasma, *IEEE Trans. Ant.Prop.*, AP-23, 83, 1975.
4. Kalluri, D. K., On reflection from a suddenly created plasma half-space: Transient solution, *IEEE Trans. Plasma Sci.*, 16, 11, 1988.
5. Wilks, S. C., Dawson, J. M., and Mori, W. B., Frequency up-conversion of electromagnetic radiation with use of an overdense plasma, *Phys. Rev. Lett.*, 61, 337, 1988.
6. Auld, B. A., Collins, J. H., and Zapp, H. R., Signal processing in a nonperiodically time-varying magnetoelastic medium, *Proc. IEEE*, 56, 258, 1968.
7. Kalluri, D. K., Green's function for a switched plasma medium and a perturbation technique for the study of wave propagation in a transient plasma with a small rise time, In: *Conf. Rec. Abstracts, IEEE Int. Conf. Plasma Sci.*, Boston, MA, 1996.
8. Huang, T. T., Lee, J. H., Kalluri, D. K., and Groves, K. M., Wave propagation in a transient plasma: Development of a Green's function, *IEEE Trans. Plasma Sci.*, 1, 19, 1998.
9. Felsen, L. B. and Whitman, G. M., Wave propagation in time-varying media, *IEEE Trans. Ant. Prop.*, AP-18, 242, 1970.
10. Ishimaru, A., *Electromagnetic Wave Propagation, Radiation, and Scattering*, Prentice-Hall, Englewood Cliffs, NJ, 1991.
11. Lekner, J., *Theory of Reflection*, Kluwer, Boston, 1987.
12. Triezenberg, D. G., Capillary waves in a diffuse liquid–gas interface, PhD thesis, University of Maryland, MD, 1973.

13. Abramowitz, M. and Stegun, I. A., *Handbook of Mathematical Functions*, Dover Publications, New York, 1965.
14. Kalluri, D. K., Frequency upshifting with power intensification of a whistler wave by a collapsing plasma medium, *J. Appl. Phys.*, 79, 3895, 1996.
15. Kalluri, D. K. and Chen, J. M., Comparison identities for wave propagation in a time-varying plasma medium, *IEEE Trans. Ant. Prop.*, 57(9), 2698–2705, 2009.

4

Switched Plasma Half-Space:
A and B Waves*

4.1 Introduction

In Chapter 2, we considered the incidence of a source wave on a step discontinuity in the dielectric properties of a time-invariant medium. The resulting reflected and transmitted waves have the same frequency as the source wave and these will be labeled as A waves. In Chapter 3, we considered the incidence of a source wave on a temporal step discontinuity in the dielectric properties of a space-invariant plasma medium. The resulting reflected and transmitted waves have an upshifted frequency [1–4] and these will be labeled as B waves. Step profiles are mathematical approximations for fast profiles and provide insight into the physical processes as well as serve as *reference solutions* for a perturbation technique (see Sections 3.5 through 3.10).

We will consider next a problem that involves simultaneous consideration of the effects of a temporal discontinuity and a spatial discontinuity.

4.2 Steady-State Solution

Figure 4.1 shows the electron density $N(t)$ of a typical transient plasma and its approximation by a step profile. Figure 4.2 shows the geometry of the problem. A plane wave of frequency ω_0 is traveling in free space in the z-direction when, at $t = 0$, a semi-infinite plasma of electron density N_0 is created in the upper half of the plane $z > 0$. The reflected wave will have two components denoted by subscripts A and B. There is no special significance to the choice of these letters for the subscripts.

The A component is due to reflection at the spatial discontinuity at $z = 0$. Subscript S will be used to indicate scattering at a spatial boundary. The

* © IEEE. Reprinted from Kalluri, D. K., *IEEE Trans. Plasma Sci.*, 16, 11–16, 1988. With permission. Chapter 4 is an adaptation of the reprint.

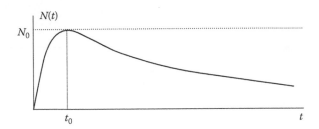

FIGURE 4.1   Electron density $N(t)$ versus time of a typical transient plasma. (From Kalluri, D. K., *IEEE Trans. Plasma Sci.*, 16, 11–16, 1988. With permission.)

corresponding reflection coefficient is

$$R_A = R_S = \frac{\eta_{p0} - \eta_0}{\eta_{p0} + \eta_0}, \tag{4.1}$$

where

$$\eta_{p0} = \frac{\eta_0}{\sqrt{1 - \omega_p^2/\omega_0^2}}, \tag{4.2}$$

$$\omega_p = \sqrt{\frac{N_0 q^2}{m\varepsilon_0}}. \tag{4.3}$$

Here, η_0 is the characteristic impedance of free space, η_{p0} is the characteristic impedance of the plasma medium, and the other symbols have the usual meanings [5].

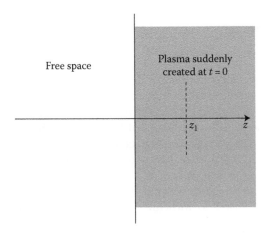

FIGURE 4.2   Geometry of the steady-state problem. (From Kalluri, D. K., *IEEE Trans. Plasma Sci.*, 16, 11–16, 1988. With permission.)

The B component arises due to the wave reflected by the temporal discontinuity at $t = 0$ and say $z = z_1$. Jiang [3] showed that this wave is of a new frequency ω,

$$\omega = \sqrt{\omega_0^2 + \omega_p^2},\tag{4.4}$$

and has a relative amplitude (relative to the incident wave amplitude)

$$R_t = \frac{\omega - \omega_0}{2\omega}.\tag{4.5}$$

The subscript t indicates scattering due to the temporal discontinuity. The wave will travel along the negative z-axis. When it reaches the spatial boundary at $z = 0$, a part of it will be transmitted into free space. The corresponding transmission coefficient is

$$T_S = \frac{2\eta_0}{\eta_p + \eta_0},\tag{4.6}$$

where

$$\eta_p = \frac{\eta_0}{\sqrt{1 - \omega_p^2/\omega^2}}.\tag{4.7}$$

The B component of the reflected wave in free space has a relative amplitude

$$R_B = R_t T_S.\tag{4.8}$$

In considering the steady-state values ($t \rightarrow \infty$), the B component can be ignored since this component will be damped out in traveling from $z = z_1 = \infty$ to $z = 0$, even if the plasma is only slightly lossy.

A quantitative idea of the damping of the B wave will now be given by calculating the damping constants for a low-loss plasma. Let ν be the collision frequency, and let $(\nu/\omega_p) \ll 1$. The attenuation constant α of the B wave of frequency ω is given by [5, pp. 7–9]

$$\alpha = \frac{(\omega/c)(\omega_p^2/2\omega^3)}{\sqrt{(1 - \omega_p^2/\omega^2)}}, \quad \nu \ll \omega_p.\tag{4.9}$$

After simplifying the algebra,

$$\alpha = \frac{\nu\omega_p^2}{2\omega_0\omega c}, \quad \nu \ll \omega_p.\tag{4.10}$$

Equation 4.10 shows that α is large for small values of ω_0. When the incident wave frequency ω_0 is low, the frequency ω of the B wave is only slightly larger than the plasma frequency ω_p, and the B wave is heavily attenuated.

Due to attenuation, R_B is now modified as

$$R_B = R_t T_S \exp(-\alpha z_1). \tag{4.11}$$

A damping distance constant z_p can now be defined:

$$z_p = \frac{1}{\alpha}. \tag{4.12}$$

In traveling this distance z_p, the B wave attenuates to e^{-1} of its original value. The attenuation can be expressed in terms of time rather than distance by noting that

$$t_p = \frac{z_p}{v_g}, \tag{4.13}$$

where t_p is the damping time constant of the B wave and v_g is the group velocity of propagation of the B wave given by

$$v_g = c\sqrt{1 - \frac{\omega_p^2}{\omega^2}} = c\frac{\omega_0}{\omega}. \tag{4.14}$$

From Equations 4.10 and 4.12 through 4.14,

$$t_p = \left(\frac{2\omega^2}{v\omega_p^2}\right), \quad v \ll \omega_p. \tag{4.15}$$

The amplitude of the B wave reduces to e^{-1} of its original value in time t_p. Numerical results are discussed in terms of normalized values. Let

$$\Omega_c = \frac{v}{\omega_p}, \tag{4.16}$$

$$\Omega_0 = \frac{\omega_0}{\omega_p}. \tag{4.17}$$

In terms of these variables, from Equations 4.1, 4.5, 4.6, 4.12, and 4.13, Equations 4.18 through 4.23 are obtained:

$$R_S = 1, \quad \Omega_0 < 1, \tag{4.18}$$

$$R_S = \frac{\Omega_0 - \sqrt{\Omega_0^2 - 1}}{\Omega_0 + \sqrt{\Omega_0^2 - 1}}, \quad \Omega_0 > 1, \tag{4.19}$$

$$R_t = \frac{\sqrt{\Omega_0^2 + 1} - \Omega_0}{2\sqrt{\Omega_0^2 + 1}}, \tag{4.20}$$

$$T_S = \frac{2\Omega_0}{\Omega_0 + \sqrt{\Omega_0^2 + 1}}, \tag{4.21}$$

$$\omega_p t_p = \frac{2(\Omega_0^2 + 1)}{\Omega_c}, \tag{4.22}$$

$$\frac{z_\rho}{\lambda_\rho} = \frac{\Omega_0\sqrt{\Omega_0^2 + 1}}{\pi\Omega_c}, \tag{4.23}$$

where λ_p is the free-space wavelength corresponding to the plasma frequency:

$$\lambda_\rho = \frac{2\pi c}{\omega_\rho}. \tag{4.24}$$

Figure 4.3 shows $R_A = R_S, R_t, T_S, R_B = R_t T_S$ versus Ω_0. The graph of R_t shows that the amplitude of the reflected wave generated in the plasma by the temporal discontinuity decreases as the incident wave frequency ω_0 increases (plasma frequency ω_p being kept constant). This is to be expected since the plasma behaves like free space when $\omega_0 \gg \omega_p$. On the other hand, the transmission coefficient T_S at the spatial discontinuity increases with ω_0. Thus, the reflection coefficient of the B wave increases, attains a maximum, and then decreases as ω_0 is increased. The variation of the two reflection coefficients (R_A and R_B) is shown in Figure 4.3 and is discussed next.

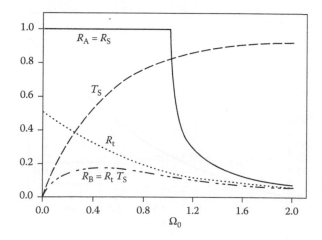

FIGURE 4.3 Reflection coefficients versus Ω_0 (normalized incident wave frequency). (From Kalluri, D. K., *IEEE Trans. Plasma Sci.*, 16, 11–16, 1988. With permission.)

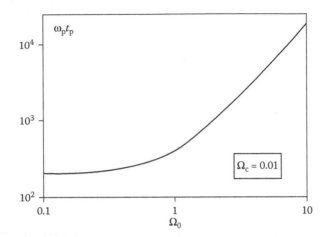

FIGURE 4.4 Damping time constant t_p versus Ω_0. (From Kalluri, D. K., *IEEE Trans. Plasma Sci.*, 16, 11–16, 1988. With permission.)

For $\Omega_0 < 1$, $|R_A|$ is 1 and R_B reaches a peak value of about 17% at about $\Omega_0 = 0.5$. For $\Omega_0 > 1$, R_A falls quickly and R_B falls slowly, both reaching about 5% at $\Omega_0 = 2$. For $\Omega_0 > 2$, $R_B/R_A \approx 1$. When the incident wave frequency is large compared to the plasma frequency, the total reflection coefficient is low, but each component contributes significantly to this low value.

Figure 4.4 shows the damping time constant t_p for $\Omega_c = 0.01$. Log–log scale is used here. From this graph it is evident that for large ω_0, the B component persists for a long time in low-loss plasma. For small ω_0, this component is damped out rather quickly. Figure 4.5 shows the damping distance constant z_p.

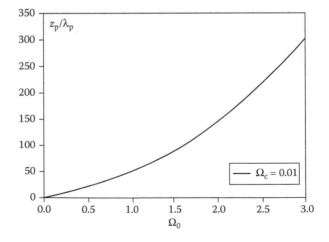

FIGURE 4.5 Damping distance constant z_p versus Ω_0. (From Kalluri, D. K., *IEEE Trans. Plasma Sci.*, 16, 11–16, 1988. With permission.)

4.3 Transient Solution

4.3.1 Formulation and Solution

Figure 4.6 shows the geometry of the problem. A perpendicularly polarized plane wave is propagating in the z-direction when, at $t = 0$, a semi-infinite plasma of particle density N_0 is created in the upper half of the z plane.

Let the electric field of the incident wave be

$$\mathbf{E}(\mathbf{r}, t) = E_y(z, t)\hat{y}, \quad t < 0, \ -\infty < z < \infty, \tag{4.25}$$

$$E_y(z, t) = E_0 \cos(\omega_0 t - k_0 z) = \text{Re}[E_0 \exp\{j(\omega_0 t - k_0 z)\}], \tag{4.26}$$

and $k_0 = \omega_0/c$. Hereinafter, Re will be omitted and assumed understood. The equations satisfied by E_y are

$$\frac{\partial^2 E_{y1}}{\partial z^2} - \frac{1}{c^2}\frac{\partial^2 E_{y1}}{\partial t^2} = 0, \quad (t > 0, \ z < 0) \tag{4.27}$$

and, from Equation 1.24,

$$\frac{\partial^2 E_{y2}}{\partial z^2} - \frac{1}{c^2}\frac{\partial^2 E_{y2}}{\partial t^2} - \frac{\omega_p^2}{c^2} E_{y2} = 0, \quad (t > 0, \ z > 0). \tag{4.28}$$

Here, ω_p is the plasma frequency defined in Equation 4.3. Subscript 1 indicates fields in the lower half of the x–z plane ($z < 0$) and subscript 2 indicates fields in the upper half ($z > 0$). Let the Laplace transform of E_y be E_y:

$$\mathcal{L}\{E_y(z, t)\} = E_y(z, s). \tag{4.29}$$

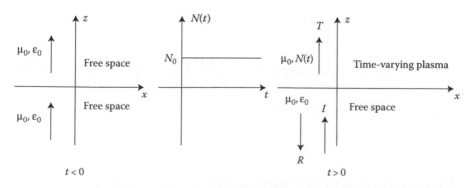

FIGURE 4.6 Geometry of the transient problem. (From Kalluri, D. K., *IEEE Trans. Plasma Sci.,* 16, 11–16, 1988. With permission.)

Equation 4.28 is transformed into (here, $' = \partial/\partial t$)

$$\frac{d^2 E_{y2}}{dz^2} - \frac{1}{c^2}(s^2 + \omega_p^2)E_{y2} + \frac{s}{c^2}E_{y2}(z,0) + \frac{1}{c^2}E'_{y2}(z,0) = 0. \qquad (4.30)$$

From the initial conditions,

$$E_{y1}(z,0) = E_{y2}(z,0) = E_0 \exp(-jk_0 z), \qquad (4.31)$$

$$E'_{y1}(z,0) = E'_{y2}(z,0) = j\omega_0 E_0 \exp(-jk_0 z). \qquad (4.32)$$

In the above equations, it is assumed that at $t = 0^+$ the newly created electrons and ions of the plasma are stationary so that the tangential components of the electric and magnetic fields are the same at $t = 0^-$ as at $t = 0^+$ [3]. From Equations 4.30 through 4.32,

$$\left[\frac{d^2}{dz^2} + \frac{s^2 + \omega_p^2}{c^2}\right]E_{y2}(z,s) = -\left[\frac{s + j\omega_0}{c^2}\right]E_0 \exp(-jk_0 z). \qquad (4.33)$$

The solution of this ordinary differential equation is given by

$$E_{y2}(z,s) = A_2(s)\exp(-q_2 z) + \left[\frac{s + j\omega_0}{s^2 + \omega_0^2 + \omega_p^2}\right]E_0 \exp(-jk_0 z), \qquad (4.34)$$

where

$$q_2 = \sqrt{s^2 + \frac{\omega_p^2}{c}}. \qquad (4.35)$$

Similarly, from Equations 4.27, 4.31, and 4.32,

$$E_{y1}(z,s) = A_1(s)\exp(q_1 z) + \left[\frac{s + j\omega_0}{s^2 + \omega_0^2}\right]E_0 \exp(-jk_0 z), \qquad (4.36)$$

where

$$q_1 = \frac{s}{c}. \qquad (4.37)$$

The second term on the right-hand side of Equation 4.36 is the Laplace transform of the incident electric field. Therefore, the first term on the right-hand side of Equation 4.36 is due to the reflected electric field. The undetermined constants $A_1(s)$ and $A_2(s)$ can be obtained from the boundary conditions of continuity of the tangential E and H, Equations 4.38 and 4.39:

$$E_{y1}(0,s) = E_{y2}(0,s), \tag{4.38}$$

$$\left[\frac{\partial E_{y1}(z,s)}{\partial z}\right]_{z=0} = \left[\frac{\partial E_{y2}(z,s)}{\partial z}\right]_{z=0}, \tag{4.39}$$

$$\frac{A_1(s)}{E_0} = \left[\frac{j\omega_0 - \sqrt{s^2+\omega_P^2}}{s+\sqrt{s^2+\omega_P^2}}\right]\left[\frac{(s+j\omega_0)\omega_P^2}{(s^2+\omega_0^2+\omega_P^2)(s^2+\omega_0^2)}\right]. \tag{4.40}$$

The reflected electric field is given by

$$\frac{E_{yr}(z,t)}{E_0} = \mathcal{L}^{-1}\left[\left\{\frac{A_{1R}(s)}{E_0}\right\}\{\exp(q_1 z)\}\right], \tag{4.41}$$

where $A_{1R}(s)$ is the real part of $A_1(s)$:

$$\left[\frac{A_{1R}(s)}{E_0}\right] = \left[\frac{\omega_P^2(-\omega_0^2 - s\sqrt{s^2+\omega_P^2})}{(s^2+\omega_0^2)(s^2+\omega_0^2+\omega_P^2)(s+\sqrt{s^2+\omega_P^2})}\right]. \tag{4.42}$$

After some algebraic manipulation, Equation 4.42 can be written as the sum of five terms:

$$\text{term1} = \frac{2\omega_0^2 - \omega_P^2}{\omega_P^2}\frac{s}{s^2+\omega_0^2}, \tag{4.43}$$

$$\text{term2} = \frac{2\omega_0^2(\omega_0^2 - \omega_P^2)}{\omega_P^2}\frac{1}{\sqrt{s^2+\omega_P^2}(s^2+\omega_0^2)}, \tag{4.44}$$

$$\text{term3} = \frac{1}{\sqrt{s^2+\omega_P^2}}, \tag{4.45}$$

$$\text{term4} = \frac{-2\omega_0^2}{\omega_P^2}\frac{s}{s^2+\omega_0^2+\omega_P^2}, \tag{4.46}$$

$$\text{term5} = \frac{-(2\omega_0^2 + \omega_P^2)\omega_0^2}{\omega_P^2}\frac{1}{\sqrt{s^2+\omega_P^2}(s^2+\omega_0^2+\omega_P^2)}. \tag{4.47}$$

The Laplace inverses of Equations 4.43, 4.45, and 4.46 are known [6]: these are term1, term3, and term4 given in Equations 4.50, 4.52, and 4.53, respectively, but the inversion of Equations 4.44 and 4.47 requires further work. The author has developed an algorithm [7] for the numerical Laplace inversion and computation of a class of Bessel-like functions of the type

$$h_{mn}(a, b, t) = \pounds^{-1} \frac{1}{\left[(s^2 + a^2)^{(n+1/2)} (s^2 + b^2)^m\right]}. \tag{4.48}$$

In terms of this notation and from the well-known Laplace-transform pairs [6],

$$\frac{E_{yr}(0, t)}{E_0} = term1 + term2 + term3 + term4 + term5, \tag{4.49}$$

where

$$term1 = \frac{2\omega_0^2 - \omega_p^2}{\omega_p^2} \cos \omega_0 t, \tag{4.50}$$

$$term2 = \frac{2\omega_0^2(\omega_0^2 - \omega_p^2)}{\omega_p^2} h_{10}(\omega_p, \omega_0, t), \tag{4.51}$$

$$term3 = J_0(\omega_p t), \tag{4.52}$$

$$term4 = -\frac{2\omega_0^2}{\omega_p^2} \cos\left(\sqrt{\omega_0^2 + \omega_p^2} t\right), \tag{4.53}$$

$$term5 = \frac{-(2\omega_0^2 + \omega_p^2)\omega_0^2}{\omega_p^2} h_{10}\left(\omega_p, \sqrt{\omega_0^2 + \omega_p^2}, t\right). \tag{4.54}$$

Here, $E_{yr}(z, t)/E_0$ is obtained by replacing t by $(t + z/c)$ in Equations 4.50 through 4.54.

4.3.2 Steady-State Solution from the Transient Solution

The steady-state value of $h_{10}(a, b, t)$ can be shown to be [7]

$$\lim_{t \to \infty} h_{10}(a, b, t) = \frac{1}{b\sqrt{a^2 - b^2}} \sin(bt), \quad b < a, \tag{4.55}$$

$$\lim_{t \to \infty} h_{10}(a, b, t) = -\frac{1}{b\sqrt{b^2 - a^2}} \cos(bt), \quad b > a. \tag{4.56}$$

From Equations 4.55 and 4.56, it can be shown that E_{LA}, the limiting value (value as t tends to infinity) of {term1 + term2} in Equation 4.49, gives

Equations 4.57 and 4.58:

$$E_{LA} = \left[2\frac{\omega_0^2 - \omega_P^2}{\omega_P^2} \right] \cos(\omega_0 t) - \left[2\omega_0 \frac{\sqrt{\omega_P^2 - \omega_0^2}}{\omega_P^2} \right] \sin(\omega_0 t), \quad \omega_0 < \omega_P,$$

(4.57)

$$E_{LA} = \left[2\omega_0^2 - \omega_P^2 - 2\omega_0 \frac{\sqrt{\omega_0^2 - \omega_P^2}}{\omega_P^2} \right] \cos(\omega_0 t), \quad \omega_0 > \omega_P.$$

(4.58)

The steady-state solution $t \to \infty$, in the presence of the space boundary alone, is easily shown to be

$$\left. \frac{E_{yr}(0,t)}{E_0} \right|_{\substack{\text{steady-state} \\ \text{space boundary}}} = \mathrm{Re}\left[\left\{ \frac{j\omega_0 - \sqrt{\omega_P^2 - \omega_0^2}}{j\omega_0 + \sqrt{\omega_P^2 - \omega_0^2}} \right\} \exp(j\omega_0 t) \right].$$

(4.59)

The term in the curly brackets, { }, is the reflection coefficient R_A of Equation 4.2. The right-hand sides of Equations 4.57 through 4.59 are the same and, hence, E_{LA} gives the steady-state value of the reflected electric field due to the spatial discontinuity.

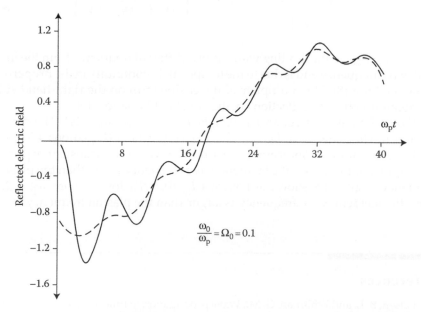

FIGURE 4.7 Reflected electric field versus $\omega_P t$. The dotted curve shows the plot of the electric field E_L. Here, $\Omega_0 = \omega_0/\omega_P = 0.1$. (From Kalluri, D. K., *IEEE Trans. Plasma Sci.*, 16, 11–16, 1988. With permission.)

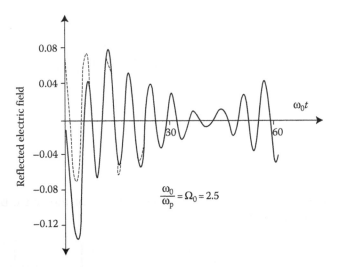

FIGURE 4.8   Reflected electric field versus $\omega_0 t$. The dotted curve shows E_L. Here, $\Omega_0 = \omega_0/\omega_p = 2.5$. (From Kalluri, D. K., *IEEE Trans. Plasma Sci.*, 16, 11–16, 1988. With permission.)

The limiting value (E_{LB}) of {term3 + term4 + term5} in Equation 4.49 can be shown to be

$$E_{LB} = \frac{\omega_0}{\sqrt{\omega_0^2 + \omega_p^2}} \left[\frac{\sqrt{\omega_0^2 + \omega_p^2} - \omega_0}{\omega_p} \right]^2 \cos\left(\sqrt{\omega_0^2 + \omega_p^2}\, t\right). \qquad (4.60)$$

This term E_{LB}, in the limiting value, is of a different frequency from the incident wave frequency and is due to the temporal discontinuity in the properties of the medium [8]. The multiplier of the cosine term on the right-hand side of Equation 4.60 is the reflection coefficient R_B of Equation 4.8.

Figure 4.7 shows $\omega_p t$ versus $E_{yr}(0, t)/E_0$ for $\Omega_0 = \omega_0/\omega_p = 0.1$. The dotted curve is the limiting value $E_L = E_{LA} + E_{LB}$ of this field (Equations 4.57 plus 4.60) and is shown to indicate the period for which the transient aspect is important. For this curve, the incident wave frequency is less than the plasma frequency. Figure 4.8 shows $\omega_0 t$ versus $E_{yr}(0, t)/E_0$ for $\Omega_0 = \omega_0/\omega_p = 2.5$. Here, the incident wave frequency is larger than the plasma frequency.

References

1. Felsen, B. L. and Whitman, G. M., Wave propagation in time-varying media, *IEEE Trans. Ant. Prop.*, AP-18, 242, 1970.
2. Fante, R. L., Transmission of electromagnetic waves into time-varying media, *IEEE Trans. Ant. Prop.*, AP-19, 417, 1971.

3. Jiang, C. L., Wave propagation and dipole radiation in a suddenly created plasma, *IEEE Trans. Ant. Prop.*, AP-23, 83, 1975.
4. Kalluri, D. K. and Prasad, R. C., Reflection of an electromagnetic wave from a suddenly created plasma half-space, In *Conf. Rec. Abstracts, IEEE Int. Conf. Plasma Sci.*, San Diego: CA, p. 46, 1983.
5. Heald, M. A. and Wharton, C. B., *Plasma Diagnostics with Microwaves*, Wiley, New York, 1965.
6. Roberts, R. E. and Kaufman, H., *Table of Laplace Transforms*, Saunders, Philadelphia, 1966.
7. Kalluri, D. K., Numerical Laplace inversion of $(s^2 + a^2)^{-(n+1/2)} (s^2 + a^2)^{-m}$, *Int. J. Comput. Math.*, 19, 327, 1986.
8. Auld, B. A., Collins, J. H., and Zapp, H. R., Signal processing in a nonperiodically time-varying magnetoelastic medium, *Proc. IEEE*, 56, 258, 1968.

5

Switched Plasma Slab: B Wave Pulses*

5.1 Introduction

In the previous chapter, we considered the switching of a plasma half-space and established that the reflected wave will have an A wave and a B wave. If the plasma is assumed to be lossless, the B wave at the upshifted frequency is present at $t = \infty$. In this chapter, we examine the case of sudden creation of a lossless finite extent plasma slab. The transient solution is obtained through the use of Laplace transform. The solution is broken into two components: an A component that in steady state has the same frequency as the incident wave frequency ω_0 and a B component that has an upshifted frequency:

$$\omega_1 = \sqrt{\omega_0^2 + \omega_p^2}. \tag{5.1}$$

Here ω_p is the plasma frequency of the switched plasma slab. Numerical results are presented to show the effect of the slab width on the reflected and transmitted waves.

For a finite slab width d, the B component decays and dies ultimately. This will be the case even if we assume that the created plasma slab is lossless. The conversion into B waves occurs at $t = 0$. After the medium change, there is no more wave conversion and therefore there is no continuous energy flow from the original wave to the new plasma modes. In a lossless plasma, the B wave lasts longer and longer as d is increased. However, if the plasma is lossy, the amplitude of the B wave that emerges in free space after traveling in the lossy plasma will decrease as the slab width d increases.

A proper choice of ω_0, ω_p, and the slab width will yield a transmitted electromagnetic pulse of significant strength at the upshifted frequency.

---

* © 1992 American Institute of Physics. Reprinted from Kalluri, D. K. and Goteti, V. R., *J. Appl. Phys.*, 72, 4575–4580, 1992. With permission. Chapter 5 is an adaptation of the reprint.

5.2 Development of the Problem

The geometry of the problem is shown in Figure 5.1. Initially, for time $t < 0$, the entire space is considered to be free space. A uniform plane EMW having a frequency of ω_0 rad/s and propagating in the positive z-direction is established over the entire space. This wave is called the incident wave. Its field components can be expressed in complex notation as

$$\mathbf{E_i}(z,t) = \hat{x}\, E_0 e^{j(\omega_0 t - k_0 z)}, \quad t < 0, \tag{5.2}$$

$$\mathbf{H_i}(z,t) = \hat{y}\, H_0 e^{j(\omega_0 t - k_0 z)}, \quad t < 0. \tag{5.3}$$

Here k_0 is the free-space wave number and $H_0 = E_0/\eta_0$, η_0 being the intrinsic impedance of free space. At time $t = 0$, an isotropic plasma medium of a constant plasma frequency ω_p is suddenly created between the planes $z = 0$ and $z = d$.

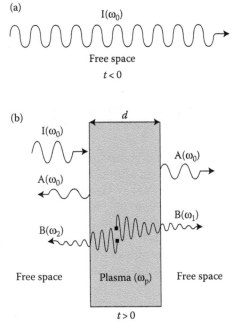

FIGURE 5.1 Effect of switching an isotropic plasma slab. Reflected and transmitted waves are sketched in the time domain to show the frequency changes. The A waves have the same frequency as the incident wave frequency (ω_0) but B waves have a new frequency $\omega_1 = (\omega_0^2 + \omega_p^2)^{1/2} = |\omega_2|$. (a) $t < 0$ and (b) $t > 0$. The waves in (b) are sketched in the time domain to show the frequency changes. See Figure 0.5b for a more accurate description. (From Kalluri, D. K. and Goteti, V. R., *J. Appl. Phys.*, 72, 4575–4580, 1992. With permission.)

The problem features a temporal discontinuity resulting from a sudden temporal change in the properties of the medium and spatial discontinuities owing to the confinement of the plasma in space. Several interesting phenomena take place due to these discontinuities. The temporal discontinuity results in the creation of two new waves in the plasma medium. These are termed the B waves. They have the same wave number, k_0, as that of the incident wave but have a frequency that differs from that of the incident wave. On the other hand, spatial discontinuity results in the creation of two new waves in the plasma medium that have the same frequency, ω_0, as the incident wave but have a different wave number. These waves are termed the A waves. Out of each set one of the two waves propagates in the positive z-direction, whereas the other propagates in the negative z-direction. The composite effect of the creation of these waves is the production of reflected waves in free space for $z < 0$ and transmitted waves in free space for $z > d$ (Figure 5.1). See Figure 0.5b for a more accurate description of the waves. The normalized (with respect to E_0) electric field of the reflected wave due to the A wave is denoted by R_A at $z = 0$ and that of the transmitted wave is denoted by T_A at $z = d$. Similarly, the symbols R_B and T_B are used to denote the contributions from the B waves.

A solution to this problem is developed here using the Laplace transforms with respect to the time variable t and ordinary differential equations with respect to the space variable z. The solution given here is restricted to the computation of R_A, R_B, T_A, and T_B.

5.3 Transient Solution

For time $t > 0$, the space is divided into three zones, namely, free space up to $z = 0$, isotropic plasma from $z = 0$ to $z = d$, and again free space for $z > d$. The scalar electric fields in the three media are denoted by $E_1(z,t), E_2(z,t)$, and $E_3(z,t)$, respectively. The partial differential equations for these fields are

$$\frac{\partial^2 E_1(z,t)}{\partial z^2} - \frac{1}{c^2}\frac{\partial^2 E_1(z,t)}{\partial t^2} = 0, \quad z < 0, \quad t > 0, \tag{5.4}$$

$$\frac{\partial^2 E_2(z,t)}{\partial z^2} - \frac{1}{c^2}\frac{\partial^2 E_2(z,t)}{\partial t^2} - \frac{\omega_p^2}{c^2}E_2(z,t) = 0, \quad 0 < z < d, \quad t > 0, \tag{5.5}$$

$$\frac{\partial^2 E_3(z,t)}{\partial z^2} - \frac{1}{c^2}\frac{\partial^2 E_3(z,t)}{\partial t^2} = 0, \quad z > d, \quad t > 0. \tag{5.6}$$

The newly created electrons are assumed to be at rest initially and are set in motion only for $t > 0$. This assumption is supported by the works of Jiang [1], Kalluri [2,3], and Goteti and Kalluri [4] and leads to the continuity of the electric and the magnetic field components over the temporal discontinuity at $t = 0$. The initial conditions are thus

$$E_1(t = 0^+) = E_2(t = 0^+) = E_3(t = 0^+) = E_i(t = 0^-) = E_0 \exp(-jk_0z), \quad (5.7)$$

$$\frac{\partial E_1}{\partial t}(t = 0^+) = \frac{\partial E_2}{\partial t}(t = 0^+) = \frac{\partial E_3}{\partial t}(t = 0^+) = \frac{\partial E_i}{\partial t}(t = 0^-)$$

$$= j\omega_0 E_0 \exp(-jk_0z). \quad (5.8)$$

Using these initial conditions, and taking the Laplace transforms of Equation 5.3, gives

$$\frac{d^2 E_1(z, s)}{dz^2} - q_1^2 E_1(z, s) = -\frac{(s + j\omega_0)}{c^2} E_0 \exp(-jk_0z), \quad (5.9)$$

$$\frac{d^2 E_2(z, s)}{dz^2} - q_2^2 E_2(z, s) = -\frac{(s + j\omega_0)}{c^2} E_0 \exp(-jk_0z), \quad (5.10)$$

$$\frac{d^2 E_3(z, s)}{dz^2} - q_1^2 E_3(z, s) = -\frac{(s + j\omega_0)}{c^2} E_0 \exp(-jk_0z), \quad (5.11)$$

where $E_1(z, s), E_2(z, s)$, and $E_3(z, s)$ are the Laplace transforms of $E_1(z, t)$, $E_2(z, t)$, and $E_3(z, t)$, respectively. Further,

$$q_1 = \frac{s}{c}, \quad (5.12)$$

$$q_2 = \frac{\sqrt{(s^2 + \omega_p^2)}}{c}. \quad (5.13)$$

The solution to the ordinary differential equations given in Equations 5.4 through 5.6 can be written as

$$E_1(z, s) = A_1 \exp(q_1z) + \frac{(s + j\omega_0)}{(s^2 + \omega_0^2)} E_0 \exp(-jk_0z), \quad z < 0, \quad (5.14)$$

$$E_2(z, s) = A_2 \exp(q_2z) + A_3 \exp(-q_2z) + \frac{(s + j\omega_0)}{(s^2 + \omega_0^2 + \omega_p^2)} E_0 \exp(-jk_0z),$$

$$0 < z < d, \quad (5.15)$$

$$E_3(z, s) = A_4 \exp(-q_1z) + \frac{(s + j\omega_0)}{(s^2 + \omega_0^2)} E_0 \exp(-jk_0z), \quad z > d. \quad (5.16)$$

The quantities A_1 through A_4 in Equations 5.14 through 5.16 are constants to be determined from the boundary conditions. The tangential components of the electric field and the magnetic field must be continuous at the $z = 0$ and $z = d$ interfaces. Thus

$$E_1(0,s) = E_2(0,s), \qquad (5.17)$$

$$\frac{dE_1(0,s)}{dz} = \frac{dE_2(0,s)}{dz}, \qquad (5.18)$$

$$E_2(d,s) = E_3(d,s), \qquad (5.19)$$

$$\frac{dE_2(d,s)}{dz} = \frac{dE_3(d,s)}{dz}. \qquad (5.20)$$

Substitution of the values of the constants in Equations 5.14 through 5.16 gives the complete description of the Laplace transforms of the fields in complex exponential form. When the incident wave is a harmonic wave of the form $E_0 \cos(\omega_0 t - k_0 z)$, it becomes necessary to take the real part on the right-hand side of Equations 5.14 through 5.16 to get the correct expressions for the fields. Following this procedure, the fields in the free-space zones $z < 0$ and $z > d$ are obtained as given below:

$$E_1(z,s) = A_{1R}(s)\exp(q_1 z) + E_0 \frac{[s\cos(k_0 z) + \omega_0 \sin(k_0 z)]}{(s^2 + \omega_0^2)}, \quad z < 0, \qquad (5.21)$$

$$E_3(z,s) = A_{4R}(s)\exp[-q_1(z-d)] + E_0 \frac{[s\cos(k_0 z) + \omega_0 \sin(k_0 z)]}{(s^2 + \omega_0^2)}, \quad z > d, \qquad (5.22)$$

where A_{1R} and A_{4R} are the real parts of A_1 and A_4, respectively.

The second term on the right-hand side of Equation 5.21 is the Laplace transform of the incident wave. Hence the first term corresponds to the field of the reflected waves in the free-space zone $z < 0$. Specifically, the inverse Laplace transform of $A_{1R}(s)$ gives the time variation of the fields of the reflected waves at $z = 0$ and the fields at any distance $z < 0$ can be obtained by replacing t with $(t + z/c)$. Similarly, the inverse Laplace transform of $A_{4R}(s)$ gives the fields of the transmitted waves at $z = d$ and replacement of t with $(t - (z - d)/c)$ gives the fields for any $z > d$. The complete expressions for $A_{1R}(s)$ and $A_{4R}(s)$ are given below:

$$\frac{A_{1R}}{E_0}(s) = R_A(s) + R_B(s), \qquad (5.23)$$

$$R_A(s) = \frac{\omega_p^2 [\exp(q_2 d) - \exp(-q_2 d)]}{DR} \frac{s}{(s^2 + \omega_0^2)}, \qquad (5.24)$$

$$R_B(s) = \frac{\omega_p^2 [\exp(q_2 d) + \exp(-q_2 d) - 2\cos(k_0 d)]}{DR} \frac{\sqrt{s^2 + \omega_p^2}}{(s^2 + \omega_0^2 + \omega_p^2)}, \qquad (5.25)$$

$$\frac{A_{4R}}{E_0}(s) = T_A(s) + T_B(s) - E_0 \frac{s\cos(k_0 d) + \omega_0 \sin(k_0 d)}{s^2 + \omega_0^2}, \qquad (5.26)$$

$$T_A(s) = -\frac{4s^2 \sqrt{s^2 + \omega_p^2}}{DR(s^2 + \omega_0^2)}, \qquad (5.27)$$

$$T_B(s) = \frac{\omega_0 \sin(k_0 d)}{(s^2 + \omega_0^2 + \omega_p^2)} + \frac{\sqrt{s^2 + \omega_p^2}}{(s^2 + \omega_0^2 + \omega_p^2)}[T_{B1}(s) - T_{B2}(s)], \qquad (5.28)$$

$$T_{B1}(s) = \frac{2(2s^2 + \omega_p^2)}{DR}, \qquad (5.29)$$

$$T_{B2}(s) = \cos(k_0 d)\left(\frac{\left[s - \sqrt{(s^2 + \omega_p^2)}\right]^2 \exp(-q_2 d)}{DR}\right)$$

$$+ \cos(k_0 d)\left(\frac{\left[s + \sqrt{(s^2 + \omega_p^2)}\right]^2 \exp(q_2 d)}{DR}\right), \qquad (5.30)$$

where

$$DR = \left(s - \sqrt{s^2 + \omega_p^2}\right)^2 \exp(-q_2 d) - \left(s + \sqrt{s^2 + \omega_p^2}\right)^2 \exp(q_2 d), \qquad (5.31)$$

and q_1 and q_2 are given by Equations 5.12 and 5.13, respectively.

An observation of Equation 5.24 shows that $R_A(s)$ corresponds to the field of the reflected wave that has a frequency ω_0 in steady state. This component arises due to reflection of the incident wave from the space boundary at $z = 0$. The term $R_B(s)$ refers to the transient effects involved in the partial transmission of the negatively propagating B wave across the $z = 0$ space boundary into the free-space zone $z < 0$. Similarly, in Equation 5.26, $T_A(s)$ refers to the electric field of the transmitted wave at the $z = d$ space boundary. Its steady-state frequency is ω_0. $T_A(s)$ is due to the spatial discontinuity in the properties of the medium at $z = d$. $T_B(s)$ refers to the partial transmission of the positively propagating B wave across the $z = d$ space boundary into the free-space zone $z > d$. Thus, based on the observation of the location of the complex poles in Equations 5.24 and 5.27, it can be concluded that quantities with the subscript A refer to the effects of spatial discontinuities and those with the subscript B refer to the effects of temporal discontinuity in the properties of the medium. This identification is further confirmed from an analysis of the steady-state solution given in Section 5.5.

5.4 Degenerate Case

The case of imposition of a semi-infinite plasma (d tending to infinity) is examined here. When the slab width d tends to infinity, the fields of the reflected waves given in Equations 5.24 and 5.25 can be shown to reduce to the following:

$$R_A(s) = -\frac{\omega_p^2}{\left(s + \sqrt{s^2 + \omega_p^2}\right)^2} \frac{s}{(s^2 + \omega_0^2)}, \tag{5.32}$$

$$R_B(s) = -\frac{\omega_p^2}{\left(s + \sqrt{s^2 + \omega_p^2}\right)^2} \frac{\sqrt{s^2 + \omega_p^2}}{(s^2 + \omega_0^2 + \omega_p^2)}. \tag{5.33}$$

These expressions agree with the results obtained by Kalluri [2] in his analysis of wave propagation due to a suddenly imposed semi-infinite plasma medium.

5.5 A Component from Steady-State Solution

Based on the assumption of the following form of waves in the three media, this steady-state boundary value problem is solved:
Medium 1 ($z < 0$): free space:
  Positive-going wave:

$$E_i(z, t) = E_0 \exp j(\omega_0 t - k_0 z). \tag{5.34}$$

Negative-going wave:

$$E_R(z, t) = B_1 \exp j(\omega_0 t + k_0 z). \tag{5.35}$$

Medium 2 ($0 < z < d$): isotropic plasma:
  Transmitted wave:

$$E_p^+(z, t) = B_2 \exp[j(\omega_0 t - \beta z)]. \tag{5.36}$$

Reflected wave:

$$E_p^-(z, t) = B_3 \exp[j(\omega_0 t + \beta z)]. \tag{5.37}$$

Medium 3 ($z > d$): free space:
  Transmitted wave:

$$E_T(z,t) = B_4 \exp[j(\omega_0 t - k_0 z)],\tag{5.38}$$

where β is the propagation constant in the plasma medium and is given by

$$\beta = \frac{\sqrt{(\omega_0^2 - \omega_p^2)}}{c}.\tag{5.39}$$

Using the boundary conditions at $z = 0$ and at $z = d$, the following expressions for B_1 and B_4 can be obtained:

$$\frac{B_1}{E_0} = \frac{(1 - a^2)[\exp(j\beta d) - \exp(-j\beta d)]}{(1 + a)^2 \exp(j\beta d) - (1 - a)^2 \exp(-j\beta d)},\tag{5.40}$$

$$\frac{B_4}{E_0} = \frac{4a \exp(jk_0 d)}{(1 + a)^2 \exp(j\beta d) - (1 - a)^2 \exp(-j\beta d)},\tag{5.41}$$

where

$$a = \frac{\sqrt{(\omega_0^2 - \omega_p^2)}}{\omega_0}.\tag{5.42}$$

Here B_1 and B_4 obtained in Equations 5.40 through 5.41 are the frequency response coefficients of the fields. From this steady-state solution, it is possible to obtain the Laplace transforms of the transient fields of the reflected wave and the transmitted wave in free space when the incident wave is of the form $E_0 \cos(\omega_0 t - k_0 z)$. The steps are as follows:

1. Replace $j\omega_0$ in Equations 5.40 through 5.41 with s.
2. Multiply the resulting expression with $s/(s^2 + \omega_0^2)$, the Laplace transform of $\cos(\omega_0 t)$.
3. Multiply B_1 with the operator $\exp(q_1 z)$ and B_4 with the operator $\exp(-q_1 z)$ to account for the propagation of the reflected and transmitted waves in free space.

This procedure gives the following expressions:

$$\frac{E_R}{E_0}(z,s) = \frac{\omega_p^2[\exp(q_2 d) - \exp(-q_2 d)]}{DR} \frac{s}{(s + \omega_0^2)} \exp(q_2 z),\tag{5.43}$$

$$\frac{E_T}{E_0}(z,s) = \frac{4s^2 \sqrt{(s^2 + \omega_p^2)}}{DR(s^2 + \omega_0^2)} \exp[-q_1(z - d)],\tag{5.44}$$

where DR, q_1, and q_2 are given in Equation 5.31. Comparison of Equation 5.43 with Equation 5.24 and of Equation 5.44 with Equation 5.27 shows that $R_A(s)$ and $T_A(s)$ are indeed the components resulting from the effects of spatial discontinuities in the properties of the medium. Thus the splitting up given in Equations 5.23 and 5.26 into components arising out of spatial discontinuities and temporal discontinuity is correct and complete. Hence, the inverse Laplace transforms of $R_A(s)$ and $T_A(s)$ can be expected to merge into the steady-state solution given in Equations 5.40 and 5.47. Proceeding on the same lines, the steady-state behavior of $R_B(s)$ and $T_B(s)$ can be obtained. This requires assumption of two waves in the plasma medium having the same frequency ω_1 but propagating in opposite directions and analyzing the reflection and transmission aspects into free space across the $z = 0$ and $z = d$ space boundaries. The amplitudes of these waves in free space can be shown to be zero in steady state for finite width of the plasma slab. Thus the steady-state values of $R_B(t)$ and $T_B(t)$ can be expected to be zero for finite values of d.

Numerical Laplace inversion [5] of the expressions given in Equations 5.24, 5.25, 5.27, and 5.28 is performed and these observations have been confirmed in Section 5.6.

5.6 Numerical Results

The numerical results are presented in normalized form. The frequency variables are normalized by taking $\omega_p = 1$ and d is normalized with respect to

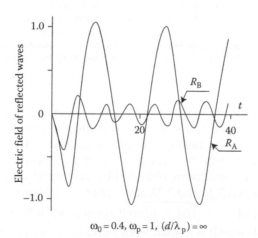

$$\omega_0 = 0.4, \ \omega_p = 1, \ (d/\lambda_p) = \infty$$

FIGURE 5.2 Electric field of reflected waves versus t for the case of switching a lossless plasma half-space. The results are presented in normalized form. The frequency variables are normalized by taking $\omega_p = 1$. $R_A(t)$ and $R_B(t)$ are the electric fields at $z = 0$ normalized with respect to the strength of the electric field (E_0) of the incident wave. $R_B(t)$ has a steady-state frequency $\omega_1 = (\omega_0^2 + \omega_p^2)^{1/2}$ and a nonzero steady-state amplitude. (From Kalluri, D. K. and Goteti, V. R., *J. Appl. Phys.*, 72, 4575–4580, 1992. With permission.)

(a)

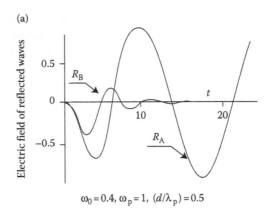

$\omega_0 = 0.4,\ \omega_p = 1,\ (d/\lambda_p) = 0.5$

(b)

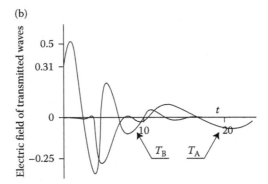

FIGURE 5.3   Electric field of (a) reflected waves and (b) transmitted waves for the case of switching a lossless plasma slab of width d. The width is normalized with respect to λ_p (free-space wavelength corresponding to plasma frequency). $T_A(t)$ and $T_B(t)$ are normalized electric field variables at $z = d$. The B waves decay even if the plasma slab is assumed to be lossless. (From Kalluri, D. K. and Goteti, V. R., *J. Appl. Phys.*, 72, 4575–4580, 1992. With permission.)

λ_p (free-space wavelength corresponding to plasma frequency). The electric fields of the reflected and transmitted waves are normalized with respect to the strength of the electric field (E_0) of the incident wave.

Figure 5.2 presents reflection coefficients at $z = 0$ versus t for the semi-infinite problem. The results are obtained by performing numerical Laplace inversion of Equations 5.32 and 5.33. $R_A(t)$ shows a frequency of ω_0 and $R_B(t)$ shows a frequency of $\omega_1 = \sqrt{\omega_0^2 + \omega_p^2}$ as it approaches infinity. The parameter ω_0 is chosen as 0.4 since it is known from previous work [2] that R_B will have maximum amplitude at about this value of ω_0. In this idealized model of lossless plasma, R_B will persist forever and its amplitude has a nonzero steady-state value. This phenomena can be physically explained in the following way [2]; one of the B waves generated at $t = 0$ and $z = \infty$ propagates in the negative z-direction and emerges into medium 1 at $t = \infty$.

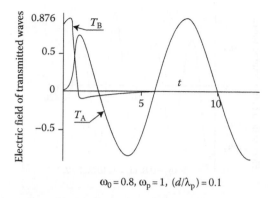

$$\omega_0 = 0.8, \ \omega_p = 1, \ (d/\lambda_p) = 0.1$$

FIGURE 5.4 $T_A(t)$ and $T_B(t)$ versus t for $d/\lambda_p = 0.1$. Since $\omega_0 < \omega_p$, T_A is due to tunneling of the incident wave through the plasma slab. It will become significantly weaker as the slab width increases. (From Kalluri, D. K. and Goteti, V. R., *J. Appl. Phys.*, 72, 4575–4580, 1992. With permission.)

For a finite value of d, the B component should decay since the conversion into B waves occurs at $t = 0$. After the medium change, there is no more wave conversion; therefore there is no continuous energy flow from the original wave to the new plasma modes. Figure 5.3 illustrates this point for the parameter $d/\lambda_p = 0.5$. The B wave that is generated at $t=0$ and traveling along the positive z-direction emerges into medium 3 and gives rise to T_B. Since $\omega_0 < \omega_p$, the incident wave frequency is in the stop band of the ordinary wave. T_A is due to tunneling (see Section 2.5) of the incident wave through the plasma slab and will become significantly weaker as the slab width increases. Medium 3 will then not have signals at the frequency of the incident wave. The above point is illustrated by progressively increasing d/λ_p. The results for the transmitted waves are shown in Figures 5.4 through 5.6. The parameter ω_0 is chosen as 0.8. This frequency is still in the stop band but results in a

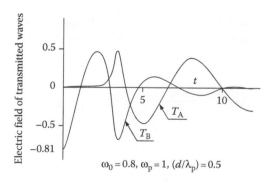

$$\omega_0 = 0.8, \ \omega_p = 1, \ (d/\lambda_p) = 0.5$$

FIGURE 5.5 $T_A(t)$ and $T_B(t)$ versus t for $d/\lambda_p = 0.5$. (From Kalluri, D. K. and Goteti, V. R., *J. Appl. Phys.*, 72, 4575–4580, 1992. With permission.)

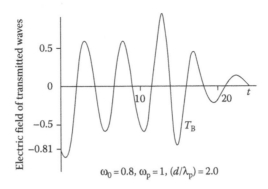

FIGURE 5.6 $T_B(t)$ versus t for $d/\lambda_p = 2.0$. $T_A(t)$ is negligible. (From Kalluri, D. K. and Goteti, V. R., *J. Appl. Phys.*, 72, 4575–4580, 1992. With permission.)

stronger B component in the transmitted wave. Figure 5.6 shows only a B component in the transmitted wave. The A component has negligible amplitude for $d/\lambda_p = 2.0$. Thus it is possible to obtain an upshifted signal [6] and suppress the original signal [6] by rapid creation of a plasma slab.

References

1. Jiang, C. L., Wave propagation and dipole radiation in a suddenly created plasma, *IEEE Trans. Ant. Prop.*, AP-23, 83, 1975.
2. Kalluri, D. K., On reflection from a suddenly created plasma half-space: Transient solution, *IEEE Trans. Plasma Sci.*, 16, 11, 1988.
3. Kalluri, D. K., Effect of switching a magnetoplasma medium on a traveling wave: Longitudinal propagation, *IEEE Trans. Ant. Prop.*, AP-37, 1638, 1989.
4. Goteti, V. R. and Kalluri, D. K., Wave propagation in a switched magnetoplasma medium: Transverse propagation, *Radio Sci.*, 25, 61, 1990.
5. *IMSL Library Reference Manual*, Vol. 2, IMSL Inc., Houston, 1982, FLINV-1.
6. Wilks, S. C., Dawson, J. M., and Mori, W. B., Frequency up-conversion of electromagnetic radiation with use of an overdense plasma, *Phys. Rev. Lett.*, 61, 337, 1988.

6

Magnetoplasma Medium: L, R, O, and X Waves

6.1 Introduction

Plasma medium in the presence of a static magnetic field behaves like an anisotropic dielectric. Therefore the rules of EMW propagation are the same as those of the light waves in crystals. Of course, in addition we have to take into account the dispersive nature of the plasma medium.

The well-established *magnetoionic theory* [1–3] is concerned with the study of wave propagation of an arbitrarily polarized plane wave in a cold anisotropic plasma, where the direction of phase propagation of the wave is at an arbitrary angle to the direction of the static magnetic field. As the wave travels in such a medium, the polarization state continuously changes. However, there are specific normal modes of propagation in which the state of polarization is unaltered. Waves with the left (L wave) or the right (R wave) circular polarization are the normal modes in the case of phase propagation along the static magnetic field. Such a propagation is labeled as *longitudinal propagation*. Ordinary wave (O wave) and extraordinary wave (X wave) are the normal modes for *transverse propagation*. The properties of these waves are explored in Sections 6.3 through 6.5.

6.2 Basic Field Equations for a Cold Anisotropic Plasma Medium

In the presence of a static magnetic field $\mathbf{B}_0$, the force Equation 1.3 needs modification due to the additional magnetic force term:

$$m\frac{d\mathbf{v}}{dt} = -q\left[\mathbf{E} + \mathbf{v} \times \mathbf{B}_0\right]. \qquad (6.1)$$

The corresponding modification of Equation 1.10 is given by

$$\frac{d\mathbf{J}}{dt} = \varepsilon_0 \omega_p^2(\mathbf{r}, t)\mathbf{E} - \mathbf{J} \times \boldsymbol{\omega}_b, \qquad (6.2)$$

where

$$\omega_b = \frac{q\mathbf{B}_0}{m} = \omega_b \hat{\mathbf{B}}_0. \tag{6.3}$$

In the above equation, $\hat{\mathbf{B}}_0$ is a unit vector in the direction of $\mathbf{B}_0$ and ω_b is the absolute value of the electron gyrofrequency.

Equations 1.1 and 1.2 are repeated here for convenience:

$$\nabla \times \mathbf{E} = -\mu_0 \frac{\partial \mathbf{H}}{\partial t}, \tag{6.4}$$

$$\nabla \times \mathbf{H} = \varepsilon_0 \frac{\partial \mathbf{E}}{\partial t} + \mathbf{J}, \tag{6.5}$$

and Equation 6.2 is the basic equation that will be used in discussing the EMW transformation by a magnetized cold plasma.

Taking the curl of Equation 1.1 and eliminating $\mathbf{H}$, the wave equation for $\mathbf{E}$ can be derived:

$$\nabla^2 \mathbf{E} - \nabla(\nabla \cdot \mathbf{E}) - \frac{1}{c^2}\frac{\partial^2 \mathbf{E}}{\partial t^2} - \frac{1}{c^2}\omega_p^2(\mathbf{r}, t)\mathbf{E} + \mu_0 \mathbf{J} \times \omega_b = 0. \tag{6.6}$$

Similar efforts will lead to a wave equation for the magnetic field:

$$\nabla^2 \dot{\mathbf{H}} - \frac{1}{c^2}\frac{\partial^2 \dot{\mathbf{H}}}{\partial t^2} - \frac{1}{c^2}\omega_p^2(\mathbf{r}, t)\dot{\mathbf{H}} + \varepsilon_0 \nabla \omega_p^2(\mathbf{r}, t) \times \mathbf{E} + \nabla \times (\mathbf{J} \times \mathbf{E}) = 0, \tag{6.7}$$

where

$$\dot{\mathbf{H}} = \frac{\partial \mathbf{H}}{\partial t}. \tag{6.8}$$

If ω_p^2 and ω_b vary only with t, Equation 6.7 becomes

$$\nabla^2 \dot{\mathbf{H}} - \frac{1}{c^2}\frac{\partial^2 \dot{\mathbf{H}}}{\partial t^2} - \frac{1}{c^2}\omega_p^2(t)\dot{\mathbf{H}} + \varepsilon_0 \omega_b(t)(\nabla \cdot \dot{\mathbf{E}}) = 0. \tag{6.9}$$

Equations 6.6 and 6.7 involve more than one field variable. It is possible to convert them into higher-order equations in one variable. In any case it is difficult to obtain meaningful analytical solutions to these higher-order vector partial differential equations. The equations in this section are useful in developing numerical methods.

We will consider next, like in the isotropic case, simple solutions to particular cases where we highlight one parameter or one aspect at a time. These solutions will serve as building blocks for the more involved problems.

6.3 One-Dimensional Equations: Longitudinal Propagation: L and R Waves

Let us consider the particular case where (1) the variables are functions of one spatial coordinate only, say the z coordinate; (2) the electric field is circularly polarized; (3) the static magnetic field is z-directed; and (4) the variables are denoted by

$$\mathbf{E} = (\hat{x} \mp j\hat{y})E(z,t), \tag{6.10}$$

$$\mathbf{H} = (\pm j\hat{x} + \hat{y})H(z,t), \tag{6.11}$$

$$\mathbf{J} = (\hat{x} \mp j\hat{y})J(z,t), \tag{6.12}$$

$$\omega_p^2 = \omega_p^2(z,t), \tag{6.13}$$

$$\boldsymbol{\omega}_b = \hat{z}\omega_b(z,t). \tag{6.14}$$

The basic equations for $E, H,$ and J take the following simple form:

$$\frac{\partial E}{\partial z} = -\mu_0 \frac{\partial H}{\partial t}, \tag{6.15}$$

$$-\frac{\partial H}{\partial z} = \varepsilon_0 \frac{\partial E}{\partial t} + J, \tag{6.16}$$

$$\frac{dJ}{dt} = \varepsilon_0 \omega_p^2(z,t)E \pm j\omega_b(z,t)J, \tag{6.17}$$

$$\frac{\partial^2 E}{\partial z^2} - \frac{1}{c^2}\frac{\partial^2 E}{\partial t^2} - \frac{1}{c^2}\omega_p^2(z,t)E \mp \frac{j}{c^2}\omega_b(z,t)\frac{\partial E}{\partial t} \mp j\mu_0\omega_b(z,t)\frac{\partial H}{\partial z} = 0, \tag{6.18}$$

$$\frac{\partial^2 \dot{H}}{\partial z^2} - \frac{1}{c^2}\frac{\partial^2 \dot{H}}{\partial t^2} - \frac{1}{c^2}\omega_p^2(z,t)\dot{H} + \varepsilon_0\frac{\partial}{\partial z}\omega_b(z,t)E \mp j\omega_b(z,t)\frac{\partial^2 H}{\partial z^2}$$
$$\pm \frac{1}{c^2}j\omega_b(z,t)\frac{\partial^2 H}{\partial t^2} \mp j\frac{\partial\omega_b(z,t)}{\partial z}\frac{\partial H}{\partial z} \mp j\varepsilon_0\frac{\partial\omega_b(z,t)}{\partial z}\frac{\partial E}{\partial t} = 0. \tag{6.19}$$

If ω_p and ω_b are functions of time only, Equation 6.19 reduces to

$$\frac{\partial^2 \dot{H}}{\partial z^2} - \frac{1}{c^2}\frac{\partial^2 \dot{H}}{\partial t^2} - \frac{1}{c^2}\omega_p^2(t)\dot{H} \mp j\omega_b(t)\frac{\partial^2 H}{\partial z^2} \pm \frac{1}{c^2}j\omega_b(t)\frac{\partial^2 H}{\partial t^2} = 0. \tag{6.20}$$

Let us next look for the plane wave solutions in a homogeneous, time-invariant unbounded magnetoplasma medium, that is,

$$f(z,t) = \exp\left[j(\omega t - kz)\right], \tag{6.21}$$

$$\omega_p^2(z,t) = \omega_p^2, \tag{6.22}$$

$$\omega_b(z,t) = \omega_b, \tag{6.23}$$

where f stands for any of the field variables $E, H,$ or J. From Equations 6.16 and 6.17 it is shown that

$$jkH = j\omega\varepsilon_0\varepsilon_{pR,L}E, \tag{6.24}$$

where

$$\varepsilon_{pR,L} = 1 - \frac{\omega_p^2}{\omega(\omega \mp \omega_b)}. \tag{6.25}$$

This shows clearly that the magnetized plasma, in this case, can be modeled as a dielectric with the dielectric constant given by Equation 6.25. The dispersion relation is obtained from

$$k^2 = \frac{\omega^2}{c^2}\varepsilon_{pR,L}, \tag{6.26}$$

which when expanded yields

$$\omega^3 \mp \omega_b\omega^2 - (k^2c^2 + \omega_p^2)\omega \pm k^2c^2\omega_b = 0. \tag{6.27}$$

The expression for the dielectric constant can be written in an alternative fashion in terms of ω_{c1} and ω_{c2}, which are called cutoff frequencies: the dispersion relation can also be recast in terms of the cutoff frequencies.

$$\varepsilon_{pR,L} = \frac{(\omega \pm \omega_{c1})(\omega \mp \omega_{c2})}{\omega(\omega \mp \omega_b)}, \tag{6.28}$$

$$\omega_{c1,c2} = \mp\frac{\omega_b}{2} + \sqrt{\left(\frac{\omega_b}{2}\right)^2 + \omega_p^2}, \tag{6.29}$$

$$k_{R,L}^2 c^2 = \omega^2\varepsilon_{pR,L} = \frac{\omega(\omega \pm \omega_{c1})(\omega \mp \omega_{c2})}{(\omega \mp \omega_b)}. \tag{6.30}$$

In the above equations, the top sign is for the right circular polarization (R wave) and the bottom sign is for the left circular polarization. Figure 6.1 gives the ε_p versus ω diagram and Figure 6.2 gives the ω–k diagram for the R wave. Figures 6.3 and 6.4 do the same for the L wave. R and L waves propagate in the direction of the static magnetic field, without any change in the polarization state of the wave, and are called the characteristic waves of longitudinal propagation. For these waves, the medium behaves like an isotropic plasma except that the dielectric constant ε_p is influenced by the strength of the static magnetic field. Particular attention is drawn to Figure 6.1, which shows that the dielectric constant $\varepsilon_p > 1$ for the R wave in the frequency band $0 < \omega < \omega_b$. This mode of propagation is called *whistler mode* in the literature on ionospheric physics and helicon mode in the literature on solid-state plasma. Chapter 9 deals with the transformation of the

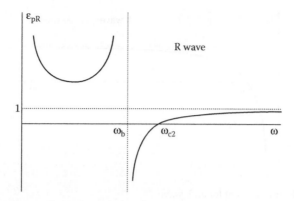

FIGURE 6.1 Dielectric constant for an R wave.

whistler wave by a transient magnetoplasma and the consequences of such a transformation. Note that an isotropic plasma does not support a whistler wave.

The longitudinal propagation of a linearly polarized wave is accompanied by the *Faraday rotation* of the plane of polarization. This phenomenon is easily explained in terms of the propagation of the R and L characteristic waves. A linearly polarized wave is the superposition of R and L waves, each of which propagates without change of its polarization state but each with a different phase velocity. (Note from Equation 6.25 that ε_p for a given ω, ω_p, and ω_b is different for R and L waves due to the sign difference in the denominator.) The result of combining the two waves after traveling a distance d is a linearly polarized wave again but with the plane of polarization rotated by an angle Ψ given by

$$\Psi = \frac{1}{2}(k_L - k_R)d. \tag{6.31}$$

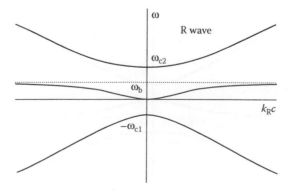

FIGURE 6.2 ω–k diagram for an R wave.

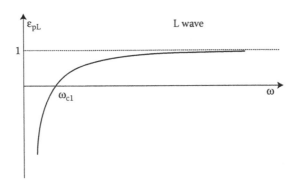

FIGURE 6.3   Dielectric constant for an L wave.

The high value of ε_p in the lower-frequency part of the whistler mode can be demonstrated by the following calculation given in Ref. [2]: for $f_p = 10^{14}$ Hz, $f_b = 10^{10}$ Hz, and $f = 10$ Hz, $\varepsilon_p = 9 \times 10^{16}$. The wavelength of the signal in the plasma is only 10 cm. See Refs. [4–6] for published work on the associated phenomena.

The resonance of the R wave at $\omega = \omega_b$ is of special significance. Around this frequency it can be shown that even a low-loss plasma strongly absorbs the energy of a source EMW and heats the plasma. This effect is the basis of radiofrequency heating of *fusion plasmas* [7]. It is also used to experimentally determine the effective mass of an electron in a crystal [8].

For the L wave, Figure 6.3 does not show any resonance effect. In fact the L wave has resonance at the ion gyrofrequency. In the Lorentz plasma model, the ion motion is neglected. A simple modification, by including the ion equation of motion, will extend the EMW transformation theory to a low-frequency source wave [2,9–11].

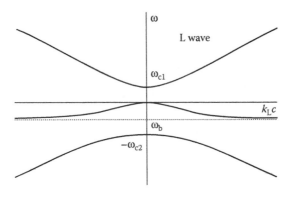

FIGURE 6.4 ω–k diagram for an L wave.

6.4 One-Dimensional Equations: Transverse Propagation: O Wave

Let us next consider that the electric field is linearly polarized in the y-direction, that is,

$$\mathbf{E} = \hat{y}E(z,t), \tag{6.32}$$

$$\mathbf{H} = -\hat{x}H(z,t), \tag{6.33}$$

$$\mathbf{J} = \hat{y}J(z,t), \tag{6.34}$$

$$\omega_p^2 = \omega_p^2(z,t), \tag{6.35}$$

$$\boldsymbol{\omega}_b = \hat{y}\omega_b(z,t). \tag{6.36}$$

The last term on the right-hand side of Equation 6.2 is zero since the current density and the static magnetic field are in the same direction. The equations are then no different from those of the isotropic case, and the static magnetic field has no effect. The electrons move in the direction of the electric field and give rise to a current density in the plasma in the same direction as that of the static magnetic field. In such a case, the electrons do not experience any magnetic force, and their orbit is not bent and they continue to move in the direction of the electric field. The one-dimensional solution for a plane wave in such a medium is called an O wave. Its characteristics are the same as those discussed in the previous chapters. In this case, unlike the case considered in Section 6.2, the direction of phase propagation is perpendicular to the direction of the static magnetic field and hence this case comes under the label of transverse propagation.

6.5 One-Dimensional Solution: Transverse Propagation: X Wave

The more difficult case of the transverse propagation is when the electric field is normal to the static magnetic field. The trajectories of the electrons that start moving in the direction of the electric field get altered and bent due to the magnetic force. Such a motion gives rise to an additional component of the current density in the direction of phase propagation, and to obtain a self-consistent solution we have to assume that a component of the electric field exists also in the direction of phase propagation. Let

$$\mathbf{E} = \hat{x}E_x(z,t) + \hat{z}E_z(z,t), \tag{6.37}$$

$$\mathbf{H} = \hat{y}H(z,t), \tag{6.38}$$

$$\mathbf{J} = \hat{x}J_x(z,t) + \hat{z}J_z(z,t), \tag{6.39}$$

$$\omega_p^2 = \omega_p^2(z,t), \tag{6.40}$$

$$\boldsymbol{\omega}_b = \hat{y}\omega_b(z,t). \tag{6.41}$$

The basic equations for E, H, and J take the following form:

$$\frac{\partial E_x}{\partial z} = -\mu_0 \frac{\partial H}{\partial t}, \tag{6.42}$$

$$-\frac{\partial H}{\partial z} = \varepsilon_0 \frac{\partial E_x}{\partial t} + J_x, \tag{6.43}$$

$$\varepsilon_0 \frac{\partial E_z}{\partial t} = -J_z, \tag{6.44}$$

$$\frac{dJ_x}{dt} = \varepsilon_0 \omega_p^2(z,t)E_x + \omega_b(z,t)J_z, \tag{6.45}$$

$$\frac{dJ_z}{dt} = \varepsilon_0 \omega_p^2(z,t)E_z - \omega_b(z,t)J_x. \tag{6.46}$$

Let us look again for the plane wave solution in a homogeneous, time-invariant, unbounded magnetoplasma medium by applying Equation 6.21 to the set of Equations 6.42 through 6.46:

$$-jkE_x = -\mu_0 j\omega H, \tag{6.47}$$

$$jkH = j\omega\varepsilon_0 E_x + J_x, \tag{6.48}$$

$$\varepsilon_0 j\omega E_z = -J_z, \tag{6.49}$$

$$j\omega J_x = \varepsilon_0 \omega_p^2 E_x + \omega_b J_z, \tag{6.50}$$

$$j\omega J_z = \varepsilon_0 \omega_p^2 E_z - \omega_b J_x. \tag{6.51}$$

From Equations 6.49 through 6.51, we obtain the following relation between J_x and E_x:

$$J_x = \varepsilon_0 \frac{\omega_p^2}{\left[1 - (\omega_b^2/(\omega^2 - \omega_p^2))\right]} \frac{E_x}{j\omega}. \tag{6.52}$$

Substituting Equation 6.52 into 6.48, we obtain

$$jkH = j\omega\varepsilon_0 \varepsilon_{pX} E_x, \tag{6.53}$$

$$\varepsilon_{pX} = 1 - \frac{\omega_p^2/\omega^2}{\left[1 - (\omega_b^2/(\omega^2 - \omega_p^2))\right]}. \tag{6.54}$$

An alternative expression for ε_{pX} can be written in terms of ω_{c1}, ω_{c2}, and ω_{uh}:

$$\varepsilon_{pX} = \frac{(\omega^2 - \omega_{c1}^2)(\omega^2 - \omega_{c2}^2)}{\omega^2(\omega^2 - \omega_{uh}^2)}. \tag{6.55}$$

The cutoff frequencies ω_{c1} and ω_{c2} are defined earlier and ω_{uh} is the upper hybrid frequency:

$$\omega_{uh}^2 = \omega_p^2 + \omega_b^2. \tag{6.56}$$

The dispersion relation $k^2 = (\omega^2/c^2)\varepsilon_{pX}$, when expanded, yields

$$\omega^4 - (k^2 c^2 + \omega_b^2 + 2\omega_p^2)\omega^2 + \left[\omega_p^4 + k^2 c^2(\omega_b^2 + \omega_p^2)\right] = 0. \tag{6.57}$$

In obtaining Equation 6.57, the expression for ε_{pX} given by Equation 6.54 is used. For the purpose of sketching the ω–k diagram, it is more convenient to use Equation 6.54 for ε_{pX} and write the dispersion relation as

$$k^2 c^2 = \frac{(\omega^2 - \omega_{c1}^2)(\omega^2 - \omega_{c2}^2)}{(\omega^2 - \omega_{uh}^2)}. \tag{6.58}$$

Figures 6.5 and 6.6 sketch ε_{pX} versus ω and ω versus kc, respectively. Substituting Equations 6.49 and 6.52 into Equation 6.51, we can find the ratio of E_z to E_x:

$$\frac{E_z}{E_x} = -j\omega\frac{\omega_b}{\omega^2 - \omega_p^2}(\varepsilon_{pX} - 1). \tag{6.59}$$

This shows that the polarization in the x–z plane, whether linear, circular, or elliptic, depends on the source frequency and the plasma parameters.

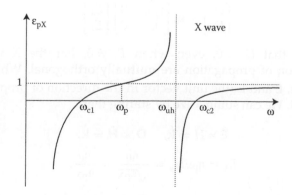

FIGURE 6.5 Dielectric constant for an X wave.

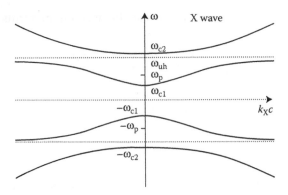

FIGURE 6.6 ω–k diagram for an X wave.

Moreover, using Equations 6.52 and 6.49, the relation between **J** and **E** can be written as

$$\mathbf{J} = \bar{\sigma} \cdot \mathbf{E}, \tag{6.60}$$

$$\begin{bmatrix} J_x \\ J_z \end{bmatrix} = j\omega\varepsilon_0 \begin{bmatrix} (\varepsilon_{pX} - 1) & 0 \\ 0 & -1 \end{bmatrix} \begin{bmatrix} E_x \\ E_z \end{bmatrix}, \tag{6.61}$$

where $\bar{\sigma}$ is the conductivity tensor for the case under consideration. The dielectric modeling of the plasma can be deduced from

$$\mathbf{D} = \varepsilon_0 \bar{\mathbf{K}} \cdot \mathbf{E}, \tag{6.62}$$

where the dielectric tensor $\bar{\mathbf{K}}$ is related to $\bar{\sigma}$ by

$$\bar{\mathbf{K}} = \bar{\mathbf{I}} + \frac{\bar{\sigma}}{j\omega\varepsilon_0}, \tag{6.63}$$

which gives, for this case,

$$\begin{bmatrix} D_x \\ D_z \end{bmatrix} = \varepsilon_0 \begin{bmatrix} \varepsilon_{pX} & 0 \\ 0 & 0 \end{bmatrix} \begin{bmatrix} E_x \\ E_z \end{bmatrix}. \tag{6.64}$$

This shows that $D_z = 0$, even when $E_z \neq 0$. For the X wave **D**, **H** and the direction of propagation are mutually orthogonal. While **E** and **H** are orthogonal, **E** and $\hat{\mathbf{k}}$ (the unit vector in the direction of propagation) are not orthogonal. We can summarize by stating that

$$\hat{\mathbf{E}} \times \hat{\mathbf{H}} \neq \hat{\mathbf{k}}, \quad \hat{\mathbf{D}} \times \hat{\mathbf{H}} = \hat{\mathbf{k}}, \tag{6.65}$$

$$E_x = \eta_{pX} H_y = \frac{\eta_0}{\sqrt{\varepsilon_{pX}}} = \frac{\eta_0}{n_{pX}}, \tag{6.66}$$

$$\mathbf{E} \cdot \hat{\mathbf{k}} \neq 0, \quad \mathbf{D} \cdot \hat{\mathbf{k}} = 0. \tag{6.67}$$

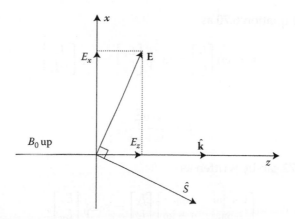

FIGURE 6.7 Geometrical sketch of the directions of various components of the X wave.

In the above equation, n_{pX} is the refractive index. Figure 6.7 is a geometrical sketch of the directions of various components of the X wave. The direction of power flow is given by the Poynting vector $\mathbf{S} = \mathbf{E} \times \mathbf{H}$, which in this case is not in the direction of phase propagation. The result that $D_z = 0$ for the plane wave comes from the general equation

$$\nabla \cdot \mathbf{D} = 0 \tag{6.68}$$

in a sourceless medium. In an anisotropic medium, it does not necessarily follow that the divergence of the electric field is also zero because of the tensorial nature of the constitutive relation. For example, from the expression

$$D_z = \varepsilon_0 \left[K_{zx} E_x + K_{zy} E_y + K_{zz} E_z \right], \tag{6.69}$$

D_z can be zero without E_z being zero.

6.6　Dielectric Tensor of a Lossy Magnetoplasma Medium

The constitutive relation for a lossy plasma with collision frequency v (rad/s) is obtained by modifying Equation 6.2 further:

$$\frac{d\mathbf{J}}{dt} + v\mathbf{J} = \varepsilon_0 \omega_p^2(\mathbf{r}, t)\mathbf{E} - \mathbf{J} \times \boldsymbol{\omega}_b. \tag{6.70}$$

Assuming time harmonic variation $\exp(j\omega t)$ and taking the z-axis as the direction of the static magnetic field,

$$\boldsymbol{\omega}_b = \hat{z}\omega_b, \tag{6.71}$$

we can write Equation 6.70 as

$$(\nu + j\omega) \begin{bmatrix} J_x \\ J_y \end{bmatrix} = \varepsilon_0 \omega_p^2 \begin{bmatrix} E_x \\ E_y \end{bmatrix} - j\omega_b \begin{bmatrix} J_y \\ -J_x \end{bmatrix} \tag{6.72}$$

and

$$J_z = \frac{\varepsilon_0 \omega_p^2}{\nu + j\omega} E_z. \tag{6.73}$$

Equation 6.72 can be written as

$$\begin{bmatrix} \nu + j\omega & \omega_b \\ -\omega_b & \nu + j\omega \end{bmatrix} \begin{bmatrix} J_x \\ J_y \end{bmatrix} = \varepsilon_0 \omega_p^2 \begin{bmatrix} E_x \\ E_y \end{bmatrix}, \tag{6.74}$$

$$\begin{bmatrix} J_x \\ J_y \end{bmatrix} = \varepsilon_0 \omega_p^2 \begin{bmatrix} \nu + j\omega & \omega_b \\ \omega_b & \nu + j\omega \end{bmatrix}^{-1} \begin{bmatrix} E_x \\ E_y \end{bmatrix}. \tag{6.75}$$

Combining Equation 6.75 with Equation 6.73, we obtain

$$\begin{bmatrix} J_x \\ J_y \\ J_z \end{bmatrix} = \bar{\sigma} \cdot \mathbf{E}, \tag{6.76}$$

where

$$\bar{\sigma} = \begin{bmatrix} \sigma_\perp & -\sigma_H & 0 \\ \sigma_H & \sigma_\perp & 0 \\ 0 & 0 & \sigma_{||} \end{bmatrix}. \tag{6.77}$$

In the above equation, $\sigma_\perp$, σ_H, and $\sigma_{||}$ are called perpendicular conductivity, Hall conductivity, and parallel conductivity, respectively. From Equation 6.63, the dielectric tensor $\bar{\mathbf{K}}$ can now be obtained:

$$\bar{\mathbf{K}} = \begin{bmatrix} \varepsilon_{r\perp} & -j\varepsilon_{rH} & 0 \\ j\varepsilon_{rH} & \varepsilon_{r\perp} & 0 \\ 0 & 0 & \varepsilon_{r||} \end{bmatrix}, \tag{6.78}$$

where

$$\varepsilon_{r\perp} = 1 - \frac{(\omega_p/\omega)^2 (1 - j\nu/\omega)}{(1 - j\nu/\omega)^2 - (\omega_b/\omega)^2}, \tag{6.79}$$

$$\varepsilon_{rH} = \frac{(\omega_p/\omega)^2 (\omega_b/\omega)}{(1 - j\nu/\omega)^2 - (\omega_b/\omega)^2}, \tag{6.80}$$

$$\varepsilon_{r||} = 1 - \frac{(\omega_p/\omega)^2}{1 - j\nu/\omega}. \tag{6.81}$$

6.7 Periodic Layers of Magnetoplasma

Layered semiconductor–dielectric periodic structures in a dc magnetic field can be modeled as periodic layers of magnetoplasma. EMW propagating in such structures can be investigated by using the dielectric tensor derived in Section 6.6. The solution is obtained by extending the theory of Section 2.7 to the case where the plasma layer is magnetized and hence has the anisotropy complexity; see Appendix J. Brazis and Safonova [12–14] investigated the resonance and absorption bands of such structures.

6.8 Surface Magnetoplasmons

The theory of Section 2.5.2 can be extended to the propagation of surface waves at a semiconductor–dielectric interface in the presence of a static magnetic field. The semiconductor in a dc magnetic field can be modeled as an anisotropic dielectric tensor. Wallis [15] discussed at length the properties of surface magnetoplasmons on semiconductors.

6.9 Surface Magnetoplasmons in Periodic Media

Refs. [16,17] discuss the propagation of surface magnetoplasmons in truncated superlattices. This model is a combination of the models used in Sections 6.7 and 6.8.

References

1. Heald, M. A. and Wharton, C. B., *Plasma Diagnostics with Microwaves*, Wiley, New York, 1965.
2. Booker, H. G., *Cold Plasma Waves*, Kluwer, Hingham, MA, 1984.
3. Swanson, D. G., *Plasma Waves*, Academic Press, New York, 1989.
4. Steele, M. C., *Wave Interactions in Solid State Plasmas*, McGraw-Hill, New York, 1969.
5. Bowers, R., Legendy, C., and Rose, F., Oscillatory galvanometric effect in metallic sodium, *Phys. Rev. Lett.*, 7, 339, 1961.
6. Aigrain, P. R., In: *Proceeding of the International Conference on Semiconductor Physics*, Czechoslovak Academy Sciences, Prague, p. 224, 1961.
7. Miyamoto, K., *Plasma Physics for Nuclear Fusion*, The MIT Press, Cambridge, MA, 1976.

8. Solymar, L. and Walsh, D., *Lectures on the Electrical Properties of Materials*, fifth edition, Oxford University Press, Oxford, 1993.
9. Madala, S. R. V. and Kalluri, D. K., Longitudinal propagation of low frequency waves in a switched magnetoplasma medium, *Radio Sci.*, 28, 121, 1993.
10. Dimitrijevic, M. M. and Stanic, B. V., EMW transformation in suddenly created two-component magnetized plasma, *IEEE Trans. Plasma Sci.*, 23, 422, 1995.
11. Madala, S. R. V., Frequency shifting of low frequency electromagnetic waves using magnetoplasmas, Doctoral thesis, University of Massachusetts Lowell, Lowell, 1993.
12. Brazis, R. S. and Safonova, L. S., Resonance and absorption band in the classical magnetoactive semiconductor–insulator superlattice, *Int. J. Infrared Millim. Waves*, 8, 449, 1987.
13. Brazis, R. S. and Safonova, L. S., Electromagnetic waves in layered semiconductor–dielectric periodic structures in DC magnetic fields, *Proc. SPIE*, 1029, 74, 1988.
14. Brazis, R. S. and Safonova, L. S., In-plane propagation of millimeter waves in periodic magnetoactive semiconductor structures, *Int. J. Infrared Millim. Waves*, 18, 1575, 1997.
15. Wallis, R. F., Surface magnetoplasmons on semiconductors, In: A. D. Boardman (Ed.), *Electromagnetic Surface Modes*, Chapter 2, Wiley, New York, 1982.
16. Wallis, R. F., Szenics, R., Quihw, J. J., and Giuliani, G. F., Theory of surface magnetoplasmon polaritons in truncated superlattices, *Phys. Rev. B*, 36, 1218, 1987.
17. Wallis, R. F. and Quinn, J. J., Surface magnetoplasmon polaritons in truncated superlattices, *Phys. Rev. B*, 38, 4205, 1988.

7

Switched Magnetoplasma Medium

7.1 Introduction

The effect of converting free space into an anisotropic medium by the creation of a plasma medium in the presence of a static magnetic field is considered in this chapter. In Section 7.2, the higher-order differential equations for fields in a time-varying magnetoplasma medium are developed. These equations are the basis of adiabatic analysis discussed in Chapter 9.

The main effect of switching the medium is the splitting of the original wave (incident wave) into new waves whose frequencies are different from the frequency of the incident wave. The strength and the direction of the static magnetic field affect the number of waves generated, their frequencies, damping rates, and their power densities. These aspects for L, R, O, and X waves are discussed in Sections 7.3 through 7.9.

7.2 One-Dimensional Equations: Longitudinal Propagation

Let us first consider the propagation of R and L plane waves in a time-varying space-invariant unbounded magnetoplasma medium, that is,

$$\omega_p^2 = \omega_p^2(t), \tag{7.1}$$

$$\omega_b = \omega_b(t), \tag{7.2}$$

$$F(z,t) = F(t)\exp(-jkz), \tag{7.3}$$

$$\frac{\partial}{\partial z} = -jk. \tag{7.4}$$

The relevant equations are obtained from Section 6.3:

$$\mu_0 \frac{\partial H}{\partial t} = jkE, \tag{7.5}$$

$$\varepsilon_0 \frac{\partial E}{\partial t} = jk\,H - J, \tag{7.6}$$

$$\frac{dJ}{dt} = \varepsilon_0 \omega_p^2(t)E \pm j\omega_b(t)J. \tag{7.7}$$

From Equation 6.20 or from Equations 7.5 through 7.7, the equation for H can be obtained:

$$\frac{d^3H}{dt^3} \mp j\omega_b(t)\frac{d^2H}{dt^2} + \left[k^2c^2 + \omega_p^2(t)\right]\frac{dH}{dt} \mp j\omega_b(t)k^2c^2H = 0. \tag{7.8}$$

The equation for E from Equation 6.18 turns out to be more complicated than the equation for H. However, if ω_p^2 alone is a function of time and ω_b is a constant, we have a third-order differential equation for E involving the first derivative of $\omega_p^2(t)$:

$$\frac{d^3E}{dt^3} \mp j\omega_b\frac{d^2E}{dt^2} + \left[k^2c^2 + \omega_p^2(t)\right]\frac{dE}{dt} + (g \mp j\omega_b k^2c^2)E = 0, \tag{7.9}$$

where

$$g = \frac{d\omega_p^2(t)}{dt}. \tag{7.10}$$

No such simplification is possible if ω_p^2 is a constant but ω_b is varying with time. In such a case, it is still possible to write a third-order equation for J. However, it includes the first and second derivatives of $\omega_b(t)$ denoted by $\dot{\omega}_b$ and $\ddot{\omega}_b$, respectively:

$$\frac{d^3J}{dt^3} \mp j\omega_b(t)\frac{d^2J}{dt^2} + \left[k^2c^2 + \omega_p^2(t) \mp 2j\dot{\omega}_b(t)\right]\frac{dJ}{dt} \mp j\left(k^2c^2\omega_b(t) + \ddot{\omega}_b(t)\right)J = 0. \tag{7.11}$$

The equation for the isotropic case should be obtained by substituting $\omega_b = 0$ in the above equations. Equation 7.8 indeed reduces to Equation 3.13. Equation 7.9 remains third order in E even if we substitute $\omega_b = 0$, whereas the equation for the isotropic case, Equation 3.11, is of second order. However, Equation 7.9 is obtained by differentiating Equation 3.11 with respect to t.

7.3 Sudden Creation: Longitudinal Propagation [1,2]

Let us next solve the problem of sudden creation of the plasma in the presence of a static magnetic field in the z-direction. The geometry of the problem is given in Figure 7.1. We shall assume that the source wave is a circularly polarized wave propagating in the z-direction. Solution for R and L waves can be obtained simultaneously.

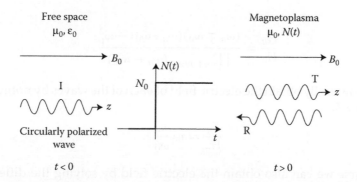

FIGURE 7.1 Geometry of the problem. Magnetoplasma medium is switched on at $t = 0$.

This problem will be similar to that of Section 3.2 if we assume that the source wave is an R or L wave. We shall again use the Laplace transform technique. We can use the state variable technique, but to add to the variety, we shall solve the higher-order differential equation for H rather than solve the state variable equations. Assuming that the source wave medium is free space and its frequency is $\omega_0 = kc$, we have

$$\frac{d^3H}{dt^3} \mp j\omega_b \frac{d^2H}{dt^2} + \left[\omega_0^2 + \omega_p^2\right]\frac{dH}{dt} \mp j\omega_b\omega_0^2 H = 0. \qquad (7.12)$$

The initial conditions on E, H, and J given by Equations 3.25 through 3.27 can be converted into the initial conditions on H and its derivatives at the origin:

$$H(0) = H_0, \qquad (7.13)$$

$$\dot{H}(0) = j\omega_0 H_0, \qquad (7.14)$$

$$\ddot{H}(0) = -\omega_0^2 H_0. \qquad (7.15)$$

After some algebra, we obtain

$$\frac{H(s)}{H_0} = \frac{(s \mp j\omega_b)(s + j\omega_0) + \omega_p^2}{(s - j\omega_1)(s - j\omega_2)(s - j\omega_3)}, \qquad (7.16)$$

where ω_1, ω_2, and ω_3 are the roots of the polynomial equation:

$$\omega^3 \mp \omega_b\omega^2 - (\omega_0^2 + \omega_p^2)\omega \pm \omega_0^2\omega_b = 0. \qquad (7.17)$$

By evaluating the residues at the poles, we can invert $H(s)$. The solution for $H(t)$ can be expressed in compact form

$$\frac{H(t)}{H_0} = \sum_{n=1}^{3} \frac{H_n}{H_0} \exp(j\omega_n t), \qquad (7.18)$$

where

$$\frac{H_n}{H_0} = \frac{(\omega_n \mp \omega_b)(\omega_n + \omega_0) - \omega_p^2}{\prod_{m=1, m \neq n}^{3} (\omega_n - \omega_m)}. \tag{7.19}$$

We can now easily obtain the electric field of each of the waves by noting from Equation 7.5,

$$\frac{E_n}{H_n} = \eta_0 \frac{\omega_n}{\omega_0}. \tag{7.20}$$

Of course we can also obtain the electric field by solving the differential Equation 7.9. One has to take care in interpreting g for the suddenly created plasma. The $\omega_p^2(t)$ profile is expressed in terms of a step profile

$$\omega_p^2(t) = \omega_p^2 u(t), \tag{7.21}$$

and $g(t)$, its derivative, can be expressed in terms of an impulse function $\delta(t)$:

$$g(t) = \omega_p^2 \delta(t). \tag{7.22}$$

Equation 7.9 becomes

$$\frac{d^3 E}{dt^3} \mp j\omega_b \frac{d^2 E}{dt^2} + \left[\omega_0^2 + \omega_p^2 \right] \frac{dE}{dt} + \left[\omega_p^2 \delta(t) \mp j\omega_b \omega_0^2 \right] E = 0. \tag{7.23}$$

Equation 7.23 can now be solved subject to the initial conditions

$$E(0) = E_0, \tag{7.24}$$

$$\dot{E}(0) = j\omega_0 E_0, \tag{7.25}$$

$$\ddot{E}(0) = -\omega_0^2 E_0. \tag{7.26}$$

Noting that

$$\pounds \{\delta(t)E(t)\} = \int_0^\infty \delta(t) \exp(-st)E(t) \, dt = E(0), \tag{7.27}$$

we obtain

$$\frac{E(s)}{E_0} = \frac{(s \mp j\omega_b)(s + j\omega_0)}{(s - j\omega_1)(s - j\omega_2)(s - j\omega_3)} \tag{7.28}$$

and

$$\frac{E_n}{E_0} = \frac{(\omega_n \mp \omega_b)(\omega_n + \omega_0)}{\prod_{m=1, m \neq n}^{3} (\omega_n - \omega_m)}. \tag{7.29}$$

After a little algebra, it can be shown that E_n obtained from Equation 7.19 is the same as that given by Equation 7.29. Moreover, we can further show that

$$\frac{E_n}{H_n} = \sqrt{\frac{\mu_0}{\varepsilon_0 \varepsilon_{pn}}} = \frac{\eta_0}{n_{pn}},$$ (7.30)

where

$$\varepsilon_{pn} = 1 - \frac{\omega_p^2}{\omega_n(\omega_n \mp \omega_b)}.$$ (7.31)

Here ε_{pn} is the dielectric constant and n_{pn} is the refractive index at the frequency ω_n. The product of Equation 7.29 with Equation 7.19 gives the ratio of the power in the nth wave to the power in the incident wave, called S_n. In general,

$$S_n = \frac{E_n}{E_0} \frac{H_n}{H_0}.$$ (7.32)

A negative value for S_n indicates that the nth wave is a reflected wave, further confirmed by the value of ω_j. The ω–k diagram discussed in the next paragraph confirmed that the group velocity of these waves has the same algebraic sign as the phase velocity, thus ruling out any backward waves [3, p. 257]. The sum of the numerical values of S_n gives the ratio of total power in the created waves to the power in the incident wave.

$$S = \sum_{n=1}^{3} |S_n|.$$ (7.33)

The effect of switching the magnetoplasma medium is the creation of three waves with new frequencies (frequencies other than the incident wave frequency) given by Equation 7.17. These new frequencies are functions of ω_0, ω_b, and ω_p. Their variation is conveniently studied with the use of normalized variables defined in Equations 7.34 through 7.37.

$$\Omega = \frac{\omega}{\omega_p},$$ (7.34)

$$\Omega_0 = \frac{\omega_0}{\omega_p},$$ (7.35)

$$\Omega_n = \frac{\omega_n}{\omega_p}, \quad n = 1, 2, 3,$$ (7.36)

$$\Omega_b = \frac{\omega_b}{\omega_p}.$$ (7.37)

In terms of these normalized variables, Equation 7.17 becomes

$$\Omega^3 \mp \Omega_b \Omega^2 - (1 + \Omega_0^2)\Omega \pm \Omega_0^2 \Omega_b = 0, \tag{7.38}$$

where the lower sign is to be used when the incident wave is an L wave and the upper sign when the incident wave is an R wave. Sketches of the curves Ω versus Ω_0 with Ω_b as a parameter are given for the two cases in Figures 7.2 and 7.3, respectively. It can be noted that the curve Ω versus Ω_0 is also the ω–k diagram.

$$\Omega_0 = \frac{\omega_0}{\omega_p} = \frac{k_0 c}{\omega_p} = \frac{kc}{\omega_p}. \tag{7.39}$$

In the above equation, k_0 is the wave number of the incident wave, which is also the wave number k of the newly created waves since the wave number is conserved at a time discontinuity. As expected, they are similar to the curves of Figures 6.4 and 6.2, respectively. The phase velocity v_{ph} of the waves is given by

$$v_{ph} = \frac{\omega}{k} = c\frac{\Omega}{\Omega_0}. \tag{7.40}$$

The group velocity v_{gr} is given by

$$v_{gr} = c\frac{d\Omega}{d\Omega_0}. \tag{7.41}$$

The branch marked Ω_1 in Figure 7.2 is the dispersion curve for the first L wave. This is a transmitted wave. The asymptote to this branch at 45° shows that the phase and group velocities of this wave approach those of the incident wave as $\Omega_0 = \omega_0/\omega_p$ becomes large. When the frequency of the incident wave is much larger than the plasma frequency, the plasma has little effect on the incident wave. The branch marked Ω_2 describes the reflected wave whose asymptote is at −45°. The branch marked Ω_3 describes the third wave, which is a reflected wave. Its asymptote is a horizontal line $\Omega = -\Omega_b$. Its phase and group velocities are negative. The third wave disappears in the isotropic case ($\Omega_b = 0$).

Figure 7.3 describes the dispersion for the three waves when the incident wave is the R wave. In this case, the third wave is a transmitted wave with an asymptote $\Omega = \Omega_b$.

The following differences between the case studied here and the isotropic case ($\Omega_b = 0$) reported in the literature [4,5] can be noted:

1. The first and the second waves do not have the same frequencies.

2. The third wave does not exist in the isotropic case. Instead, a static but spatially varying magnetic field exists [4]. The same result, that is, $\omega_3 = 0$, $E_3/E_0 = 0$, and $H_3/H_0 = \omega_p^2/(\omega_0^2 + \omega_p^2)$, is obtained, when $\omega_b = 0$ is substituted in Equations 7.17, 7.29, and 7.19.

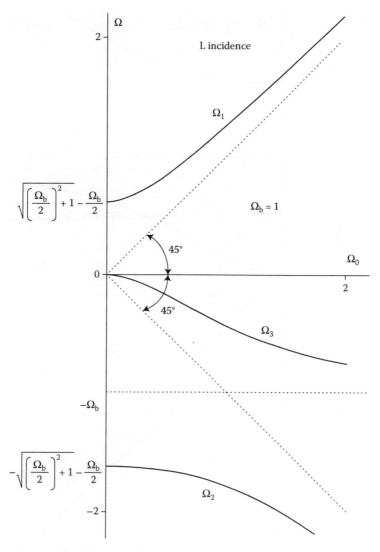

FIGURE 7.2 $\omega-k$ diagram for the case of incident L wave (L incidence). The results are presented in terms of normalized variables. The horizontal axis is $\Omega_0 = \omega_0/\omega_p$, where ω_p is the plasma frequency of the switched medium. Ω_0 is also proportional to the wave number of the created waves. The vertical axis is $\Omega = \omega/\omega_p$, where ω is the frequency of the created waves. The branch marked Ω_n describes the nth created wave. The parameter is $\Omega_b = \omega_b/\omega_p$, where ω_b is the gyrofrequency of the switched medium.

7.4 Numerical Results: Longitudinal Propagation

In this section, the characteristics of the three waves mentioned above are studied in detail. The horizontal axis is the normalized frequency (normalized

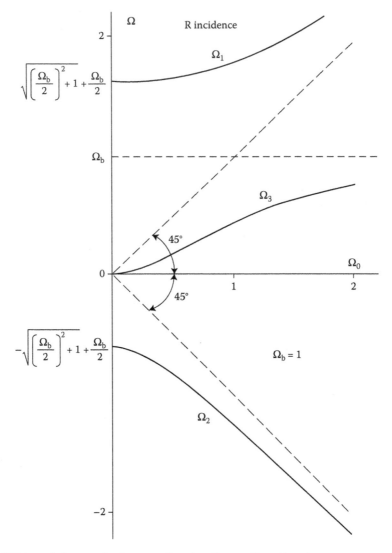

FIGURE 7.3 ω–k diagram for the case of incident R wave (R incidence).

with respect to the plasma frequency) of the incident wave. The normalized gyrofrequency (Ω_b) is used as a parameter. The case of an incident L wave (for brevity this case will be referred to as L incidence) is shown in Figures 7.4 through 7.7 and the case of an incident R wave (R incidence) is shown in Figures 7.8 through 7.11.

Figure 7.4 describes the first wave for L incidence that is a transmitted wave. In Figure 7.4a, the variation of the normalized frequency of the first wave is shown. The isotropic case is described by the curve $\Omega_b = 0$. For strong static

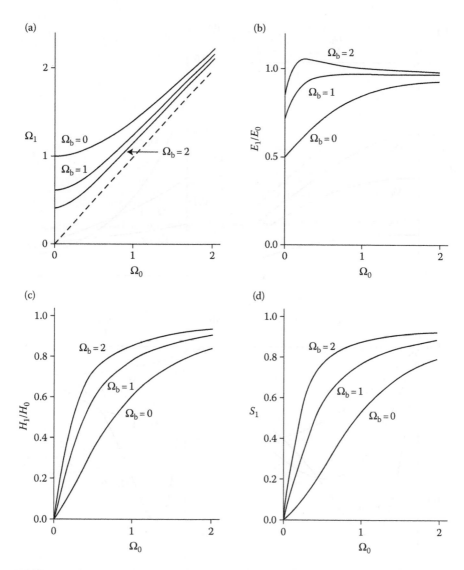

FIGURE 7.4 Characteristics of the first wave for L incidence: (a) frequency, (b) electric field, (c) magnetic field, and (d) power density.

magnetic fields (high Ω_b) and low values of the incident wave frequencies, the frequency of the first wave can be made smaller than the plasma frequency. In Figure 7.4b, the electric field is shown; it shows a peak value of 1.07 at about $\Omega_0 = 0.27$ for $\Omega_b = 2$. For a very large value of Ω_b ($\Omega_b = 10$), the peak value is about 1.2 occurring at $\Omega_0 = 0.05$ (not shown in the figure). In Figure 7.4c the wave magnetic field and in Figure 7.4d the time-averaged power density are shown. The static magnetic field improves the power level of this wave. For a

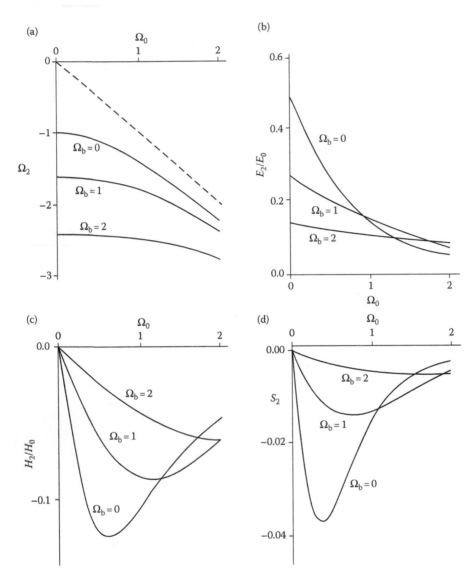

FIGURE 7.5 Characteristics of the second wave for L incidence: (a) frequency, (b) electric field, (c) magnetic field, and (d) power density.

large value of Ω_0, the frequency of the first wave asymptotically approaches the incident wave frequency. Plasma has very little effect.

Figure 7.5 describes the second wave that is a reflected wave. Its frequency is always greater than the incident wave frequency. Its power level peaks to 4% at about $\Omega_0 = 0.35$ and $\Omega_b = 0$. The static magnetic field reduces the power level of this reflected wave.

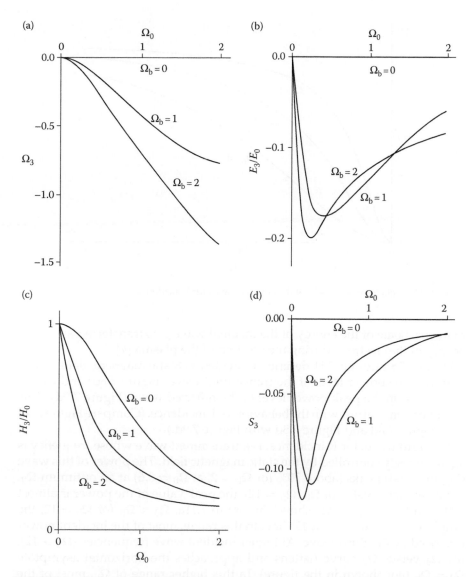

FIGURE 7.6 Characteristics of the third wave for L incidence: (a) frequency, (b) electric field, (c) magnetic field, and (d) power density.

Figure 7.6 describes the third wave, which in this case (L incidence) is a reflected wave. This wave is due to the presence of the static magnetic field and is strongly influenced by it. The frequency of this wave is less than Ω_b. The power level peaks to about 12% (at $\Omega_0 = 0.1$ for $\Omega_b = 0.1$) and the peaking point is influenced by Ω_b. Figure 7.7 shows the ratio of the total power in the newly created waves to the power in the incident wave. It is always less

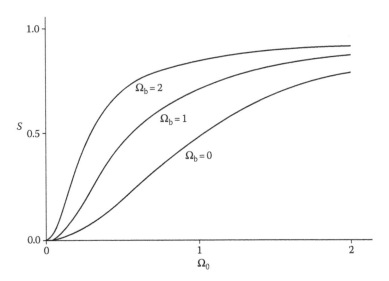

FIGURE 7.7   Total power density of the created waves for L incidence.

than one. Some of the energy of the incident wave gets transferred as kinetic energy of the electrons during the creation of the plasma [4].

Figures 7.8 through 7.11 describe R incidence. Static magnetic field causes a drop in the power level of the transmitted wave (Figure 7.8d) and causes an increase in the peak power level of the reflected wave (Figure 7.9d). This behavior is in contrast with the behavior for L incidence. (Compare Figure 7.4d with Figure 7.8d and Figure 7.5d with Figure 7.9d.)

The third wave, for R incidence, is a transmitted wave whose frequency is again strongly controlled by the static magnetic field. The power of this wave (Figure 7.10d) peaks (about 70% for $\Omega_b = 2$, at $\Omega_0 = 0.6$) at an optimum Ω_0. For a very large value of Ω_b ($\Omega_b = 12$), the peak value of the power is almost 100%. In the range of Ω_0 shown in Figure 7.11a, $\Omega_3 \approx \Omega_0$ for $\Omega_b = 12$; the graph of Ω_3 versus Ω_0 is a 45° line. In this range, most of the incident power is carried by the third wave. At higher incident wave frequencies ($\Omega_0 > 12$), the Ω_3 versus Ω_0 curve flattens and approaches the horizontal asymptote $\Omega_3 = \Omega_b$ (not shown in the figure). In this higher range of Ω_0, most of the incident power will be carried by the first wave.

A physical interpretation of the three waves can be given in the following way: the electric field of the incident wave accelerates the electrons in the newly created magnetoplasma, which in turn radiates new waves. The frequency (ω) of the new waves is constrained by the requirements that the wave number is conserved and the refractive index (n) is that applicable to longitudinal propagation in magnetoplasma.

$$\omega = \beta v_{ph} = k v_{ph} = \frac{\omega_0}{n}, \tag{7.42}$$

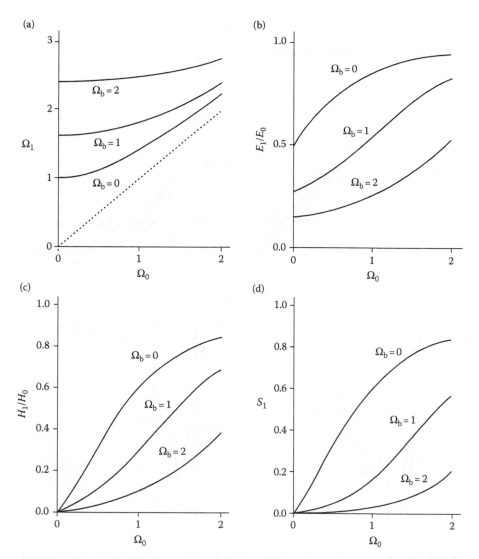

FIGURE 7.8 Characteristics of the first wave for R incidence: (a) frequency, (b) electric field, (c) magnetic field, and (d) power density.

where

$$n = \sqrt{1 - \frac{\omega_p^2}{\omega(\omega \mp \omega_b)}}. \tag{7.43}$$

Equation 7.42, when simplified, reduces to Equation 7.17.

The propagation transverse to the imposed magnetic field results in ordinary or extraordinary waves (X waves). The ordinary wave case is the same as the isotropic case. The X wave case is discussed in Section 7.6.

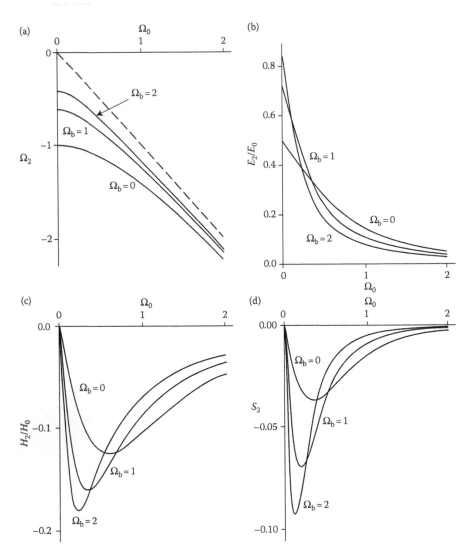

FIGURE 7.9 Characteristics of the second wave for R incidence: (a) frequency, (b) electric field, (c) magnetic field, and (d) power density.

7.5 Damping Rates: Longitudinal Propagation

We have ignored in our analysis the collision frequency v. Assuming a low-loss plasma medium where $v/\omega_p \ll 1$, Kalluri and Goteti [6] give expressions for the damping time constant and attenuation length for longitudinal propagation of the B wave in a magnetoplasma. A reprint of Ref. [6] is given in this book as Appendix B.

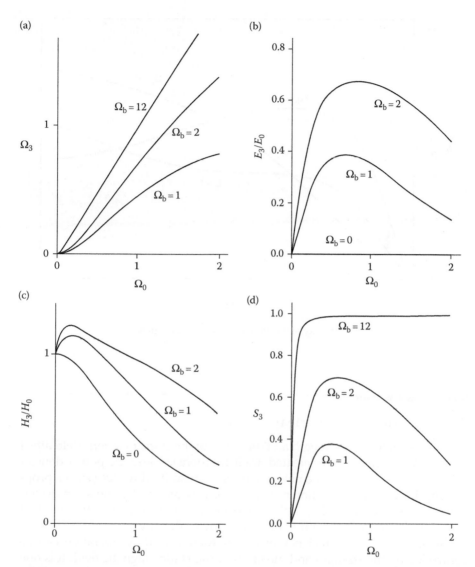

FIGURE 7.10 Characteristics of the third wave for R incidence: (a) frequency, (b) electric field, (c) magnetic field, and (d) power density.

7.6 Sudden Creation: Transverse Propagation: X wave

The complete theory of "wave propagation in a switched magnetoplasma medium: transverse propagation" is discussed in Ref. [7]; a reprint of this paper is given in this book as Appendix C.

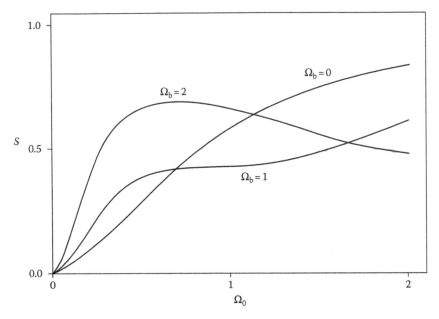

FIGURE 7.11   Total power density of the created waves for R incidence.

7.7   Additional Numerical Results

We have seen that the strength and direction of the static magnetic field affect
the number of new waves created, their frequencies, and the power density
of these waves. Some of the new waves are transmitted waves (waves propagating in the same direction as the incident wave) and some are reflected
waves (waves propagating in the opposite direction to that of the incident
wave). Reflected waves tend to have less power density.

The shift ratio and efficiency of the frequency shifting operation can be
controlled by the strength and direction of the static magnetic field. It is one
of the two aspects that are discussed in this section.

Figure 7.12 presents the results for the L1 wave. It has a frequency $\Omega_1 > \Omega_0$
and is a transmitted wave with considerable power in it. Figure 7.12a and b
shows the variation of Ω_1 and S_1 with Ω_0 for various values of Ω_b. Figure 7.12c
is constructed from Figure 7.12a and b and shows the frequency upshift ratio
Ω_1/Ω_0 versus Ω_0. The solid line and the broken line curves have percentage
power and the electron gyrofrequency (Ω_b) as parameters, respectively. The
beneficial effect of the strength of the static magnetic field in increasing the
power for a given shift ratio is evident from Figure 7.12.

In Figures 7.13 and 7.14, the cases of R1 and R3 waves are considered,
respectively. R1 is upshifted, whereas R3 is downshifted.

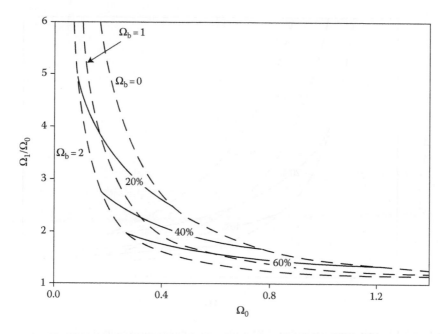

FIGURE 7.12 Frequency upshifting of L1 wave. Frequency shift ratio as a function of the incident wave frequency (ω_0). The solid line and the broken line curves have S_1 expressed in percent and the electron gyrofrequency (ω_b) as parameters, respectively. All frequency variables are normalized with respect to the plasma frequency (ω_p): $\Omega_b = \omega_b/\omega_p$, $\Omega_0 = \omega_0/\omega_p$, and $\Omega_1 = \omega_1/\omega_p$. The power density is normalized with respect to that of the source wave.

Figures 7.15 and 7.16 show the frequency shifts of X1 and X2 waves. X1 wave is upshifted, while X2 wave is upshifted for $\Omega_0 < 1$ and downshifted for $\Omega_0 > 1$.

The waves generated in the switched medium are damped as they travel in the medium if it is lossy. The formulas for the skin depth (d_{pn}) and the damping time constant (t_{pn}) were derived earlier (see Appendices B and C).

Table 7.1 is prepared from the results presented in Figures 7.4 through 7.16 and lists the parameters of the switched magnetoplasma medium for a given shift ratio. From the shift ratio $f_n/f_0 = \Omega_n/\Omega_0$, the variables Ω_0 and S_n/S_0 are found from Figures 7.4 through 7.16 for $\Omega_b = 0, 1$, and 2. For an assumed value of f_0, Ω_b and ω_p are determined and these are converted into B_0 and N_0. It can be noted that R3 and X2 waves do not exist for the isotropic case ($\Omega_b = 0$). The numbers for B_0 and N_0 in Table 7.1 are computed assuming $f_0 = 10\,\text{GHz}$. One can obtain these numbers for other values of f_0 by noting that N_0 scales as f_0^2 and B_0 scales as f_0. The damping constants are calculated assuming a low-loss plasma with $\nu/\omega_p = 0.01$, where ν is the collision frequency. The last column in Table 7.1 lists the answer to the question whether the source frequency ω_0 falls in the stop band of the particular mode. An incident wave whose frequency ω_0 falls in the stop band is totally reflected by the space boundary

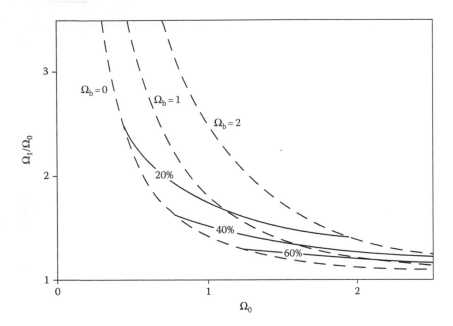

FIGURE 7.13   Frequency upshifting of R1 wave. The symbols are explained in Figure 7.12.

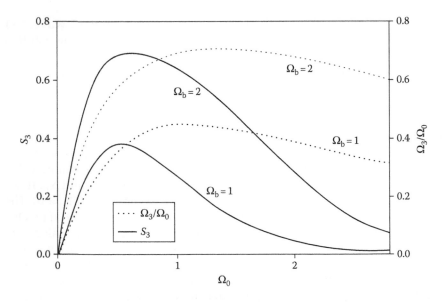

FIGURE 7.14   Frequency downshifting of R3 wave. The symbols are explained in Figure 7.12.

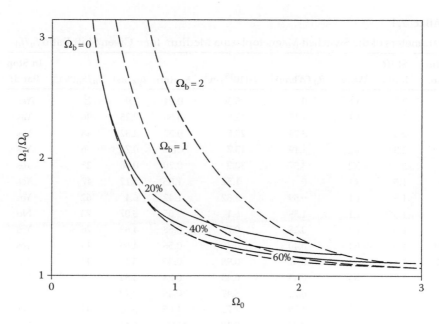

FIGURE 7.15 Frequency upshifting of X1 wave. The symbols are explained in Figure 7.12.

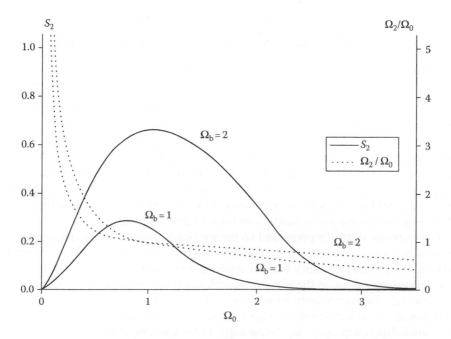

FIGURE 7.16 Frequency shifting of X2 wave. The symbols are explained in Figure 7.12.

TABLE 7.1

Parameters of the Switched Magnetoplasma Medium for a Given Shift Ratio f_n/f_0

Line No.	Shift Ratio	Wave	B_0 (Wb/m)^2	$N_0(10^{18}/m^3)$	δ_{pn}(m)	t_{pn}(ns)	S_n/S_0(%)	In Stop Band?
1	2.3	O	0	5.36	0.24	1.9	22	Yes
2	2.3*	L1	1.15	12.8	0.38	2.25	36	Yes
3	2.3**	L1	3.25	25.5	0.57	2.82	48	Yes
4	2.3*	X2	1.19	13.7	N.D.[a]	0.74	6	Yes
5	2.3**	X2	3.57	30.9	0.20	0.88	25	Yes
6	1.5	O	0	1.52	1.05	5.12	47	Yes
7	1.5*	L1	0.59	3.32	1.42	6.4	62	Yes
8	1.5**	L1	1.59	6.1	2.05	8.07	73	No
9	1.5*	R1	0.28	0.76	0.68	4.63	26	Yes
10	1.5**	R1	0.40	0.38	0.56	4.69	14	Yes
11	1.5*	X1	0.32	0.98	0.36	3.4	11	No
12	1.5**	X1	0.43	0.45	0.26	3.84	4	No
13	1.5*	X2	0.75	5.47	0.45	3.42	17	Yes
14	1.5**	X2	1.79	7.72	1.05	6.35	41	Yes
15	0.7**	X2	0.20	0.10	N.D.[a]	6.41	1	No
16	0.6**	R3	0.26	0.16	0.55	6.28	7.5	No
17	0.6**	R3	1.28	3.94	2.47	10.3	69	No
18	0.4*	R3	0.19	0.34	0.25	4.4	5.5	No
19	0.4*	R3	0.58	3.21	0.84	5.2	38	No
20	0.4**	R3	2.75	18.26	2.23	10.48	55	No

Incident wave frequency: $f_0 = 10\,\text{GHz}$. Normalized collision frequency: $\nu/\omega_p = 0.01$.
*$\Omega_B = 1$, **$\Omega_B = 1$.
[a] Not determined.

of the medium. The free space on the forward side of the magnetoplasma medium then will not have signals at the frequency of the incident wave. From the results in Table 7.1 it appears that the L1 wave is best suited for upshifting and the R3 wave is best suited for downshifting if there are no constraints on B_0 and N_0. In both cases the strength of the magnetic field improves the power density of the frequency shifted wave. A stronger static magnetic field requires a stronger ionization to achieve the same shift ratio. A stronger static magnetic field gives larger time constant.

The numerical results presented so far are normalized with reference to the plasma frequency. For the convenience of the experimentalists who preferred to keep the source frequency fixed and vary the plasma frequency by adjusting the vacuum pressure in the plasma vessel, we have generated the numerical results and graphs normalized with reference to the source frequency.

It has also become possible to give explicit expressions for the frequencies and the fields when $\omega_b \ll \omega_0$. These aspects are discussed in Ref. [8]; a reprint of this is given in this book as Appendix D.

7.8 Sudden Creation: Arbitrary Direction of the Static Magnetic Field

The switching of the plasma medium in the presence of an arbitrarily directed magnetic field creates many more frequency-shifted modes. This aspect is discussed by Stanic et al. [9].

7.9 Frequency Shifting of Low-Frequency Waves

The results given so far need modification to be applicable to very low values of ω_0. The ion motion has to be introduced into the analysis and suitable approximations corresponding to longitudinal and transverse propagation of Alfven waves can be made. These aspects are discussed in Refs. [9–11].

References

1. Kalluri, D. K., Effect of switching a magnetoplasma medium on a traveling wave: Longitudinal propagation, *IEEE Trans. Ant. Prop.*, AP-37, 1638, 1989.
2. Kalluri, D. K., Frequency-shifting using magnetoplasma medium, *IEEE Trans. Plasma Sci.*, 21, 77, 1993.
3. Ramo, S., Whinnery, J. R., and Duzer, T. V., *Fields and Waves in Communication Electronics*, Wiley, New York, 1965.
4. Jiang, C. L., Wave propagation and dipole radiation in a suddenly created plasma, *IEEE Trans. Ant. Prop.*, AP-23, 83, 1975.
5. Kalluri, D. K., On reflection from a suddenly created plasma half-space: Transient solution, *IEEE Trans. Plasma Sci.*, 16, 11, 1988.
6. Kalluri, D. K. and Goteti, V. R., Damping rates of waves in a switched magnetoplasma medium: Longitudinal propagation, *IEEE Trans. Plasma Sci.*, 18, 797, 1990.
7. Goteti, V. R. and Kalluri, D. K., Wave propagation in a switched magnetoplasma medium: Transverse propagation, *Radio Sci.*, 25, 61, 1990.
8. Kalluri, D. K., Frequency shifting using magnetoplasma medium: Flash ionization, *IEEE Trans. Plasma Sci.*, 21, 77, 1993.
9. Madala, S. R. V. and Kalluri, D. K., Longitudinal propagation of low frequency waves in a switched magnetoplasma medium, *Radio Sci.*, 28, 121, 1993.
10. Dimitrijevic, M. M. and Stanic, B. V., EMW transformation in suddenly created two-component magnetized plasma, *IEEE Trans. Plasma Sci.*, 23, 422, 1995.
11. Madala, S. R. V., Frequency shifting of low frequency electromagnetic waves using magnetoplasmas, Doctoral thesis, University of Massachusetts Lowell, Lowell, 1993.

8

Longitudinal Propagation in a Magnetized Time-Varying Plasma*

8.1 Introduction

The technique of solving the problem of longitudinal propagation in a magnetized time-varying plasma is laid down in Sections 3.5 through 3.9. The anisotropic case considered here is a little more difficult. In preparation for its assimilation, the reader is urged to revisit Sections 3.5 through 3.9 and Section 7.3.

Wave propagation in a transient magnetoplasma has been considered by a number of authors [1–10]. At one extreme, when the source wave period is much larger than the rise time T_r, the problem can be solved as an initial value problem using the sudden-switching approximation ($T_r = 0$) [1]. One effect of switching the medium is creation of three frequency-shifted waves. The power scattering coefficients depend on the ratio of the source frequency to the plasma frequency and gyrofrequency. The power reflection coefficient falls rapidly with an increase of T_r.

At the other extreme, the switching is considered sufficiently slow that an adiabatic analysis can be used [2]. In this approximation, two of the scattering coefficients are negligible and the main forward-propagating wave is the modified source wave (MSW).

In this chapter, we consider the wave propagation in a magnetized transient plasma for the case of a rise time comparable to the period of the source wave [8]. Geometry of the problem in the z–t plane is sketched in Figure 8.1. A right circularly polarized source wave, $R_s(\omega_1)$, of angular frequency ω_1 is propagating in an unbounded magnetized plasma medium whose parameters are the plasma frequency ω_{p1} and the gyrofrequency ω_b. The direction of wave propagation, which is also the direction of the static magnetic field, is assumed to be along the z-axis (longitudinal propagation). The plasma frequency undergoes a transient change in the interval $-T_{r1} < t < T_{r2}$ attaining its final value of ω_{p2} at $t = T_{r2}$. The transient change in the medium properties gives rise to three frequency-shifted waves, designated R_1, R_2, and R_3.

* © IEEE. Reprinted from Kalluri, D. K. and Huang, T. T., *IEEE Trans. Plasma Sci.*, 26, 1022–1030, 1998. With permission. Chapter 8 is an adaptation of the reprint.

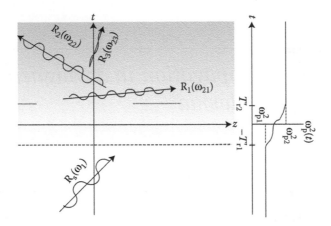

FIGURE 8.1 Geometry of the problem. A right circularly polarized source wave propagating in a magnetized time-varying plasma medium is considered. The direction of wave propagation, which is also the direction of the static magnetic field, is assumed to be along the z-axis.

The computation of the instantaneous frequencies and the final asymptotic frequencies of the waves is quite straightforward [2]. Enforcing the conservation of the wave number on the dispersion relation of the magnetized plasma results in a cubic equation

$$\omega_{2n}^3 - \omega_{2n}^2 \omega_b - \omega_{2n}[k^2 c^2 + \omega_p^2(t)] + k^2 c^2 \omega_b = 0, \tag{8.1}$$

where

$$k^2 c^2 = \omega_1^2 \left[1 - \frac{\omega_{p1}^2}{\omega_1(\omega_1 - \omega_b)} \right], \tag{8.2}$$

and k is the conserved wave number. The roots of Equation 8.1 are the instantaneous frequencies and are sketched in Figure 8.2.

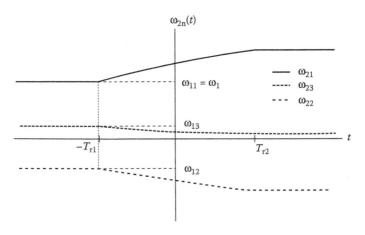

FIGURE 8.2 Instantaneous frequencies.

On the other hand, there is no exact solution for the amplitudes of the waves. A perturbation technique for the computation of the scattering coefficients is the focus of this chapter. A causal Green's function is developed as the basis for the perturbation. The method gives a closed-form expression for the scattering coefficients for a general temporal profile for which an exact solution is not available. The extension of the results for the case of a left circularly polarized source wave (L_s wave) is quite straightforward and is not discussed here.

8.2 Perturbation from Step Profile

The governing field equations for the problem, with the assumption of a lossless cold plasma, are

$$\nabla \times \mathbf{E} = -\mu_0 \frac{\partial \mathbf{H}}{\partial t}, \tag{8.3}$$

$$\nabla \times \mathbf{H} = \varepsilon_0 \frac{\partial \mathbf{E}}{\partial t} + \mathbf{J}, \tag{8.4}$$

$$\frac{\partial \mathbf{J}}{\partial t} = \varepsilon_0 \omega_p^2(t)\mathbf{E} - \mathbf{J} \times \boldsymbol{\omega}_b, \tag{8.5}$$

where $\mathbf{E}$, $\mathbf{H}$, and $\mathbf{J}$ have the usual meaning [2]. For the case of the longitudinal propagation of an R wave, that is, $\mathbf{H}(z, t) = (\hat{x} - j\hat{y})H(t)e^{-jkz}$, and so on [2], the magnetic field satisfies

$$\frac{d^3 H(t)}{dt^3} - j\omega_b \frac{d^2 H(t)}{dt^2} + [k^2 c^2 + \omega_p^2(t)]\frac{dH(t)}{dt} - jk^2 c^2 \omega_b H(t) = 0. \tag{8.6}$$

The solution of Equation 8.4, where $\omega_p^2(t) = \tilde{\omega}_p^2(t)$ is a step function, can be obtained by using the Laplace transform technique [1]:

$$H_0(t) = e^{j\omega_1 t}, \quad t < 0, \tag{8.7}$$

$$H_0(t) = H_{21}^{(0)} e^{j\omega_{21} t} + H_{22}^{(0)} e^{j\omega_{22} t} + H_{23}^{(0)} e^{j\omega_{23} t}, \quad t > 0, \tag{8.8}$$

where

$$H_{2n}^{(0)} = \frac{(\omega_{2n} - \omega_b)(\omega_{2n} + \omega_1) - k^2 c^2 + \omega_1^2 - \omega_{p2}^2}{\prod_{p=1, p \neq n}^{3}(\omega_{2n} - \omega_{2p})}, \quad n = 1, 2, 3. \tag{8.9}$$

ω_1 is the source wave frequency, and ω_{2n} are the roots of

$$\omega_{2n}^3 - \omega_{2n}^2 \omega_b - \omega_{2n}(k^2 c^2 + \omega_{p2}^2) + k^2 c^2 \omega_b = 0. \tag{8.10}$$

For a general profile (Figure 8.1) of $\omega_p^2(t)$, we seek to have a perturbation solution by using the solution for the step profile $\tilde{\omega}_p^2(t)$ as a reference solution. The formulation is developed below:

$$\frac{d^3 H(t)}{dt^3} - j\omega_b \frac{d^2 H(t)}{dt^2} + [k^2 c^2 + \omega_p^2(t)] \frac{dH(t)}{dt} - jk^2 c^2 \omega_b H(t) = 0 \qquad (8.11)$$

$$\frac{d^3}{dt^3}[H_0(t) + H_1(t)] - j\omega_b \frac{d^2}{dt^2}[H_0(t) + H_1(t)]$$

$$+ [k^2 c^2 + \tilde{\omega}_p^2(t) + \Delta\omega_p^2(t)] \frac{d}{dt}[H_0(t) + H_1(t)] - jk^2 c^2 \omega_b [H_0(t) + H_1(t)] = 0.$$

$$(8.12)$$

In Equation 8.12,

$$\Delta\omega_p^2(t) = \omega_p^2(t) - \tilde{\omega}_p^2(t). \qquad (8.13)$$

The first-order perturbation H_1 can be estimated by dropping the second-order term $-\Delta\omega_p^2(t)\, d[H_1(t)]/\, dt$ from Equation 8.12:

$$\frac{d^3 H_1(t)}{dt^3} - j\omega_b \frac{d^2 H_1(t)}{dt^2} + [k^2 c^2 + \tilde{\omega}_p^2(t)] \frac{dH_1(t)}{dt} - jk^2 c^2 \omega_b H_1(t)$$

$$= -\Delta\omega_p^2(t) \frac{dH_0(t)}{dt}. \qquad (8.14)$$

Treating the right-hand side of Equation 8.14 as the driving function $F(t)$, we can write the solution of H_1 as

$$H_1(t) = G(t, \tau) * F(t) = -\int_{-\infty}^{t} G(t, \tau) \Delta\omega_p^2(\tau) \frac{dH_0(\tau)}{d\tau}\, d\tau, \qquad (8.15)$$

where $G(t, \tau)$ is the Green's function that satisfies the third-order differential equation

$$\frac{d^3 G(t, \tau)}{dt^3} - j\omega_b \frac{d^2 G(t, \tau)}{dt^2} + [k^2 c^2 + \tilde{\omega}_p^2(t)] \frac{dG(t, \tau)}{dt} - jk^2 c^2 \omega_b G(t, \tau) = \delta(t - \tau).$$

$$(8.16)$$

The approximate one-iteration solution for the original time-varying differential equation takes the form

$$H(t) \approx H_0(t) + H_1(t) = H_0(t) - \int_{-\infty}^{t} G(t, \tau) \Delta\omega_p^2(\tau) \frac{dH_0(\tau)}{d\tau}\, d\tau. \qquad (8.17)$$

More iterations can be made until the desired accuracy is reached. The general N-iterations solution is

$$H(t) = H_0(t) + \sum_{n=1}^{N} H_n(t), \tag{8.18}$$

where

$$H_n(t) = - \int_{-\infty}^{t} G(t, \tau) \Delta \omega_p^2(\tau) \frac{\mathrm{d}}{\mathrm{d}\tau} H_{n-1}(\tau) \, \mathrm{d}\tau. \tag{8.19}$$

8.3 Causal Green's Function for Temporally-Unlike Magnetized Plasma Media

The technique of constructing the Green's function for this case is similar, in principle, to that used in Ref. [7], but is more difficult to construct since the case deals with a third-order system. The frequency of the upshifted forward-propagating wave is different from that of the backward-propagating wave. Also, there is a second forward-propagating wave with a downshifted frequency. The appropriate Green's function in the domain $-\infty < t < 0$ is built up of functions $e^{j\omega_{1m}t}$, where ω_{1m} are the roots of

$$\omega_{1m}^3 - \omega_{1m}^2 \omega_b - \omega_{1m}(k^2 c^2 + \omega_{p1}^2) + k^2 c^2 \omega_b = 0. \tag{8.20}$$

One of the roots of the cubic Equation 8.20 is ω_1, say

$$\omega_{11} = \omega_1. \tag{8.21}$$

In the region $0 < t < \infty$, the Green's function is built up from the functions $e^{j\omega_{2n}t}$, where ω_{2n} are the roots of Equation 8.10.

From the causality requirement,

$$G(t, \tau) = 0, \quad t < \tau. \tag{8.22}$$

For $t > \tau$, $G(t, \tau)$ will be determined from the impulse source conditions for a third-order system, that is,

$$G(\tau^+, \tau) = G(\tau^-, \tau), \tag{8.23}$$

$$\frac{\partial G(\tau^+, \tau)}{\partial t} = \frac{\partial G(\tau^-, \tau)}{\partial t}, \tag{8.24}$$

$$\frac{\partial^2 G(\tau^+, \tau)}{\partial t^2} - \frac{\partial^2 G(\tau^-, \tau)}{\partial t^2} = 1. \tag{8.25}$$

From Equation 8.22, one obtains $G(\tau^-, \tau) = 0$, $\partial G(\tau^-, \tau)/\partial t = 0$, and $\partial^2 G(\tau^-, \tau)/\partial t^2 = 0$. Thus, Equations 8.23 through 8.25 reduce to

$$G(\tau^+, \tau) = 0, \tag{8.26}$$

$$\frac{\partial G(\tau^+, \tau)}{\partial t} = 0, \tag{8.27}$$

$$\frac{\partial^2 G(\tau^+, \tau)}{\partial t^2} = 1. \tag{8.28}$$

A geometrical interpretation of the steps involved in constructing the Green's function is given in Figure 8.3, where the horizontal axis is the spatial coordinate z, along which the waves are propagating. The horizontal axis ($t = 0$) is also the temporal boundary between the two magnetized temporally-unlike plasma media.

In Figure 8.3a, we consider the case $\tau > 0$. The impulse source is located in the second medium. From Equation 8.16, the Green's function can be written as

$$G(t, \tau) = c_{B1}(\tau)e^{j\omega_{21}t} + c_{B2}(\tau)e^{j\omega_{22}t} + c_{B3}(\tau)e^{j\omega_{23}t}, \quad t > \tau > 0, \tag{8.29}$$

and c_{B1}, c_{B2}, and c_{B3} are determined from Equations 8.26 through 8.28:

$$c_{B1}(\tau) = H_{B1}e^{-j\omega_{21}\tau} = \frac{-1}{(\omega_{21} - \omega_{22})(\omega_{21} - \omega_{23})}e^{-j\omega_{21}\tau}, \tag{8.30}$$

$$c_{B2}(\tau) = H_{B2}e^{-j\omega_{22}\tau} = \frac{-1}{(\omega_{22} - \omega_{21})(\omega_{22} - \omega_{23})}e^{-j\omega_{22}\tau}, \tag{8.31}$$

$$c_{B3}(\tau) = H_{B3}e^{-j\omega_{23}\tau} = \frac{-1}{(\omega_{23} - \omega_{21})(\omega_{23} - \omega_{22})}e^{-j\omega_{23}\tau}. \tag{8.32}$$

In Figure 8.3b, we consider the case $\tau < 0$. The impulse source is located in the first medium. Thus $G(t, \tau)$, in the interval $\tau < t < 0$ can be written as

$$G(t, \tau) = c_{A1}(\tau)e^{j\omega_{11}t} + c_{A2}(\tau)e^{j\omega_{12}t} + c_{A3}(\tau)e^{j\omega_{13}t}, \quad \tau < t < 0, \tag{8.33}$$

where c_{A1}, c_{A2}, and c_{A3} are once again determined by Equations 8.26 through 8.28:

$$c_{A1}(\tau) = H_{A1}e^{-j\omega_{11}\tau} = \frac{-1}{(\omega_{11} - \omega_{12})(\omega_{11} - \omega_{13})}e^{-j\omega_{11}\tau}, \tag{8.34}$$

$$c_{A2}(\tau) = H_{A2}e^{-j\omega_{12}\tau} = \frac{-1}{(\omega_{12} - \omega_{11})(\omega_{12} - \omega_{13})}e^{-j\omega_{12}\tau}, \tag{8.35}$$

$$c_{A3}(\tau) = H_{A3}e^{-j\omega_{13}\tau} = \frac{-1}{(\omega_{13} - \omega_{11})(\omega_{13} - \omega_{12})}e^{-j\omega_{13}\tau}. \tag{8.36}$$

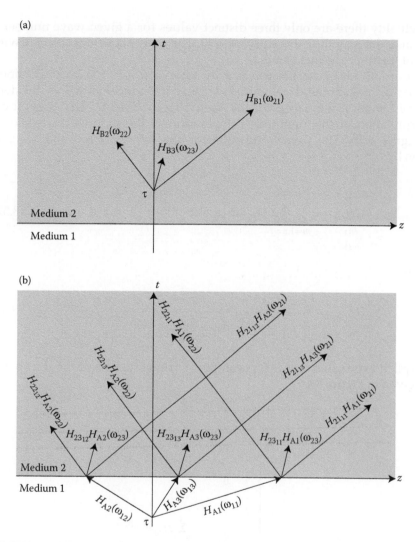

FIGURE 8.3 (a) Geometrical interpretation of the Green's function for magnetized temporally-unlike plasma media for the case $\tau > 0$. The frequencies are shown in parentheses. (b) Geometrical interpretation of the Green's function for magnetized temporally-unlike plasma media for the case $\tau < 0$. The frequencies are shown in parentheses.

Referring to Figure 8.3b, we note that the three waves launched by the impulse source at $t = \tau$ travel along the positive z-axis (two forward-propagating waves) and along the negative z-axis (one backward-propagating wave). They reach the temporal interface at $t = 0$, where the medium undergoes a temporal change. Each of these waves gives rise to three waves at the shifted frequencies ω_{21}, ω_{22}, and ω_{23}. It can be noted that as n varies from 1 to 3 and m varies from 1 to 3, ω_{mn}, will have nine values.

In actuality there are only three distinct values for a given wave number k as constrained by the cubic equation 8.10. The other six values coincide with one of the three distinct values.

The amplitude of the nine waves in the interval $0 < t < \infty$ can be obtained by taking into account the amplitude of each of the incident waves (incident on the temporal interface) and the scattering coefficients. The geometrical interpretation of the amplitude computation of the nine waves is shown in Figure 8.3b. The Green's function for the interval $t > 0$ and $\tau < 0$ is given by

$$G(t, \tau) = \sum_{m=1}^{3} \sum_{n=1}^{3} H_{Am} H_{2n_{1m}} e^{j(\omega_{2n} t - \omega_{1m} \tau)}, \quad \tau < 0 < t, \tag{8.37}$$

where

$$H_{2n_{1m}} = \frac{(\omega_{2n} - \omega_b)(\omega_{2n} + \omega_{1m}) - k^2 c^2 + \omega_{1m} - \omega_{p2}^2}{\prod_{p=1, p \neq n}^{3} (\omega_{2n} - \omega_{2p})}. \tag{8.38}$$

Explicit expressions for $G(t, \tau)$ valid for various regions of the (t, τ) plane are given in Figure 8.4.

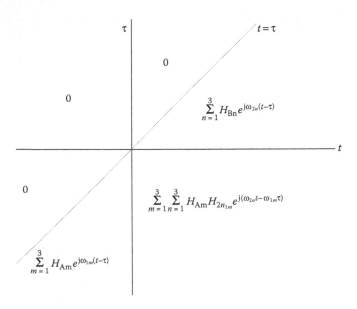

FIGURE 8.4 Causal Green's function for magnetized temporally-unlike plasma media.

8.4 Scattering Coefficients for a General Profile

The approximate amplitude of the electric field after one iteration ($N = 1$) can be obtained from Equation 8.17:

$$H(t) = H_0(t) - \int_0^{+t} d\tau \sum_{m=1}^{3} H_{Bm} e^{j\omega_{2m}(t-\tau)} \Delta\omega_p^2(\tau) \sum_{n=1}^{3} j\omega_{2n} H_{2n}^{(0)} e^{j\omega_{21}t}$$

$$- \int_{-\infty}^{0} d\tau \sum_{m=1}^{3} \sum_{n=1}^{3} H_{Am} H_{2n_{1m}} e^{j(\omega_{2n}t - \omega_{1m}\tau)} \Delta\omega_p^2(\tau) j\omega_1 e^{j\omega_1\tau}. \quad (8.39)$$

In the asymptotic limit of $t \to \infty$, $H_1(t)$ can be written as

$$H_1(t) = H_{21}^{(1)} e^{j\omega_{21}t} + H_{22}^{(1)} e^{j\omega_{22}t} + H_{23}^{(1)} e^{j\omega_{23}t}, \quad (8.40)$$

where $H_{21}^{(1)}$, $H_{22}^{(1)}$, and $H_{23}^{(1)}$ are the first-order correction terms of the scattering coefficients:

$$H_{21}^{(1)} = - \int_0^{\infty} d\tau H_{B1} e^{-j\omega_{21}\tau} \Delta\omega_p^2(\tau) \sum_{n=1}^{3} j\omega_{2n} H_{2n}^{(0)} e^{j\omega_{2n}\tau}$$

$$- \int_{-\infty}^{0} d\tau \sum_{m=1}^{3} H_{Am} H_{21_{1m}} e^{-j\omega_{1m}\tau} \Delta\omega_p^2(\tau) j\omega_1 e^{j\omega_1\tau}, \quad (8.41)$$

$$H_{22}^{(1)} = - \int_0^{\infty} d\tau H_{B2} e^{-j\omega_{22}\tau} \Delta\omega^2(\tau) \sum_{n=1}^{3} j\omega_{2n} H_{2n}^{(0)} e^{j\omega_{2n}\tau}$$

$$- \int_{-\infty}^{0} d\tau \sum_{m=1}^{3} H_{Am} H_{22_{1m}} e^{-j\omega_{1m}\tau} \Delta\omega^2(\tau) j\omega_1 e^{j\omega_1\tau}, \quad (8.42)$$

$$H_{23}^{(1)} = - \int_0^{\infty} d\tau H_{B3} e^{-j\omega_{23}\tau} \Delta\omega_p^2(\tau) \sum_{n=1}^{3} j\omega_{2n} H_{2n}^{(0)} e^{j\omega_{2n}\tau}$$

$$- \int_{-\infty}^{0} d\tau \sum_{m=1}^{3} H_{Am} H_{23_{1m}} e^{-j\omega_{1m}\tau} \Delta\omega_p^2(\tau) j\omega_1 e^{j\omega_1\tau}. \quad (8.43)$$

Higher-order correction terms ($N > 1$) can be obtained by using more iterations.

8.5 Scattering Coefficients for a Linear Profile

As a particular example, we use Equations 8.41 through 8.43 to compute scattering coefficients for a profile of $\omega_p^2(t)$ rising linearly from 0 (at $t = -T_r/2$) to ω_{p2}^2 (at $t = T_r/2$) with a slope of ω_{p2}^2/T_r. Here T_r is the rise time of the profile. The function $\Delta\omega_p^2(t)$ takes the form of $\omega_{p2}^2(-1/2 + t/T_r)$ for $0 < t < T_r/2$ and $\omega_{p2}^2(1/2 + t/T_r)$ for $-T_r/2 < t < 0$. It is zero on the rest of the real line (Figure 8.5):

$$
\begin{aligned}
H_{21}^{(1)} = j\omega_{p2}^2 \Bigg[&+ \frac{\omega_{21}H_{21}^{(0)}H_{B1}T_r}{8} - \omega_{21}H_{22}^{(0)}H_{B1}\frac{e^{j(\omega_{22}-\omega_{21})T_r/2} - 1 - j(\omega_{22} - \omega_{21})T_r/2}{T_r(\omega_{22} - \omega_{21})^2} \\
&- \omega_{21}H_{23}^{(0)}H_{B1}\frac{e^{j(\omega_{23}-\omega_{21})T_r/2} - 1 - j(\omega_{23} - \omega_{21})T_r/2}{T_r(\omega_{23} - \omega_{21})^2} \\
&- \frac{\omega_{11}H_{21_{11}}H_{A1}T_r}{8} + \omega_{12}H_{21_{12}}H_{A2}\frac{e^{-j(\omega_{11}-\omega_{12})T_r/2} - 1 + j(\omega_{11} - \omega_{12})T_r/2}{T_r(\omega_{11} - \omega_{12})^2} \\
&+ \omega_{13}H_{21_{13}}H_{A3}\frac{e^{-j(\omega_{11}-\omega_{13})T_r/2} - 1 + j(\omega_{11}-\omega_{13})T_r/2}{T_r(\omega_{11} - \omega_{13})^2} \Bigg],
\end{aligned} \tag{8.44}
$$

$$
\begin{aligned}
H_{22}^{(1)} = j\omega_{p2}^2 \Bigg[&- \omega_{22}H_{21}^{(0)}H_{B2}\frac{e^{j(\omega_{21}-\omega_{22})T_r/2} - 1 - j(\omega_{21} - \omega_{22})T_r/2}{T_r(\omega_{21} - \omega_{22})^2} + \frac{\omega_{22}H_{22}^{(0)}H_{B2}T_r}{8} \\
&- \omega_{22}H_{23}^{(0)}H_{B2}\frac{e^{j(\omega_{23}-\omega_{22})T_r/2} - 1 - j(\omega_{23} - \omega_{22})T_r/2}{T_r(\omega_{23} - \omega_{22})^2} \\
&+ \omega_{12}H_{22_{12}}H_{A2}\frac{e^{-j(\omega_{11}-\omega_{12})T_r/2} - 1 + j(\omega_{11} - \omega_{12})T_r/2}{T_r(\omega_{11} - \omega_{12})^2} - \frac{\omega_{11}H_{22_{11}}H_{A1}T_r}{8} \\
&+ \omega_{13}H_{22_{13}}H_{A3}\frac{e^{-j(\omega_{11}-\omega_{13})T_r/2} - 1 + j(\omega_{11} - \omega_{13})T_r/2}{T_r(\omega_{11} - \omega_{13})^2} \Bigg],
\end{aligned} \tag{8.45}
$$

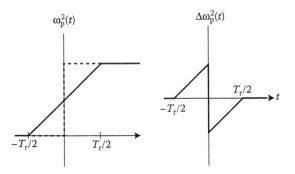

FIGURE 8.5   Linear profile. The broken line curve is the step profile. $\tilde{\omega}_p^2(t)$.

$$
H_{23}^{(1)} = j\omega_{p2}^2 \left[-\omega_{21}H_{21}^{(0)}H_{B3}\frac{e^{j(\omega_{21}-\omega_{23})T_r/2} - 1 - j(\omega_{21}-\omega_{23})T_r/2}{T_r(\omega_{21}-\omega_{23})^2} \right.
$$

$$
-\omega_{22}H_{22}^{(0)}H_{B3}\frac{e^{j(\omega_{22}-\omega_{23})T_r/2} - 1 - j(\omega_{22}-\omega_{23})T_r/2}{T_r(\omega_{22}-\omega_{23})^2} + H_{B3}\frac{\omega_{23}H_{23}^{(0)}T_r}{8}
$$

$$
+\omega_{11}H_{23_{12}}H_{A2}\frac{e^{-j(\omega_{11}-\omega_{12})T_r/2} - 1 + j(\omega_{11}-\omega_{12})T_r/2}{T_r(\omega_{11}-\omega_{12})^2}
$$

$$
\left. +\omega_{11}H_{23_{13}}H_{A3}\frac{e^{-j(\omega_{11}-\omega_{13})T_r/2} - 1 + j(\omega_{11}-\omega_{13})T_r/2}{T_r(\omega_{11}-\omega_{13})^2} - H_{A1}\frac{\omega_{11}H_{23_{11}}T_r}{8} \right].
$$

$$
\text{(8.46)}
$$

The solution for other profiles can be obtained similarly. However, to perform the integrations in Equations 8.41 through 8.43 for a general profile, it may be necessary to expand the exponential in Equations 8.41 through 8.43 into a power series and then do the integration. For a given T_r, it is possible to choose the relative positioning of $\omega_p^2(t)$ and $\tilde{\omega}_p^2(t)$ such that

$$
\int_{-\infty}^{+\infty} \Delta\omega_p^2(\tau)\, d\tau = 0. \tag{8.47}
$$

Such a choice will improve the accuracy of the scattering coefficients when we approximate the exponential terms in Equations 8.41 through 8.43 by keeping only the dominant terms.

8.6 Numerical Results

Each of the scattering coefficients after one iteration is determined by adding to Equation 8.9 the corresponding first-order correction terms given by Equations 8.43 through 8.46. Figures 8.6 through 8.8 depict the power scattering coefficients; the scattering coefficients are first computed from the equations and their magnitude squares are multiplied by the ratio $|(\omega_1/\omega_{2n})|$ to get the scattering power density. The numerical results for a linear profile are presented in a normalized form by taking $f_1 = \omega_1/2\pi = 1.0$. The other parameters are $f_{p1} = \omega_{p1}/2\pi = 0$ and $f_{p2} = \omega_{p2}/2\pi = 1.2$.

From these figures, one can see that the effect of the rise time is less prominent when the longitudinal magnetic field is strong. From Figure 8.8, it can be noted that the power scattering coefficient of the R_3 wave increases with the strength of the static magnetic field.

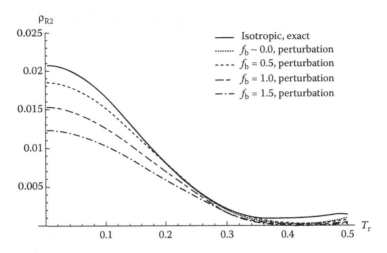

FIGURE 8.6 Power density ρ_{R2} of the backward-propagating wave R_2 versus rise time for a linear profile.

8.7 Wiggler Magnetic Field

Note that in the $f_b = 0$ case, the perturbation solutions for the R_1 and R_2 waves track the exact isotropic analytical solutions up to $T_r = 0.25$ [7], as expected. The solution for the R_3 wave has a zero power as shown in Figure 8.8. From the sudden-switching analysis [1], $T_r = 0$, we know that the electric field

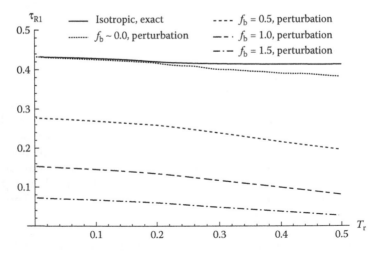

FIGURE 8.7 Power density τ_{R1} of the forward-propagating wave R_1 versus rise time for a linear profile.

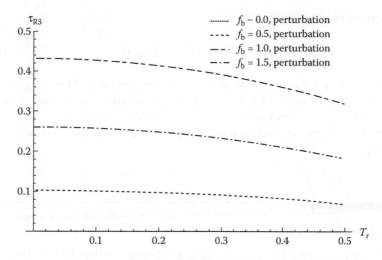

FIGURE 8.8 Power density τ_{R3} of the forward-propagating wave R_3 versus rise time for a linear profile.

for the third wave $E_{23}^{(0)}$ is zero when $f_b = 0$. All the subsequent perturbation correction terms for the electric field are also zero. The magnetic field however is not zero [5]. The wave reduces to a magnetic wiggler field. The perturbation solution for the wiggler field can be calculated from Equation 8.46 by substituting $f_b = 0$. Adding this value to that of $H_{23}^{(0)}$ from Equation 8.9, we obtain the magnetic wiggler field H_{23} after one iteration. Thus, the magnetic

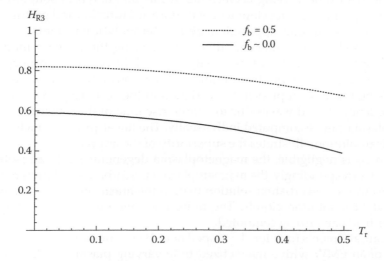

FIGURE 8.9 Magnetic wiggler field versus rise time.

wiggler field H_w is given by

$$H_w = H_{23}^{(0)} + H_{23}^{(1)}, \quad f_b = 0. \tag{8.48}$$

Figure 8.9 shows the variation of H_{R3} as a function of T_r. The wiggler magnetic field H_w is obtained by choosing a very small value f_b and is shown as the solid line. The broken line curve shows H_{R3} for $f_b = 0.5$. This is the magnetic field of the second forward-propagating wave with a downshifted frequency.

8.8   *E*-Formulation

The problem can also be solved using an *E*-formulation [9,10]. The time-varying differential equation for this formulation is given by Equation 7.9. The results obtained by such a formulation agree with the results shown in Section 8.7.

8.9   Summary

Wave propagation in a transient magnetoplasma with a rise time comparable to the source wave period is a challenging problem. The underlining difficulty is in finding the solution to the third-order governing differential equation with time-varying coefficients. It has no analytical solution even for a linear profile. The development of the Green's function and perturbation series provides a unique approach in giving the analytical insights.

The method has the inherent advantage of using the intrinsic modes of wave propagation, the forward-propagating and the backward-propagating waves, and thus is superior to the numerical solution of the initial value problem. Using this new approach, the variation of the scattering coefficients of the frequency-shifted waves due to a finite rise time of the transient magnetized plasma can be computed individually. The linear profile, which has no analytical solution, illustrates the superiority of the approach.

When ω_b is negligible, the magnetoplasma degenerates into an isotropic plasma. Correspondingly, the magnetoplasma perturbation solution reduces to the isotropic perturbation solution that, in the linear profile case, matches the analytical solution closely. The finite rise time effect on the magnetic wiggler field can also be computed.

The causal Green's function developed here, being fundamental to the interaction of an EMW with a magnetized time-varying plasma, can be used in other analyses besides computation of scattering coefficients.

References

1. Kalluri, D. K., Effect of switching a magnetoplasma medium on a traveling wave: Longitudinal propagation, *IEEE Trans. Ant. Prop.*, AP-37, 1638, 1989.
2. Kalluri, D. K., Goteti, V. R., and Sessler, A. M., WKB solution for wave propagation in a time-varying magnetoplasma medium: Longitudinal propagation, *IEEE Trans. Plasma Sci.*, 21, 70, 1993.
3. Goteti, V. R. and Kalluri, D. K., Wave propagation in a switched magnetoplasma medium: Transverse propagation, *Radio Sci.*, 25, 61, 1990.
4. Dimitrijevic, M. M. and Stanic, B. V., EMW transformation in suddenly created two-component magnetized plasma, *IEEE Trans. Plasma Sci.*, 23, 422, 1995.
5. Kalluri, D. K., Frequency upshifting with power intensification of a whistler wave by a collapsing plasma medium, *J. Appl. Phys.*, 79, 3895, 1996.
6. Lai, C. H., Katsouleas, T. C., Mori, W. B., and Whittum, D., Frequency upshifting by an ionization front in a magnetized plasma, *IEEE Trans. Plasma Sci.*, 21, 45, 1993.
7. Huang, T. T., Lee, J. H., Kalluri, D. K., and Groves, K. M., Wave propagation in a transient plasma: Development of a Green's function, *IEEE Trans. Plasma Sci.*, 1, 19–25, 1998.
8. Kalluri, D. K. and Huang, T. T., Longitudinal propagation in a magnetized time-varying plasma: Development of Green's function, *IEEE Trans. Plasma Sci.*, 26, 1022–1030, 1998.
9. Huang, T. T., Wave propagation in a time-varying magnetoplasma: development of a Green's function, Doctoral thesis, University of Massachusetts Lowell, Lowell, 1997.
10. Huang, T. T. and Kalluri, D. K., Effect of finite rise time on the strength of the magnetic field generated in a switched plasma, *Int. J. Infrared Millim. Waves*, 19(7), 977–992, 1998.

9

Adiabatic Analysis of the MSW in a Transient Magnetoplasma

9.1 Introduction

In Chapter 7, we considered a step profile for the electron density. In Chapter 8, we improved the model by allowing a small rise time T_r. In this chapter, we look at the other limit of approximation of a large rise time. We assume that the parameters of the magnetoplasma medium change slowly enough that an *adiabatic analysis* can be made. The objective is to determine the modifications of the source wave caused by the time-varying parameters. We shall illustrate in the next section such an analysis for a source R wave. A similar analysis can be done for the L wave by associating the appropriate algebraic sign with ω_b.

9.2 Adiabatic Analysis for R Wave

The starting point for the adiabatic analysis of the R wave in a time-varying space-invariant magnetoplasma medium is Equation 7.8, which is repeated here for convenience:

$$\frac{d^3 H}{dt^3} - j\omega_b(t)\frac{d^2 H}{dt^2} + \left[k^2 c^2 + \omega_p^2(t)\right]\frac{dH}{dt} - j\omega_b(t)k^2 c^2 H = 0. \quad (9.1)$$

Note that this equation for H does not involve the time derivatives of $\omega_b(t)$ or $\omega_p^2(t)$. The technique [1–4] used in this book in obtaining an adiabatic solution of this equation is as follows. A complex instantaneous frequency function is defined such that

$$\frac{dH}{dt} = p(t)H(t) = \left[\alpha(t) + j\omega(t)\right]H(t). \quad (9.2)$$

Here ωt is the instantaneous frequency. Substituting Equation 9.2 into Equation 9.1 and neglecting α and all derivatives, a zero-order solution can be obtained. The solution gives a cubic in ω giving the instantaneous frequencies of three waves created by the switching action,

$$\omega^3 - \omega_b(t)\omega^2 - \left[k^2c^2 + \omega_p^2(t)\right]\omega + k^2c^2\omega_b(t) = 0. \tag{9.3}$$

The cubic has two positive real roots and one negative real root. At $t = 0$, one of the positive roots, say ω_m, has a value ω_0:

$$\omega_m(0) = \omega_0. \tag{9.4}$$

The mth wave is the MSW. An equation for α can now be obtained by substituting Equation 9.2 into Equation 9.1 and equating the real part to zero. In the adiabatic analysis, the derivatives and powers of α are neglected:

$$\alpha = \dot{\omega} \frac{3\omega - \omega_b(t)}{k^2c^2 + \omega_p^2(t) - 3\omega^2 + 2\omega\omega_b(t)}. \tag{9.5}$$

The approximate solution to Equation 9.1 can now be written as

$$H(t) = H_m(t) \exp\left[j \int_0^t \omega_m(\tau)\, d\tau\right], \tag{9.6}$$

where

$$H_m(t) = H_0 \exp\left[\int_0^t \alpha_m(\tau)\, d\tau\right]. \tag{9.7}$$

The amplitude of the other two modes (waves other than the mth mode) are of the order of the slopes at the origin of $\omega_b(t)$ and $\omega_p^2(t)$ and hence are neglected in the adiabatic analysis. The integral in Equation 9.7 can be evaluated numerically, but in the case of only one of the parameters varying with time, it can be evaluated analytically.

In the case of ω_b being constant and ω_p^2 varying with time, we can eliminate $\omega_p^2(t)$ from Equation 9.5, by using Equation 9.3 and simplifying

$$\omega_p^2(t) = \frac{(\omega^2 - k^2c^2)(\omega - \omega_b)}{\omega}, \tag{9.8}$$

$$\alpha = -\dot{\omega} \left[\frac{3\omega^2 - \omega\omega_b}{2\omega^3 - \omega^2\omega_b - k^2c^2\omega_b}\right], \quad \omega_b = \text{constant}. \tag{9.9}$$

Since the numerator in the bracketed fraction is one-half of the derivative of the denominator with respect to ω in the bracketed fraction, Equation 9.9 can be written as

$$\alpha\, dt = -\frac{1}{2}\frac{dr}{r}, \tag{9.10}$$

where

$$r = 2\omega^3 - \omega^2\omega_b - k^2c^2\omega_b, \quad \omega_b = \text{constant}. \tag{9.11}$$

Equation 9.7 reduces to

$$\frac{H_m(t)}{H_0} = \left[\frac{2\omega_0^3 - \omega_b\omega_0^2 - k^2c^2\omega_b}{2\omega_m^3(t) - \omega_b\omega_m^2(t) - k^2c^2\omega_b} \right]^{1/2}, \quad \omega_b = \text{constant}. \tag{9.12}$$

In the case of ω_p^2 being constant and ω_b varying with time, we can eliminate $\omega_b(t)$ from Equation 9.5 by using Equation 9.3 and simplifying

$$\omega_b(t) = \omega\left(1 - \frac{\omega_p^2}{\omega^2 - k^2c^2}\right), \tag{9.13}$$

$$\alpha = -\dot\omega \frac{\omega(2\omega^2 - 2k^2c^2 + \omega_p^2)}{(\omega^2 - k^2c^2)^2 + \omega_p^2\omega^2 + k^2c^2\omega_p^2}. \tag{9.14}$$

As in the previous case we can integrate analytically and write

$$\int_0^t \alpha(\tau)\, d\tau = \ln\left(\frac{s[\omega(0)]}{s[\omega(t)]}\right)^{1/2}, \tag{9.15}$$

where $s(\omega)$ is the denominator in the fraction on the right-hand side of Equation 9.14. From Equation 9.7,

$$\frac{H_m(t)}{H_0} = \left[\frac{(\omega_0^2 - k^2c^2)^2 + \omega_p^2\omega_0^2 + k^2c^2\omega_p^2}{(\omega_m^2(t) - k^2c^2)^2 + \omega_p^2\omega_m^2(t) + k^2c^2\omega_p^2} \right]^{1/2}. \tag{9.16}$$

The electric field $E_m(t)$ is easily obtained from $H_m(t)$ by using the wave impedance concept:

$$E_m(t) = \eta_{pm}H_m(t), \tag{9.17}$$

$$\eta_{pm} = \frac{\eta_0}{n_m} = \frac{\eta_0\omega_m}{kc}. \tag{9.18}$$

In the above equation, n_m is the refractive index of the medium when the frequency of the signal in the plasma is ω_m.

9.3 Modification of the Source Wave by a Slowly Created Plasma

Assume that an R wave of frequency ω_0 is propagating in free space. At $t = 0$, an unbounded homogeneous slowly varying transient plasma with an exponential profile is created:

$$\omega_p^2(t) = \omega_{p0}^2 \left[1 - \exp(-bt)\right]. \tag{9.19}$$

The fields are given by Equation 9.12 with $kc = \omega_0$. Figure 9.1 shows the variation of the normalized instantaneous frequency and the normalized electric and magnetic fields as a function of time. The parameters are $b = 0.1$ and $\omega_0 = 1.0$, and ω_{p0} is chosen as 0.4. A detailed derivation of the equations based on a slightly different formulation is given in Ref. [1].

9.4 Modification of the Whistler Wave by a Collapsing Plasma Medium

Section 6.3 mentioned the whistler mode of propagation of the R wave in the frequency band $0 < \omega < \omega_b$. In this band, the refractive index is greater than one. This section looks at the adiabatic transformation of the whistler wave by a collapsing plasma medium. If n_0 is the refractive index and ω_0 is the frequency of the whistler wave before the collapse begins, the wave magnetic field $H_m(t)$ during the collapse is given by Equation 9.12, where $k^2 c^2 = \omega_0^2 n_0^2$ and the electric field is obtained from Equations 9.17 and 9.18. Denoting the first wave as the MSW, that is, $m = 1$ and $\omega_m(0) = \omega_1(0) = \omega_0$,

$$\frac{H_1(t)}{H_0} = \left[\frac{2\omega_0^3 - \omega_b\omega_0^2 - n_0^2\omega_0^2\omega_b}{2\omega_1^3(t) - \omega_b\omega_1^2(t) - n_0^2\omega_0^2\omega_b}\right]^{1/2}, \tag{9.20}$$

$$\frac{E_1(t)}{E_0} = \frac{\omega_1(t)}{\omega_0}\frac{H_1(t)}{H_0}, \qquad \omega_b = \text{constant}. \tag{9.21}$$

It has been mentioned before, see Section 6.3, that the refractive index could be quite large when $\omega_0 \ll \omega_b$. In such a case,

$$n_0 \approx n_w = \frac{\omega_p}{\sqrt{\omega_0\omega_b}} \gg 1, \tag{9.22}$$

$$\omega_1(t \to \infty) = n_w\omega_0, \tag{9.23}$$

$$\frac{E_1(t \to \infty)}{E_0} \approx \frac{n_w}{\sqrt{2}}, \tag{9.24}$$

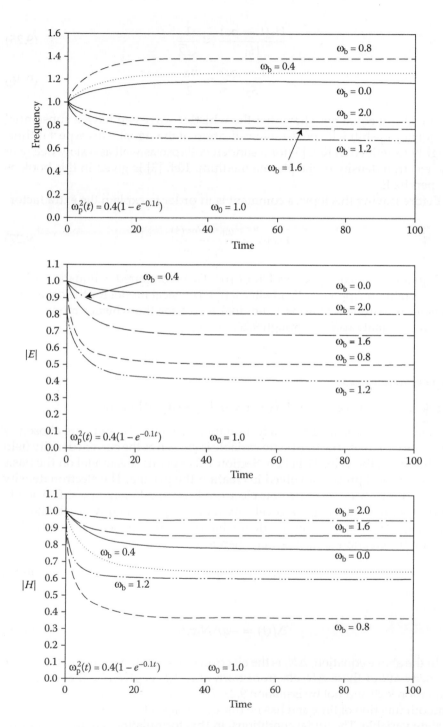

FIGURE 9.1 Frequency and fields of the modified source R wave.

$$\frac{H_1(t \to \infty)}{H_0} \approx \frac{1}{\sqrt{2}}, \tag{9.25}$$

$$\frac{S_1(t \to \infty)}{S_0} \approx \frac{n_w}{2}. \tag{9.26}$$

The above equations show that the whistler wave can have a substantial frequency upshift with power intensification of the signal. This aspect is thoroughly examined in Ref. [3] for a sudden collapse as well as a slow decay of the electron density of the plasma medium. Ref. [3] is given in this book as Appendix E.

Before leaving this topic, a comment is in order to explain the extra factor

$$\text{Factor} = \frac{\omega_b - \omega_1(t)}{\omega_b - \omega_0} \tag{9.27}$$

in Equation E.14a of Appendix E for $E_1(t)/E_0$ as compared to Equation 9.21 of this section. In Appendix E, an alternative physical model for the collapsing plasma is assumed. However, the final result for the whistler mode is still given by Equations 9.22 through 9.26.

9.5 Alternative Model for a Collapsing Plasma

Equation 9.1 is obtained from a formulation where E, H, and J are chosen as the state variables. The third state variable is chosen as the current density field J rather than the velocity field v. Section 1.2 explains this model on the basis of the physical process involved in creating the plasma. The electron density $N(t)$ increases because of the new electrons born at different times. The newly born electrons start with zero velocity and are subsequently accelerated by the fields. Thus all the electrons do not have the same velocity at a given time during the creation of the plasma. Therefore

$$J(t) \neq -qN(t)v(t), \tag{9.28}$$

but

$$\Delta J(t) = -q\Delta N_i v_i(t). \tag{9.29}$$

In the above equation, ΔN_i is the electron density added at t_i and $v_i(t)$ is the velocity at t of these ΔN_i electrons born at t_i. $J(t)$ is given by the integral of Equation 9.29 and not by Equation 9.28. $J(t)$ obtained from Equation 9.29 is a smooth function of time and has no discontinuities. It fits the requirements of a state variable. The initial conditions, in this formulation, are the continuity of E, H, and J at $t = 0$. This is a good model for a building up plasma. We can ask the following questions regarding the modeling of a transient plasma. Is this the only model or are there other models? Is this a good model for the collapsing plasma? A lot depends on the physical processes responsible for the temporal change in the electron density [5].

Consider the following model for the collapse of the plasma. The decrease in electron density takes place by a process of sudden removal of $\Delta N(t)$ electrons; the velocities of all of the remaining electrons are unaffected by this capture and have the same instantaneous value $v(t)$. The initial conditions in this model are the continuity of E, H, and v; these field variables can be used as the state variables. The state variable equations are

$$\mu_0 \frac{dH}{dt} = jkE, \tag{9.30}$$

$$\varepsilon_0 \frac{dE}{dt} = jkH + \varepsilon_0 \omega_p^2(t)u, \tag{9.31}$$

$$\frac{du}{dt} = -E + j\omega_b(t)u, \tag{9.32}$$

where

$$u = e\frac{v}{m}, \tag{9.33}$$

and the simplest higher-order differential equation, when ω_b is a constant, is in the variable u and is given by

$$\frac{d^3u}{dt^3} - j\omega_b \frac{d^2u}{dt^2} + \left[k^2c^2 + \omega_p^2(t)\right]\frac{du}{dt} + \left[g(t) - j\omega_b k^2 c^2\right]u = 0, \quad \omega_b = \text{constant}, \tag{9.34}$$

where

$$g(t) = \frac{d\omega_p^2(t)}{dt}. \tag{9.35}$$

Equation 9.34 is slightly different from Equation 9.1. The last term on the left-hand side of Equation 9.34 has an extra parameter $g(t)$. Consequently, while the equation for ω remains the same as Equation 9.3, the equation for α is a modified version of Equation 9.5 and is given by Equation 9.36:

$$\alpha = \frac{-g(t) + \dot{\omega}(3\omega - \omega_b)}{k^2c^2 + \omega_p^2(t) - 3\omega^2 + 2\omega\omega_b}, \quad \omega_b = \text{constant}. \tag{9.36}$$

Substituting Equation 9.8 for $\omega_p^2(t)$ into Equation 9.36,

$$\alpha = \frac{g(t)\omega - \dot{\omega}(3\omega^2 - \omega_b\omega)}{2\omega^3 - \omega^2\omega_b - \omega_b k^2 c^2}, \quad \omega_b = \text{constant}. \tag{9.37}$$

An expression for $g(t)$ can be obtained by differentiating Equation 9.8:

$$g(t) = \frac{d\omega_p^2(t)}{dt} = \frac{\dot{\omega}}{\omega}\left[2\omega^3 - \omega^2\omega_b - k^2c^2\omega_b\right], \quad \omega_b = \text{constant.} \quad (9.38)$$

Substituting Equation 9.38 into Equation 9.37,

$$\alpha = \frac{\dot{\omega}}{\omega} - \dot{\omega}\frac{\omega^2 - \omega\omega_b}{2\omega^3 - \omega^2\omega_b - k^2c^2\omega_b}, \quad \omega_b = \text{constant.} \quad (9.39)$$

The velocity field of the MSW is obtained, by analytical integration, as before:

$$\frac{u_1(t)}{u_0} = \exp\left[\int_0^t \alpha(\tau)d\tau\right] = \frac{\omega_1(t)}{\omega_0}\left[\frac{2\omega_0^3 - \omega_0^2\omega_b - n_0^2\omega_0^2\omega_b}{2\omega_1^3 - \omega_1^2\omega_b - n_0^2\omega_0^2\omega_b}\right]^{1/2}. \quad (9.40)$$

The electric field Equation 9.43 is now obtained by noting from Equation 9.32 that

$$E_0 = j(\omega_b - \omega_0)u_0, \quad (9.41)$$

$$E_1 = j(\omega_b - \omega_1)u_1, \quad (9.42)$$

$$\frac{E_1(t)}{E_0} = \left[\frac{\omega_b - \omega_1(t)}{\omega_b - \omega_0}\right]\left[\frac{\omega_1(t)}{\omega_0}\right]\left[\frac{2\omega_0^3 - \omega_0^2\omega_b - n_0^2\omega_0^2\omega_b}{2\omega_1^3(t) - \omega_1^2(t)\omega_b - n_0^2\omega_0^2\omega_b}\right]^{1/2},$$

$$\omega_b = \text{constant.} \quad (9.43)$$

When $\omega_0 \ll \omega_b \sim \omega_p$, $n_0 \approx n_w$ is large and the first bracketed term on the right-hand side of Equation 9.43 can be approximated by

$$\left[\frac{\omega_b - \omega_1(t)}{\omega_b - \omega_0}\right] \approx \left[1 + \frac{\omega_0}{\omega_b} - \frac{\omega_{p0}}{\omega_b}\sqrt{\frac{\omega_0}{\omega_b}}\right] \approx 1. \quad (9.44)$$

9.6 Modification of the Whistler Wave by a Collapsing Magnetic Field

Equation 9.16 can be directly applied to this case. The third mode is denoted as the MSW, that is, $\omega_3(0) = \omega_0$ and $kc = n_0\omega_0$

$$\frac{H_3(t)}{H_0} = \sqrt{\frac{(\omega_0^2 - n_0^2\omega_0^2)^2 + \omega_p^2\omega_0^2 + n_0^2\omega_0^2\omega_p^2}{(\omega_3^2(t) - n_0^2\omega_0^2)^2 + \omega_p^2\omega_3^2(t) + n_0^2\omega_0^2\omega_p^2}}. \quad (9.45)$$

In this case, it is immaterial whether we choose J or v as the state variable since the continuity of v ensures the continuity of J, the electron density N being a constant.

For the case when $\omega_0 \ll \omega_b \sim \omega_p$, $n_0 \approx n_w$ is large and it can be shown that

$$H_3(\infty) \approx H_0, \tag{9.46}$$

$$E_3(\infty) \approx 0, \tag{9.47}$$

$$\omega_3(\infty) \approx 0. \tag{9.48}$$

The whistler wave is converted into a wiggler magnetic field. This aspect is discussed at length in Ref. [4] and is given in this book as Appendix F.

9.7   Adiabatic Analysis for the X Wave

In this chapter, we have discussed the adiabatic analysis of the R wave in a transient magnetoplasma. A similar analysis can be performed for the X wave [2] and is given in this book as Appendix H.

References

1. Kalluri, D. K., Goteti, V. R., and Sessler, A. M., WKB solution for wave propagation in a time-varying magnetoplasma medium: Longitudinal propagation, *IEEE Trans. Plasma Sci.*, 21, 70, 1993.
2. Lee, J. H. and Kalluri, D. K., Modification of an electromagnetic wave by a time-varying switched mangetoplasma medium: Transverse propagation, *IEEE Trans. Plasma Sci.*, 26, 1–6, 1998.
3. Kalluri, D. K., Frequency upshifting with power intensification of a whistler wave by a collapsing plasma medium, *J. Appl. Phys.*, 79, 3895, 1996.
4. Kalluri, D. K., Conversion of a whistler wave into a controllable helical wiggler magnetic field, *J. Appl. Phys.*, 79, 6770, 1996.
5. Stepanov, N. S., Dielectric constant of unsteady plasma, *Soviet Radiophys. Quant. Electr.*, 19, 683, 1976.

10

Miscellaneous Topics

10.1 Introduction

The first nine chapters dealt with topics in an orderly fashion progressing step by step from one kind of complexity of the medium to the other. In the process, the reader, it is hoped, has developed a feel for the theory of *Frequency Shifting by a Transient Magnetoplasma Medium*.

In this chapter, we will consider miscellaneous topics connected with the general theme of the book. They include references to the contributions of some of the research groups with which the author is familiar.

10.2 Proof of Principle Experiments

We considered earlier two distinct cases:

a. Case 1. *Time-invariant space-varying medium*: We solved the problem by imposing the requirement that ω is conserved (Chapter 2). In the ω–k diagram, ω = constant is a horizontal line.

b. Case 2. *Space-invariant time-varying medium*: The conserved quantity now is k and k = constant is a vertical line in the ω–k diagram.

Experimental realization of case 2 is not easy. We have to ionize uniformly a large region of space at a given time. Joshi et al. [1], Kuo [2], Kuo and Ren [3], and Rader and Alkexeff [4] developed ingenious experimental techniques to achieve these conditions and demonstrated the principle of frequency shifting using an isotropic plasma. One of the earliest experimental evidence for frequency shifting quoted in the literature is the seminal paper by Yablanovitch [5]. See Part IV for a more detailed description of the experiments.

10.3 Moving Ionization Front (Figures 10.1 through 10.4)

The Doppler frequency shift associated with reflection of an electromagnetic signal by a moving object (medium) is familiar to all. In this case, the medium

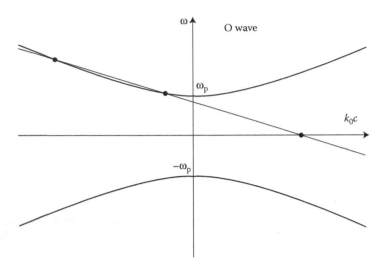

FIGURE 10.1 ω–k diagram of the O wave for an ionization front problem.

along with the boundary is moving. An alternative is to create a moving boundary between two media and the two media themselves do not move. Imagine an electromagnetic signal propagating through an unionized gas. At $t = 0$, a powerful laser pulse is injected into the gas. As the pulse travels with the speed of light through the gas, say from right to left, it ionizes the gas and creates a plasma. At any given time there exists a boundary between unionized gas, which is like free space, and the plasma. The boundary itself is moving with the speed of light. Such a medium is called an ionization front.

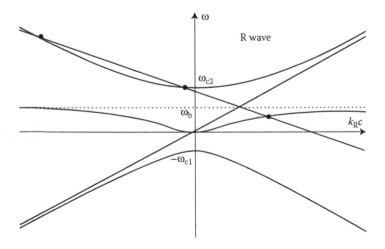

FIGURE 10.2 ω–k diagram of the R wave for an ionization front problem.

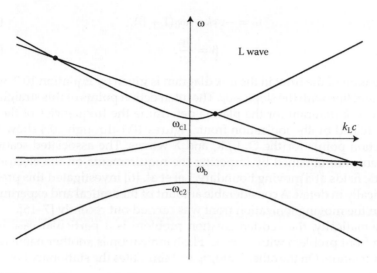

FIGURE 10.3 ω–k diagram of the L wave for an ionization front problem.

Figure 10.1 shows the geometry of the ionization front where we used the velocity of the front as v_F. In space–time the front is described by the equation

$$z = -v_F t. \tag{10.1}$$

Let ω_0 and k_0 be the frequency and the wave number in free space to the left of the front and ω and k the corresponding entities in the plasma to the right of the front. From the phase invariance principle,

$$(\omega_0 t - k_0 z)|_{z=-v_F t} = (\omega t - kz)|_{z=-v_F t}, \tag{10.2}$$

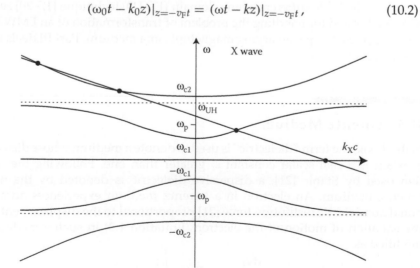

FIGURE 10.4 ω–k diagram of the X wave for an ionization front problem.

$$\omega = -v_F k + \omega_0(1 + \beta), \tag{10.3}$$

$$\beta = \frac{v_F}{c}. \tag{10.4}$$

The equation of the front in the ω–k diagram is given by Equation 10.3, which is a straight line with the slope $-v_F$. The intersection points of this straight line with the ω–k diagram for the plasma determine the frequencies of the new waves created by the ionization front. Figures 10.1 through 10.4 show these intersection points for the O, L, R, and X waves. The associated scattering coefficients are obtained by imposing the requirements on the electric and magnetic fields at a moving boundary. Lai et al. [6] investigated this problem theoretically in detail. A considerable amount of theoretical and experimental work on the moving ionization front was carried out recently [7–16].

Mathematically, the sudden creation problem is a particular case of the moving front problem with $v_F = \infty$. Flash ionization is another name for the sudden creation. On the other hand, $v_F = 0$ simulates the stationary boundary problem.

10.4   The Finite-Difference Time-Domain Method

In the previous chapters, we chose suitable models and made approximations that permitted us to obtain the exact or an approximate analytical solution. These solutions are useful in giving us a qualitative picture of the effects of each type of complexity in the properties of the medium. Numerical methods are useful for studying the effects of simultaneous presence of several complexities. The finite-difference time-domain (FDTD) technique [17–20] seems to be well suited for handling the problem of transformation of an EMW by a time-varying and space-varying magnetoplasma medium. Part III deals with this aspect in detail.

10.5   Lorentz Medium

In this book, the term "dielectric" is used to denote a medium whose dielectric constant is a real scalar constant ε_r greater than one. Following the notation used by Stanic [21], a dispersive dielectric is denoted by the name Lorentz medium. An electron in a Lorentz medium experiences an additional quasielastic restitution force [21] proportional to the displacement and the equation of motion of the electron (Equation 1.3) in such a medium is modified as

$$m\frac{d\mathbf{v_L}}{dt} = -e\mathbf{E} - \alpha \int_0^t \mathbf{v_L}\, d\tau. \tag{10.5}$$

The dielectric constant of such a medium denoted by ε_L is given by

$$\varepsilon_L = 1 - \frac{\omega_p^2}{\omega^2 + \omega_L^2}, \tag{10.6}$$

where

$$\omega_L = \left(\frac{\alpha}{m}\right)^{1/2}. \tag{10.7}$$

For $\alpha = 0$, $\omega_L = 0$, the electron is free and the Lorentz medium becomes a cold plasma.

Stanic [21] and Stanic and Draganic [22] extended our analysis of Chapters 3 and 7 to a Lorentz medium and gave results for the frequencies, fields, and power densities of various modes. The techniques developed in this book for the time-varying media can be applied to more complex media with multiple Lorentz resonances [17,18].

10.6 Mode Conversion of X Wave

It is known from the magnetoionic theory that the X wave has a resonance at the upper hybrid frequency:

$$\omega_{UH} = \sqrt{\omega_p^2 + \omega_b^2}. \tag{10.8}$$

However, using warm-electron gas equations, it can be shown that in place of the resonance at ω_{UH}, there is a transition from a basically transverse wave to a basically longitudinal wave at this frequency. The ω–k diagram for such a medium can be obtained from the known expression [23] for the refractive index of the X wave:

$$n_X^2 = \frac{(\omega^2 - \omega_p^2)(\omega^2 - \omega_{UH}^2 - \delta^2 k^2 c^2) - \omega_p^2 \omega_b^2}{(\omega^2 - \omega_{UH}^2 - \delta^2 k^2 c^2)}, \tag{10.9}$$

and the relation

$$k^2 c^2 = \omega^2 n_X^2. \tag{10.10}$$

In the above equation,

$$\delta^2 = \frac{c_e^2}{c^2} = \frac{\gamma k_B T_e}{m} \frac{1}{c^2}, \tag{10.11}$$

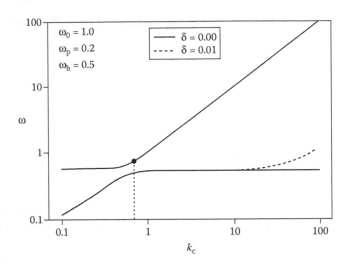

FIGURE 10.5   Mode conversion problem. $\omega–k$ diagram of the X wave for $\omega_p = 0.2$.

where γ, k_B, T_e, and m have the usual meaning [23] and c_e is the electron thermal speed. From Equation 10.9 it is obvious that the upper hybrid resonance is eliminated by the thermal term.

Figure 10.5 shows the $\omega–k$ diagram for $\delta = 0.01$, $\omega_0 = 1$, $\omega_\pi = 0.2$, and $\omega_\beta = 0.5$. In Figures 10.5 through 10.8 the frequencies are normalized with respect to the source frequency ω_0. The solid line is the diagram for the cold plasma approximation. Figures 10.6 through 10.8 show the $\omega–k$ diagram for other values of the parameters ω_p and ω_b. These figures offer qualitative

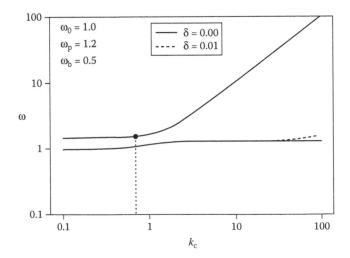

FIGURE 10.6   Mode conversion problem. $\omega–k$ diagram of the X wave for $\omega_p = 1.2$.

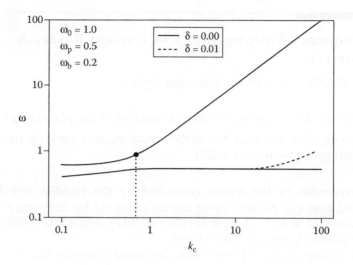

FIGURE 10.7 Mode conversion problem. ω–k diagram of the X wave for $\omega_b = 0.2$.

proof on the possibilities of mode conversion in a time-varying magneto-plasma medium. See Appendix H for a discussion on the possibilities of mode conversion, where the modification of an X wave by the slowly varying magnetoplasma parameter is considered.

No work has been done till now on the mode conversion in a time-varying magnetoplasma medium. However, some literature exists on the mode conversion in a space-varying magnetoplasma medium [23]. Some literature also exists on such a phenomenon in a magnetoelastic medium [24–25].

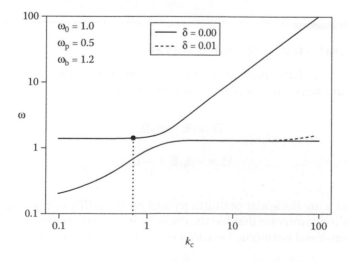

FIGURE 10.8 Mode conversion problem. ω–k diagram of the X wave for $\omega_b = 1.2$.

10.7 Frequency-Shifting Topics of Current Research Interest

Kuo et al. [26,27] dealt with the following topics:

a. Frequency downshifting due to collisionality of the plasma [26]
b. Frequency downshifting due to the switching of a periodic medium consisting of plasma layers [27]

The frequencies of the waves generated by the sudden switching of unbounded periodic plasma layers can be obtained by drawing a vertical line $k = $ constant in the ω–β diagram (Figure 2.12). Appendix J gives results when the plasma layers are magnetized.

Mendonca and Silba [28] dealt with the mode coupling theory due to a time-varying and space-varying plasma in a cavity.

Kapoor and Dillon [29,30] extended the work of Chapter 5 on switching of the plasma slab to determining the fields in the plasma region in the isotropic case as well as the case of transverse propagation in a magnetized plasma.

Nerukh [31] considered an evolutionary approach in transient electrodynamic problems.

Muggli et al. [32] recently conducted a proof of principle experiment on radiation generation by an ionization front in a gas-filled capacitor array. Yoshii et al. [33] discussed the generation of tunable microwaves from Cerenkov wakes in a magnetized plasma.

10.8 Chiral Media: R and L Waves

Chiral materials have molecules of helical structure. They are bi-isotropic materials and have the complex constitutive relations,

$$\mathbf{D} = \varepsilon\mathbf{E} - j\xi_c\mathbf{B}, \tag{10.12}$$

$$\mathbf{H} = -j\xi_c\mathbf{E} + \frac{1}{\mu}\mathbf{B}, \tag{10.13}$$

where ε and μ are the scalar permittivity and permeability of the media and ξ_c is the chirality parameter that has the dimensions of admittance. By assuming a one-dimensional harmonic variation of field variables $f(z, t)$,

$$f(z, t) = \exp[j(\omega t - k_c z)], \tag{10.14}$$

we can show that the R and L waves are the natural modes with the wave number [34]

$$k_c = k_{R,L} = \pm\omega\mu\xi_c + [k^2 + (\omega\mu\xi_c)^2]^{1/2}, \tag{10.15}$$

$$k = \omega\sqrt{\mu\varepsilon}, \tag{10.16}$$

where the upper sign is for the R wave and the lower sign is for the L wave. The wave impedance is given by

$$\eta_c = \frac{E_x}{H_y} = \left(\frac{\mu}{\varepsilon + \mu\xi_c^2}\right)^{1/2}. \tag{10.17}$$

Since the R and L waves have different wave numbers, the propagation of a linearly polarized wave is accompanied by the phenomenon of rotation of the plane of polarization. Liquids such as sugar solutions and crystals such as quartz and sodium chlorate [34] exhibit *natural rotation* at optical frequencies due to *optical activity*. Artificial chiral materials that exhibit such rotations at microwave frequencies are commercially available [35]. The natural rotation is similar in principle to the Faraday rotation (magnetic rotation) discussed in Section 6.3. However, the phenomena have the following important difference. In the case of natural rotation, the sign of the angle of rotation is dependent on whether the wave is a forward-propagating wave or a backward-propagating wave. In the magnetic rotation, it depends on the direction of the static magnetic field and does not depend on whether the wave is a forward-propagating wave or a backward-propagating wave. Therefore the total angle of rotation for a round trip travel of the wave is zero for the natural rotation. On the other hand, the total angle of rotation for a round trip travel of the wave for the magnetic rotation is double that of a single trip travel. An optical isolator is constructed using these rotators [36].

The theory developed in Section 2.7 can be used to study the wave propagation in an unbounded layered chiral medium. Equation 2.127 can be used with the replacement of n by the equivalent refractive index n_c of the chiral medium. Kalluri and Rao [37] used such a technique to study the filter characteristics of periodic chiral layers. Recent work in designing artificial chiral materials using conical inclusions is discussed in Ref. [38].

10.9 Solitons

This topic deals with a desirable aspect of the EMW transformation by the balancing act of two complexities, each of which by itself produces an undesirable effect. The balancing of the wave breaking feature of a nonlinear medium complexity with the flattening feature of the dispersion leads to a soliton solution [39].

10.10 Astrophysical Applications

Yablonovitch [40] considered an EMW in a rapidly growing plasma and connected it with the Unruh–Davies–Fulling–DeWitt radiation and the nonadiabatic Casimir effect. He showed that the sudden ionization of a gas or a semiconductor crystal to generate plasma on a subpicosecond timescale can produce a reference frame accelerating at about 10^{20} g relative to an inertial frame.

10.11 Virtual Photoconductivity

Yablonovitch et al. [41] showed that a semiconductor photoconductor "will exhibit a strong reactive response to optical radiation tuned to the transparent region, just below the band gap. This response is due to the excitation of a virtual electron–hole gas, readily polarized by the dc electric field, which can contribute to a major change (>1) in the low-frequency dielectric constant." They suggest that this reactive response can aid the laboratory detection of the Unruh–Davies–Fulling–DeWitt radiation.

References

1. Joshi, C. S., Clayton, C. E., Marsh, K., Hopkins D. B., Sessler, A., and Whittum. D., Demonstration of the frequency upshifting of microwave radiation by rapid plasma creation, *IEEE Trans. Plasma Sci.*, 18, 814, 1990.
2. Kuo, S. P., Frequency up-conversion of microwave pulse in a rapidly growing plasma, *Phys. Rev. Lett.*, 65, 1000, 1990.
3. Kuo, S. P. and Ren, A., Experimental study of wave propagation through a rapidly created plasma, *IEEE Trans. Plasma Sci.*, 21, 53, 1993.
4. Rader, M. and Alexeff, I., Microwave frequency shifting using photon acceleration, *Int. J. Infrared Millim. Waves*, 12(7), 683, 1991.
5. Yablonavitch, E., Spectral broadening in the light transmitted through a rapidly growing plasma, *Phys. Rev. Lett.*, 31, 877, 1973.
6. Lai, C. H., Katsouleas, T. C., Mori, W. B., and Whittum D., Frequency upshifting by an ionization front in a magnetized plasma, *IEEE Trans. Plasma Sci.*, 21, 45, 1993.
7. Lampe, M., Ott, E., Manheimer, W. M., and Kainer, S., Submillimeter-wave production by upshifted reflection from a moving ionization front, *IEEE Trans. Microw. Theory Tech.*, 25, 556, 1991.
8. Esarey, E., Joyce, G., and Sprangle, P., Frequency up-shifting of laser pulses by copropagating ionization fronts, *Phys. Rev. A*, 44, 3908, 1991.
9. Esarey, E., Ting, A., and Sprangle, P., Frequency shifts induced in laser pulses by plasma waves, *Phys. Rev. A*, 42, 3526, 1990.
10. Lampe, M., Ott, E., and Walker, J. H., Interaction of electromagnetic waves with a moving ionization front, *Phys. Fluids*, 21, 42, 1978.

11. Mori, W. B., Generation of tunable radiation using an underdense ionization front, *Phys. Rev. A*, 44, 5121, 1991.
12. Savage, Jr., R. L., Joshi, C., and Mori, W. B., Frequency up-conversion of electromagnetic radiation upon transmission into an ionization front, *Phys. Rev. Lett.*, 68, 946, 1992.
13. Wilks, S. C., Dawson, J. M., and Mori W. B., Frequency up-conversion of electromagnetic radiation with use of an overdense plasma, *Phys. Rev. Lett.*, 63, 337, 1988.
14. Gildenburg, V. B., Pozdnyakova, V. I., and Shereshevski, I. A., Frequency self-upshifting of focused electromagnetic pulse producing gas ionization, *Phys. Lett. A*, 43, 214, 1995.
15. Xiong, C., Yang Z., and Liu, S., Frequency shifts by beam-driven plasma wake waves, *Phys. Lett. A*, 43, 203, 1995.
16. Savage, Jr., R. L., Brogle, R. P., Mori, W. B., and Joshi, C., Frequency upshifting and pulse compression via underdense relativistic ionization fronts, *IEEE Trans. Plasma Sci.*, 21, 5, 1993.
17. Taflove, A., *Computational Electrodynamics*, Artech House, Boston, MA, 1995.
18. Kunz, K. S. and Luebbers, R. J., *Finite Difference Time Domain Method for Electromagnetics*, CRC Press, Boca Raton, FL, 1993.
19. Jelanek, A. and Stanic, B. V., FDTD simulation of frequency upshifting from suddenly created magnetoplasma slab, *Conf. Rec. Abstracts IEEE Int. Conf. Plasma Sci.*, 141, 1995.
20. Kalluri, D. K. and Lee, J. H., Numerical simulation of electromagnetic wave transformation in a dynamic magnetized plasma, *Final Report*, Summer Research Extension Program AFOSR, December 1997.
21. Stanic, B. V., Electromagnetic waves in a suddenly created Lorentz medium and plasma, *J. Appl. Phys.*, 70, 1987, 1991.
22. Stanic, B. V. and Draganic, I. N., Electromagnetic waves in a suddenly created magnetized Lorentz medium (transverse propagation), *IEEE Trans. Ant. Prop.*, AP-44, 1394, 1996.
23. Swanson, D. G., *Plasma Waves*, Academic Press, New York, 1989.
24. Auld, B. A., Collins, J. H., and Japp, H. R., Signal processing in a nonperiodically time-varying magnetoclassic medium, *Proc. IEEE*, 56, 258, 1968.
25. Rezende. S. M., Magnetoelastic and magnetostatic waves in a time-varying magnetic fields, PhD dissertation, MIT, Cambridge, MA, 1967.
26. Kuo, S. P., Ren, A., and Schmidt, G., Frequency downshift in a rapidly ionizing media, *Phys. Rev. E*, 49, 3310, 1994.
27. Kuo, S. P. and Faith, J., Interaction of an electromagnetic wave with a rapidly created spatially periodic plasma, *Phys. Rev. E*, 56, 1, 1997.
28. Mendonca, J. T. and Silva, O. E., Mode coupling theory of flash ionization in a cavity, *IEEE Trans. Plasma Sci.*, 24, 147, 1996.
29. Kapoor, S., Wave propagation in a bounded suddenly created isotropic plasma medium: Transverse propagation, MS thesis, Tuskegee University, 1992.
30. Dillon, W. J., The effects of a sudden creation of a magnetoplasma slab on wave propagation: transverse propagation, MS Thesis, Tuskegee University, 1996.
31. Nerukh, A. G, Evolutionary approach in transient electrodynamic problems, *Radio Sci.*, 30, 481, 1995.

32. Muggli, P., Liou, R., Lai, C. H., and Katsouleas, T. C., Radiation generation by an ionization front in a gas-filled capacitor array, In: *Conf. Rec. Abstracts, IEEE Int. Conf. Plasma Sci.*, San Diego, CA, p. 157, May 19–22, 1997.
33. Yoshii, J., Lai, C. H., and Katsouleas, T. C., Tunable microwaves from Cerenkov wakes in magnetized plasma, In: *Conf. Rec. Abstracts, IEEE Int. Conf. Plasma Sci.*, San Diego, CA, p. 157, 1997.
34. Ishimaru, A., *Electromagnetic Wave Propagation, Radiation, and Scattering*, Prentice-Hall, Englewood Cliffs, NJ, 1991.
35. Lindell, I. V., Sihvola, A. H., Tretyakov, S. A., and Viitanen, A. J., *Electromagnetic Waves in Chiral and Bi-Isotropic Media*, Artech House, Boston, MA, 1994.
36. Saleh, B. E. A. and Teich, M. C., *Fundamentals of Photonics*, Wiley, New York, 1991.
37. Kalluri, D. K. and Rao, T. C. K., Filter characteristics of periodic chiral layers, *Pure Appl. Opt.*, 3, 231, 1994.
38. DeMartinis, G. *Chiral* Media using conical coil wire inclusions, Doctoral thesis, University of Massachusetts Lowell, Lowell, 2008.
39. Hirose, A. and Longren, K. E., *Introduction to Wave Phenomena*, Wiley, New York, 1985.
40. E. Yablonovitch, Accelerating reference wave for electromagnetic waves in a rapidly growing plasma: Unruh–Davis–Fulling–DeWitt radiation and the nonadiabatic Casimir effect, *Phys. Rev. Lett.*, 62, 1742–1745, 1989.
41. E. Yablonovitch, J. P. Heritage, D. E. Aspnes, and Y. Yafet, Virtual photoconductivity, *Phys. Rev. Lett.*, 63, 976–979, 1989.

Additional References

1. Nerukh, A. G., Scherbatko, I. V., and Rybin, O. N., The direct numerical calculation of an integral Volterra equation for an electromagnetic signal in a time-varying dissipative medium, *J. Electromagn. Waves Appl.*, 12, 163, 1998.
2. Nerukh, A. G., Evolutionary approach in transient electromagnetic problems, *Radio Sci.*, 30, 48, 1995.
3. Nerukh, A. G. and Khizhnyak, N. A., Enhanced reflection of an electromagnetic wave from a plasma cluster moving in a waveguide, *Microwave Opt. Technol. Lett.*, 17, 267, 1998.
4. Nerukh, A. G., Scherbatko, I. V., and Nerukh, D. A., Using evolutionary recursion to solve an electromagnetic problem with time-varying parameters, *Microwave Opt. Technol. Lett.*, 14, 31, 1997.
5. Nerukh, A. G., Splitting of an electromagnetic pulse during a jump in the conductivity of a bounded medium (English transl.), *Sov. Phys.—Tech. Phys.*, 37, 543, 1992.
6. Nerukh, A. G., On the transformation of electromagnetic waves by a nonuniformly moving boundary between two media (English transl.), *Sov. Phys.—Tech. Phys.*, 34, 281, 1989.
7. Nerukh, A. G., Electromagnetic waves in dielectric layers with time-dependent parameters (English transl.), *Sov. Phys.—Tech. Phys.*, 32, 1258, 1987.

Appendix A

Constitutive Relation for a Time-Varying Plasma Medium

The electron density profile function in Figure A.1a $N(\mathbf{r}, t)$ can be written as the sum of steady density $N_0(\mathbf{r})$ and the time-varying density $N_1(\mathbf{r}, t)$.

$$N(\mathbf{r}, t) = N_0(\mathbf{r}) + N_1(\mathbf{r}, t). \tag{A.1}$$

Let $\mathbf{J}_1(\mathbf{r}, t)$ be the current density due to $N_1(\mathbf{r}, t)$. Note that $N_1(\mathbf{r}, 0) = 0$ as shown in Figure A.1b. From Equations 1.4 through 1.6,

$$\Delta \mathbf{J}_1(\mathbf{r}, t) = \frac{q^2}{m} \left\{ \frac{\partial N_1}{\partial t} \bigg|_{t=t_i} \right\} \Delta t_i \int_{t_i}^{t} \mathbf{E}(\mathbf{r}, \alpha) \, d\alpha, \tag{A.2}$$

$$\mathbf{J}_1(\mathbf{r}, t) = \frac{q^2}{m} \int_{0}^{t} \frac{\partial N_1(\mathbf{r}, \tau)}{\partial \tau} \, d\tau \int_{\tau}^{t} \mathbf{E}(\mathbf{r}, \alpha) \, d\alpha. \tag{A.3}$$

Let

$$\mathbf{f}(\mathbf{r}, t) = \int_{0}^{t} \frac{\partial N_1(\mathbf{r}, \tau)}{\partial \tau} \, d\tau \int_{\tau}^{t} \mathbf{E}(\mathbf{r}, \alpha) \, d\alpha, \tag{A.4}$$

and

$$\mathbf{p} = \int \mathbf{E}(\mathbf{r}, \alpha) \, d\alpha, \tag{A.5}$$

$$\int_{\tau}^{t} \mathbf{E}(\mathbf{r}, \alpha) \, d\alpha = \mathbf{p}(\mathbf{r}, t) - \mathbf{p}(\mathbf{r}, \tau), \tag{A.6}$$

$$\mathbf{f}(\mathbf{r}, t) = \mathbf{p}(\mathbf{r}, t) \int_{0}^{t} \frac{\partial N_1(\mathbf{r}, \tau)}{\partial \tau} \, d\tau - \int_{0}^{t} \mathbf{p}(\mathbf{r}, \tau) \frac{\partial N_1(\mathbf{r}, \tau)}{\partial \tau} \, d\tau. \tag{A.7}$$

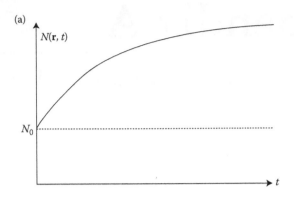

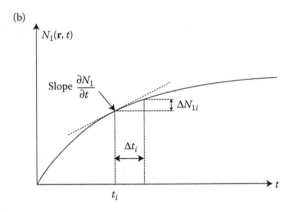

FIGURE A.1 (a) Time-varying electron density profile. N_0 is the initial value and (b) N_1 is the time-varying component.

Integrating the second integral on the right-hand side of Equation A.7 by parts gives

$$\mathbf{f}(\mathbf{r},t) = \mathbf{p}(\mathbf{r},t)\left[N_1(\mathbf{r},t) - N_1(\mathbf{r},0)\right] - \mathbf{p}(\mathbf{r},\tau)N_1(\mathbf{r},\tau)\big|_0^t + \int_0^t N_1(\mathbf{r},\tau)\frac{\partial \mathbf{p}(\mathbf{r},\tau)}{\partial \tau}\,d\tau.$$

$$(A.8)$$

Since $N_1(\mathbf{r},0) = 0$,

$$\mathbf{f}(\mathbf{r},t) = \int_0^t N_1(\mathbf{r},\tau)\mathbf{E}(\mathbf{r},\tau)\,d\tau. \qquad (A.9)$$

From Equations A.2, A.3, and A.9,

$$\mathbf{J}_1(\mathbf{r},t) = \varepsilon_0 \int_0^t \omega_{p1}^2(\mathbf{r},\tau)\mathbf{E}(\mathbf{r},\tau)\,d\tau, \qquad (A.10)$$

where

$$\omega_{p1}^2(\mathbf{r}, t) = \frac{q^2 N_1(\mathbf{r}, t)}{m\varepsilon_0}. \tag{A.11}$$

The current density $\mathbf{J}_0(\mathbf{r}, t)$ due to the steady density $N_0(\mathbf{r}, t)$ is easily computed since the initial velocity at $t = 0$ of all the electrons is the same and is given by $\mathbf{v}_0(\mathbf{r}, 0)$:

$$\mathbf{J}_0(\mathbf{r}, t) = -qN_0(\mathbf{r})\mathbf{v}_0(\mathbf{r}, t)$$

$$= \frac{q^2 N_0(\mathbf{r})}{m} \int_0^t \mathbf{E}(\mathbf{r}, \tau)\, d\tau - qN_0(\mathbf{r})\mathbf{v}_0(\mathbf{r}, 0)$$

$$= \varepsilon_0 \omega_{p0}^2(\mathbf{r}) \int_0^t \mathbf{E}(\mathbf{r}, \tau)\, d\tau + \mathbf{J}_0(\mathbf{r}, 0), \tag{A.12}$$

where

$$\mathbf{J}_0(\mathbf{r}, 0) = -qN_0(\mathbf{r})\mathbf{v}_0(\mathbf{r}, 0). \tag{A.13}$$

Adding Equations A.12 and A.10, we obtain the total current $\mathbf{J}(\mathbf{r}, t)$:

$$\mathbf{J}(\mathbf{r}, t) = \varepsilon_0 \int_0^t \omega_p^2(\mathbf{r}, \tau)\mathbf{E}(\mathbf{r}, \tau)\, d\tau + \mathbf{J}(\mathbf{r}, 0), \tag{A.14}$$

where

$$\omega_p^2(\mathbf{r}, t) = \omega_{p0}^2(\mathbf{r}) + \omega_{p1}^2(\mathbf{r}, t), \tag{A.15}$$

$$\mathbf{J}(\mathbf{r}, 0) = \mathbf{J}_0(\mathbf{r}, 0). \tag{A.16}$$

Appendix B

Damping Rates of Waves in a Switched Magnetoplasma Medium: Longitudinal Propagation*

Dikshitulu K. Kalluri

B.1 Introduction

The study of the interaction between EMWs and plasmas is of considerable interest [1–3]. The real-life plasma can be bounded, having time-varying and space-varying (inhomogeneous) parameters. The general problem when all the features are simultaneously present has no analytic solution. At best, one can obtain a numerical solution. In some problems it is possible to neglect one or more of the features and concentrate on the dominant parameter.

Shock waves and controlled-fusion containment experiments are typical examples of time-varying plasmas [1]. The effect of a rapid rise of the electron density to a peak value can be modeled as a sudden creation of a plasma medium; that is, the plasma medium is suddenly switched on at $t = 0$, creating a temporal discontinuity in the properties of the medium.

Kalluri [4] has discussed the reflection of a traveling wave when an isotropic plasma of plasma frequency ω_p is switched on only over the $z > 0$ half-space. He has shown that the reflected field in the free space comprises two components, A and B. The A component is due to reflection at the spatial discontinuity at $z = 0$. Its frequency is the same as that of the incident wave. The temporal discontinuity gives rise to waves, called the B waves, and the one propagating in the negative z-direction undergoes partial transmission into free space. This is designated as the B component. Its frequency is different from that of the incident wave. Further, it is shown that the B component will be damped out even if the plasma is only slightly lossy. The B component is shown [3] to carry a maximum of 3.7% of the power in the incident wave when the incident wave frequency ω is $0.35\omega_p$. For this value of the incident wave frequency, the damping time constant is shown [4] to be $224.5/\omega_p$ s

---

* © 1990 IEEE. Reprinted from *IEEE Trans. Plasma Sci.*, 18, 797–801, 1990. With permission.

when the collision frequency ν is 1% of ω_p. The B component is thus weak and is damped out rapidly.

Kalluri [5] has studied the effect on a traveling wave of switching on an unbounded magnetoplasma medium. The direction of propagation of the incident (circularly polarized) wave is taken to be the same as that of the static magnetic field (longitudinal propagation [6]). He has shown that the original wave splits into three B waves. One of these waves is strongly influenced by the strength of the static magnetic field and carries a significant portion of the incident power for a value of $\omega = 0.67\omega_p$ and the gyrofrequency $\omega_b = 2\omega_p$. If this wave lasts long enough, there is a good chance of detecting its presence. This has motivated us to investigate the damping rates of these waves. The results given in Ref. [5] and the results obtained in this paper are used in suggesting a suitable experiment to detect these waves.

B.2 Frequencies and Velocities of the New Waves

Figure B.1 shows the geometry of the problem. Initially, for time $t < 0$, a circularly polarized plane wave is assumed to be propagating in free space in the positive z-direction. The incident wave is designated by

$$\bar{e}(z,t) = (\hat{x} \mp j\hat{y})E_0 \exp(j(\omega_0 t - k_0 z)), \tag{B.1}$$

$$\bar{h}(z,t) = (\pm j\hat{x} + \hat{y})H_0 \exp(j(\omega_0 t - k_0 z)). \tag{B.2}$$

In Equations B.1 and B.2, the upper sign corresponds to right circular polarization, designated as R-incidence, and the lower sign corresponds to left circular polarization designated as L-incidence. This convention will be

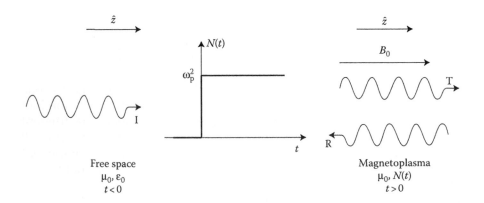

FIGURE B.1 Geometry of the problem. Magnetoplasma medium is switched on at time $t = 0$.

followed throughout the rest of this appendix text. Also,

$$H_0 = \frac{E_0}{\eta_0}, \quad \eta_0 = \sqrt{\frac{\mu_0}{\varepsilon_0}}, \quad k_0 = \omega_0 \sqrt{\mu_0 \varepsilon_0}, \tag{B.3}$$

where μ_0 and ε_0 are, respectively, the free space permeability and permittivity.

At time $t = 0$, the entire space is converted to a magnetoplasma medium. The plasma is assumed to consist of a constant number density N_0, together with a collision frequency v, and a static magnetic field of strength B_0 directed along the positive z-direction. These values correspond to a plasma frequency ω_p ($= \{N_0 q^2 / m\varepsilon_0\}^{1/2}$) and a gyrofrequency ω_b ($= qB_0/m$). The analysis is based on the conservation of the wave number of the plasma waves produced due to the temporal discontinuity [1–4]. The plasma fields can be described by

$$\bar{e}(z,t) = (\hat{x} \mp j\hat{y})e(t)\exp(-jk_0 z), \quad t > 0, \tag{B.4a}$$

$$\bar{h}(z,t) = (\pm j\hat{x} + \hat{y})h(t)\exp(-jk_0 z), \quad t > 0, \tag{B.4b}$$

$$\bar{v}(z,t) = (\hat{x} \mp j\hat{y})v(t)\exp(-jk_0 z), \quad t > 0. \tag{B.4c}$$

In Equation B.4c, v refers to the velocity field. The plasma fields under this condition satisfy the following equations:

$$\frac{d}{dt}(\eta_0 h) = j\omega_0 e, \quad t > 0, \tag{B.5a}$$

$$\frac{d}{dt}(e) = j\omega_0(\eta_0 h) + \frac{N_0 q}{\varepsilon_0} v, \quad t > 0, \tag{B.5b}$$

$$\frac{d}{dt}(v) = \frac{q}{m} e \pm j\omega_0 v - vv, \quad t > 0. \tag{B.5c}$$

In the absence of collisions ($v = 0$), the real frequencies of the newly created waves can be obtained by replacing (d/dt) by $j\omega$ in Equation B.5. This will yield the cubic given in Equation B.6:

$$D(\omega) = \omega^3 \mp \omega_b \omega^2 - (\omega_0^2 + \omega_p^2)\omega \pm \omega_0^2 \omega_b = 0. \tag{B.6}$$

Positive values of the frequencies in Equation B.6 correspond to the frequencies of the transmitted waves and negative values correspond to the frequencies of the reflected waves. A double subscript notation is used wherever necessary to indicate the nature of circular polarization (first subscript) of the incident wave and to indicate the wave being considered (second subscript). The three waves were studied [5], and it was found that: (1) For R-incidence, one of the three waves propagates in the negative z-direction; this wave has a frequency ω_{R2}; and (2) for L-incidence, two of the waves propagate in the negative z-direction; these waves have frequencies ω_{L2} and ω_{L3}. The group velocity u_{gn} of each of the newly created waves is obtained as

$$\frac{u_{gn}}{c} = \left[\frac{d\omega_n}{d\omega_0} \right] = \frac{2\omega_0(\omega_n \mp \omega_b)}{3\omega_n^2 \mp 2\omega_n\omega_b - \omega_0^2 - \omega_p^2}, \quad n = 1, 2, 3. \tag{B.7}$$

The velocity of energy transport u_{en} is defined as the ratio of the time-averaged Poynting vector to the total energy density. The total energy density comprises the energy stored in the electric field, the energy stored in the magnetic field, and the kinetic energy of the moving electrons. The energy velocity represents the velocity at which energy is carried by an EMW. The energy velocities of the three waves under consideration are given by

$$\frac{u_{en}}{c} = \frac{2\omega_0\omega_n(\omega_n \mp \omega_b)^2}{\left[(\omega_n \mp \omega_b)^2(\omega_n^2 + \omega_0^2) + \omega_n^2\omega_p^2 \right]}, \quad n = 1, 2, 3. \tag{B.8}$$

Equation B.8, when simplified with the help of Equation B.6, reduces to Equation B.7. Thus it can be concluded that the group velocity of the waves adequately characterizes the velocity of energy transfer by the three waves.

From these equations and the ω–β diagrams discussed by Kalluri [5], the following conclusions are drawn with regard to the dependence of the energy velocity on the gyrofrequency:

1. The energy velocity decreases for the wave having frequency ω_{R1} or ω_{L2}
2. The energy velocity increases for the wave having frequency ω_{R2} or ω_{L1}
3. The energy velocity increases for the wave having frequency ω_{R3} or ω_{L3}

These conclusions are useful in explaining the variation of the damping constants of the waves with the gyrofrequency and the incident wave frequency (Figures B.2 and B.3).

B.3   Damping Rates of the New Waves

In the ideal case of a lossless plasma, the B waves continue to propagate at all times. They ultimately damp out if the magnetoplasma is even slightly lossy. The possibility of detection of the B waves before they damp out in a lossy magnetoplasma is investigated in this section. To this end, the decay time constant and the attenuation length of the B waves will be obtained. The plasma is considered to be slightly lossy ($\nu \ll \omega_p$) so that approximation methods can be used for obtaining these parameters.

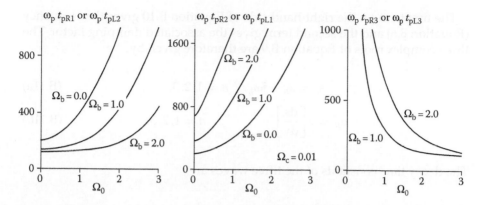

FIGURE B.2   Decay of waves with time at a given point in space.

When collisions are included, each of the three frequencies will be associated with a damping term, thus forming a complex frequency. Therefore, in order to obtain the complex frequencies, it is necessary to replace (d/dt) in Equation B.5 by the complex frequency variable s. The characteristic equation $D(s)$ is obtained from Equation B.5:

$$D(s) = s^3 + s^2(v \mp j\omega_b) + s(\omega_0 + \omega_p^2) + \omega_0^2(v \mp j\omega_b) = 0. \qquad (B.9)$$

The imaginary parts of the roots of $D(s)$ give the real frequencies, while their real parts indicate the nature of their exponential decay with time. Taylor's series expansion of s about $v = 0$ gives the relation

$$s(v) = [s]_{v=0} + v\left[\frac{ds}{dv}\right]_{v=0}. \qquad (B.10)$$

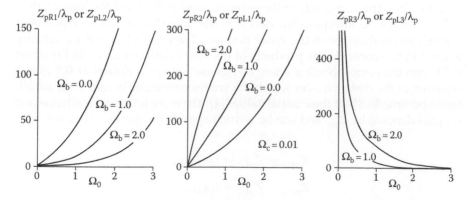

FIGURE B.3   Decay of waves in space at a given time.

The first term on the right-hand side of Equation B.10 gives the frequency (Equation B.6) and the second term gives the associated damping factor. The three complex roots of Equation B.9 are therefore given by

$$s_n = \sigma_n + j\omega_n, \quad n = 1, 2, 3,$$ (B.11a)

$$\sigma_n = v \left[\frac{ds}{dv} \right]_{v=0, s=j\omega_n}, \quad n = 1, 2, 3.$$ (B.11b)

The decay time constants of the three waves are

$$t_{pn} = -\frac{1}{\sigma_n} = \frac{(3\omega_n^2 \mp 2\omega_n\omega_b - \omega_0^2 - \omega_p^2)}{v(\omega_n^2 - \omega_0^2)}, \quad n = 1, 2, 3.$$ (B.12a)

Substitution for ω_b from Equation B.6 gives

$$t_{pn} = \frac{1}{v} \left[1 + \frac{\omega_p^2}{(\omega_n^2 - \omega_0^2)} + \frac{2\omega_p^2\omega_0^2}{(\omega_n^2 - \omega_0^2)^2} \right], \quad n = 1, 2, 3.$$ (B.12b)

The decay time constants for R-incidence and L-incidence are related:

$$t_{pR1} = t_{pL2},$$ (B.13a)

$$t_{pR2} = t_{pL1},$$ (B.13b)

$$t_{pR3} = t_{pL3}.$$ (B.13c)

The time constant associated with each frequency can be interpreted as the time in which the amplitude of the wave reduces to 36.8% of the initial value.

It is possible to interpret this decay in the time domain as an equivalent attenuation to distance in the space domain. The concept of energy velocity is used in this context. The products of the time constants given in Equation B.13 with the corresponding energy velocities given in Equation B.8 give a measure of the distance over which the waves propagate before their amplitudes become 36.8% of their initial values [4]. These are termed the attenuation lengths denoted by Z'_{pn} and can be written specifically for the three waves:

$$Z'_{pR1} = Z'_{pL2} = t_{pR1}u_{eR1},$$ (B.14a)

$$Z'_{pR2} = Z'_{pL1} = t_{pR2}u_{eR2},$$ (B.14b)

$$Z'_{pR3} = Z'_{pL3} = t_{pR3}u_{eR3}.$$ (B.14c)

An alternative way of obtaining the expression for the attenuation length is through the expression for the complex propagation constant γ_n [1] given by

$$\gamma_n = \alpha_n + j\beta_n = j\mu_n \frac{\omega_n}{c}, \quad n = 1, 2, 3, \tag{B.15a}$$

$$\mu_n^2 = 1 - \frac{\omega_p^2}{\omega_n \left(\omega_n \mp \omega_b - j\nu\right)}, \quad n = 1, 2, 3. \tag{B.15b}$$

The space attenuation of the plasma waves can be interpreted in terms of the attenuation length constant Z_p given by

$$Z_{pn} = \frac{1}{\alpha_n}, \quad n = 1, 2, 3. \tag{B.16}$$

For the three plasma waves under consideration, the attenuation lengths are given by

$$Z_{pR1} = Z_{pL2} = \frac{1}{\alpha_{R1}}, \tag{B.17a}$$

$$Z_{pR2} = Z_{pL1} = \frac{1}{\alpha_{R2}}, \tag{B.17b}$$

$$Z_{pR3} = Z_{pL3} = \frac{1}{\alpha_{R3}}. \tag{B.17c}$$

The attenuation lengths calculated from Equation B.14 are found to be very close to those obtained from Equation B.17 when the collision frequency is small. The decay time constants and attenuation lengths are computed for various values of the incident wave frequency and the gyrofrequency. The results are discussed in the following.

B.4 Numerical Results

In this analysis, the frequencies are normalized with respect to the plasma frequency. Thus,

$$\text{Frequency of the incident wave: } \Omega_0 = \frac{\omega_0}{\omega_p}, \tag{B.18a}$$

$$\text{Gyrofrequency: } \Omega_b = \frac{\omega_b}{\omega_p}, \tag{B.18b}$$

$$\text{Collision frequency: } \Omega_c = \frac{\nu}{\omega_p}, \tag{B.18c}$$

$$\text{Frequency of the plasma waves: } \Omega_n = \frac{\omega_n}{\omega_p}, \quad n = 1, 2, 3. \tag{B.18d}$$

Figure B.2 shows the variation of the damping time constant of the plasma waves for a collision frequency of $\Omega_c = 0.01$. In these plots the values of $\omega_p t_p$ are plotted as functions of Ω_0. Figure B.2a shows the damping rate of the waves having frequency Ω_{R1}. The same plot holds for the wave having frequency Ω_{L2}. For all values of the incident wave frequency, this wave decays faster as the strength of the static magnetic field is increased. The decay is more pronounced for $\Omega_0 > 1.0$ than for lower frequencies. For lower values of Ω_0, the static magnetic field has little effect (when $\Omega_0 = 0.5$, $\omega_p t_{pR1}$ is 150 for $\Omega_b = 1.0$, while it is 120 for $\Omega_b = 2.0$). Also, for a given Ω_b, high values of Ω_0 result in longer persistence of this wave in the plasma. Figure B.2b shows the decay rates of the wave having frequency Ω_{R2}. The behavior of this wave with Ω_b is different from that of the first wave. As Ω_b is reduced, this wave decays faster. For a specific Ω_b, as Ω_0 is increased, this wave persists longer in the plasma. Figure B.2c shows the time-domain decay of the third wave. For a given value of the incident wave frequency, an increase in the value of Ω_b results in a longer time of persistence of this wave. In the region where $\Omega_0 < 1.0$, the rate of decay of this wave with Ω_0 is much larger than its rate of decay for higher values of Ω_0. Figure B.3 shows the normalized attenuation lengths of the plasma waves in the space domain for a normalized collision frequency of $\Omega_c = 0.01$. Here the attenuation length Z_p is normalized with respect to λ_p ($= 2\pi c/\omega_p$), which is the wavelength in free space corresponding to plasma frequency. Figure B.3a shows the normalized attenuation length of the wave having frequency Ω_{R1} (or Ω_{L1}). Z_{pR1}/λ_p starts at a small value and increases to a large value with Ω_0 for any specific Ω_b.

The stronger the imposed magnetic field, the shorter the distance of propagation of this wave. Figure B.3b shows the space behavior of the wave having frequency Ω_{R2} (or Ω_{L1}). The stronger the imposed magnetic field, the longer the distance of propagation of this wave in the plasma. Figure B.3c shows the attenuation of the third wave having frequency Ω_{R3} (or Ω_{L3}). This wave is not present in the isotropic case and is the distinguishing feature of longitudinal propagation in a switched-on plasma medium. A small increase in Ω_0 brings about an appreciable reduction in the attenuation length of this wave when ω_0 is less than the plasma frequency ($\Omega_0 < 1.0$). When Ω_0 tends to infinity, the attenuation length of this wave becomes zero. When the collision frequency ν is small, the attenuation length of the waves can be shown to be approximately given by

$$\frac{Z_{pn}}{\lambda_p} \cong \frac{1}{\pi\Omega_c} \frac{(\Omega_n \mp \Omega_b)^2 \Omega_0}{\Omega_n}, \quad n = 1, 2, 3. \tag{B.19}$$

The damping constants, t_{pn} and Z_{pn}, increase with the energy velocity (group velocity). This is the physical explanation for the variation of the damping constants with Ω_b given in Figures B.2 and B.3.

TABLE B.1

Optimum Parameters for a Suitable Experiment

Rise Time (t_0)	Incident Frequency ($\omega_0/2\pi$)	Plasma Frequency ($\omega_p/2\pi$)	Gyro-frequency ($\omega_b/2\pi$)	Collision Frequency ($v/2\pi$)	Wave to be Observed (Third Wave)	
					Frequency ($\omega_3/2\pi$)	Time Delay (Constant t_p)
1 μs	10 kHz	15 kHz	30 kHz	150 s^{-1}	6.37 kHz	10 ms
1 ns	10 MHz	15 kHz	30 MHz	150×10^3s^{-1}	6.37 MHz	10 μs

B.5 Suggested Experiment

The characteristic wave, the third wave, in the case of R-incidence, carries considerable power [5] and lasts long (Figure B.3c) for a value of $\Omega_0 = 0.67$ and $\Omega_b = 2.0$. Based on the assumption that the plasma can be considered as suddenly created if the time period of the incident wave is at least 100 times the rise time t_0 of the plasma electron density, the incident wave frequency is given by $\omega_0 = (2\pi/100t_0)$. The other frequency parameters are, therefore, $\omega_p = (3\pi/100t_0)$, $\omega_b = (6\pi/100t_0)$, and $v = (3\pi/10,000t_0)$ for $\Omega_c = 0.01$. The decay time constant of the wave is found to be $t_p = 10,080t_0$. Table B.1 shows the required values of the parameters for two values of t_0.

B.6 Conclusions

The sudden imposition of a magnetoplasma medium on a traveling wave causes the creation of three new waves having frequencies different from that of the incident wave. Two of these waves are transmitted, and one is a reflected wave when the incident wave has right-hand circular polarization. The situation reverses when the incident wave has left-hand circular polarization. The energy velocity of each wave is the same as its group velocity ($v = 0$). The creation of the third wave is the distinguishing feature of the longitudinal propagation.

The attenuation of the B waves in the case of the imposition of a lossy magnetoplasma medium can be estimated through the computation of their decay time constants. The energy velocity can be used to relate the decay time constant to the attenuation length. An alternative means of estimating the attenuation length is through the consideration of the complex propagation constant computed at the new frequencies.

References

1. Heald, M. A. and Wharton, C. B., *Plasma Diagnostics with Microwaves*, Wiley, New York, 1965.
2. Auld, B. A., Collins, J. H., and Zapp, H. R., Signal processing in a nonperiodically time-varying magnetoelastic medium, *Proc. IEEE*, 56, 258, 1968.
3. Jiang, C. L., Wave propagation and dipole radiation in a suddenly created plasma, *IEEE Trans. Ant. Prop.*, AP-23, 83, 1975.
4. Kalluri, D.K., On reflection from a suddenly created plasma half-space: Transient solution, *IEEE Trans. Plasma Sci.*, 16, 11, 1988.
5. Kalluri, D. K., Effect of switching a magnetoplasma medium on a traveling wave: Longitudinal propagation, *IEEE Trans. Ant. Prop.*, AP-37, 1638, 1989.
6. Booker, H. G., *Cold Plasma Waves*, Kluwer, Hingham, MA, 1984.

Appendix C

Wave Propagation in a Switched Magnetoplasma Medium: Transverse Propagation*

Venkata R. Goteti and Dikshitulu K. Kalluri

C.1 Introduction

The study of interactions between electromagnetic radiation and plasmas having time-varying parameters is of considerable interest [1–6]. The interaction is characterized by the creation of new waves of frequencies different from that of the incident wave. The nature of these new waves (amplitudes, frequencies, and power carried) can be studied through a theoretical model of a suddenly created plasma. In this model it is assumed that a plane wave of frequency ω_0 is traveling in free space. At time $t = 0$ the free electron density in the medium increases suddenly from zero to some constant value N_0, that is, a plasma of frequency ω_p, where

$$\omega_p = \left(\frac{N_0 q^2}{m \varepsilon_0} \right)^{1/2} \tag{C.1}$$

is switched on at $t = 0$. In Equation C.1, q is the numerical value of the charge of an electron, m is the mass of an electron, and ε_0 is the permittivity of free space.

In the absence of a static magnetic field the created plasma behaves like an isotropic medium. It is known [4] that in the isotropic case, two new waves are created of frequencies

$$\omega_{1,2} = \pm(\omega_0^2 + \omega_p^2)^{1/2}. \tag{C.2}$$

In Equation C.2, the negative value for the frequency indicates a reflected wave.

In the presence of the static magnetic field, the switched medium is anisotropic. The case of longitudinal propagation (the static magnetic field in the direction of wave propagation and the incident wave circularly polarized) was recently investigated by Kalluri [6].

This paper investigates the effect of switching on a magnetoplasma medium for the case of transverse propagation. The incident wave (the wave existing prior to switching the medium) is assumed to be linearly polarized and propagating in the positive z-direction. The static magnetic field is assumed to be along the positive y-direction (transverse propagation [7]). The case of electric field along the y-direction results in ordinary wave propagation. It is the same as the isotropic case and is already discussed in detail by one of the authors, Kalluri [5]. The case of the incident electric field along the x-axis results in an extraordinary mode [7] of propagation and forms the subject of this paper.

In Section C.2, the basic nature of the interaction is studied in terms of the frequencies, amplitudes of the fields, and the power carried by the new waves. For this purpose the switched medium is assumed to occupy the entire space for $t > 0$. The fields of these waves will be attenuated if the plasma is lossy. The damping rates in time and space of these waves are examined in Section C.3. A more realistic situation of switching on a finite extent magnetoplasma is modeled in Section C.4 by considering the switched medium as a magnetoplasma half-space. The steady-state solution to this problem is developed along the lines of Kalluri [5], who considered the isotropic case. The transient solution to the switched lossy magnetoplasma half-space problem is developed in Section C.5 on the lines of Kalluri [5]. Numerical Laplace inversion of the solution confirms the damping rates of Section C.3 and the steady-state solution of Section C.4. In Section C.6, the numerical results are discussed. In the following analysis, ion motion is neglected (radio approximation).

C.2 Frequencies and Power Content of the New Waves

The geometry of the problem is shown in Figure C.la. Initially, a plane wave is assumed to be propagating in free space in the positive z-direction. Thus, for $t < 0$, the fields are described by

$$e(z, t) = \hat{x} E_0 \exp\left[j(\omega_0 t - k_0 z)\right], \quad t < 0, \tag{C.3}$$

$$h(z, t) = \hat{y} H_0 \exp\left[j(\omega_0 t - k_0 z)\right], \quad t < 0, \tag{C.4}$$

where ω_0 is the incident wave frequency, k_0 is the free-space wave number, μ_0 and ε_0 are the free-space parameters, and

$$H_0 = \frac{E_0}{\eta_0}, \quad \eta_0 = \sqrt{\frac{\mu_0}{\varepsilon_0}}. \tag{C.5}$$

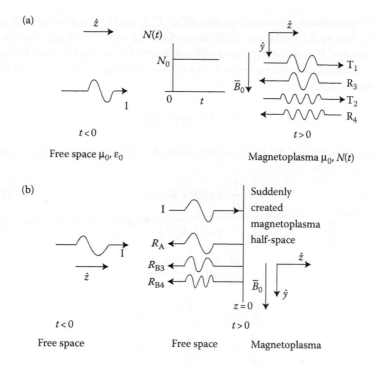

FIGURE C.1 Geometry of the problem: Switching of (a) an unbounded magnetoplasma medium and (b) a magnetoplasma half-space. The electric field of the incident wave is normal to the static magnetic field B_0.

At time $t = 0$, the entire space is converted to a magnetoplasma. The plasma is characterized by a constant number density N_0 and an imposed static magnetic field of strength B_0 along the y-direction. The fields satisfy the following equations under this condition (for $t > 0$):

$$\nabla \times \mathbf{e} = -\mu_0 \frac{\partial \mathbf{h}}{\partial t}, \quad t > 0, \tag{C.6}$$

$$\nabla \times \mathbf{h} = \varepsilon_0 \frac{\partial \mathbf{e}}{\partial t} - N_0 q \mathbf{v}, \quad t > 0, \tag{C.7}$$

$$\frac{\partial \mathbf{v}}{\partial t} = -\frac{q}{m} \mathbf{e} - \frac{q}{m} \mathbf{v} \times \mathbf{B}_0, \quad t > 0, \tag{C.8}$$

where v denotes the velocity field.

The waves in the plasma will be extraordinary waves [7]. So the plasma fields can be written as

$$\mathbf{e} = \hat{x} e_x + \hat{z} e_z, \tag{C.9a}$$

$$\mathbf{h} = \hat{y} h_y, \tag{C.9b}$$

$$\mathbf{v} = \hat{x} v_x + \hat{z} v_z. \tag{C.9c}$$

It is now possible to express Equations C.6, C.7, and C.8 as partial differential equations in z and t variables. With regard to the wave propagation in time-varying media, it is shown in the works of Fante [2], Jiang [4], and Kalluri [5] that the wave number is conserved. Therefore the z variation of the fields can be written in the form

$$f(z,t) = f(t)\exp(-jk_0 z), \tag{C.10}$$

Based on Equation C.10, the field equations C.6 through C.9 can be written as

$$\frac{d}{dt}(\eta_0 h_0) = j\omega_0 e_x, \tag{C.11a}$$

$$\frac{d}{dt}(e_x) = j\omega_0(\eta_0 h_y) + \frac{N_0 q}{\varepsilon_0}v_x, \tag{C.11b}$$

$$\frac{d}{dt}(e_z) = \frac{N_0 q}{\varepsilon_0}v_z, \tag{C.11c}$$

$$\frac{d}{dt}(v_x) = -\frac{q}{m}e_x + \omega_b v_z, \tag{C.11d}$$

$$\frac{d}{dt}(v_z) = -\frac{q}{m}e_z - \omega_b v_x, \tag{C.11e}$$

where ω_b is the gyrofrequency,

$$\omega_b = \frac{q B_0}{m}. \tag{C.11f}$$

The solution of Equation C.11 with appropriate initial conditions gives the time-domain description of the fields for $t = 0$.

Assume that the newly created electrons are stationary at $t = 0$ [4] and that they are set in motion only for $t > 0$. The free current density in the medium thus remains at zero value over the time discontinuity. Therefore the tangential electric and magnetic fields are continuous over $t = 0$. Further, it is assumed that the longitudinal component of the electric field, e_z, is present only for $t > 0$. The initial conditions are thus

$$v_x(z,0) = 0, \tag{C.12a}$$

$$v_z(z,0) = 0, \tag{C.12b}$$

$$e_x(z,0^-) = e_x(z,0^+) = E_0\exp(-jk_0 z), \tag{C.12c}$$

$$h_y(z,0^-) = h_y(z,0^+) = H_0\exp(-jk_0 z), \tag{C.12d}$$

$$e_z(z,0) = 0. \tag{C.12e}$$

Equations C.11 are solved through the use of Laplace transforms together with the initial conditions specified in Equation C.12. The Laplace transform of $f(t)$ is defined as

$$F(s) = \int_0^\infty f(t)e^{-st}\,dt. \tag{C.13}$$

The Laplace transforms of the field quantities are obtained as

$$E_x(z,s) = \frac{N_1(s)}{D(s)} E_0 \exp(-jk_0 z), \tag{C.14a}$$

$$E_z(z,s) = \frac{N_2(s)}{sD(s)} E_0 \exp(-jk_0 z), \tag{C.14b}$$

$$H_y(z,s) = \frac{1}{\eta_0} \left(j\omega_0 \frac{N_1(s)}{sD(s)} + \frac{1}{s} \right) E_0 \exp(-jk_0 z), \tag{C.14c}$$

where

$$N_1(s) = (s + j\omega_0)(s^2 + \omega_b^2 + \omega_p^2), \tag{C.15a}$$

$$N_2(s) = (s + j\omega_0)\omega_b\omega_p^2, \tag{C.15b}$$

$$D(s) = (s^2 + \omega_0^2)(s^2 + \omega_b^2 + \omega_p^2) + \omega_p^2(s^2 + \omega_p^2). \tag{C.15c}$$

Observation of Equations C.14 and C.15 shows that the poles of $E_x(z,s)$, $E_z(z,s)$, and $H_y(z,s)$ are all on the $j\omega$-axis. They are the zeros of $D(s)$. By setting $s = j\omega$ in Equation C.15, the zeros of $D(s)$ can be obtained as the roots of the equation

$$\omega^4 - \omega^2(\omega_0^2 + \omega_b^2 + 2\omega_p^2) + [\omega_p^4 + \omega_0^2(\omega_b^2 + \omega_p^2)] = 0. \tag{C.16}$$

Equation C.16 is a quadratic in ω^2 and can be readily solved to give four frequencies:

$$\omega_1 = (A + \sqrt{A^2 - B})^{1/2}, \tag{C.17a}$$

$$\omega_2 = (A - \sqrt{A^2 - B})^{1/2}, \tag{C.17b}$$

$$\omega_3 = -\omega_1, \tag{C.17c}$$

$$\omega_4 = -\omega_2, \tag{C.17d}$$

where

$$A = \frac{(\omega_0^2 + \omega_b^2 + 2\omega_p^2)}{2}, \tag{C.18a}$$

$$B = \omega_p^4 + \omega_0^2(\omega_b^2 + \omega_p^2). \tag{C.18b}$$

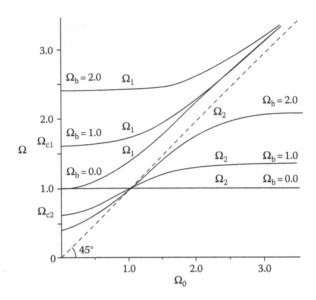

FIGURE C.2 Frequencies of the plasma waves, Ω, versus incident wave frequency, Ω_0. The results are presented in terms of normalized variables. The horizontal axis is $\Omega_0 = \omega_0/\omega_p$ where ω_0 is the frequency of the incident wave and ω_p is the plasma frequency of the switched medium. The vertical axis is $\Omega = \omega/\omega_p$, where ω is the frequency of the created waves. The branch Ω_n describes the nth created wave. The parameter $\Omega_b = \omega_b/\omega_p$, where ω_b is the gyrofrequency of the switched medium.

The important conclusion that can be drawn from Equations C.17 and C.18 is that ω_1 and ω_2 are real and positive and are the frequencies of the transmitted waves. The frequencies ω_3 and ω_4 are the frequencies of the corresponding reflected waves. Thus the incident wave of frequency ω_0 splits into four waves because of the imposition of the magnetoplasma. All these four waves will have the same wave number as that of the incident wave. The variation of these frequencies with the incident wave frequency is discussed in Section C.6 (Figure C.2).

Referring again to Equations C.14b and C.14c, it is evident that in addition to these four waves there is a purely space-varying component present in e_z and h_y. This corresponds to the residue of the pole at the origin of $E_z(s)$ and $H_y(s)$. The complete expression for the individual frequency components of $e_x(z,t)$, $e_z(z,t)$, and $h_y(z,t)$ can be obtained as

$$\frac{e_{xn}(z,t)}{E_0} = \frac{(\omega_n + \omega_0)(\omega_n^2 - \omega_p^2 - \omega_b^2)\exp[j(\omega_n t - k_0 z)]}{\prod_{m=1,m\neq n}^{4}(\omega_n - \omega_m)}, \quad n = 1,2,3,4,$$

$$\text{(C.19)}$$

$$\frac{e_{zn}(z,t)}{E_0} = j\frac{(\omega_n + \omega_0)\omega_b\omega_p^2\exp[j(\omega_n t - k_0 z)]}{\omega_n \prod_{m=1,m\neq n}^{4}(\omega_n - \omega_m)}, \quad n = 1,2,3,4, \quad \text{(C.20a)}$$

$$\frac{e_{z0}(z,t)}{E_0} = j\frac{\omega_0 \omega_b \omega_p^2}{\omega_1 \omega_2 \omega_3 \omega_4} \exp(-jk_0 z), \tag{C.20b}$$

$$h_{yn}(z,t) = \frac{1}{\eta_0}\frac{\omega_0}{\omega_n}e_{xn}(z,t), \quad n = 1,2,3,4, \tag{C.21a}$$

$$\frac{h_{y0}(z,t)}{H_0} = \frac{\omega_p^4}{\omega_1 \omega_2 \omega_3 \omega_4} \exp(-jk_0 z). \tag{C.21b}$$

The active power associated with e_x and h_y components is along the direction of propagation. No active power is associated due to e_z and h_y component interaction. The relative power content of the four waves expressed as a ratio of the time-averaged Poynting vector for each frequency component to that of the incident wave is given by

$$S_n = \frac{\langle e_{xn} h_{yn} \rangle}{E_0 H_0 / 2}, \quad n = 1,2,3,4. \tag{C.22}$$

The relative power content of the waves with frequencies ω_1 and ω_2 is found to be positive, confirming transmission, whereas that for ω_3 and ω_4 is found to be negative, confirming reflection. The variation of the power content of these waves with the incident wave frequency is discussed in Section C.6 (Figure C.3).

C.3 Damping Rates for the New Waves

The effects of switching on of a lossy plasma on wave propagation are examined in this section by introducing a collision frequency v in the plasma field equations. The momentum Equation C.8 gets modified as

$$\frac{d\mathbf{v}}{dt} = -\frac{q}{m}\mathbf{e} - \frac{q}{m}\mathbf{v} \times \mathbf{B}_0 - v\mathbf{v}. \tag{C.23}$$

Analysis can be carried out as in Section C.2 and the expression for the fields can be obtained. These are given by Equations C.14 and C.15, where ω_p is replaced by $c(s)$ and

$$c(s) = \frac{\omega_p^2[s(s+v) + \omega_p^2]}{[s(s+v)^2 + \omega_p^2(s+v) + \omega_b^2 s]}. \tag{C.24}$$

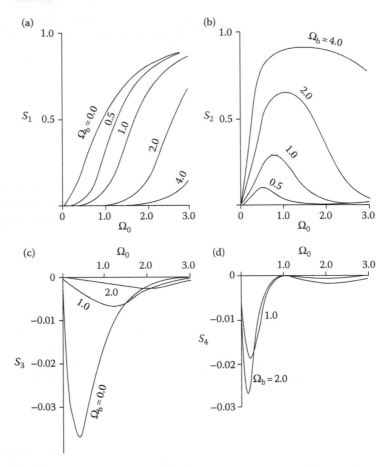

FIGURE C.3   Relative power content of the plasma waves versus incident wave frequency.

After simplifying one obtains

$$E_x(z,s) = E_x(s)E_0 \exp(-jk_0z), \qquad\qquad \text{(C.25a)}$$

$$E_x(s) = \frac{NR(s)}{DR(s)}, \qquad\qquad \text{(C.25b)}$$

where

$$NR(s) = (s + j\omega_0)\,[s(s + v)^2 + \omega_p^2(s + v) + \omega_b^2 s], \qquad \text{(C.26a)}$$

$$DR(s) = s^5 + a_4s^4 + a_3s^3 + a_2s^2 + a_1s + a_0, \qquad \text{(C.26b)}$$

$$a_4 = 2v, \qquad\qquad \text{(C.26c)}$$

$$a_3 = (\omega_0^2 + v^2 + \omega_b^2 + 2\omega_p^2), \qquad\qquad \text{(C.26d)}$$

$$a_2 = 2\nu(\omega_0^2 + \omega_p^2), \tag{C.26e}$$

$$a_1 = \omega_0^2(\nu^2 + \omega_b^2 + \omega_p^2) + \omega_p^4, \tag{C.26f}$$

$$a_0 = \nu\omega_0^2\omega_p^2, \tag{C.26g}$$

Further, it is possible to write Equation C.26b as

$$\text{DR}(s) = (s + s_0)[(s + s_1)^2 + \omega_{11}^2][(s + s_2)^2 + \omega_{21}^2]. \tag{C.27}$$

The basis for the assumption of a solution of this form is the fact that due to collisions, the fields produced in the plasma due to a temporal discontinuity die out ultimately [4,5]. Therefore the introduction of a collision frequency term contributes to the following changes in the nature of the plasma waves. First is a modification of the frequencies from ω_1 to ω_{11} and from ω_2 to ω_{21}. When ν is small, ω_{11} and ω_{21} will be close to ω_1 and ω_2. Second is the decay of these waves. Waves having frequency ω_{11} decay with a time constant of $1/s_1$, and those with frequency ω_{21} have a decay time constant of $1/s_2$. Third is an addition of a pure exponential decay term with a time constant of $1/s_0$. In the absence of collisions, this term is not present. Thus the x-component of the electric field can be written as

$$e_x(z,t) = \sum_{n=0}^{4} e_{xn}(t)\exp(-jk_0z), \tag{C.28a}$$

$$e_{x0}(t) = E_{p0}\exp\left(\frac{-t}{t_{p0}}\right), \tag{C.28b}$$

$$e_{xn}(t) = E_{pn}\exp(-t/t_{pn})\exp(-j\omega_{n1}t), \quad n = 1,2,3,4, \tag{C.28c}$$

where

$$t_{p0} = \frac{1}{s_0}, \tag{C.29a}$$

$$t_{p1} = t_{p3} = \frac{1}{s_1}, \tag{C.29b}$$

$$t_{p2} = t_{p4} = \frac{1}{s_2}, \tag{C.29c}$$

$$\omega_{31} = -\omega_{11}, \tag{C.29d}$$

$$\omega_{41} = -\omega_{21}. \tag{C.29e}$$

In Equation C.28, E_{p0}, E_{p1}, E_{p2}, E_{p3}, and E_{p4} are the residues of $E_x(s)$ given in Equation C.25 at the five poles of $E_x(s)$. For specified values of ω_0, ω_p, ω_b, and v, the plasma electric field is determined, and the amplitude and decay of each of the five component terms are examined.

While the foregoing analysis is concerned with the time-domain decay of the plasma waves, a second way of interpreting the decay is possible [5]. This concerns attenuation of the waves in the space domain as they propagate in the plasma. The complex propagation constant of the plasma waves is given by Heald and Wharton [8],

$$\gamma_n = \alpha_n + j\beta_n = j\mu_n \frac{\omega_n}{c}, \tag{C.30a}$$

where

$$\mu_n^2 = 1 - \frac{\omega_p^2/\omega_n^2}{1 - (j\nu/\omega_n) - (\omega_b^2/\omega_n^2)[1 - (j\nu/\omega_n) - (\omega_p^2/\omega_n^2)]^{-1}}. \tag{C.30b}$$

In this analysis, the wave propagation is characterized by the term $\exp(-\gamma_n z)$, indicating thereby that as the nth wave propagates in the plasma, it will be attenuated in space. The space attenuation constant can be defined as

$$Z_{pn} = \frac{1}{\alpha_n}. \tag{C.31}$$

Thus it is possible to explain the decay of plasma waves in two ways: in terms of a decay with time by the decay time constants t_{p1}, t_{p2}, t_{p3}, and t_{p4} and in terms of an attenuation with distance by the space attenuation constants Z_{p1}, Z_{p2}, Z_{p3}, and Z_{p4}. The variation of the damping rates with the incident wave frequency is discussed in Section C.6 (Figures C.6 and C.7).

C.4   Magnetoplasma Half-Space: Steady-State Solution

The switched-on magnetoplasma is considered to be confined to the $z > 0$ half-space in this analysis (Figure C.1b). This problem presents many interesting features. It presents a sharp discontinuity in space in addition to a sharp discontinuity in time. Therefore, in addition to an initial value problem in the time domain, it poses a boundary value problem in the space domain. The distinguishing feature of propagation in a time discontinuity and in a space discontinuity lies in the fact that time discontinuity results in the conservation of the wave number while space discontinuity results in the conservation of the frequency. Thus the plasma waves produced as a result of time discontinuity will consist of a set of four waves having a wave number identical to that of the incident wave. Two of these waves will propagate in the positive z-direction, while the remaining two will propagate in the negative z-direction. Added to these waves will be one transmitted wave and one reflected wave produced as a result of the space discontinuity. These waves will have the same frequency as the incident wave frequency but with a different wave number. In free space ($z < 0$), there will be three reflected waves. One will be a wave having relative amplitude R_A and frequency ω_0 produced by reflection from the $z = 0$ space boundary. The other two waves are due to those produced in the plasma by the time discontinuity and traveling in the negative z-direction. These waves will be partially transmitted into free space when they encounter the space boundary at $z = 0$. If the plasma is lossless, the two waves continue to propagate in steady state [5]. Their frequencies are already designated as ω_3 and ω_4, and let the relative amplitudes be R_{B3} and R_{B4}, respectively. Therefore the total free-space reflected field in steady state is given by a superposition of the three waves. In other words, when the plasma is lossless and the incident wave is $E_0 \cos(\omega_0 t - k_0 z)$, the total reflected electric field in steady state at the interface, $z = 0$, can be written as

$$A_{1Rss}(t) = R \cos \omega_0 t - X \sin \omega_0 t + R_{B3} \cos \omega_3 t + R_{B4} \cos \omega_4 t, \quad (C.32a)$$

where

$$R_A = R + jX = \frac{(\eta_{p0} - \eta_0)}{(\eta_{p0} + \eta_0)}, \quad (C.32b)$$

$$\eta_{p0} = \eta_0 (\varepsilon_{r0})^{-1/2}, \quad (C.32c)$$

$$\varepsilon_{r0} = 1 - \frac{\omega_p^2 / \omega_0^2}{1 - (\omega_b^2 / \omega_0^2)[1 - (\omega_p^2 / \omega_0^2)]^{-1}}, \quad (C.32d)$$

$$R_{B3} = R_{T3} T_{S3}, \quad (C.32e)$$

$$R_{B4} = R_{T4} T_{S4}, \quad (C.32f)$$

$$R_{T3} = \frac{(\omega_3 + \omega_0)(\omega_3^2 - \omega_p^2 - \omega_b^2)}{(\omega_3 - \omega_1)(\omega_3 - \omega_2)(\omega_3 - \omega_4)}, \quad (C.32g)$$

$$R_{T4} = \frac{(\omega_4 + \omega_0)(\omega_4^2 - \omega_p^2 - \omega_b^2)}{(\omega_4 - \omega_1)(\omega_4 - \omega_2)(\omega_4 - \omega_3)}, \quad (C.32h)$$

$$T_{Sn} = 2\eta_0 (\eta_{pn} + \eta_0)^{-1}, \quad n = 3, 4, \quad (C.32i)$$

$$\eta_{pn} = \eta_0 \frac{|\omega_n|}{\omega_0}, \quad n = 3, 4. \quad (C.32j)$$

In Equation C.32, R_A is the reflection coefficient from free space to plasma, η_{p0} is the plasma intrinsic impedance at frequency ω_0, R_{T3} and R_{T4} are the relative amplitudes of the negatively propagating waves given by Equation C.19 in Section C.2, and T_{S3} and T_{S4} are the transmission coefficients from

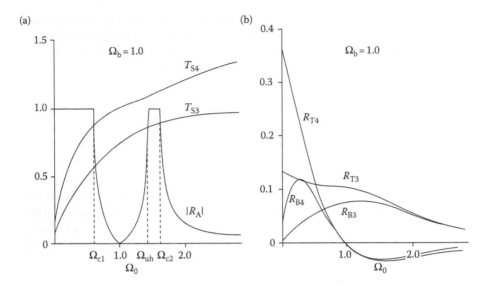

FIGURE C.4 Reflection coefficients versus Ω_0.

plasma to free space at frequencies ω_3 and ω_4, respectively. Finally, R_{B3} and R_{B4} are the relative amplitudes of the transmitted parts of R_{T3} and R_{T4} into free space. The frequencies ω_1, ω_2, ω_3, and ω_4 are obtained in Equation C.17. The variation of the various reflection and transmission coefficients (mentioned earlier) with the incident wave frequency is discussed in Section C.6 (Figure C.4). The fields at any distance z from the interface in free space can be obtained by replacing t with $(t + z/c)$ in Equation C.32.

When the plasma is lossy, even slightly, the steady-state reflected field is only that due to R and X in Equation C.32a since R_{B3} and R_{B4} components die out in steady state. To obtain the steady-state fields when the plasma is lossy, it is necessary to include the effect of the collision frequency ν in the expression for the dielectric constant, and it is given by [8]

$$\varepsilon_{r0} = 1 - \frac{\omega_p^2/\omega_0^2}{1 - (j\nu/\omega_0) - (\omega_b^2/\omega_0^2)[1 - (j\nu/\omega_0) - (\omega_p^2/\omega_0^2)]^{-1}}, \quad \nu \neq 0. \quad \text{(C.33)}$$

C.5 Magnetoplasma Half-Space: Transient Solution

The transient electric field in the plasma and free space can be obtained by formulating the problem as an initial value problem in time domain and as a boundary value problem in the space domain. This involves the solution of partial differential equations for the fields. The electric field of the incident wave is written as

$$\mathbf{e}(z,t) = \hat{x}\,\mathrm{Re}\left\{E_0\,\exp[j(\omega_0 t - k_0 z)]\right\}, \quad t < 0. \tag{C.34}$$

At time $t = 0$, the entire $z > 0$ half-space is converted to magnetoplasma. The partial differential equations describing the fields can be written as

$$\frac{\partial^2 e_{xf}}{\partial z^2} - \frac{1}{c^2}\frac{\partial^2 e_{xf}}{\partial t^2} = 0, \quad t > 0, \quad z < 0, \tag{C.35}$$

$$\frac{\partial^2 e_{xp}}{\partial z^2} - \frac{1}{c^2}\frac{\partial^2 e_{xp}}{\partial t^2} + N_0 q \mu_0 \frac{\partial v_{xp}}{\partial t} = 0, \quad t > 0, \quad z > 0, \tag{C.36a}$$

$$\frac{\partial v_{xp}}{\partial t} = -\left(\frac{q}{m}\right) e_{xp} + \omega_b v_{zp}, \quad t > 0, \quad z > 0, \tag{C.36b}$$

$$\frac{\partial v_{zp}}{\partial t} = -\left(\frac{q}{m}\right) e_{zp} - \omega_b v_{xp}, \quad t > 0, \quad z > 0, \tag{C.36c}$$

$$\frac{\partial e_{zp}}{\partial t} = -\left(\frac{N_0 q}{\varepsilon_0}\right) v_{zp}, \quad t > 0, \quad z > 0. \tag{C.36d}$$

Equation C.35 describes the free-space electric field e_{xf}, and Equation C.36 describes the fields in the plasma. Taking Laplace transforms of Equations C.35 and C.36 with respect to t and imposing the initial conditions,

$$e_{xf}(z,0) = e_{xp}(z,0) = e(z,0) = E_0\,\exp(-jk_0 z), \tag{C.37a}$$

$$\frac{\partial e_{xf}(z,0)}{\partial t} = \frac{\partial e_{xp}(z,0)}{\partial t} = \frac{\partial e(z,0)}{\partial t} = j\omega_0 E_0\,\exp(-jk_0 z), \tag{C.37b}$$

$$e_{zp}(z,0) = 0, \tag{C.37c}$$

$$v_{xp}(z,0) = 0, \tag{C.37d}$$

$$v_{zp}(z,0) = 0, \tag{C.37e}$$

one obtains

$$\frac{d^2 E_{xf}(z,s)}{dz^2} - q_1^2 E_{xf}(z,s) = -\frac{(s+j\omega_0)}{c^2} E_0\,\exp(-jk_0 z), \tag{C.38}$$

$$\frac{d^2 E_{xp}(z,s)}{dz^2} - q_2^2 E_{xp}(z,s) = -\frac{(s+j\omega_0)}{c^2} E_0\,\exp(-jk_0 z), \tag{C.39}$$

where

$$q_1 = \frac{s}{c}, \tag{C.40a}$$

$$q_2 = \frac{1}{c}\left(s^2 + \frac{\omega_p^2(s^2 + \omega_p^2)}{(s^2 + \omega_p^2 + \omega_b^2)}\right)^{1/2}. \tag{C.40b}$$

The solution to these ordinary differential equations can be obtained as

$$E_{xf}(z,s) = A_1(s)\exp(q_1 z) + \frac{(s + j\omega_0)}{(s^2 + \omega_0^2)} E_0 \exp(-jk_0 z), \tag{C.41}$$

$$E_{xp}(z,s) = A_2(s)\exp(-q_2 z) + \frac{(s + j\omega_0)}{c^2(q_2^2 + k_0^2)} E_0 \exp(-jk_0 z), \tag{C.42}$$

The second term on the right-hand side of Equation C.41 is the Laplace transform of the incident electric field. The first term should therefore correspond to the reflected field. The undetermined coefficients $A_1(s)$ and $A_2(s)$ can be obtained from the continuity condition of tangential electric and magnetic fields at $z = 0$. Thus

$$E_{xf}(0,s) = E_{xp}(0,s), \tag{C.43a}$$

$$\frac{dE_{xf}}{dz}(0,s) = \frac{dE_{xp}}{dz}(0,s). \tag{C.43b}$$

Using these conditions, $A_1(s)$ can be shown to be given by

$$\frac{A_1(s)}{E_0} = \frac{(q_2 - jk_0)}{(q_1 + q_2)} \left[\frac{(s + j\omega_0)}{c^2(q_2^2 + k_0^2)} - \frac{(s + j\omega_0)}{(s^2 + \omega_0^2)} \right]. \tag{C.44}$$

The reflected electric field thus becomes

$$\frac{e_{xR}}{E_0} = \mathcal{L}^{-1} \left\{ \frac{A_{1R}(s)}{E_0} \exp(q_1 z) \right\}, \tag{C.45}$$

where $A_{1R}(s)$, the real part of $A_1(s)$, is given by

$$\frac{A_{1R}(s)}{E_0} = \left[\frac{-\omega_p^2(sq_2 + \omega_0 k_0)(s^2 + \omega_p^2)}{(q_1 + q_2)(s^2 + \omega_0^2)\left[(s^2 + \omega_0^2)(s^2 + \omega_p^2 + \omega_b^2) + \omega_p^2(s^2 + \omega_p^2)\right]} \right]. \tag{C.46}$$

Numerical Laplace inversion of this irrational function was performed, and it is shown that for large t the total reflected field $A_{1Rss}(t)/E_0$ agrees with that obtained from Equation C.32. See Figure C.5 of Section C.6.

When a loss term v is introduced, it is possible to show that the reflected field $A_{1R}(s)/E_0$ can be obtained by replacing ω_p by ω_{p1} in Equation C.46, where ω_{p1} is given by Equation C.24. Numerical Laplace inversion then gives the reflected transient electric field of the lossy plasma. In steady state this should agree with the expression given by R_A. Numerical computations verified that this is indeed the case.

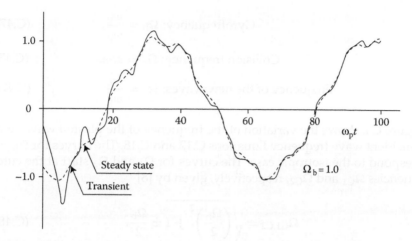

FIGURE C.5 Steady-state and transient free-space reflected electric field versus normalized time $\omega_p t$.

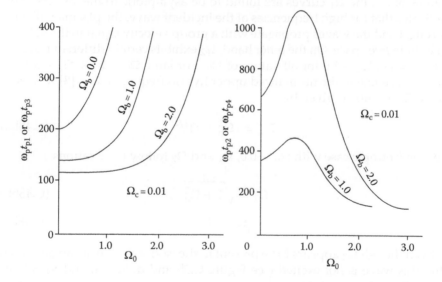

FIGURE C.6 Normalized damping time constant $\omega_p t_p$ of the plasma waves versus Ω_0.

C.6 Numerical Results and Discussion

In the following analysis, frequencies are normalized with respect to the plasma frequency. The normalized variables are denoted as

$$\text{Incident wave frequency: } \Omega_0 = \frac{\omega_0}{\omega_p}, \qquad \text{(C.47a)}$$

$$\text{Gyrofrequency: } \Omega_b = \frac{\omega_b}{\omega_p}, \tag{C.47b}$$

$$\text{Collision frequency: } \Omega_c = \nu/\omega_p, \tag{C.47c}$$

$$\text{Frequency of the new waves: } \Omega_n = \frac{\omega_n}{\omega_p}. \tag{C.47d}$$

Figure C.2 shows the variation of the frequency of the plasma waves with the incident wave frequency Equations C.17 and C.18. The curves for $\Omega_b = 0$ correspond to the isotropic case. The curves for Ω_1 and Ω_2 start at the cutoff frequencies Ω_{C1} and Ω_{C2}, respectively, given by [8]

$$\Omega_{C1,C2} = \sqrt{\left(\frac{\Omega_b}{2}\right)^2 + 1} \pm \frac{\Omega_b}{2}. \tag{C.48a}$$

Here the upper sign refers to Ω_{C1} and the lower sign refers to Ω_{C2}. For large values of Ω_0, the Ω_1 curves are found to be asymptotic to the 45° line. This indicates that for high frequencies of the incident wave, the plasma will have no effect and the waves propagate with a group velocity equal to the velocity of light in free space. On the other hand, Ω_2 exhibits certain different features. At $\Omega_0 = 1.0$, Ω_2 is 1.0 for all values of Ω_b. For large Ω_0 values, Ω_2 saturates at a value equal to the normalized upper hybrid frequency Ω_{uh} [8] with zero slope. This value is given by

$$\Omega_{uh} = (\Omega_b^2 + 1)^{1/2}. \tag{C.48ba}$$

For the isotropic case with $\Omega_b = 0.0$, Ω_1 and Ω_2 follow the relations

$$\Omega_1 = \sqrt{1 + \Omega_0^2}, \tag{C.48bb}$$

$$\Omega_2 = 1. \tag{C.48bc}$$

Even though Ω_2 appears to be present in the isotropic case, it can be shown that this wave is not excited (see Figure C.3b and d; $S_2 = 0$ and $S_4 = 0$ for $\Omega_b = 0$).

Figure C.3 shows the dependence of the relative power content of the four plasma waves on Ω_0. In Figure C.3a the data for the first transmitted wave are shown, and in Figure C.3c the data for the first reflected wave are shown. Both these waves have the same frequency (Ω_1 or $|\Omega_3|$). The curves in Figure C.3b and d correspond to the second pair of waves with frequency Ω_2 or $|\Omega_4|$. S_1 and S_2, are positive, indicating propagation in the positive z-direction, while S_3 and S_4 being negative indicate propagation in the negative z-direction. When $\Omega_b = 0.0$, only S_1 and S_3 are present. S_1 is observed to increase monotonically from zero to unity for a given Ω_b. For a given Ω_0, an increase in Ω_b results in a reduction in S_1.

The S_2 curve is observed to start from zero at $\Omega_b = 0$, increase up to a peak value, and finally reduce to zero as Ω_0 increases for a given Ω_b. An increase in Ω_b increases the value of S_2 for a specific Ω_0. S_2 possesses a peak value of 0.275 for $\Omega_0 = 0.8$ and $\Omega_b = 1.0$, while it is 0.64 at $\Omega_0 = 1.05$ and $\Omega_b = 2.0$.

The relative power content in the first reflected wave is shown by the curve S_3 in Figure C.3c. Its power content is much smaller than that of the first transmitted wave. S_3 has a peak of about 0.036 (about 3.6% only). The stronger the imposed magnetic field, the weaker this wave is. The peak power for $\Omega_b = 1.0$ is found to be 0.0068 at $\Omega_0 = 1.2$.

The second reflected wave has a power content S_4 shown in Figure C.3d. S_4 is zero at $\Omega_0 = 1.0$ for all Ω_b. This is because the corresponding electric and magnetic fields are both zero at this frequency. Even though S_4 is small compared to S_2, it is comparable with S_3. S_4 starts from zero at $\Omega_0 = 0.0$, increases to a negative peak value, becomes zero at $\Omega_0 = 1.0$, goes through a second negative peak, and finally becomes zero as Ω_0 increases for a given value of Ω_b. This wave becomes stronger with the static magnetic field.

Figure C.4 relates to the components of the steady-state free-space reflected fields for the case of the switched-on magnetoplasma half-space discussed in Section C.4. In Figure C.4a, R_A corresponds to the reflection coefficient from free space to plasma. R_A is 1.0 for $\Omega_0 < \Omega_{C1}$, is zero at $\Omega_0 = 1.0$, and increases to 1.0 at $\Omega_0 = \Omega_{uh}$. It is unity up to $\Omega_0 = \Omega_{C2}$ and from thereon decreases to zero. The cutoff frequencies are 0.618 and 1.618 while the upper hybrid frequency is 1.414 for $\Omega_b = 1.0$. T_{S3} refers to the transmission coefficient from plasma to free space for the first negatively propagating wave generated in the plasma. Its frequency is Ω_1. T_{S3} is zero at $\Omega_0 = 0.0$ and increases to unity as Ω_0 increases. T_{S4} corresponds to the transmission coefficient of the second negatively propagating wave having frequency Ω_2. It also starts at 0.0 at $\Omega_0 = 0.0$, but increases to 2.0 as Ω_0 increases. The relative amplitudes of the electric fields of the two waves mentioned above are shown by the curves R_{T3} and R_{T4} in Figure C.4b. R_{T3} starts at 0.13 for $\Omega_0 = 0.0$ and ultimately becomes zero. R_{T4} starts at 0.36 and reduces to zero at $\Omega_0 = 1.0$. R_{T4} exhibits a negative peak when $\Omega_0 > 1.0$ but becomes zero as Ω_0 becomes large. The relative amplitudes of the corresponding free-space fields are shown by R_{B3} and R_{B4}. Both these curves start at zero and ultimately become zero as Ω_0 increases. But while R_{B3} is always positive, R_{B4} goes through a positive peak, becomes zero at $\Omega_0 = 1.0$, and goes through a negative peak afterwards. R_{B3} is found to have a peak value of 0.08 at $\Omega_0 = 1.2$, $\Omega_b = 1.0$. R_{B4} has a positive peak of 0.12 at $\Omega_0 = 0.25$, $\Omega_b = 1.0$, and a negative peak of 0.03 at $\Omega_0 = 1.6$, $\Omega_b = 1.0$.

Figure C.5 compares the transient reflected electric field $A_{1R}(t)/E_0$ with the superposition of R_A, R_{B3}, and R_{B4} components. The initial peaks in the transient solution are due to the switching on of the magnetoplasma half-space. It is found that these switching transients die out quickly, within $\omega_p t = 40$ for $\Omega_b = 1.0$ and from then on there is fair agreement between $A_{1R}(t)$ and the superposition of the three steady-state components. The low-frequency part of these curves corresponds to the R_A component. Superposed on this are the two R_B components having frequencies $|\Omega_3|$ and $|\Omega_4|$. Also, the major contribution to the reflected field is from the steady-state space boundary R_A since R_{B3} and R_{B4} are only a fraction of R_A.

When a collision frequency $\Omega_c = 0.01$ is introduced, the fields produced because of the time discontinuity are shown to decay exponentially with time. Figure C.6 shows the dependence of the decay time constants $\omega_p t_p$ of the plasma waves with Ω_0. Figure C.6a shows the variation of $\omega_p t_{p3}$ with Ω_0 for the first reflected wave of frequency Ω_1. It is found that the stronger the imposed magnetic field, the quicker the decay of this wave, for any incident wave frequency. Figure C.6b shows the decay of the second reflected wave having frequency Ω_2. This wave is absent for $\Omega_b = 0.0$. For a given Ω_0 and Ω_b this wave is observed to persist longer in time than the first reflected wave. Also as Ω_0 increases, $\omega_p t_{p2}$ (or $\omega_p t_{p4}$) attains a peak value and then finally decreases to a specific value. This final value is about 134 for $\Omega_b = 1.0$. The peak value corresponds to the incident wave frequency that gives the longest time of persistence of these waves. The peak value is about 480 at $\Omega_0 = 0.6$ for $\Omega_b = 1.0$. Also the decay of this wave is found to be faster with weaker magnetic fields.

Figure C.7 explains the attenuation of the waves with distance as they propagate in the lossy plasma. Here the attenuation depth Z_p is normalized with the free-space wavelength corresponding to the plasma frequency λ_p ($= 2\pi c/\omega_p$). The Z_{p3}/λ_p curve of Figure C.7a is for the first reflected wave, and the Z_{p4}/λ_p curve is for the second reflected wave. The dependence of Z_p/λ_p in these curves appears to follow the variation of the corresponding $\omega_p t_p$ curve. The first reflected wave is observed to attenuate over shorter distances as Ω_b is increased (Figure C.7a), while the second reflected wave (Figure C.7b) is observed to attenuate over shorter distances as Ω_b is reduced.

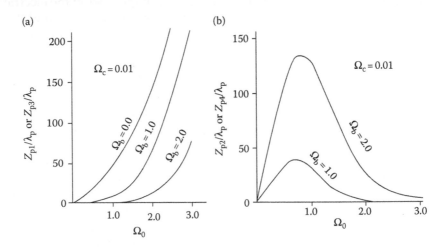

FIGURE C.7 Normalized attenuation length Z_p/λ_p of the plasma waves versus Ω_0. The free-space wavelength λ_p corresponds to the plasma frequency.

TABLE C.1

Optimum Parameters for a Suitable Experiment

Rise Time (t_0)	Incident Wave Frequency $(\omega_0/2\pi)$	Plasma Frequency $(\omega_p/2\pi)$	Gyrofrequency $(\omega_b/2\pi)$	Collision Frequency $(v/2\pi)$	Frequency[a] (ω_2/t_0)	Time Decay Constant[a] $(t_p/2\pi)$
1 μs	10 kHz	15 kHz	15 kHz	150 s^{-1}	12.45 kHz	5.6 ms
1 ns	10 MHz	15 MHz	15 MHz	150×10^3 s^{-1}	12.45 MHz	5.6 μs

[a] Value of the second wave.

Also a peak attenuation depth is a characteristic of the Z_{p4}/λ_p curve. When $\Omega_b = 1.0$, the peak value of Z_{p4}/λ_p is 40 occurring at $\Omega_0 = 0.8$.

C.7 Conclusions

It is shown that the sudden creation of a magnetoplasma medium (supporting transverse propagation) splits an existing wave (incident wave) into four propagating waves having frequencies different from the incident wave frequency. Two of the newly created waves are transmitted waves and two are reflected waves. The newly created waves are attenuated and damped when the magnetoplasma is lossy. The frequencies, the damping rates, and the power carried by the new waves are governed by the magnetoplasma parameters and the incident wave parameters.

It appears that the second wave has the best chance of being observed in an experiment since it carries considerable power (Figure C.3b) and lasts for a long time (Figure C.6b) for a value of $\Omega_0 \approx 0.67$. The optimum parameters for a suitable experiment can now be specified in terms of the rise time t_0 of a rapidly rising particle density. Assuming that the period of the incident wave should at least be 100 times t_0 in order for the plasma to be considered as suddenly created, the incident wave frequency can be obtained as $\omega_0 = (2\pi/100t_0)$. Therefore the plasma frequency should be $\omega_p = (3\pi/100t_0)$, and with the choice $\Omega_b = 1.0$, the gyrofrequency should be $\omega_b = (3\pi/100t_0)$. For this choice of these parameters the frequency of the second wave, from Equation C.I7b, is $\omega_2 = (2.482\pi/100t_0)$, and for $\Omega_c = 0.01$ the collision frequency is $v = (3\pi \times 10^{-4}/t_0)$. From Figure C.6, the time decay constant of the second wave is $t_{p2} = 5623t_0$. Table C.1 shows the required parameters for two values of the rise time t_0.

References

1. Auld, B. A., Collins, J. H., and Zapp, H. R., Signal processing in a nonperiodically time-varying magnetoelastic medium, *Proc. IEEE*, 56, 258, 1968.

2. Fante, R. L., Transmission of electromagnetic waves into time-varying media, *IEEE Trans. Ant. Prop.*, AP-19, 417, 1971.
3. Felsen, L. B. and Whitman, G. M., Wave propagation in time-varying media, *IEEE Trans. Ant. Prop.*, AP-18, 242, 1970.
4. Jiang, C. L., Wave propagation and dipole radiation in a suddenly created plasma, *IEEE Trans. Ant. Prop.*, AP-23, 83, 1975.
5. Kalluri, D. K., On Reflection from a suddenly created plasma half-space: Transient solution, *IEEE Trans. Plasma Sci.*, 16, 11, 1988.
6. Kalluri, D. K., Effect of switching a magnetoplasma medium on a traveling wave: Longitudinal propagation, *IEEE Trans. Ant. Prop.*, AP-37, 1638, 1989.
7. Booker, H. G., *Cold Plasma Waves*, p. 349, Kluwer, Hingham, MA, 1984.
8. Heald, M. A. and Wharton, C. B., *Plasma Diagnostics with Microwaves*, Wiley, New York, 1965.

Appendix D

Frequency Shifting Using Magnetoplasma Medium: Flash Ionization*

Dikshitulu K. Kalluri

D.1 Introduction

Jiang [1] investigated the effect of switching suddenly an unbounded isotropic cold plasma medium. Since the plasma was assumed to be unbounded, the problem was modeled as a pure initial value problem. The wave propagation in such a medium is governed by the conservation of the wave number across the time discontinuity and consequent change in the wave frequency. Rapid creation of the medium can be approximated as a sudden switching of the medium. The main effect of switching the medium is the splitting of the original wave (henceforth called incident wave, in the sense of incidence on a time discontinuity in the properties of the medium) into new waves whose frequencies are different from the incident wave. Recent experimental work [2–4] and computer simulation [5] demonstrated the frequency upshifting of microwave radiation by rapid plasma creation.

In any practical situation, the plasma is bounded and the problem can no longer be modeled as a pure initial value problem. Kalluri [6] dealt with this aspect by considering the reflection of a traveling wave when an isotropic cold plasma is switched on only over $z > 0$ half-space. Kalluri has shown that the reflected field in free space comprises two components, A and B. The A component is due to reflection at the spatial discontinuity at $z = 0$ formed at $t = 0$. Its frequency is the same as that of the incident wave. The temporal discontinuity gives rise to two additional waves in the plasma. These are called B waves. One of them propagates in the negative z-direction, and it is this wave that undergoes partial transmission into free space. The frequency of the B component is different from that of the incident wave.

Kalluri and Goteti [7] recently brought the solved problem closer to the practical situation by considering the switched plasma medium as a slab of width d (Figure D.1). A qualitative picture of the reflected and transmitted waves is given in Figure D.1. If the incident wave frequency ω_0 is less than the

---

* © 1993 IEEE. Reprinted from *IEEE Trans. Plasma Sci.*, 21, 77–81, 1993. With permission.

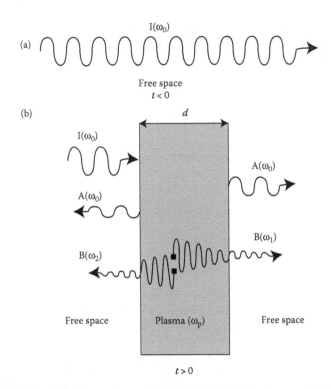

FIGURE D.1 Effect of switching an isotropic plasma slab. (a) Incident wave. Reflected and transmitted waves are sketched. The A waves have the same frequency as the incident wave frequency, but B waves have a new frequency $\omega_1 = (\omega_0^2 + \omega_p^2)^{1/2} = |\omega_2|$. The waves in (b) are sketched in the time domain to show the frequency changes. See Figure 0.5b for a more accurate description.

plasma wave frequency ω_p, that is, ω_0 is in the stop band of the medium, the transmitted wave will be a B wave of the new [1] frequency $\omega_1 = \sqrt{\omega_0^2 + \omega_p^2}$.

The solution of the slab problem was complex, but it would be more complex if the posed problem was to be brought even closer to the practical situation, that is, the problem of switching a space-varying and time-varying plasma of finite extent. However, since the B waves are generated in the plasma, their basic nature in terms of the frequency shift, power, and energy in them and how they are damped can all be studied qualitatively by considering the switching of the unbounded medium.

D.2 Wave Propagation in a Switched Unbounded Magnetoplasma Medium

It is known [8] that the presence of a static magnetic field will influence the frequency shift, and so on. Rapid creation of a plasma medium in the presence of a static magnetic field can be approximated as the sudden switching of a magnetoplasma medium. An idealized mathematical description of the problem is given next. A uniform plane wave is traveling in free space along the z-direction and it is assumed that at $t = 0^-$, the wave occupies the entire space. A static magnetic field of strength $\mathbf{B_0}$ is assumed to be present throughout. At $t = 0$, the entire space is suddenly ionized, thus converting the medium from free space to a magnetoplasma medium. The incident wave frequency is assumed to be high enough that the effects of ion motion can be neglected.

The strength and direction of the static magnetic field affect the number of new waves created, their frequencies, and the power density of these waves. Some of the new waves are transmitted waves (waves propagating in the same direction as the incident wave) and some are reflected waves (waves propagating in the opposite direction to that of the incident wave). Reflected waves tend to have less power density.

A physical interpretation of the waves can be given in the following way. The electric and magnetic fields of the incident wave and the static magnetic field accelerate the electrons in the newly created magnetoplasma, which in turn radiate new waves. The frequencies of the new waves and their fields can be obtained by adding contributions from the many electrons whose positions and motions are correlated by the collective effects supported by the magnetoplasma medium. Such a detailed calculation of the radiated fields seems to be quite involved. A simple, but less accurate, description of the plasma effect is obtained by modeling the magnetoplasma as a dielectric medium whose refractive index is computed through magnetoionic theory [9]. The frequencies of the new waves are constrained by the requirements that the wave number (k_0) is conserved over the time discontinuity, and the refractive index n is that applicable to the type of wave propagation in the magnetoplasma. This gives a conservation [10] law $k_0 c = \omega_0 = n(\omega)\omega$ from which ω can be determined. The solution of the associated electromagnetic initial value problem gives the electric and magnetic fields of the new waves. Using this approach, the general aspects of wave propagation in a switched magnetoplasma medium were discussed earlier [8,11,12] for the cases of L-, R-, and X-incidence [9].

D.3 L and R Waves

L-incidence is an abbreviation used to describe the case where the incident wave has left-hand circular polarization and the static magnetic field is in the z-direction (Figure D.2a). Three new waves with left-hand circular polarization labeled as L1, L2, and L3 are generated by the medium switching. L1 is a transmitted wave. R-incidence refers to the case where the incident wave has right-hand circular polarization and the static magnetic field is in the z-direction (Figure D.2b). The medium switching, in this case, creates three

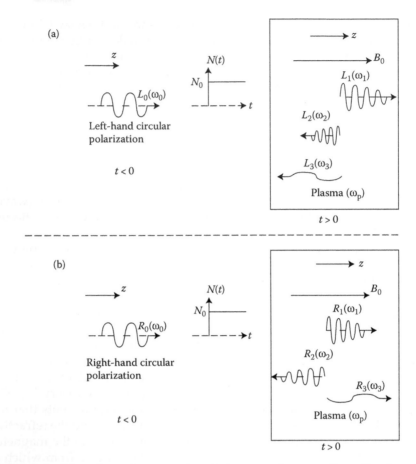

FIGURE D.2 Effect of switching an unbounded magnetoplasma medium. Sketches of the B waves generated in the plasma are given in the time domain to show the frequencies: (a) L-incidence and (b) R-incidence. The waves are sketched in time domain to show the change in frequencies.

R waves labeled as R1, R2, and R3. R1 and R3 are transmitted waves. The frequencies of the waves are the roots of the characteristic equation [9].

$$\omega^3 \mp \omega_b \omega^2 - (\omega_0^2 + \omega_p^2)\omega \pm \omega_0^2 \omega_b = (\omega - \omega_1)(\omega - \omega_2)(\omega - \omega_3) = 0, \quad \text{(D.1)}$$

where the lower sign is for the L waves and the upper sign for the R waves. The electric and magnetic fields of the nth wave are given in Equations D.2 and D.3.

$$\frac{E_n}{E_0} = \frac{(\omega_n \mp \omega_b)(\omega_n + \omega_0)}{\prod_{m=1, m \neq n}^{3} (\omega_n - \omega_m)}, \quad n = 1, 2, 3, \quad \text{(D.2)}$$

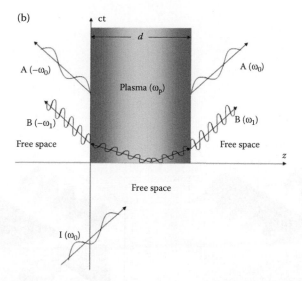

FIGURE 0.5 Effect of switching an isotropic plasma slab. A waves have the same frequency as the incident wave frequency (ω_0), and B waves have upshifted frequency $\omega_1 = \sqrt{\left(\omega_0^2 + \omega_p^2\right)} = -\omega_2$. (b) Shows a more accurate description.

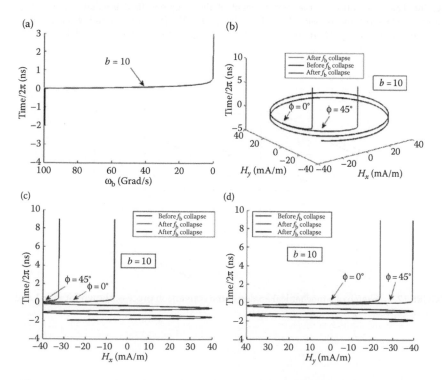

FIGURE 12.7 Evolution of the magnetic field intensity as f_b collapses under constant plasma density. $b = 10$, $\omega_0 = 150\,\text{Grad/s}$, $\omega_{b0} = 150\,\text{Grad/s}$, $m = 1$, and $z \approx 0$. (a) ω_b profile, (b) H_x–H_y variation in time, (c) H_x-variation in time, and (d) H_y-variation in time.

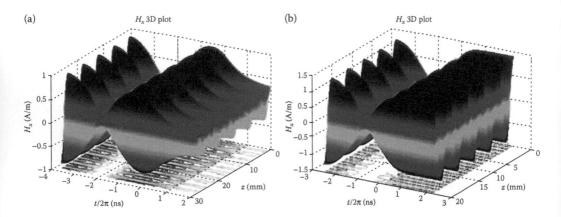

FIGURE 12.8 Evolution of the x-component of the magnetic field intensity as f_b collapses under constant plasma density. $b = 10$, $\phi = 45°$, $m = 10$, (a) $\omega_0 = 1$ Grad/s, $\omega_{p0} = 1000/s$, $\omega_{b0} = 10$ Grad/s, $d = 2.83$ cm, (b) $\omega_0 = 1$ Grad/s, $\omega_{p0} = 150$ Grad/s, $\omega_{b0} = 1.1$ Grad/s, and $d = 2$ cm.

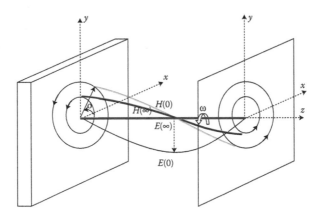

FIGURE 12.9 Sketch of a mechanical analog of the conversion process, shown for $m = 1$ and E_0 being negative.

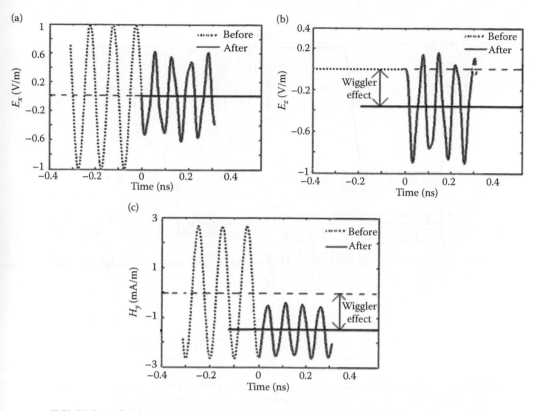

FIGURE O.5 Total electric and magnetic fields before and after switching the medium: Parameters: $f_0 = 10\,\text{GHz}$, $\varphi_0 = 90°$, $f_p = 17.63\,\text{GHz}$, $f_b = 20\,\text{GHz}$, $n = 1$, $E_0 = 1\,\text{V/m}$, $z = d/2$ for (a) and (b), and $z = 0$ for (c). See Table O.2 for more information. (a) Total electric field of x-component; (b) total electric field of z-component (including wiggler field); and (c) total magnetic field of y-component (including wiggler magnetic).

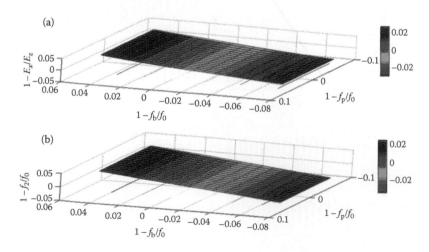

FIGURE O.6 Effect of small changes in f_b and f_p on circular polarization of mode 2. The parameters are as specified in Table O.3. (a) Variation of the ratio of the electric field components and (b) variation of the frequency of the second mode.

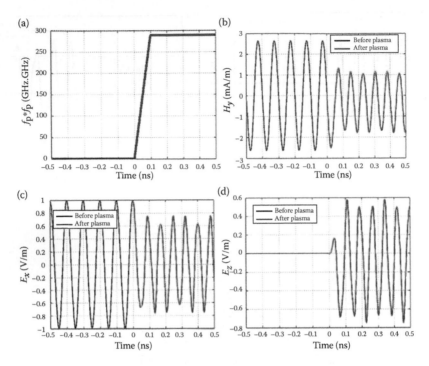

FIGURE P.3 Evolution of the electric and magnetic fields for the case of $T_r = 0.1$ ns. (a) Electron density profile, (b) H_y, (c) E_x, and (d) E_z. The parameters are $f_0 = 10$ GHz, $f_p = 17.63$ GHz, $f_b = 20$ GHz, $\phi_0 = 90°$, $\nu = 0$, $n = 1$, $z = d/2$ for E, and $z \approx 0$ for H. See Equation P.16 for the relation between the imposed magnetic flux density B_0 and the electron gyrofrequency ω_b in rad/s.

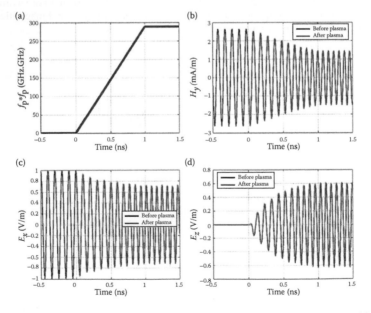

FIGURE P.4 Evolution of the electric and magnetic fields for the case of $T_r = 1$ ns. (a) Electron density profile, (b) H_y, (c) E_x, and (d) E_z. The other parameters are the same as in Figure P.3.

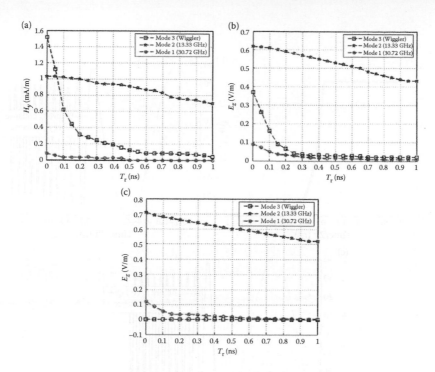

FIGURE P.6 Amplitudes of the three modes versus rise time. (a) H_y, (b) E_z, and (c) E_x. Mode 3 is the wiggler. The other parameters are the same as those in Figure P.3.

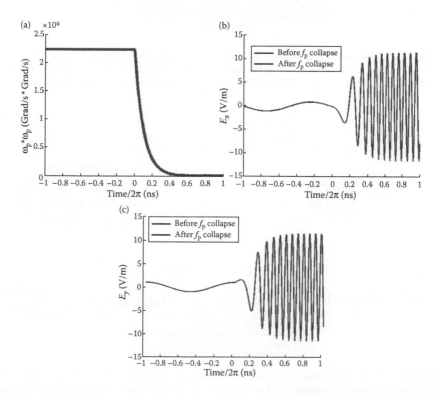

FIGURE Q.4 Evolution of the electric field intensity as plasma collapses under constant external magnetic field. (a) Plasma profile, (b) E_x variation in time, and (c) E_y variation in time. $b = 10$, $\phi_0 = 90°$, $\omega_0 = 1$ Grad/s, $\omega_{p0} = 150$ Grad/s, $\omega_{b0} = 100$ Grad/s, $m = 1$, $z = d/2$, $\nu/\omega_{p0} = 0$, and $E_0 = 1$ V/m.

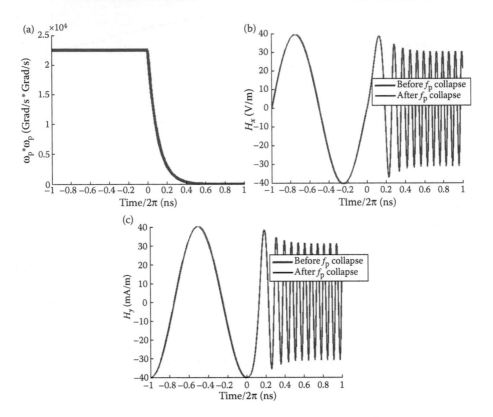

FIGURE Q.5 Evolution of the magnetic field intensity as plasma collapses under constant external magnetic field. (a) Plasma profile, (b) H_x variation in time, and (c) H_y variation in time. $b = 10$, $\phi_0 = 90°$, $\omega_0 = 1$ Grad/s, $\omega_{p0} = 150$ Grad/s, $\omega_{b0} = 100$ Grad/s, $m = 1$, $z \approx 0$, and $v/\omega_{p0} = 0$.

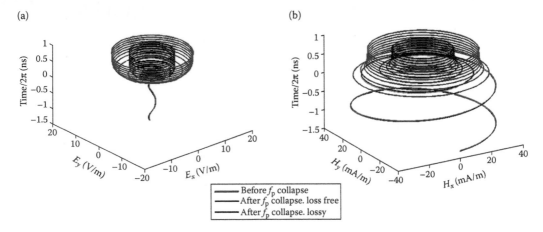

FIGURE Q.6 Evolution of the electric and magnetic field intensities as lossy plasma collapses under constant external magnetic field in the cavity. (a) E_x–E_y variation in time and (b) H_x–H_y variation in time. $b = 10$, $\phi_0 = 45°$, $\omega_0 = 1$ Grad/s, $\omega_{p0} = 150$ Grad/s, $\omega_{b0} = 100$ Grad/s, and $k_v = v/\omega_{p0} = 0.01$.

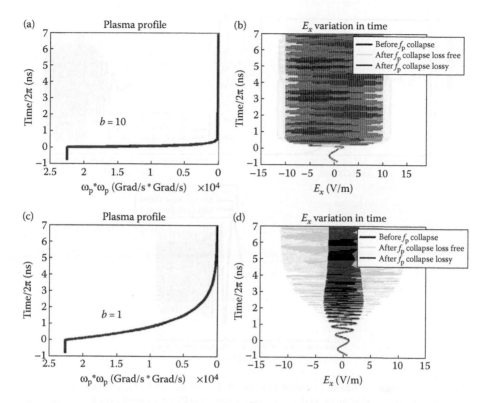

FIGURE Q.7 Evolution of the electric field intensity as lossy plasma collapses under constant external magnetic field in the cavity. $\phi_0 = 45°$, $\omega_0 = 1$ Grad/s, $\omega_{p0} = 150$ Grad/s, $\omega_{b0} = 100$ Grad/s, and $\nu/\omega_{p0} = 0.01$. (a) $b = 10$ and (c) $b = 1$. (b) E_x variation for $b = 10$ and (d) E_x variation for $b = 1$.

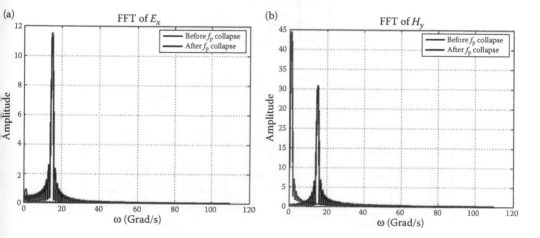

FIGURE Q.8 FFT of the (a) electric and (b) magnetic fields before and after the collapse of the plasma under a steady magnetic field. $b = 10$, $\phi_0 = 45°$, $\omega_0 = 1$ Grad/s, $\omega_{p0} = 150$ Grad/s, $\omega_{b0} = 100$ Grad/s, $m = 1$, $z = d/2$ (for E), and $z \approx 0$ (for H).

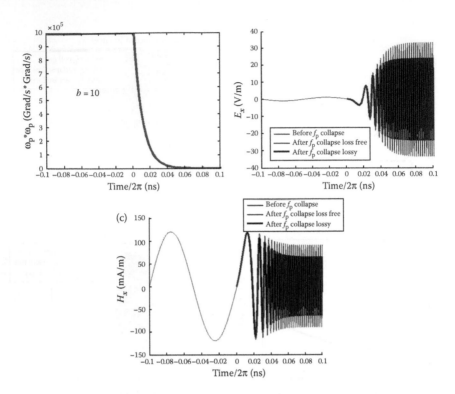

FIGURE Q.9 Higher frequency-upshift ratio of 50: $b = 10$, $\phi_0 = 90°$, $\omega_0 = 10$ Grad/s, $\omega_{p0} = 1000$ Grad/s, $\omega_{b0} = 50$ Grad/s, and $k_\nu = \nu/\omega_{p0} = 0.01$. (a) Plasma profile, (b) E_x variation in time, and (c) H_x variation in time.

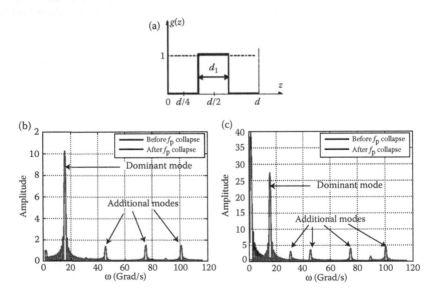

FIGURE Q.10 FFT of the electric and magnetic fields before and after the collapse of the plasma space profile under constant magnetic field. $b = 10$, $\phi_0 = 45°$, $\omega_0 = 1$ Grad/s, $\omega_{p0} = 150$ Grad/s, $\omega_{b0} = 100$ Grad/s, $m = 1$, $z = d/4$ (for E), $z \approx d/4$ (for H), and $d_1/d = 0.3$. (a) Spatial profile, (b) FFT of E_x, and (c) FFT of H_y.

$$\frac{H_n}{H_0} = \frac{(\omega_n \mp \omega_b)(\omega_n + \omega_0) - \omega_p^2}{\prod_{m=1,m\neq n}^{3}(\omega_n - \omega_m)}, \quad n = 1,2,3. \tag{D.3}$$

In the above equation, ω_b is the electron gyrofrequency and the incident wave variables are denoted by subscript zero. A negative value for ω_n [8] indicates that the wave is a reflected wave, that is, it propagates in a direction opposite to that of the source (incident) wave. If the source wave is right-going, then the reflected wave will be left-going.

The following approximations can be made when $\omega_b \ll \omega_0$:

$$\omega_1 = \sqrt{\omega_0^2 + \omega_p^2} \pm \frac{\omega_b}{2}\frac{\omega_p^2}{\omega_0^2 + \omega_p^2}, \quad \omega_b \ll \omega_0, \tag{D.4a}$$

$$\omega_2 = -\sqrt{\omega_0^2 + \omega_p^2} \pm \frac{\omega_b}{2}\frac{\omega_p^2}{\omega_0^2 + \omega_p^2}, \quad \omega_b \ll \omega_0, \tag{D.4b}$$

$$\omega_3 = \pm\,\omega_b\frac{\omega_0^2}{\omega_0^2 + \omega_p^2}, \quad \omega_b \ll \omega_0, \tag{D.4c}$$

$$\frac{E_3}{E_0} = \pm\,\omega_b\frac{\omega_0\omega_p^2}{(\omega_0^2 + \omega_p^2)^2}, \quad \omega_b \ll \omega_0, \tag{D.5a}$$

$$\frac{H_3}{H_0} = \frac{E_3}{E_0} + \frac{\omega_p^2}{(\omega_0^2 + \omega_p^2)}, \quad \omega_b \ll \omega_0, \tag{D.5b}$$

$$\frac{S_3}{S_0} = \frac{E_3 H_3}{E_0 H_0} = \pm\frac{\omega_0\omega_b\omega_p^4}{(\omega_0^2 + \omega_p^2)^3}, \quad \omega_b \ll \omega_0. \tag{D.5c}$$

The first terms on the right-hand side of Equations D.4a and D.4b are the frequencies for the isotropic case. The approximations are quite good (error less than 5%) for ω_b/ω_0 as high as 0.5.

D.4 X Waves

X-incidence refers to the case where the electric field of the incident wave is in the x-direction, and the static magnetic field is in the y-direction. The medium switching in this case creates four extraordinary waves labeled as X1, X2, X3, and X4. X1 and X2 are transmitted waves. The frequencies of these waves are

$$\omega_1 = \sqrt{A + \sqrt{A^2 - B}}, \tag{D.6a}$$

$$\omega_2 = \sqrt{A - \sqrt{A^2 - B}},$$
(D.6b)

$$\omega_3 = -\omega_1,$$
(D.6c)

$$\omega_4 = -\omega_2,$$
(D.6d)

where

$$A = \frac{(\omega_0^2 + \omega_b^2 + 2\omega_p^2)}{2},$$
(D.6e)

$$B = \omega_p^4 + \omega_0^2(\omega_p^2 + \omega_b^2).$$
(D.6f)

The fields for this case are given in Ref. [11]. The following approximations can be made when $\omega_b \ll \omega_0$:

$$\omega_1 = \sqrt{\omega_0^2 + \omega_p^2} + \frac{\omega_b^2}{2} \frac{\omega_p^2}{\omega_0^2 \sqrt{\omega_0^2 + \omega_p^2}}, \qquad \omega_b \ll \omega_0,$$
(D.7a)

$$\omega_2 = \sqrt{\omega_b^2 + \omega_p^2} - \frac{\omega_b^2}{2} \frac{\omega_p^2}{\omega_0^2 \sqrt{\omega_b^2 + \omega_p^2}}, \qquad \omega_b \ll \omega_0.$$
(D.7b)

The approximations are quite good (error less than 5%) for ω_b/ω_0 as high as 0.5. A further approximation for ω_2 can be made if ω_b is also much less than ω_p

$$\omega_2 = \omega_p + \frac{1}{2} \frac{\omega_b^2}{\omega_0^2} \frac{\omega_0^2 - \omega_p^2}{\omega_p}, \qquad \omega_b \ll \omega_0 \quad \text{and} \quad \omega_b \ll \omega_p.$$
(D.7c)

The first terms on the right-hand side of Equations D.7a and D.7c are the frequencies for the isotropic case.

D.5 Frequency-Shifting Characteristics of Various Waves

The shift ratio and the efficiency of the frequency-shifting operation can be controlled by the strength and the direction of the static magnetic field. It is this aspect that is emphasized in this paper. The results are presented by normalizing all frequency variables with respect to the incident wave frequency (source frequency) ω_0. This normalization is achieved by taking $\omega_0 = 1$ in numerical computations. Here, $\omega_b = qB_0/m$ is the electron gyrofrequency, ω_0 is the frequency (angular) of the source (incident) wave, ω_n is the frequency

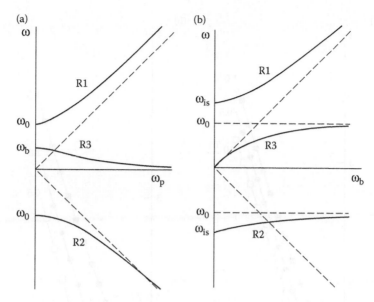

FIGURE D.3 Frequency shifting of R waves. Sketch of (a) ω versus ω_p and (b) ω versus ω_b. $\omega_{is} = (\omega_0^2 + \omega_p^2)^{1/2}$. Here ω_0 is the incident wave frequency, ω_p is the plasma frequency, and ω_b is the electron gyrofrequency.

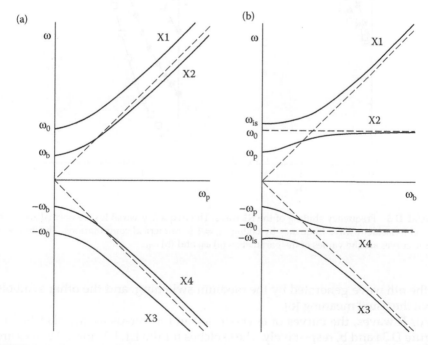

FIGURE D.4 Frequency shifting of the X wave. Sketch of (a) ω versus ω_p and (b) ω versus ω_b. The symbols are explained in Figure D.3.

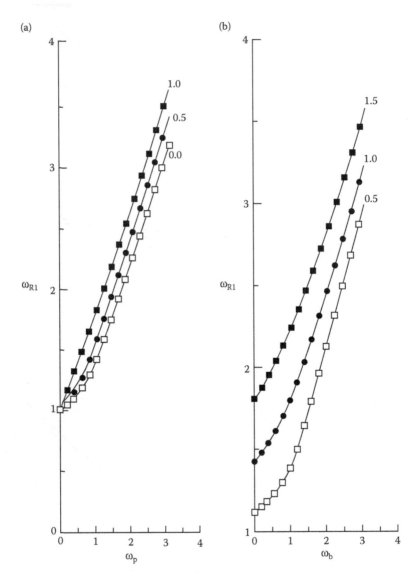

FIGURE D.5 Frequency shifting of the R1 wave. The frequency variables are normalized with respect to incident wave frequency by taking $\omega_0 = 1$ in numerical computations. The numbers on the curves are the values of the parameters (a) ω_b and (b) ω_p.

of the nth wave generated by the medium switching, and the other symbols have the usual meaning [8].

For R waves, the curves of ω versus ω_p and ω versus ω_b are sketched in Figure D.3a and b, respectively. The sketches for the L1, L2, and L3 waves are mirror images (with respect to the horizontal axis) of those of R2, R1, and R3 waves, respectively. Figure D.4 shows the sketches for the X waves.

TABLE D.1

R1 Wave Shift Ratio and Power Density for Two Sets of (ω_p, ω_b)

ω_0	ω_p	ω_b	E_1/E_0	H_1/H_0	$S_1/S_0(\%)$	ω_{R1}/ω_0
1	0.5	0.5	0.83	0.69	57.3	1.20
1	2.0	2.0	0.39	0.18	7.0	3.33

In Figure D.5, the results are presented for the R1 wave: the values on the vertical axis give the frequency-shift ratio since the frequency variables are normalized with respect to ω_0. This is an upshifted wave and both ω_p and ω_b improve the shift ratio. From this figure it appears that, by a suitable choice

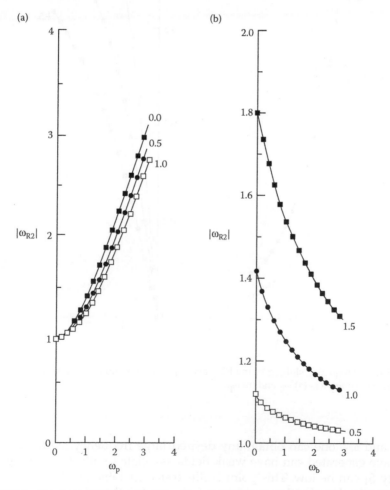

FIGURE D.6 Frequency shifting of the R2 wave, $\omega_0 = 1$. This is a reflected wave and the vertical axis gives the magnitude of ω_{R2}. The numbers on the curves are the values of the parameters (a) ω_b and (b) ω_p.

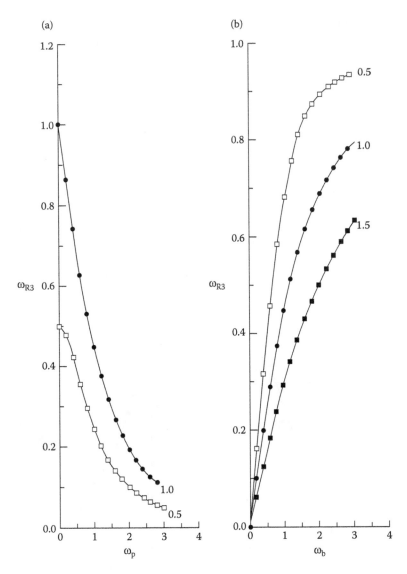

FIGURE D.7 Frequency shifting of the R3 wave, $\omega_0 = 1$. The numbers on the curves are the values of the parameters (a) ω_b and (b) ω_p.

of ω_p and ω_b, one can obtain any desired large frequency shift. However, the wave generated can have weak fields associated with it and the power density S_1 can be low. This point is illustrated in Table D.1 by considering two sets of values for the parameters (ω_p, ω_b). For the set (0.5, 0.5), the shift ratio is 1.2, but the power density ratio S_1/S_0 is 57.3% where, as for the set (2.0, 2.0), the shift ratio is 3.33, but the power density ratio is only 7%. Similar

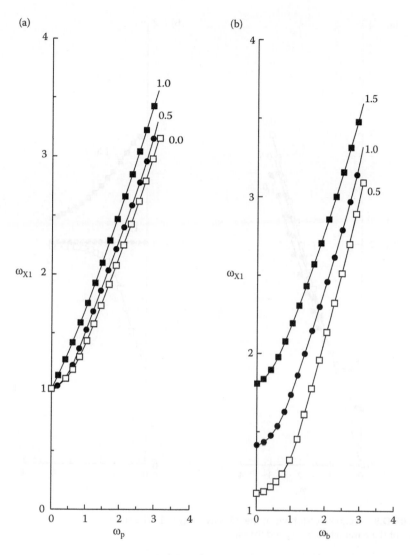

FIGURE D.8 Frequency shifting of the X1 wave. $\omega_0 = 1$. The numbers on the curves are the values of the parameters (a) ω_b and (b) ω_p.

remarks apply to other waves. A detailed report on this aspect can be found elsewhere.*

Figure D.6 shows the magnitude of the frequency of the R2 wave. This is a reflected wave. The L1 wave is a transmitted wave with the same frequency characteristics. This is an upshifted wave, and the shift ratio increases with ω_p but decreases with ω_b.

---

* See Chapter 7 of this book.

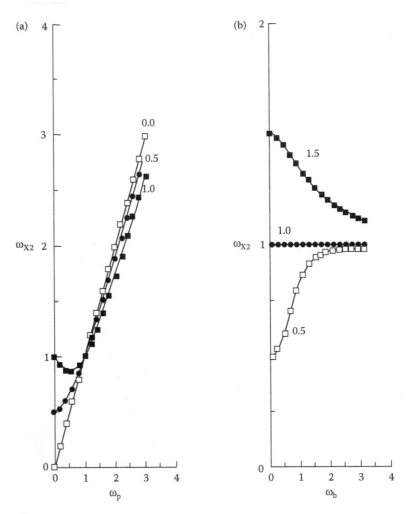

FIGURE D.9 Frequency shifting of the X2 wave. $\omega_0 = 1$. The numbers on the curves are the values of the parameters (a) ω_b and (b) ω_p.

The R3 wave (Figure D.7) is a transmitted wave that is downshifted. The shift ratio decreases with ω_p and increases with ω_b. When $\omega_b = 0$, ω_{R3} becomes zero. The electric field E_3 becomes zero, and the magnetic field degenerates to the wiggler magnetic field [8].

Figures D.8 and D.9 describe the characteristics of X waves. The X1 wave is a transmitted wave, and the shift ratio increases with ω_p and ω_b. The X2 wave is downshifted for $\omega_p < \omega_0$ and upshifted for $\omega_p > \omega_0$. The shift ratio increases with ω_b for $\omega_p < \omega_0$ and decreases with ω_b for $\omega_p > \omega_0$. Attention is drawn to the curve with the parameter $\omega_b = 0$ in Figure D.9a. For this case, $\omega_{X2} = \omega_p$. However, it can be shown [11] that all the fields associated with this wave are zero. From Figure D.9b, the shift ratio is unity when $\omega_p = \omega_0$ for all values of ω_b, but the power density increases [11] with ω_b.

References

1. Jiang, C. L., Wave propagation and dipole radiation in a suddenly created plasma, *IEEE Trans. Ant. Prop.*, AP-23, 83, 1975.
2. Rader, M., Dyer, F., Matas, A., and Alexeff, I., Plasma-induced frequency shifts in microwave beams, In: *Conf. Rec. Abstracts, IEEE Int. Conf. Plasma Sci.*, Oakland, CA, p. 171, 1990.
3. Kuo, S. P., Zhang, Y. S., and Ren, A. Q., Frequency up-conversion of microwave pulse in a rapidly growing plasma, In: *Conf. Rec. Abstracts, IEEE Int. Conf. Plasma Sci.*, Oakland, CA, p. 171, 1990.
4. Joshi, C. S., Clayton, C. E., Marsh, K., Hopkins, D. B., Sessler, A., and Whittum, D., Demonstration of the frequency upshifting of microwave radiation by rapid plasma creation, *IEEE Trans. Plasma Sci.*, 18, 814, 1990.
5. Wilks, S. C., Dawson, J. M., and Mori, W. B., Frequency up-conversion of electromagnetic radiation with use of an overdense plasma, *Phys. Rev. Lett.*, 61, 337, 1988.
6. Kalluri, D. K., On eflection from a suddenly created plasma half-space: Transient solution, *IEEE Trans. Plasma Sci.*, 16, 11, 1988.
7. Kalluri, D. K. and Goteti, V. R., Frequency shifting of electromagnetic radiation by sudden creation of a plasma slab, *J. Appl. Phys.*, 72, 4575, 1992.
8. Kalluri, D. K., Effect of switching a magnetoplasma medium on a traveling wave: Longitudinal propagation, *IEEE Trans. Ant. Prop.*, AP-37, 1638, 1989.
9. Booker, H. G., *Cold Plasma Waves*, Kluwer, Hingham, MA, 1984.
10. Kalluri, D. K., Effect of switching a magnetoplasma medium on a travelling wave: Conservation law for frequencies of newly created waves, In: *Conf. Rec. Abstracts, IEEE Int. Conf. Plasma Sci.*, Oakland, CA, p. 129, 1990.
11. Goteti, V. R. and Kalluri, D. K., Wave propagation in a switched magnetoplasma medium: Transverse propagation, *Radio Sci.*, 25, 61, 1990.
12. Kalluri, D. K. and Goteti, V. R., Damping rates of waves in a switched magnetoplasma medium: Longitudinal propagation, *IEEE Trans. Plasma Sci.*, 18, 797, 1990.

Appendix E

*Frequency Upshifting with Power Intensification of a Whistler Wave by a Collapsing Plasma Medium**

Dikshitulu K. Kalluri

E.1 Introduction

Frequency upshifting using time-varying *magnetoplasmas* has been investigated [1–6] recently. In all these cases, the electron density profile was a monotonically increasing function of time. A fast profile was modeled as sudden creation of a plasma medium. The slow profile problem was solved using the WKB method. The calculations showed that the power density in the frequency-shifted waves was less than the power density of the source wave.

It is known [7] that across a temporal discontinuity in the properties of the medium, while the energy is conserved, the sum of the power densities of the new waves may be less than or greater than the power density of the source wave. The case of "greater power density in the new wave" with upshifted frequency seems to occur under the following circumstances: A right circularly polarized EMW (R wave) is propagating in the whistler mode in an unbounded magnetoplasma medium. At $t = 0$, the electron density starts decreasing, and for $t > 0$, the electron density profile is a monotonically decreasing function of time.

E.2 Sudden Collapse

An R wave is assumed to be propagating in the positive z-direction in a spatially unbounded magnetoplasma medium. The electric and magnetic fields

---

* © 1996 American Institute of Physics. Reprinted from *J. Appl. Phys.*, 79, 3895–3899, 1996. With permission.

of this wave, named as the source wave, are

$$\bar{e}_i(z,t) = (\hat{x} - j\hat{y})E_0 \exp[j(\omega_0 t - \beta_0 z)], \qquad (E.1a)$$

$$\bar{h}_i(z,t) = (j\hat{x} + \hat{y})H_0 \exp[j(\omega_0 t - \beta_0 z)], \qquad (E.1b)$$

where ω_0 is the frequency of the incident wave and

$$\beta_0 = \frac{\omega_0 n_0}{c}, \qquad (E.2a)$$

$$n_0 = \sqrt{1 - \frac{\omega_{p0}^2/\omega_0^2}{1 - \omega_b/\omega_0}}. \qquad (E.2b)$$

Here, c is the velocity of light in free space, ω_{p0} is the plasma frequency, ω_b is the electron gyrofrequency, and n_0 is the refractive index of the magnetoplasma medium existing prior to $t = 0$. Assume that at $t = 0$ the electron density undergoes a step change and the corresponding new plasma frequency for $t > 0$ is ω_{p1}. The new refractive index n_1 is given by

$$n_1 = \sqrt{1 - \frac{\omega_{p1}^2/\omega^2}{1 - \omega_b/\omega}}, \qquad (E.3)$$

where ω is the frequency of the new waves created by the switching.

E.2.1 Frequencies of the New Waves

Since the wave number is conserved across the temporal discontinuity [7,8], we obtain

$$\omega^2 n_1^2 = \omega_0^2 n_0^2. \qquad (E.4)$$

Equation E.4 is a cubic in ω. Let us denote the frequencies of the three new waves generated by the switching action by ω_1, ω_2, and ω_3. A negative value for ω indicates a reflected wave traveling in a direction opposite to that of the source wave.

A fast collapse of the plasma may be idealized as a sudden collapse and the frequencies of the new waves in this case are obtained by substituting $\omega_{p1} = 0$ in Equation E.4, giving

$$\omega_1 = n_0 \omega_0, \quad \omega_2 = -n_0 \omega_0, \quad \text{and} \quad \omega_3 = \omega_b. \qquad (E.5)$$

E.2.2 Fields of the New Waves

The fields of the new waves may be determined by imposing the initial conditions that the E and H fields are continuous over the temporal discontinuity:

$$E_1 + E_2 + E_3 = E_0, \tag{E.6a}$$

$$H_1 + H_2 + H_3 = H_0. \tag{E.6b}$$

The third equation, which should come from the initial condition of continuity of the velocity field, is not obvious since it is assumed that the plasma has suddenly collapsed. We have recently completed a study wherein is assumed a gradual collapse of the electron density; that is, $\omega_p(t)$ is a monotonically decreasing function of time. The study clearly showed that the frequency of the third wave tends to ω_b and the fields E_3 and H_3 tend to 0 as t goes to infinity. The third wave that is supported only by a magnetoplasma medium vanishes in free space. Thus take $E_3 = H_3 = 0$ in Equation E.6. Noting that $H_1 = E_1/120\,\pi$, $H_2 = -E_2/120\,\pi$, and $H_0 = E_0 n_0/120\,\pi$, we obtain

$$\frac{E_1}{E_0} = \frac{(1+n_0)}{2}, \qquad \frac{E_2}{E_0} = \frac{(1-n_0)}{2}, \tag{E.7a}$$

$$\frac{H_1}{H_0} = \frac{(1+n_0)}{2n_0}, \qquad \frac{H_2}{H_0} = \frac{-(1-n_0)}{2n_0}, \tag{E.7b}$$

$$\frac{S_1}{S_0} = \frac{(1+n_0)^2}{4n_0}, \qquad \frac{S_2}{S_0} = \frac{-(1-n_0)^2}{4n_0}. \tag{E.7c}$$

Here the symbol S stands for the power density in the wave and the subscript indicates the appropriate wave.

E.2.3 Choice of the Parameters

In the whistler mode [9,10], that is, $\omega_0 < \omega_b$, n_0 is greater than 1 and therefore $\omega_1/\omega_0 > 1$, $E_1/E_0 > 1$, and $S_1/S_0 > 1$. Thus a frequency-upshifted wave with power intensification is obtained.

The frequency upshift is large near cyclotron resonance, that is, when ω_0 is approximately equal to ω_b, but in this neighborhood the effect of collisions cannot be ignored. There is a tradeoff in the choices of ω_p/ω_0 and ω_b/ω_0. For a low-loss magnetoplasma of collision frequency ν, the refractive index is complex and given by [10]

$$\tilde{n}_0 = n_0 - j\chi_0 = \sqrt{1 - \frac{\omega_{p0}^2/\omega_0^2}{1 - \omega_b/\omega_0 - j\nu/\omega_0}}. \tag{E.8}$$

As ω_b/ω_0 approaches 1, n_0 increases, but the attenuation constant $\alpha = \chi_0(\omega_0/c)$ also increases and therefore it will be difficult to maintain the source wave before $t = 0$. For a sample calculation, let us take $v/\omega_0 = 0.01$ and $\omega_p/\omega_0 = 0.8$. For $\omega_b/\omega_0 = 1.2$, $n_0 = 2.05$ and $\alpha = 0.039(\omega_0/c)$. For $\omega_b/\omega_0 = 1.1$, $n_0 = 2.71$ but $\alpha = 0.117(\omega_0/c)$. For a given ω_b/ω_0, as ω_p/ω_0 increases, n_0 increases, but the attenuation constant also increases. Again for a sample calculation, let us take $v/\omega_0 = 0.01$ and $\omega_b/\omega_0 = 1.2$. For $\omega_p/\omega_0 = 0.5$, $n_0 = 1.5$ and $\alpha = 0.021(\omega_0/c)$. For $\omega_p/\omega_0 = 0.8$, $n_0 = 2.05$, but $\alpha = 0.039(\omega_0/c)$.

E.2.4 Eckersley Approximation

The refractive index is also large when $\omega_0 \ll \omega_b \sim \omega_{p0}$ and from Equation E.2b

$$n_0 \approx n_w = \frac{\omega_{p0}}{(\omega_0\omega_b)^{1/2}}, \quad \omega_0 \ll \omega_b \sim \omega_{p0}, \tag{E.9a}$$

and from Equations E.7a and E.7c

$$\frac{E_1}{E_0} \approx \frac{n_w}{2}, \quad \omega_0 \ll \omega_b \sim \omega_{p0}, \tag{E.9b}$$

$$\frac{H_1}{H_0} \approx \frac{1}{2}, \quad \omega_0 \ll \omega_b \sim \omega_{p0}, \tag{E.9c}$$

$$\frac{S_1}{S_0} \approx \frac{n_w}{4}, \quad \omega_0 \ll \omega_b \sim \omega_{p0}. \tag{E.9d}$$

In this approximation, called the Eckersley approximation [9], the phase and the group velocities are proportional to the square root of the signal frequency. Eckersley used this approximation to explain the phenomenon of whistlers. Whistlers are naturally occurring EMWs (due to lightning) having frequencies in the audio band. They propagate from one hemisphere to the other through the plasmasphere along the Earth's magnetic field. Since the group velocity of these waves increases with frequency, the low frequencies arrive later and give rise to the descending tone.

A physical explanation for the power intensification may be given based on energy consideration. The energy in the whistler mode, under the Eckersley approximation, is predominantly the magnetic energy due to plasma current and its density w_m is given by

$$w_m = \frac{1}{2}\varepsilon_0 E_0^2 n_w^2. \tag{E.10}$$

After the plasma collapses, the plasma current collapses and the magnetic energy due to plasma current are converted into wave electric and magnetic energy giving rise to the frequency-upshifted waves with enhanced electric field and power density. The energy balance equation may be written as

$$\frac{1}{2}\varepsilon_0 E_1^2 + \frac{1}{2}\mu_0 H_1^2 + \frac{1}{2}\varepsilon_0 E_2^2 + \frac{1}{2}\mu_0 H_2^2 = 2\varepsilon_0 E_1^2 = \frac{1}{2}\varepsilon_0 E_0^2 n_w^2. \qquad (E.11)$$

The intermediate step in Equation E.11 is explained as follows: the electric and magnetic field energy densities are equal for each of the waves. Moreover, the amplitudes of the electric fields E_1 and E_2 have approximately the same magnitude. Thus the result $E_1/E_0 = n_w/2$ is obtained from energy balance, in agreement with Equation E.9b.

The power intensification calculation may be made from a different viewpoint if the following least efficient way of establishing the source wave in the magnetoplasma medium is considered. Let us pose the following idealized problem. For $t < 0$, let the half-space $z < 0$ be free space in which an R wave of frequency ω_0 and fields E_{-1} and H_{-1} is propagating along the direction of the z-axis. The half-space $z > 0$ is a magnetoplasma of plasma frequency ω_{p0} and electron gyrofrequency ω_b. At $t = 0$, let the plasma suddenly collapse giving rise to the frequency-upshifted wave. It is easily shown that $E_0/E_{-1} = 2/(1 + n_0)$ and from Equation E.7a, $E_1/E_{-1} = 1$. Similarly $H_1/H_{-1} = 1$.

In the above calculation, losses are neglected but we make the point that even in the less efficient case a frequency transformation without a power reduction can be obtained.

The key to frequency upshifting using the collapse of the magnetoplasma is the establishment of the source wave propagation in the pass band of the magnetoplasma with the refractive index $n_0 > 1$. Thus one can use an X wave [10] rather than the R wave in the frequency range $\omega_{p0} < \omega_0 < \omega_{uh}$ where ω_{uh} is the upper hybrid frequency. The X wave is the extraordinary mode of transverse propagation in a magnetized plasma. The direction of phase propagation of this wave is perpendicular to the direction of the imposed magnetic field. The electrical field is elliptically polarized in a plane that contains the direction of propagation but is perpendicular to the imposed magnetic field. The wave magnetic field is parallel to the imposed magnetic field.

E.2.5 Experimental Feasibility

A sudden collapse of the plasma on a half-space is an idealization of the problem and is difficult to achieve experimentally. If a slab of finite width [11] instead of a half-space is considered, the new wave will be an intensified electromagnetic pulse at the upshifted frequency.

On the other hand, an experimental technique analogous to the ionization front [6,12,13] may be used. In this case, a deionization (recombination) front is used, that is, a process of deionization [13] that moves with a deionization front velocity V_0 close to c is created. Work is in progress in this regard.

E.3 Slow Decay

A WKB type of solution can be obtained for the case of slow decay; that is, $\omega_p^2(t)$ is a slowly decaying function of time. The technique is similar to that used in Refs. [3,14,15] and therefore much of the mathematical detail is omitted here. The frequencies of the three waves created by the switching action are the roots of the cubic:

$$\omega^3 - \omega^2\omega_b - [\omega_0^2 n_0^2 + \omega_p^2(t)]\omega + \omega_0^2\omega_b n_0^2 = 0. \tag{E.12}$$

This cubic has two positive real roots and one negative real root. The amplitudes of the fields associated with these waves are obtained by solving the initial value problem. The continuity of the electric and magnetic fields provides two initial conditions. The third initial condition is the continuity of the velocity field at $t = 0$. The decay of the plasma with time is assumed to take place in a process of sudden capture of a number of free electrons as time passes. The velocity of the remaining free electrons is assumed to be unaffected.

The amplitudes of two of the waves are of the order of the time derivative of the $\omega_p^2(t)$ curve and is thus negligible. The amplitude of the wave whose initial frequency is ω_0 is significant and this wave will be called the MSW. Its fields $e_1(z, t)$ and $h_1(z, t)$ may be expressed as

$$e_1(z, t) = E_1(t)\exp\left[j\left(\int_0^t \omega_1(t)\,dt - \beta_0 z\right)\right] \tag{E.13a}$$

and

$$h_1(z, t) = H_1(t)\exp\left[j\left(\int_0^t \omega_1(t)\,dt - \beta_0 z\right)\right], \tag{E.13b}$$

where

$$\frac{E_1(t)}{E_0} = \frac{\omega_b - \omega_1(t)}{\omega_b - \omega_0}\frac{\omega_1(t)}{\omega_0}\sqrt{\frac{(\omega_b\omega_0^2 n_0^2 + \omega_b\omega_0^2 - 2\omega_0^3)}{(\omega_b\omega_0^2 n_0^2 + \omega_b\omega_1^2 - 2\omega_1^3)}}, \tag{E.14a}$$

$$\frac{H_1(t)}{H_0} = \frac{E_1(t)}{E_0}\frac{n(t)}{n_0}, \tag{E.14b}$$

$$n(t) = \sqrt{\left(1 - \frac{\omega_p^2(t)/\omega_1^2(t)}{1 - \omega_b/\omega_1(t)}\right)}, \tag{E.14c}$$

and

$$\frac{S_1(t)}{S_0} = \frac{E_1(t)}{E_0} \frac{H_1(t)}{H_0}.$$ (E.15)

After the plasma completely collapses, that is, as $t \to \infty$, it is obtained:

$$\omega_1(t \to \infty) = n_0\omega_0,$$ (E.16a)

$$\frac{E_1(t \to \infty)}{E_0} = \frac{\omega_b - n_0\omega_0}{\omega_b - \omega_0} \sqrt{\frac{\omega_b(1 + n_0^2) - 2\omega_0}{(2\omega_b - 2n_0\omega_0)}}.$$ (E.16b)

E.3.1 Eckersley Approximation

Again, when $\omega_0 \ll \omega_b \sim \omega_{p0}, n_0 \approx n_w$, and n_w is given by Equation E.9a. The final values for the frequency and the electric field are

$$\omega_1(t \to \infty) \approx n_w\omega_0,$$ (E.17a)

$$\frac{E_1(t \to \infty)}{E_0} \approx \frac{n_w}{\sqrt{2}} \left(1 - \frac{\omega_{p0}}{2\omega_b}\sqrt{\frac{\omega_0}{\omega_b}}\right) \approx \frac{n_w}{\sqrt{2}},$$ (E.17b)

$$\frac{H_1(t \to \infty)}{H_0} \approx \frac{1}{\sqrt{2}} \left(1 - \frac{\omega_{p0}}{2\omega_b}\sqrt{\frac{\omega_0}{\omega_b}}\right) \approx \frac{1}{\sqrt{2}},$$ (E.17c)

$$\frac{S_1(t \to \infty)}{S_0} \approx \frac{n_w}{2} \left(1 - \frac{\omega_{p0}}{\omega_b}\sqrt{\frac{\omega_0}{\omega_b}}\right) \approx \frac{n_w}{2}.$$ (E.17d)

The middle set of equations in Equation E.17 is the first set of approximations and the last set of approximations in Equation E.17 are obtained by approximating the middle set further.

A physical explanation is obtained from the energy balance equation:

$$\frac{1}{2}\varepsilon_0 E_1^2 + \frac{1}{2}\mu_0 H_1^2 = \varepsilon_0 E_1^2 = \frac{1}{2}\varepsilon_0 E_0^2 n_w^2.$$ (E.18)

In the slow decay case under discussion, the fields of the second and third waves are insignificant; also the free space electric and magnetic energy densities are equal and we obtain for E_1/E_0 the approximate expression given in Equation E.17b from the energy balance Equation E.18.

E.4 An Illustrative Example

Figure E.1 illustrates the results for an exponential decay profile whose time constant is $(1/b)$:

$$\omega_p^2(t) = \omega_{p0}^2 e^{-bt}.$$ (E.19)

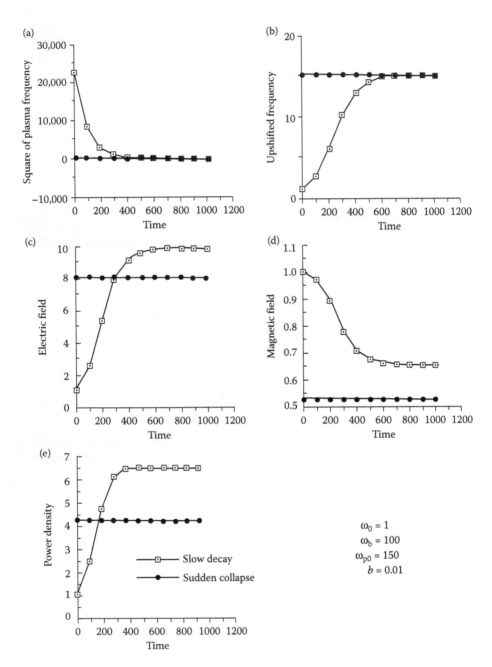

FIGURE E.1   Frequency upshifting with power intensification of a whistler wave. The values on the vertical axes are normalized with respect to source wave quantities. The source frequency ω_0 is taken as 1. The horizontal timescale is normalized with respect to the period of the source wave. The parameters are such that Eckersley approximation is applicable.

All variables are normalized with reference to the source wave quantities. The source wave frequency ω_0 is taken to be 1 so that ω_1 gives the frequency upshift ratio. The parameters ω_b and ω_{p0} are assigned the values 100 and 150, respectively. The parameters are such that the Eckersley approximation is valid. A value of 0.01 is assigned to b. The parameter describes a slow decay since $100/2\pi$ cycles of the source wave are accommodated in one time constant. The independent variable, time, is normalized with respect to the period of the source wave.

The results shown in Figure E.1 for slow decay are numerically computed from Equations E.12, E.14, and E.15 and the following values are obtained at $t = 1000$: $\omega_1/\omega_0 = 15.10$, $E_1/E_0 = 9.96$, $H_1/H_0 = 0.657$, and $S_1/S_0 = 6.57$. Approximate values computed from Equation E.17 give for $t \to \infty$: $\omega_1/\omega_0 \approx n_w \approx 15$, $E_1/E_0 \approx 9.81 \approx 10.6$, $H_1/H_0 \approx 0.654 \approx 0.707$, and $S_1/S_0 \approx 6.38 \approx 7.5$. Here the first set of approximate values are from the middle set of equations in Equation E.17 and the second set of approximate values are from the last set of equations in Equation E.17.

Figure E.1 also shows, for comparison, horizontal lines that are the results for sudden collapse. These are obtained from Equations E.2b and E.7 and are $n_0 = \omega_1/\omega_0 = 15.11$, $E_1/E_0 = 8.06$, $H_1/H_0 = 0.533$, and $S_1/S_0 = 4.294$. Approximate values obtained from Equation E.9 are: $n_w \approx 15$, $E_1/E_0 \approx 7.5$, $H_1/H_0 \approx 1/2$, and $S_1/S_0 \approx 3.75$.

The approximate results are in agreement with numerically computed results confirming that Equations E.9 and E.17 may be used as simple approximate expressions under the Eckersley approximation.

E.5 Conclusion

The following conclusions can be drawn:

1. The frequency of a source wave propagating in a whistler mode gets upshifted when the ionization collapses. The final upshifted frequency is the source frequency multiplied by the refractive index of the magnetoplasma medium before the collapse begins and is essentially independent of the rate of collapse of the ionization.

2. When the source frequency is much less than the electron gyrofrequency, the refractive index n_w is quite large and the electric field is intensified by a factor of $n_w/2$ for the case of sudden collapse and by a factor of $n_w/\sqrt{2}$ for the case of slow decay. The corresponding intensification factors for the power density are $n_w/4$ and $n_w/2$.

The principle of frequency upshifting using plasmas permits the generation of signals in easily obtainable bands and upshifts them into frequency bands not easily obtainable by other methods. In addition, the output radiation can be tuned by controlling the plasma density. However, the strength of the output radiation is reduced in the process. The technique outlined in this article, which involves using a decaying plasma in the presence of a static magnetic field, gives an upshifted as well as intensified signal. Also, the strength of the static magnetic field gives one more controlling parameter for tuning the frequency of the output radiation.

References

1. Kalluri, D. K., Effect of switching a magnetoplasma medium on a traveling wave: Longitudinal propagation, *IEEE Trans. Ant. Prop.*, AP-37, 1638, 1989.
2. Kalluri, D. K., Frequency shifting using magnetoplasma medium: Flash ionization, *IEEE Trans. Plasma Sci.*, 21, 77, 1993.
3. Kalluri, D. K., Goteti, V. R., and Sessler, A. M., WKB solution for wave propagation in a time-varying magnetoplasma medium: Longitudinal propagation, *IEEE Trans. Plasma Sci.*, 21, 70, 1993.
4. Goteti, V. R. and Kalluri, D. K., Wave propagation in a switched magnetoplasma medium: Transverse propagation, *Radio Sci.*, 25, 61, 1990.
5. Madala, S. R. V. and Kalluri, D. K., Longitudinal propagation of low-frequency waves in a switched magnetoplasma medium, *Radio Sci.*, 28, 121, 1993.
6. Lai, C. H. and Katsouleas, T. C., Frequency upshifting by an ionization front in a magnetized plasma, *IEEE Trans. Plasma Sci.*, 21, 45, 1993.
7. Auld, B. A., Collins, J. H., and Zapp, H. R., Signal processing in a nonperiodically time-varying magnetoelastic medium, *Proc. IEEE*, 56, 258, 1968.
8. Kalluri, D. K., Effect of switching a magnetoplasma medium on a traveling wave: Conservation law for frequencies of newly created waves, In: *Conf. Rec. Abstracts, IEEE Int. Conf. Plasma Sci.*, Oakland, CA, p. 129, 1990.
9. Booker, H. G., *Cold Plasma Waves*, p. 77, Kluwer, Hingham, Boston, MA, 1984.
10. Heald, M. A. and Wharton, C. B., *Plasma Diagnostics with Microwaves*, p. 12, Wiley, New York, 1965.
11. Kalluri, D. K. and Goteti, V. R., Frequency shifting of electromagnetic radiation by sudden creation of a plasma slab, *J. Appl. Phys.*, 72, 4575, 1992.
12. Savage, Jr., R. L., Brogle, R. P., Mori, W. B., and Joshi, C., Frequency upshifting and pulse compression via underdense relativistic ionization fronts, *IEEE Trans. Plasma Sci.*, 21, 5, 1993.
13. Lampe, M. and Ott, E., Interaction of electromagnetic waves with a moving ionization front, *Phy. Fluids*, 21, 42, 1978.
14. Lee, J. H. and Kalluri, D. K., Modification of an electromagnetic wave by a time-varying switched mangetoplasma medium: Transverse propagation, *IEEE Trans. Plasma Sci.*, 26, 1–6, 1998.
15. Kalluri, D. K. and Goteti, V. R., WKB solution for wave propagation in a decaying plasma medium, *J. Appl. Phys.*, 66, 3472, 1989.

Appendix F

Conversion of a Whistler Wave into a Controllable Helical Wiggler Magnetic Field*

Dikshitult K. Kalluri

F.1 Introduction

Coherent radiation generated in a free electron laser (FEL) [1] is due to the interaction of an electron beam with a wiggler magnetic field, which is a spatially varying static (zero frequency) magnetic field. Linear or helical wigglers can be constructed from electromagnets or permanent magnets. The B field of a helical wiggler can be written as

$$\bar{B}_W(z) = (j\hat{x} + \hat{y})B_W \cos(k_W z), \tag{F.1}$$

where k_W is the wave number of the wiggler. The wavelength λ of the output radiation is given by [1]

$$\lambda = \frac{\lambda_W}{2\gamma_0^2}\left(1 + \frac{K_W^2}{2}\right), \tag{F.2a}$$

where

$$\lambda_W = \frac{2\pi}{k_W}, \tag{F.2b}$$

$$K_W = \frac{qB_W}{(mck_W)}, \tag{F.2c}$$

$$\gamma_0 = \left(1 - \frac{v_0^2}{c^2}\right)^{-1/2}. \tag{F.2d}$$

* © American Institute of Physics. Reprinted from *J. Appl. Phys.*, 79, 6770–6774, 1996. With permission.

Here λ_W is the wiggler wavelength (spatial period), K_W is the wiggler strength parameter, q is the absolute value of the charge of the electron, v_0 is the velocity of the electron beam, and the other symbols have the usual meaning [1]. The tunability of the FEL comes from the variability of the parameter γ_0 through the kinetic energy of the electrons. When once an FEL is constructed, λ_W is fixed.

It is known [2–5] that part of the energy of a source wave is converted into a wiggler magnetic field when an unbounded plasma medium is suddenly switched on. If the source wave in free space is of the right circular polarization given by

$$\bar{e}_i(z,t) = (\hat{x} - j\hat{y})E_0 \exp[j(\omega_0 t - k_0 z)], \tag{F.3a}$$

$$\bar{h}_i(z,t) = (j\hat{x} + \hat{y})H_0 \exp[j(\omega_0 t - k_0 z)], \tag{F.3b}$$

$$k_0 = \frac{\omega_0}{c} = \frac{2\pi}{\lambda_0}, \tag{F.3c}$$

the wiggler field produced is a helical magnetic field given by

$$\frac{H_W}{H_0} = \frac{\omega_p^2}{\omega_0^2 + \omega_p^2} \cos(k_0 z), \tag{F.4a}$$

$$B_W = \mu_0 H_W. \tag{F.4b}$$

Here ω_0 is the angular source frequency, ω_p is the plasma frequency of the switched medium, and λ_0 is the wavelength in free space of the source wave, which is also the period of the generated wiggler field. For a fixed source wave, the wiggler period is also fixed.

This article deals with the principle of establishing a helical wiggler magnetic field with controllable λ_W even if the source wave is of fixed frequency. A whistler wave that propagates in a magnetoplasma when the source frequency ω_0 is less than the electron gyrofrequency ω_b gives rise to a wiggler magnetic field in the plasma when the static magnetic field is switched off. The strength and period of the wiggler field depend on the parameters of the magnetoplasma medium. When the parameters are such that ω_0 is much less than ω_b and ω_p, the total energy of the source wave is converted into the magnetic energy of the wiggler field.

F.2 Formulation

A wave with right circular polarization, called an R wave [6,7], is assumed to be propagating in the positive z-direction in a spatially unbounded magnetoplasma medium. The electric, magnetic, and velocity fields of this wave, named the source wave, are

$$\bar{e}_i(z,t) = (\hat{x} - j\hat{y})E_0 \exp[j(\omega_0 t - k_W z)], \tag{F.5a}$$

$$\bar{h}_i(z,t) = (j\hat{x} + \hat{y})H_0 \exp[j(\omega_0 t - k_W z)], \tag{F.5b}$$

$$\bar{v}_i(z,t) = (\hat{x} - j\hat{y})V_0 \exp[j(\omega_0 t - k_W z)], \tag{F.5c}$$

where ω_0 is the frequency of the incident wave and

$$k_W = \omega_0 \frac{n_R}{c}, \tag{F.5d}$$

$$n_R = \sqrt{1 + \frac{\omega_p^2}{\omega_0(\omega_{b0} - \omega_0)}}, \tag{F.5e}$$

$$n_0 H_0 = n_R E_0, \tag{F.5f}$$

$$(m/q)V_0 = \frac{j}{\omega_0 - \omega_b} E_0. \tag{F.5g}$$

Here, c is the velocity of light in free space, η_0 is the intrinsic impedance of free space, ω_p is the plasma frequency, $\omega_{b0} = qB_{s0}/m$ is the electron gyrofrequency, and n_R is the refractive index of the magnetoplasma medium existing prior to $t = 0$.

F.3 Sudden Switching Off of the Static Magnetic Field

Assume that at $t = 0$ the static magnetic field B_{s0} is switched off in a short time. Let us idealize the problem by considering a sudden switching off of the static magnetic field. The fields for $t > 0$ may be expressed as

$$\bar{e}(z,t) = (\hat{x} - j\hat{y})e(t) \exp[-(jk_W z)], \quad t > 0, \tag{F.6a}$$

$$\bar{h}(z,t) = (j\hat{x} + \hat{y})h(t) \exp[-(jk_W z)], \quad t > 0, \tag{F.6b}$$

$$\bar{v}(z,t) = (\hat{x} - j\hat{y})v(t) \exp[-(jk_W z)], \quad t > 0. \tag{F.6c}$$

In writing the above expressions, the well-known condition [2,8,9] that the wave number is conserved at a time discontinuity in the properties of the medium is used. From Maxwell's equations and the force equation for the velocity field, one obtains

$$\frac{d}{dt}(\eta_0 h) = (j\omega_0 n_R)e, \quad t > 0, \tag{F.7a}$$

$$\frac{d}{dt}(e) = (j\omega_0 n_R)(\eta_0 h) + \omega_p^2 u, \quad t > 0, \tag{F.7b}$$

$$\frac{d}{dt}(u) = -e, \quad t > 0, \tag{F.7c}$$

where $u = mv/q$.

F.3.1 Frequencies and the Fields of the New Waves

Equations F.7, subject to the initial values $e(0) = E_0, h(0) = H_0$, and $u(0) = mV_0/q$, are solved using the Laplace transform technique. The following solutions are obtained for the frequencies and fields of the new waves:

$$e(t) = \sum_1^3 E_n \exp(j\omega_n t), \tag{F.8a}$$

$$h(t) = \sum_1^3 H_n \exp(j\omega_n t), \tag{F.8b}$$

$$\omega_1 = \omega_{up}, \quad \omega_2 = -\omega_{up}, \quad \omega_3 = 0, \tag{F.8c}$$

$$\frac{E_1}{E_0} = \frac{(1 + \omega_0/\omega_{up})}{2}, \quad \frac{E_2}{E_0} = \frac{(1 + \omega_0/\omega_{up})}{2}, \quad \frac{E_3}{E_0} = 0, \tag{F.8d}$$

$$\frac{H_1}{H_0} = \frac{\omega_0}{\omega_{up}}\frac{E_1}{E_0}, \quad \frac{H_2}{H_0} = -\frac{\omega_0}{\omega_{up}}\frac{E_2}{E_0}, \quad \frac{H_3}{H_0} = \left(1 - \frac{\omega_0^2}{\omega_{up}^2}\right), \tag{F.8e}$$

$$\frac{S_1}{S_0} = \frac{\omega_0}{4\omega_{up}}(1 + \omega_0/\omega_{up})^2, \quad \frac{S_2}{S_0} = -\frac{\omega_0}{4\omega_{up}}(1 + \omega_0/\omega_{up})^2, \quad \frac{S_3}{S_0} = 0, \tag{F.8f}$$

where

$$\omega_{up} = \sqrt{\omega_p^2 + n_R^2 \omega_0^2}. \tag{F.8g}$$

Here a negative value for ω indicates a reflected wave traveling in a direction opposite to that of the source wave and a zero value for ω gives a wiggler field. The symbol S stands for power density equal to EH.

F.3.2 Damping Rates

The above calculations are made assuming that the plasma is lossless. The modification for a lossy plasma is derived next by a simple modification of Equation F.7c,

$$\frac{d}{dt}(u) = -e - vu, \quad t > 0. \tag{F.9}$$

Here, v is the collision frequency. For a low-loss case, v is small and an approximate solution is obtained using the Laplace transform technique [10]. The results are approximately the same as those given by Equation F.8 except that the fields are damped due to the losses in the plasma. The damping time constants for the waves are given by

$$t_n = \frac{(3\omega_n^2 - \omega_p^2 - n_R^2\omega_0^2)}{v(\omega_n^2 - n_R^2\omega_0^2)}, \quad n = 1, 2, 3. \tag{F.10}$$

The wiggler field is produced by the third wave for which $\omega_n = \omega_3 = 0$.

F.3.3 Eckersley Approximation

The refractive index $n_R = n_w$ is large when $\omega_0 \ll \omega_{b0} \sim \omega_p$ and from Equation F.5e,

$$n_R \approx n_w = \frac{\omega_{p0}}{\sqrt{\omega_0\omega_{b0}}}, \quad \omega_0 \ll \omega_{b0} \sim \omega_p, \tag{F.11a}$$

and from Equation F.8, we have

$$\omega_1 = \omega_{up} \approx \omega_p, \quad \omega_2 = -\omega_{up} \approx -\omega_p, \quad \omega_3 = 0, \tag{F.11b}$$

$$\frac{E_1}{E_0} \approx \frac{1}{2}, \quad \frac{E_2}{E_0} \approx \frac{1}{2}, \quad \frac{E_3}{E_0} = 0, \tag{F.11c}$$

$$\frac{H_1}{H_0} \approx 0, \quad \frac{H_2}{H_0} \approx 0, \quad \frac{H_3}{H_0} \approx 1, \tag{F.11d}$$

$$\frac{S_1}{S_0} \approx 0, \quad \frac{S_2}{S_0} \approx 0, \quad \frac{S_3}{S_0} \approx 0. \tag{F.11e}$$

From Equation F.10, the Eckersley approximations for the damping constants may be obtained:

$$t_{1,2} \approx \frac{2}{v(1 + \omega_0/\omega_{b0})} \approx \frac{2}{v}, \tag{F.11f}$$

$$t_3 \approx \frac{1}{v}\frac{\omega_{b0}}{\omega_0}\left(1 + \frac{\omega_0}{\omega_{b0}}\right) \approx \frac{1}{v}\frac{\omega_{b0}}{\omega_0}. \tag{F.11g}$$

The wiggler magnetic field decays in the plasma much more slowly than the wave fields. From Equation F.11 it is clear that the first and second waves with upshifted frequencies are weak but the third mode that is a wiggler magnetic field is quite strong. The energy in the whistler wave, under the Eckersley approximation, is predominantly the magnetic energy due to plasma current and its density w_m is given by [6]

$$w_m = \frac{1}{2}\varepsilon_0 E_0^2 n_w^2. \tag{F.12}$$

After the static magnetic field B_s collapses, the upshifted first and second waves are weak and do not carry much energy. Most of the energy for $t > 0$ is in the wiggler magnetic field H_3. Thus, we have

$$\frac{1}{2}\varepsilon_0 E_0^2 n_w^2 \approx \frac{1}{2}\mu_0 H_3^2, \tag{F.13a}$$

$$H_3 \approx n_w \frac{E_0}{\eta_0} = H_0. \tag{F.13b}$$

The wiggler strength parameter K_W and λ_W may now be approximated as

$$K_W \approx \frac{qE_0}{(mc\omega_0)} = \frac{586.23E_0}{\omega_0}, \quad \omega_0 \ll \omega_{b0} \sim \omega_p, \tag{F.14a}$$

$$\lambda_W \approx \frac{\lambda_0}{n_w}, \quad \omega_0 \ll \omega_{b0} \sim \omega_p. \tag{F.14b}$$

The following sample calculation illustrates the tunability that can be achieved. Let the source wave be of frequency $f_0 = 1$ GHz so that $\lambda_0 = 30$ cm and $f_p = 100$ GHz. Let the dc current in the magnet be varied so that f_{b0} can be fixed from 10 to 50 GHz. The refractive index n_w changes from $100/3$ to $100/7$ and the wiggler wavelength λ_W changes from 9 to 2.1 cm.

F.4   Slow Decay of B_s

The formulation of Section C.3 assumed the sudden collapse of B_s, which is unrealistic. If the dc current of the magnet is switched off, due to inductance, B_s will not decline suddenly but will decay over a period of time. In this section, the effect of a slow decline of B_s will be studied.

A WKB type of solution can be obtained for the case of slow decay, that is, $\omega_b(t) = qB_s(t)/m$ is a slowly decaying function of time. The differential equations satisfied by the fields h and e are given by Equations F.7a and F.7b, respectively, and the differential equation for u is given by

$$\frac{d}{dt}(u) = -e + j\omega_b(t)u, \quad t > 0. \tag{F.15}$$

From Equations F.7a, F.7b, and F.15, a third-order differential equation for h can be obtained:

$$\frac{d^3h}{dt^3} - j\omega_b(t)\frac{d^2h}{dt^2} + [\omega_0^2 n_R^2 + \omega_p^2]\frac{dh}{dt} - j\omega_0^2 n_R^2 \omega_b(t)h = 0. \tag{F.16}$$

The technique of solving this equation is similar to that used in Refs. [11–13] and therefore much of the mathematical detail is omitted here. A complex instantaneous frequency function is defined such that

$$\frac{dh(t)}{dt} = p(t)h(t) = [\alpha(t) + j\omega(t)]h(t). \tag{F.17}$$

Here, $\omega(t)$ is the instantaneous frequency. Substituting Equation F.17 into Equation F.16 and neglecting α and all derivatives, a zeroth-order solution may be obtained. The solution is a cubic in ω giving the instantaneous frequencies of three waves created by the switching action,

$$\omega^3 - \omega_b(t)\omega^2 - (n_R^2\omega_0^2 + \omega_p^2)\omega + n_R^2\omega_0^2\omega_b(t) = 0. \tag{F.18}$$

This cubic has two positive real roots and one negative real root and has the following solutions at $t = 0$ and $t = \infty$:

$$\omega_1(0) = \sqrt{\left(\frac{\omega_0 + \omega_b(0)}{2}\right)^2 + \frac{\omega_p^2\omega_b(0)}{\omega_b(0) - \omega_0}} - \frac{\omega_0 + \omega_b(0)}{2}, \tag{F.19a}$$

$$\omega_2(0) = -\sqrt{\left(\frac{\omega_0 + \omega_b(0)}{2}\right)^2 + \frac{\omega_p^2\omega_b(0)}{\omega_b(0) - \omega_0}} - \frac{\omega_0 + \omega_b(0)}{2}, \tag{F.19b}$$

$$\omega_3(0) = \omega_0, \tag{F.19c}$$

$$\omega_1(\infty) = \omega_{up}, \tag{F.19d}$$

$$\omega_2(\infty) = -\omega_{up}, \tag{F.19e}$$

$$\omega_3(\infty) = 0. \tag{F.19f}$$

An equation for α may now be obtained by substituting Equation F.17 into Equation F.16 and equating the real part to zero. In obtaining the WKB solution we neglect the derivatives and powers of α, and so on,

$$\alpha = \dot{\omega}\frac{3\omega - \omega_b}{\omega_0^2 n_R^2 + \omega_p^2 - 3\omega^2 + 2\omega\omega_b}. \tag{F.20}$$

The amplitudes of the fields associated with these waves are obtained by solving the initial value problem. The continuity of the electric and magnetic fields provides two initial conditions. The third initial condition is the continuity of the velocity field at $t = 0$. The amplitudes of the magnetic fields of two of the waves are of the order of the time derivative of the $\omega_b(t)$ curve and negligible. The amplitude of the magnetic field of the wave whose initial frequency is ω_0 is significant and this wave is called an MSW. Its magnetic field $h_3(z, t)$ may be expressed as

$$h_3(z,t) = H_3(t) \exp\left[j\left(\int_0^t \omega_3(t)\, dt - k_w z \right) \right], \qquad (F.21)$$

where

$$H_3(t) = H_0(t) \exp\left(\int_0^t \alpha_3(\tau)\, d\tau \right), \qquad (F.22)$$

The integral in Equation F.22 may be evaluated numerically but in this case further mathematical manipulation will permit us to do analytical integration. From Equation F.18,

$$\omega_b = \omega\left[1 - \frac{\omega_p^2}{\omega^2 - n_0^2 \omega_0^2} \right]. \qquad (F.23)$$

Substituting for ω_b in Equation F.20, we have

$$\alpha = \dot{\omega}\frac{-\omega(2\omega^2 - 2n_R^2\omega_0^2 + \omega_p^2)}{(\omega^2 - n_R^2\omega_0^2)^2 + \omega_p^2\omega^2 + n_R^2\omega_0^2\omega_p^2}. \qquad (F.24)$$

Denoting the denominator in Equation F.24 by a variable r,

$$\alpha = -\frac{1}{2r}\frac{dr}{d\omega}\dot{\omega}, \qquad (F.25)$$

$$\int_0^t \alpha(\tau)\, d\tau = -\frac{1}{2}\int_{r(0)}^{r(t)} \frac{dr}{r} = \ln\sqrt{\frac{r[\omega(0)]}{r[\omega(t)]}}. \qquad (F.26)$$

Therefore

$$H_3(t) = H_0\sqrt{\frac{(\omega_3^2(0) - n_R^2\omega_0^2)^2 + \omega_p^2\omega_3^2(0) + n_R^2\omega_0^2\omega_p^2}{(\omega_3^2(t) - n_R^2\omega_0^2)^2 + \omega_p^2\omega_3^2(t) + n_R^2\omega_0^2\omega_p^2}}. \qquad (F.27)$$

For the third wave $\omega_3(0) = \omega_0$ and Equation F.27 becomes

$$H_3(t) = H_0 \sqrt{\frac{(n_R^2 - 1)^2 + (\omega_p^2/\omega_0^2)(n_R^2 + 1)}{(n_R^2 - [\omega_3^2(t)/\omega_0^2])^2 + (\omega_p^2/\omega_0^2)\{n_R^2 + [\omega_3^2(t)/\omega_0^2]\}}}, \qquad \text{(F.28a)}$$

$$E_3(t) = E_0 \frac{\omega_3(t)}{\omega_0}. \qquad \text{(F.28b)}$$

Since $\omega_3(\infty) = 0$,

$$H_3(\infty) = H_0 \sqrt{\frac{(n_R^2 - 1)^2 + (\omega_p^2/\omega_0^2)(n_R^2 + 1)}{n_R^4 + (\omega_p^2/\omega_0^2)n_R^2}}, \qquad \text{(F.29a)}$$

$$E_3(\infty) = 0. \qquad \text{(F.29b)}$$

For the Eckersley approximation $n_R \gg 1$, and

$$H_3(\infty) = H_0, \qquad \omega_0 \ll \omega_{b0} \sim \omega_p, \qquad \text{(F.30a)}$$

$$E_3(\infty) = 0, \qquad \text{(F.30b)}$$

F.5   An Illustrative Example

Figure F.1 illustrates the results for an exponential decay profile whose time constant is $(1/b)$,

$$\omega_b(t) = \omega_{b0}e^{-bt}. \qquad \text{(F.31)}$$

All variables are normalized with reference to the source wave quantities. The source wave frequency ω_0 is taken to be 1. The parameters ω_{b0} and ω_p are assigned the values 100 and 1000, respectively. The parameters are such that the Eckersley approximation is valid. A value of 0.01 is assigned to b. The parameter describes a slow decay since $100/2\pi$ cycles of the source wave are accommodated in one time constant. The independent variable, time, is normalized with respect to the period of the source wave.

The results shown in Figure F.1 for slow decay are numerically computed from Equations F.18, F.28a, and F.28b and the following values are obtained at $t = 1000$: $\omega_b/\omega_0 = 4.54 \times 10^{-3}$, $\omega_3/\omega_0 = 4.54 \times 10^{-5}$, $E_3/E_0 = 4.54 \times 10^{-5}$, and $H_3/H_0 = 1.000048$. Figure F.1 shows also, for comparison, horizontal lines that are the results for sudden collapse. Irrespective of the rate of collapse of B_s, a strong wiggler field is generated.

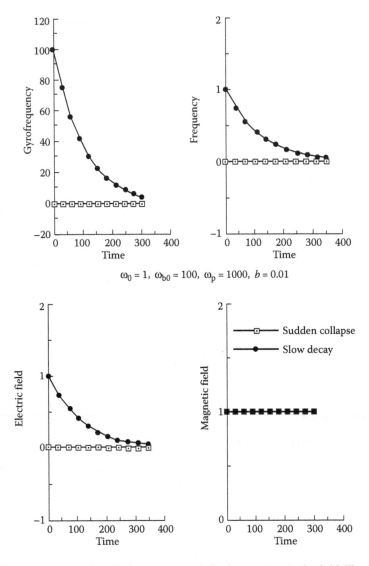

FIGURE F.1 Conversion of a whistler wave into a helical magnetic wiggler field. The values on the vertical axes are normalized with respect to source wave quantities. The source frequency ω_0 is taken as 1. The horizontal timescale is normalized with respect to the period of the source wave. The parameters are such that the Eckersley approximation is applicable.

Figure F.2 shows the results for $\omega_{b0} = 1.1$ and $\omega_p = 0.5$ with the other parameters remaining the same as those of Figure F.1. The Eckersley approximation is not valid in this case, but Equation F.28 still holds good; the third wave becomes the wiggler magnetic field with $H_3/H_0 = 0.75$; however, the damping time constant given by Equation F.10 is much less than the value for the parameters of Figure F.1.

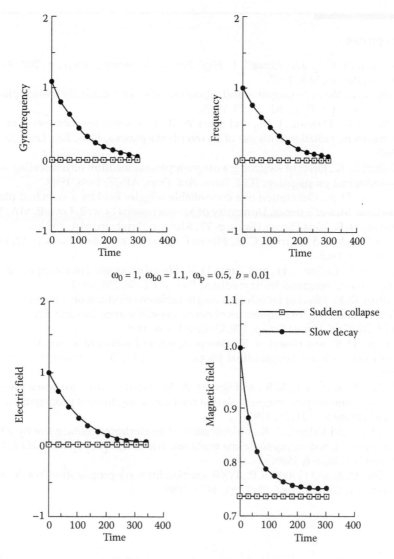

FIGURE F.2 Conversion of a whistler wave into a helical magnetic wiggler field. The parameters are such that the Eckersley approximation is not applicable.

F.6 Conclusion

The effect of switching off of the static magnetic field on a whistler wave is considered. A sudden collapse as well as a slow decay of the static magnetic field showed that the energy of the whistler wave is converted into a wiggler magnetic field.

References

1. Granastein, V. L. and Alexeff, I., *High-Power Microwave Sources*, p. 207, Artech House, Boston, MA, 1987.
2. Jiang, C. L., Wave propagation and dipole radiation in a suddenly created plasma, *IEEE Trans. Ant. Prop.*, AP-23, 83, 1975.
3. Wilks, S. C., Dawson, J. M., and Mori W. B., Frequency up-conversion of electromagnetic radiation with use of an overdense plasma, *Phys. Rev. Lett.*, 63, 337, 1988.
4. Kalluri, D. K., Effect of switching a magnetoplasma medium on a traveling wave: Longitudinal propagation, *IEEE Trans. Ant. Prop.*, AP-37, 1638, 1989.
5. Lapierre, D. A., Generation of a controllable wiggler field by a switched plasma medium, Master's thesis, University of Massachusetts Lowell, Lowell, MA, 1993.
6. Booker, H. G., *Cold Plasma Waves*, p. 77, Kluwer, Hingham, MA, 1984.
7. Heald, M. A. and Wharton, C. B., *Plasma Diagnostics with Microwaves*, p. 12, Wiley, New York, 1965.
8. Auld, B. A., Collins, J. H., and Zapp, H. R., Signal processing in a nonperiodically time-varying magnetoelastic medium, *Proc. IEEE*, 56, 258, 1968.
9. Kalluri, D. K., Effect of switching a magnetoplasma medium on a traveling wave: Conservation law for frequencies of newly created waves, In: *Conf. Rec. Abstracts, IEEE Int. Conf. Plasma Sci.*, p. 129, Oakland, CA, 1990.
10. Kalluri, D. K. and Goteti, V. R., Damping rates of waves in a switched magnetoplasma medium: Longitudinal propagation, *IEEE Trans. Plasma Sci.*, 18, 797, 1990.
11. Kalluri, D. K., Goteti, V. R., and Sessler, A. M., WKB solution for wave propagation in a time-varying magnetoplasma medium: Longitudinal propagation, *IEEE Trans. Plasma Sci.*, 21, 70, 1993.
12. Lee, J. H. and Kalluri, D. K., Modification of an electromagnetic wave by a time-varying switched mangetoplasma medium: Transverse propagation, *IEEE Trans. Plasma Sci.*, 26, 1–6, 1998.
13. Kalluri, D. K. and Goteti, V. R., WKB solution for wave propagation in a decaying plasma medium, *J. Appl. Phys.*, 66, 3472, 1989.

Appendix G

Effect of Switching a Magnetoplasma Medium on the Duration of a Monochromatic Pulse*

Dikshitulu K. Kalluri

G.1 Introduction

The study of the interaction of electromagnetic waves with time-varying plasmas has been a subject of considerable recent activity in plasma science [1] due to its potential applications in the generation of tunable electromagnetic radiation over a broad frequency range.

Jiang [2] investigated the effect of switching suddenly an unbounded isotropic cold plasma medium on a monochromatic plane electromagnetic wave. The wave propagation in such a medium is governed by the conservation of the wave number across the time discontinuity and the consequent change in the wave frequency. Rapid creation of the medium may be approximated as a sudden switching of the medium. The main effect of switching the medium is the splitting of the original wave (henceforth called the incident wave, in the sense of incidence on a time discontinuity in the properties of the medium) into new waves whose frequencies are different from the incident wave. Recent experimental work [3–5] and computer simulation [6] demonstrated the frequency upshifting of microwave radiation by rapid plasma creation.

G.2 Effect on a Monochromatic Wave: Review

It is known [7] that the presence of a static magnetic field will influence the frequency shift, and so on. Rapid creation of a plasma medium in the presence of a static magnetic field may be approximated as the sudden switching

* © 1997 Plenum Publishing Corporation. Reprint from *Int. J. Infrared Millim. Waves*, 18, 1585–1603, 1997. With permission.

of a magnetoplasma medium. An idealized mathematical description of the problem is given next. A uniform plane wave is traveling in free space along the z-direction and it is assumed that at $t = 0^-$, the wave occupies the entire space. A static magnetic field of strength B_0 is assumed to be present through-out. At $t = 0$, the entire space is suddenly ionized, thus converting the medium from free space to a magnetoplasma medium. The incident wave frequency is assumed to be high enough that the effects of ion motion may be neglected.

The strength and direction of the static magnetic field affect the number of new waves created, their frequencies, and the power density of these waves. Some of the new waves are transmitted waves (waves propagating in the same direction as the incident wave) and some are reflected waves (waves propagating in the opposite direction to that of the incident wave). Reflected waves tend to have less power density.

A physical interpretation of the waves may be given in the following way. The electric and magnetic fields of the incident wave and the static mag-netic field accelerate the electrons in the newly created magnetoplasma; the electrons in turn radiate new waves. The frequencies of the new waves and their fields can be obtained by adding contributions from the many elec-trons whose positions and motions are correlated by the collective effects supported by the magnetoplasma medium. Such a detailed calculation of the radiated fields seems to be quite involved. A simple, but less accurate, descrip-tion of the plasma effect is obtained by modeling the magnetoplasma as a dielectric medium whose refractive index is computed through magnetoionic theory [8]. The frequency of the new waves is constrained by the requirements that the wave number (k_0) is conserved over the time discontinuity and the refractive index n is that applicable to the type of wave propagation in the magnetoplasma. This gives a conservation [9] law $k_0 c = \omega_0 = n(\omega)\omega$ from which ω may be determined. The solution of the associated electromagnetic initial value problem gives the electric and magnetic fields of the new waves. Using this approach the general aspects of wave propagation in a switched magnetoplasma medium were discussed earlier [7,10,11] for the cases of L-, R-, and X-incidence [8].

L-incidence is an abbreviation used to describe the case where the incident wave has left-hand circular polarization and the static magnetic field is in the z-direction (Figure G.1a). Three new waves with left-hand circular polar-ization labeled as L1, L2, and L3 are generated by the medium switching. L1 is a transmitted wave. R-incidence refers to the case where the incident wave has right-hand circular polarization and the static magnetic field is in the z-direction. The medium switching, in this case, creates three R waves labeled as R1, R2, and R3. R1 and R3 are transmitted waves (Figure G.1b). The frequencies of the R waves are the roots of the cubic characteristic equation [7]:

$$\omega_n^3 - \omega_b \omega_n^2 - (\omega_0^2 + \omega_p^2)\omega_n + \omega_0^2 \omega_b = 0. \qquad (G.1)$$

In the above equation, ω_b is the electron gyrofrequency and the incident wave variables are denoted by subscript zero. A negative value for ω_n [7]

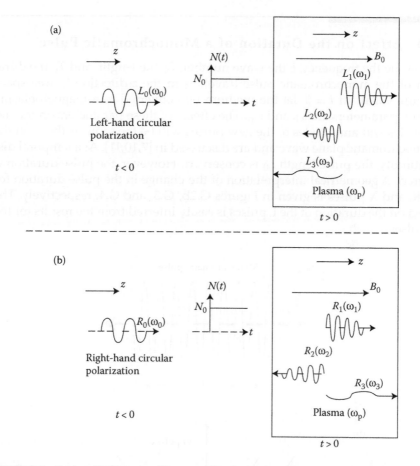

FIGURE G.1 L (a) and R (b) waves. Waves are sketched in time domain to show the frequencies.

indicates that the wave is a reflected wave, that is, it propagates in a direction opposite to that of the source (incident) wave. If the source wave is right-going, then the reflected wave will be left-going.

X-incidence refers to the case where the electric field of the incident wave is in the x-direction and the static magnetic field is in the y-direction. The medium switching in this case creates four extraordinary waves labeled as X1, X2, X3, and X4. X1 and X2 are transmitted waves. The frequencies of these waves are the roots of the characteristic equation [10]:

$$\omega_n^4 - \omega_n^2(\omega_0^2 + \omega_b^2 + 2\omega_p^2) + (\omega_p^4 + \omega_0^2(\omega_b^2 + \omega_p^2)) = 0. \qquad (G.2)$$

The source in an experiment is likely to be a monochromatic pulse of a given duration. In this paper, the effect of switching a magnetoplasma medium on the duration of such a pulse is considered.

G.3   Effect on the Duration of a Monochromatic Pulse

Let ω_0 be the frequency, k the wave number, L_0 the length, and T_0 the dura-tion of the monochromatic pulse traveling in the z-direction in free space (Figure G.2a). At $t = 0$, let the medium be converted to a magnetoplasma whose parameters are ω_b and ω_p. The effects of these parameters on the fre-quencies and amplitudes of the new pulses will be the same as those on the monochromatic plane wave and are discussed in [7,10,11]. At a temporal dis-continuity, the pulse length L_0 is conserved. However, the pulse duration is altered. A geometrical interpretation of the change in the pulse duration for O, R, and X pulses is given in Figures G.2b, G.3, and G.4, respectively. The effect on the duration of the L pulses is easily inferred from the results for the R pulses.

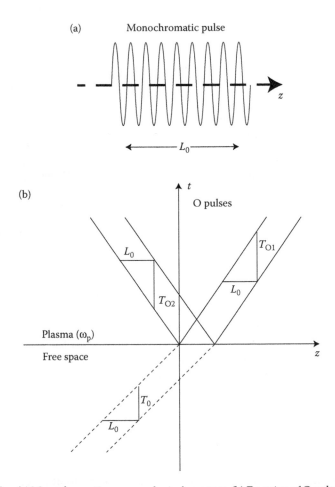

FIGURE G.2   (a) Monochromatic source pulse in free space. (b) Duration of O pulses in plasma.

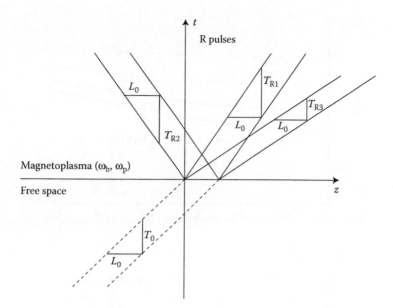

FIGURE G.3 Duration of R pulses.

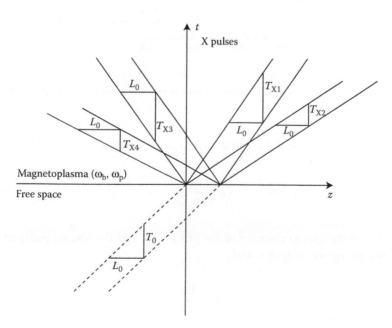

FIGURE G.4 Duration of X pulses.

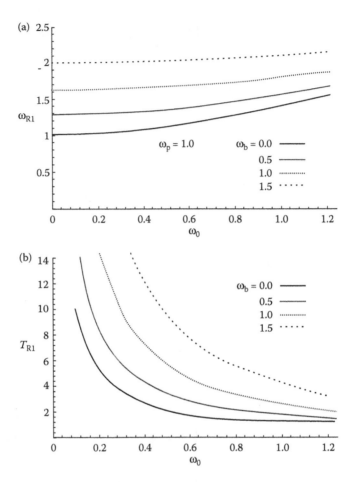

FIGURE G.5 R1 pulse: (a) pulse frequency versus source frequency and (b) pulse duration versus source frequency.

The pulse duration and the pulse length for the nth wave are related by

$$T_n = \frac{L_0}{v_{grn}}, \tag{G.3}$$

where v_{grn} is the group velocity of the nth wave. For the source pulse in free space, the group velocity is c and

$$T_0 = \frac{L_0}{c}. \tag{G.4}$$

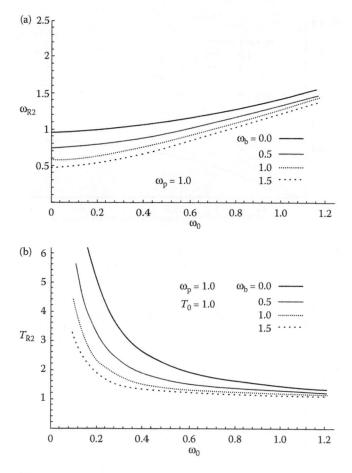

FIGURE G.6 R2 pulse: (a) pulse frequency versus source frequency and (b) pulse duration versus source frequency.

From the dispersion relation, the group velocity of the nth wave $(d\omega_n/dk)$ may be obtained. The pulse duration ratio is given by

$$\frac{T_n}{T_0} = \frac{c}{|v_{grn}|} = \frac{c}{|d\omega_n/dk|} = \frac{1}{|d\omega_n/d\omega_0|}. \tag{G.5}$$

An increase (decrease) in the numerical value of the group velocity produces a decrease (increase) in the pulse duration.

For O waves in an isotropic plasma medium, the dispersion relation is

$$\omega_n^2 = \omega_0^2 + \omega_p^2, \tag{G.6}$$

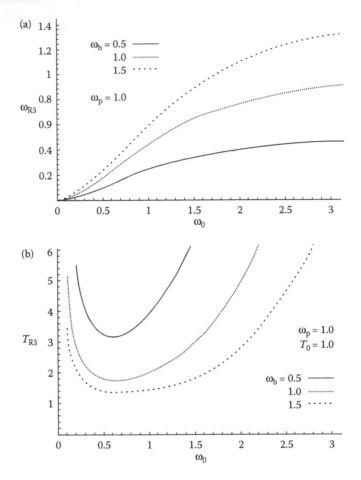

FIGURE G.7 R3 pulse: (a) pulse frequency versus source frequency and (b) pulse duration versus source frequency.

and

$$\frac{T_{O1,O2}}{T_0} = \frac{\omega_0}{|\omega_n|}.$$ (G.7)

For R waves, the dispersion relation is given by Equation G.1 and

$$\frac{d\omega_n}{d\omega_0} = \frac{2\omega_0(\omega_n - \omega_0)}{3\omega_n^2 - 2\omega_n\omega_b - \omega_0^2\omega_p^2}.$$ (G.8)

For X waves, the dispersion relation is given by Equation G.2 and

$$\frac{d\omega_n}{d\omega_0} = \frac{\omega_0}{\omega_n} \frac{\omega_n^2 - \omega_b^2 - \omega_p^2}{2\omega_n^2 - (\omega_0^2 + \omega_b^2 + 2\omega_p^2)}.$$ (G.9)

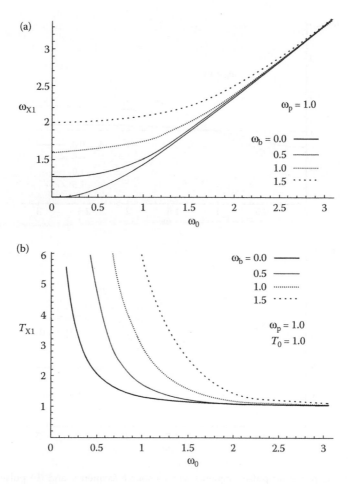

FIGURE G.8 X1 pulse: (a) pulse frequency versus source frequency and (b) pulse duration versus source frequency.

G.4 Numerical Results and Discussion

The top portions of Figures G.5 through G.9 show the variation of frequency of the pulses as a function of the frequency ω_0 of the source pulse in free space. The frequencies are normalized by choosing the plasma frequency parameter $\omega_p = 1.0$. The gyrofrequency parameter ω_b is varied from 0 to 1.5 in steps of 0.5. The slope of the pulse frequency versus the source frequency curve is proportional to the group velocity and the pulse duration is inversely proportional to the magnitude of this slope. The bottom portions of these curves describe the variation of the pulse duration of the various B pulses. Pulse duration is normalized by choosing $T_0 = 1$. The R1 and X1 curves for $\omega_b = 0$ describe the O1 pulse.

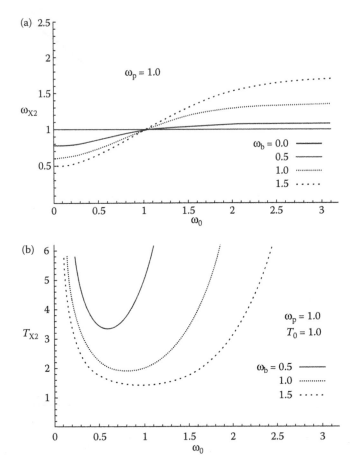

FIGURE G.9 X2 pulse: (a) pulse frequency versus source frequency and (b) pulse duration versus source frequency.

The following interesting points may be noted:

1. The pulse duration ratio for R1 and X1 pulses asymptotically reaches unit value as ω_0 tends to infinity. For large values of ω_0, the plasma has very little effect.

2. R2 is a reflected pulse and its graph for the pulse duration is qualitatively similar to the R1 graph. O2, X3, and X4 pulses are also reflected pulses and their graphs are the same as those for O1, X1, and X2 pulses, respectively.

3. R3 and X2 pulses are generated only in a switched anisotropic plasma medium and their behavior is strongly influenced by the ω_b parameter. Their pulse duration goes through a minimum and the minimum point is strongly influenced by the ω_b parameter.

It is suggested that the principle of the change in the pulse duration may be used to diagnose dynamically the time-varying parameters of an anisotropic medium.

Acknowledgment

The work was supported by Air Force Office of Scientific Research under contract AFOSR 97-0847.

References

1. Mori, W. B. (Ed.), Special issue on generation of coherent radiation using plasmas, *IEEE Trans. Plasma Sci.*, 21, 1, 1993.
2. Jiang, C. L., Wave propagation and dipole radiation in a suddenly created plasma, *IEEE Trans. Ant. Prop.*, AP-23, 83, 1975.
3. Rader, M. and Alexeff, I., Microwave frequency shifting using photon acceleration, *Int. J. Infrared Millim. Waves*, 12, 683, 1991.
4. Kuo, S. P., Zhang, Y. S., and Ren, A. Q., Frequency up-conversion of microwave pulse in a rapidly growing plasma, In: *Conf. Rec. Abstracts, IEEE Int. Conf. Plasma Sci.*, Oakland, CA, p. 171, 1990.
5. Joshi, C. S., Clayton, C. E., Marsh, K., Hopkins, D. B., Sessler, A., and Whittum, D., Demonstration of the frequency upshifting of microwave radiation by rapid plasma Creation, *IEEE Trans. Plasma Sci.*, 18, 814, 1990.
6. Wilks, S. C., Dawson, J. M., and Mori, W. B., Frequency up-conversion of electromagnetic radiation with use of an overdense plasma, *Phys. Rev. Lett.*, 61, 337, 1988.
7. Kalluri, D. K., Effect of switching a magnetoplasma medium on a traveling wave: Longitudinal propagation, *IEEE Trans. Ant. Propag.*, AP-37, 1638, 1989.
8. Booker, H. G., *Cold Plasma Waves*, Kluwer, Hingham, MA, 1984.
9. Kalluri, D. K., Effect of switching a magnetoplasma medium on a traveling wave: Conservation law for frequencies of newly created waves, In: *Conf. Rec. Abstracts, IEEE Int. Conf. Plasma Sci.*, Oakland, CA, p. 129, 1990.
10. Goteti, V. R. and Kalluri, D. K., Wave propagation in a switched magnetoplasma medium: Transverse propagation, *Radio Sci.*, 25, 61, 1990.
11. Kalluri, D. K., Frequency shifting using magnetoplasma medium: Flash ionization, *IEEE Trans. Plasma Sci.*, 21, 77, 1993.

Appendix H

Modification of an Electromagnetic Wave by a Time-Varying Switched Magnetoplasma Medium: Transverse Propagation*

Joo Hwa Lee and Dikshitulu K. Kalluri

H.1 Introduction

The main effect of switching a magnetoplasma [1–17] is the splitting of the source wave into new waves whose frequencies are different from the frequency of the source wave. When the medium properties change slowly with time due to slow change of the electron density in the plasma, one of these waves whose initial frequency is the source frequency will be the dominant new wave in the sense that its field amplitudes will be significant. Other waves have field amplitudes of the order of the initial slope of the electron density profile. The case of longitudinal propagation is already discussed in [13,16,17]. The case of transverse propagation involving X waves [18] is the topic of this paper.

H.2 Transverse Propagation

H.2.1 Development of the Problem

The geometry of the problem is shown in Figure H.1. Initially, for time $t < 0$, the entire space is considered to be free space. A uniform electromagnetic plane wave with a frequency ω_0 propagating in the positive z-direction and the electric field in the positive x-direction is established over the entire space.

---

* © 1998 IEEE. Reprinted from *IEEE Trans. Plasma Sci.*, 26, 1–6, 1998. With permission.

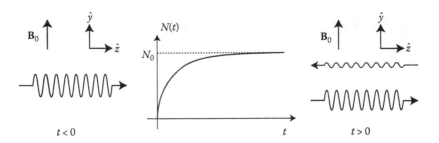

FIGURE H.1   Geometry of the problem. Modification of the source wave by a time-varying magnetoplasma medium: transverse propagation.

The fields of the source wave are given by

$$\mathbf{E}(z,t) = \hat{x}E_0 e^{j(\omega_0 t - k_0 z)}, \quad t < 0, \tag{H.1a}$$

$$\mathbf{H}(z,t) = \hat{y}H_0 e^{j(\omega_0 t - k_0 z)}, \quad t < 0, \tag{H.1b}$$

where k_0 is the free space wave number, $E_0 = \eta_0 H_0$, η_0 is the intrinsic impedance of free space, and $j = \sqrt{-1}$. A static magnetic field $\mathbf{B}_0$ is present along the y-direction.

At time $t = 0$, the entire space is converted into a time-varying plasma medium. The plasma medium is spatially homogeneous and unbounded. Therefore, the wave number k_0 remains constant [1]. The free electron density $N(t)$ is slowly increasing to a certain value N_0 with zero initial value. Therefore, the plasma medium has a time-varying plasma frequency $\omega_p(t)$ and an electron gyrofrequency ω_b given by

$$\omega_p(t) = \sqrt{\frac{N(t)q}{m\varepsilon_0}}, \tag{H.2a}$$

$$\omega_b = \frac{B_0 q}{m}, \tag{H.2b}$$

where q is the absolute value of the charge, m the mass of the electron, and B_0 the magnitude of $\mathbf{B}_0$. With these assumptions, we can write the fields for $t > 0$ as follows:

$$\mathbf{E}(z,t) = \left[\hat{x}E_x(t) + \hat{z}E_z(t)\right]e^{-jk_0 z}, \quad t > 0, \tag{H.3a}$$

$$\mathbf{H}(z,t) = \hat{y}H_y(t)e^{-jk_0 z}, \quad t > 0, \tag{H.3b}$$

$$\mathbf{v}(z,t) = \left[\hat{x}v_x(t) + \hat{z}v_z(t)\right]e^{-jk_0 z}, \quad t > 0, \tag{H.3c}$$

where $\mathbf{v}(z, t)$ is the velocity of an electron. The differential equations for these fields can be obtained by using the Maxwell equations:

$$\nabla \times \mathbf{E} = -\mu_0 \frac{\partial \mathbf{H}}{\partial t}, \tag{H.4a}$$

$$\nabla \times \mathbf{H} = \varepsilon_0 \frac{\partial \mathbf{E}}{\partial t} + \mathbf{J}. \tag{H.4b}$$

From Equations H.3 and H.4, the following differential equations are obtained:

$$\eta_0 \frac{\partial H_y}{\partial t} = j\omega_0 E_x, \tag{H.5a}$$

$$\frac{\partial E_x}{\partial t} = j\omega_0 \eta_0 H_y - \frac{1}{\varepsilon_0} J_x, \tag{H.5b}$$

$$\frac{\partial E_z}{\partial t} = -\frac{1}{\varepsilon_0} J_z. \tag{H.5c}$$

Here J_x and J_z denote the current density components. The current density vectors are closely related to the motion of the electron charges.

H.2.2 Current Density Vector

In the previous subsection, we obtained three equations with five unknowns. Two more equations are needed to determine the five unknowns. The equation of motion of the electrons is given by the Lorentz force equation

$$m \frac{d\mathbf{v}}{dt} = -q \left[\mathbf{E} + \mathbf{v} \times \mathbf{B}_0 \right]. \tag{H.6}$$

The velocity vector may be related to the current density vector. When free space is slowly converted into plasma, it is necessary to impose the requirement that the electrons born at different times have zero initial velocity [12,13]. Applying these concepts to the case of transverse propagation [14], we obtain the equations for J_x and J_z given below:

$$\frac{dJ_x}{dt} = \varepsilon_0 \omega_p^2 E_x + \omega_b J_z, \tag{H.7a}$$

$$\frac{dJ_z}{dt} = \varepsilon_0 \omega_p^2 E_z - \omega_b J_x. \tag{H.7b}$$

H.2.3 Higher-Order Differential Equation

Equations H.5 and H.7 are the first-order coupled differential equations that contain time-varying and complex coefficients. Generally, it is difficult to obtain exact solutions, but reasonable approximate solutions can be obtained. Equations H.5 and H.7 are manipulated by differentiation and substitution and a fifth-order differential equation for E_x is obtained:

$$E_x^{(5)} + AE_x^{(3)} + BE_x^{(1)} + g[5E_x^{(2)} + (2\omega_0^2 + 3\omega_p^2)E_x]$$

$$+ \frac{dg}{dt}(3E_x^{(1)} + \omega_b E_z) + \frac{d^2g}{dt^2}E_x = 0, \tag{H.8}$$

where

$$A = \omega_0^2 + 2\omega_p^2 + \omega_b^2, \tag{H.9a}$$

$$B = \omega_0^2\omega_b^2 + \omega_0^2\omega_p^2 + \omega_p^4, \tag{H.9b}$$

$$g = \frac{d\omega_p^2}{dt}. \tag{H.9c}$$

We neglect dg/dt, d^2g/dt^2, and g^2 since $\omega_p^2(t)$ is a slowly varying function of time. The equation for E_x becomes simpler and is given by

$$E_x^{(5)} + AE_x^{(3)} + BE_x^{(1)} + g[5E_x^{(2)} + (2\omega_0^2 + 3\omega_p^2)E_x] = 0. \tag{H.10}$$

H.2.4   Riccati Equation

To solve Equation H.10, the WKB method can be applied [15]. A complex instantaneous frequency function, $p(t)$, is defined such that

$$\frac{dE_x}{dt} = p(t)E_x. \tag{H.11}$$

Then,

$$E_x^{(2)} = (\dot{p} + p^2)E_x, \tag{H.12a}$$

$$E_x^{(3)} = (\ddot{p} + 3p\dot{p} + p^3)E_x, \tag{H.12b}$$

$$E_x^{(4)} = (\dddot{p} + 4p\ddot{p} + 6p^2\dot{p} + 3\dot{p}^2 + p^4)E_x, \tag{H.12c}$$

$$E_x^{(5)} = [(p^{(4)} + 5p\dddot{p} + 10(\dot{p} + p^2)\ddot{p} + 5(2p^3 + 3p\dot{p})\dot{p} + p^5)]E_x. \tag{H.12d}$$

Substitution of Equation H.12 into Equation H.10 gives

$$p^{(4)} + 5p\dddot{p} + 10(\dot{p} + p^2)\ddot{p} + 5(2p^3 + 3p\dot{p})\dot{p} + p^5$$

$$+ A(\ddot{p} + 3p\dot{p} + p^3) + Bp + g[5(\dot{p} + p^2) + (2\omega_0^2 + 3\omega_p^2)] = 0. \tag{H.13}$$

This Riccati equation can be solved to zero order, and the solution gives real frequencies of the new waves.

H.2.5 Frequencies of the New Waves

Let $p = j\omega$ in Equation H.13 and neglect higher-order terms ($\dot{p}$, $\ddot{p}$, $\dddot{p}$, $p^{(4)}$, g, and their powers); then we obtain

$$\omega^5 - A\omega^3 + B\omega = 0. \tag{H.14}$$

The solutions to this equation can be easily obtained as for $\omega_b < \omega_0$,

$$\omega_1 = \left[\frac{1}{2} \left\{ A + \sqrt{A^2 - 4B} \right\} \right]^{1/2}, \tag{H.15a}$$

$$\omega_2 = \left[\frac{1}{2} \left\{ A - \sqrt{A^2 - 4B} \right\} \right]^{1/2}, \tag{H.15b}$$

$$\omega_3 = -\omega_1, \tag{H.15c}$$

$$\omega_4 = -\omega_2, \tag{H.15d}$$

$$\omega_5 = 0, \tag{H.15e}$$

for $\omega_b > \omega_0$,

$$\omega_1 = \left[\frac{1}{2} \left\{ A - \sqrt{A^2 - 4B} \right\} \right]^{1/2}, \tag{H.16a}$$

$$\omega_2 = \left[\frac{1}{2} \left\{ A + \sqrt{A^2 - 4B} \right\} \right]^{1/2}, \tag{H.16b}$$

$$\omega_3 = -\omega_1, \tag{H.16c}$$

$$\omega_4 = -\omega_2, \tag{H.16d}$$

$$\omega_5 = 0. \tag{H.16e}$$

We split the solutions into two regions given by Equations H.15 and H.16 to obtain the same initial values at $t = 0$. Consequently, in both regions, the first wave has the same initial value ω_0 and may be considered as the MSW. The other waves may be considered as new waves created by the switching action and are of the order of $g(0) = g_0$. Equations H.15 and H.16 are the same as those that can be obtained from the consideration of conservation of the wave number of the X wave in a time-varying, unbounded, and homogeneous plasma medium [15].

Equations H.15 and H.16 show that five new waves are generated when $t > 0$ as a result of the imposition of the magnetoplasma. The first two solutions are transmitted waves, while the next two are reflected waves. Since ω_5 is zero, the fifth solution is not a wave.

H.2.6   Amplitudes of the Fields of the New Waves

Amplitudes of the fields of the new waves can be obtained by expanding the frequency function to include a real part. Let $p = \alpha + j\omega$. Neglecting higher-order terms except α, $\dot{\omega}$, and g, we can obtain the relation for α:

$$\alpha_n = \frac{g(5\omega_n^2 - 2\omega_0^2 - 3\omega_p^2) - \dot{\omega}(10\omega_n^2 - 3A)\omega_n}{5\omega_n^4 - 3A\omega_n^2 + B}. \tag{H.17}$$

H.2.7   Complete Solution

The complete solutions to the problem can be represented by the linear combination of the five new modes. Let us suppose that the field components have the following form:

$$E_x = E_0 \sum_{n=1}^{5} A_n \exp\left[\int_0^t \left(j\omega_n(\tau) + \alpha_n(\tau)\right) d\tau\right]. \tag{H.18}$$

To determine the unknown coefficients A_n, we can use the initial conditions at $t = 0$:

$$E_x(0) = E_0, \tag{H.19a}$$

$$E_z(0) = 0, \tag{H.19b}$$

$$H_y(0) = H_0, \tag{H.19c}$$

$$J_x(0) = 0, \tag{H.19d}$$

$$J_z(0) = 0. \tag{H.19e}$$

Symbolic manipulations [19] have been carried out to obtain the following coefficients:

$$A_1 = 1 - jg_0 \frac{\omega_b^4 + 14\omega_b^2\omega_0^2 + \omega_0^4}{8\omega_0(\omega_0^2 - \omega_b^2)^3}, \tag{H.20a}$$

$$A_2 = jg_0 \frac{\omega_b}{2(\omega_0 + \omega_b)(\omega_0 - \omega_b)^3}, \tag{H.20b}$$

$$A_3 = jg_0 \frac{1}{8\omega_0(\omega_0^2 - \omega_b^2)}, \tag{H.20c}$$

$$A_4 = -jg_0 \frac{\omega_b}{2(\omega_0 - \omega_b)(\omega_0 + \omega_b)^3}, \tag{H.20d}$$

$$A_5 = 0, \tag{H.20e}$$

The above expressions become singular when $\omega_0 = \omega_b$. The evaluation of the unknown constants using Equation H.18 makes it possible to describe the transmitted fields and the reflected fields completely in terms of their time-varying envelope functions and the frequency functions. The resulting expression is

$$E_x(z,t) = E_0 \sum_{n=1}^{4} A_n \exp[\beta_n(t)] \exp[j(\theta_n(t) - k_0 z)], \qquad \text{(H.21)}$$

where

$$\theta_n(t) = \int_0^t \omega_n(\tau)\, d\tau, \qquad \text{(H.22a)}$$

$$\beta_n(t) = \int_0^t \alpha_n(\tau)\, d\tau. \qquad \text{(H.22b)}$$

H.2.8 Explicit Expressions for the Amplitudes of the Fields

The expression for g in terms of ω may be obtained by differentiating Equation H.14:

$$g = \frac{2(2\omega^2 - A)\omega\dot{\omega}}{2\omega^2 - \omega_0^2 - 2\omega_p^2}. \qquad \text{(H.23)}$$

From Equations H.17 and H.23, $\alpha(t)$ can be expressed as

$$\alpha(t) = \frac{\left(\omega_0^2 A - 2(2\omega_b^2 + \omega_0^2)\omega^2\right)\dot{\omega}}{2\omega(2\omega^2\omega_b^2 - 3\omega_0^2\omega_b^2 + \omega_0^4 + 2\omega_b^2\omega_p^2)} = -\frac{2\omega\omega_b^2}{G(G - \omega_b^2)}\dot{\omega} - \frac{\omega_0^2}{2\omega G}\dot{\omega}, \qquad \text{(H.24)}$$

where

$$G = \sqrt{4\omega^2\omega_b^2 - 4\omega_0^2\omega_b^2 + \omega_0^4}, \qquad \text{(H.25a)}$$

$$\omega_p^2 = \frac{(2\omega^2 - \omega_0^2 - G)}{2}. \qquad \text{(H.25b)}$$

We can integrate Equation H.24 to obtain $\beta(t)$:

$$\beta(t) = \int_0^t \alpha(\tau)\, d\tau = -\frac{1}{2}\ln(G - \omega_b^2) + \frac{\omega_b^2}{2\sqrt{\omega_0^4 - 4\omega_0^2\omega_b^2}} \ln\frac{G + \sqrt{\omega_0^4 - 4\omega_0^2\omega_b^2}}{\omega}\Bigg|_{\omega(0)}^{\omega(t)}.$$

$$\text{(H.26)}$$

H.2.9 Adiabatic Analysis of the MSW

For the first wave, from Equation H.26,

$$F(t) = \exp\left[\beta_1(t)\right] = \sqrt{\frac{\omega_0^2 - \omega_b^2}{2\omega_1^2 - \omega_0^2 - 2\omega_p^2 - \omega_b^2}}$$

$$\times \left(\frac{\omega_0}{\omega_1} \frac{2\omega_1^2 - \omega_0^2 - 2\omega_p^2 + \sqrt{\omega_0^4 - 4\omega_b^2\omega_0^2}}{\omega_0^2 + \sqrt{\omega_0^4 - 4\omega_b^2\omega_0^2}}\right)^{\omega_0^2/2\sqrt{\omega_0^4 - 4\omega_b^2\omega_0^2}}, \qquad \text{(H.27)}$$

In Equation H.27, the factors including the last exponent are real for $\omega_0^2 > 4\omega_b^2$. For other ranges, some of these factors are complex but easily computed by using mathematical software [15]. For small g_0, and ω_0 sufficiently different from ω_b, $A_1 \cong 1$ and the other A coefficients are negligible. Only the first wave has significant amplitude and it may be labeled as the MSW. Each of the fields of the MSW may be written as a product of an instantaneous amplitude function and a frequency function

$$E_x(z,t) = E_0 e_x(t) \cos\left[\theta_1(t) - k_0 z\right], \qquad \text{(H.28a)}$$

$$H_y(z,t) = H_0 h_y(t) \cos\left[\theta_1(t) - k_0 z\right], \qquad \text{(H.28b)}$$

$$E_z(z,t) = E_0 e_z(t) \cos\left[\theta_1(t) - k_0 z + \frac{\pi}{2}\right], \qquad \text{(H.28c)}$$

where the instantaneous amplitude functions are given by

$$e_x(t) = F(t), \qquad \text{(H.28d)}$$

$$h_y(t) = \frac{\omega_0}{\omega_1(t)} F(t), \qquad \text{(H.28e)}$$

$$e_z(t) = \frac{\omega_b \omega_p^2}{\omega_1(t)\left[\omega_1^2(t) - \omega_p^2(t) - \omega_b^2\right]} F(t). \qquad \text{(H.28f)}$$

and the power density function $S(t)$ is given by

$$S(t) = \frac{\omega_0}{\omega_1(t)} \left[F(t)\right]^2. \qquad \text{(H.29)}$$

For $\omega_b \ll \omega_0$, the amplitude functions reduce to

$$e_x(t) \cong \left(\frac{\omega_0^2}{\omega_0^2 + \omega_p^2(t)} \right)^{1/4}$$

$$\times \left[1 + \frac{\left(2\omega_0^2 \left(\omega_0^2 + \omega_p^2(t) \right) \ln \left(\omega_0^2 / \left(\omega_0^2 + \omega_p^2(t) \right) \right) - \omega_p^2(t) \left(3\omega_0^2 + 2\omega_p^2(t) \right) \right) \omega_b^2}{4\omega_0^4 \left(\omega_0^2 + \omega_p^2(t) \right)} \right]. \quad \text{(H.30a)}$$

$$e_z(t) \cong \left(\frac{\omega_0^2}{\omega_0^2 + \omega_p^2(t)} \right)^{1/4} \frac{\omega_b \omega_p^2(t)}{\omega_0^2 \sqrt{\omega_0^2 + \omega_p^2(t)}}. \quad \text{(H.30b)}$$

For $\omega_b \gg \omega_0$ and $\omega_b \gg \omega_p$,

$$e_x(t) \cong 1 - \frac{3}{4} \frac{\omega_p^2(t)}{\omega_b^2}, \quad \text{(H.30c)}$$

$$e_z(t) \cong -\frac{\omega_p^2(t)}{\omega_b \omega_0}. \quad \text{(H.30d)}$$

Graphical results given in Section H.3 are obtained by computing $F(t)$ from Equation H.27. The results are verified using Equation H.22b involving a numerical integration routine.

H.3   Graphical Illustrations and Results

An exponentially varying electron density profile is used for an illustrative example. The time-varying plasma frequency is considered as

$$\omega_p^2(t) = \omega_{p0}^2 [1 - \exp(-bt)]. \quad \text{(H.31)}$$

The results for the first wave are presented in normalized form by taking $\omega_0 = 1$. The parameter b is chosen as 0.01. Figures H.2 through H.4 show the characteristics of the MSW. It has an upshifted frequency (Figure H.2) compared with the source wave frequency ω_0 when $\omega_b < \omega_0$ and a downshifted frequency when $\omega_b > \omega_0$. The amplitude function $e_x(t)$ (Figure H.3) of the MSW decreases with an increase in ω_b as long as $\omega_b < \omega_0$. However, it increases with ω_b for $\omega_b > \omega_0$. The amplitude function $h_y(t)$ follows a similar

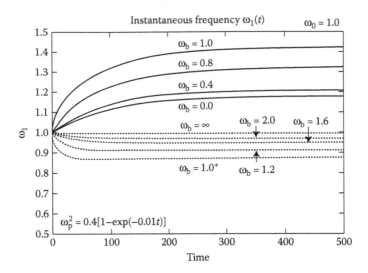

FIGURE H.2 Instantaneous frequency of the MSW.

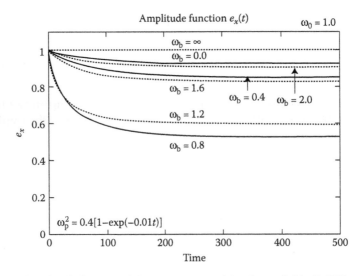

FIGURE H.3 Amplitude function of the x-component of the electric field of MSW.

pattern. The amplitude function $e_z(t)$ changes sign as the ω_b value crosses $\omega_b = \omega_0$ (Figure H.4).

Figures H.5 through H.8 show the variation of the saturation values ($t \to \infty$) of the frequency and amplitude function of the MSW. The horizontal axis is ω_b. However, in the neighborhood of $\omega_b \approx \omega_0$, the assumptions that $A_1 \approx 1$ and all other As are zero are not valid. The following validity condition is obtained by restricting the magnitude of the second term on the right-hand

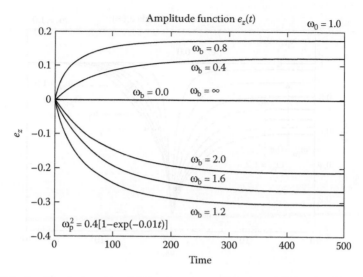

FIGURE H.4 Amplitude function of the z-component of the electric field of MSW.

side of Equation H.20a to be less than 0.1. The validity condition is

$$\left| 1 - \frac{\omega_b^2}{\omega_0^2} \right| > \left(20 \frac{b\omega_{p0}^2}{\omega_0^3} \right)^{1/3}. \tag{H.32}$$

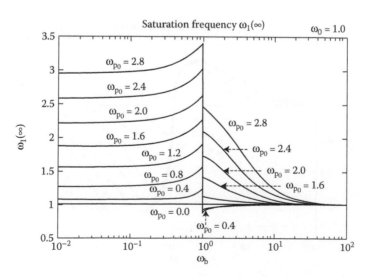

FIGURE H.5 Variation of the saturation frequency of the MSW with the electron cyclotron frequency (ω_b) and the saturation plasma frequency (ω_{p0}).

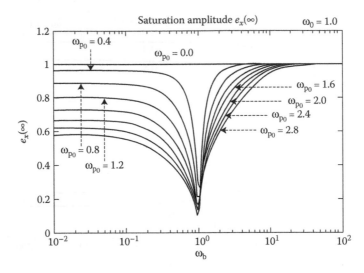

FIGURE H.6   Variation of the saturation amplitude of the x-component of the electric field of the MSW with ω_b and ω_{p0}.

In Figures H.5 through H.8, the following qualitative result may be stated: as the value of ω_b crosses ω_0, there will be a rapid change in the frequency and the fields of the source wave.

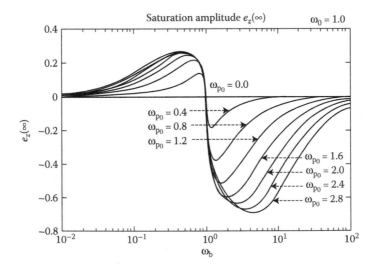

FIGURE H.7   Variation of the saturation amplitude of the z-component of the electric field of the MSW with ω_b and ω_{p0}.

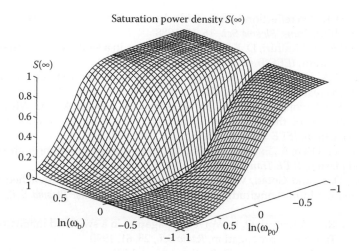

FIGURE H.8 Variation of the saturation power density of the MSW with ω_b and ω_{p0}.

H.4 Conclusion

The creation of a time-varying magnetoplasma medium splits the incident wave into four new waves having different frequencies in the case of transverse propagation. Two of the waves are the transmitted waves and the other two are the reflected waves. The first wave whose initial frequency is that of the incident wave has the strongest intensity compared with the others and is labeled as the MSW. Based on the adiabatic analysis of the MSW, the following qualitative statement may be made: as the value of ω_b crosses ω_0, there will be a rapid change in the frequency and the fields.

We still have difficulty in quantifying the behavior of the fields in the neighborhood of $\omega_b \approx \omega_0$. The accuracy of the solution degrades since $\dot{\omega}$ at $t = 0$ tends to infinity as ω_b tends toward ω_0. In this neighborhood, the higher-order terms become important. The aspect of the effect at $\omega_b = \omega_0$ is currently under investigation using a very different approach, and the result will be reported in due course.

Adiabatic solutions given in this paper can give an explicit analytical description for the frequency and amplitude of the MSW except in the neighborhood of $\omega_b \approx \omega_0$.

References

1. Jiang, C. L., Wave propagation and dipole radiation in a suddenly created plasma, *IEEE Trans. Ant. Prop.*, AP-23, 83, 1975.

2. Kalluri, D. K., On reflection from a suddenly created plasma half-space: Transient solution, *IEEE Trans. Plasma Sci.*, 16, 11, 1988.

3. Goteti, V. R. and Kalluri, D. K., Wave propagation in a switched-on time-varying plasma medium, *IEEE Trans. Plasma Sci.*, 17, 828, 1989.

4. Kalluri, D. K. and Goteti, V. R., Frequency shifting of electromagnetic radiation by sudden creation of a plasma slab, *J. Appl. Phys.*, 72, 4575, 1992.

5. Joshi, C. S., Clayton, C. E., Marsh, K., Hopkins, D. B., Sessler, A., and Whittum, D., Demonstration of the frequency upshifting of microwave radiation by rapid plasma creation, *IEEE Trans. Plasma Sci.*, 18, 814, 1990.

6. Kuo, S. P. and Ren, A., Experimental study of wave propagation through a rapidly created plasma, *IEEE Trans. Plasma Sci.*, 21, 53, 1993.

7. Kalluri, D. K. and Goteti, V. R., Damping rates of waves in a switched magnetoplasma medium: Longitudinal propagation, *IEEE Trans. Plasma Sci.*, 18, 797, 1993.

8. Goteti, V. R. and Kalluri, D. K., Wave propagation in a switched magnetoplasma medium: Transverse propagation, *Radio Sci.*, 25, 61, 1990.

9. Madala, S. R. V. and Kalluri, D. K., Frequency shifting of a wave traveling in an isotropic plasma medium due to slow switching of a magnetic field, *Plasma Sources Sci. Technol.*, 1, 242, 1992.

10. Madala, S. R. V. and Kalluri, D. K., Longitudinal propagation of low-frequency waves in a switched magnetoplasma medium, *Radio Sci.*, 28, 121, 1993.

11. Lai, C. H. and Katsouleas, T. C., Frequency upshifting by an ionization front in a magnetized plasma, *IEEE Trans. Plasma Sci.*, 21, 45, 1993.

12. Baños, Jr., A., Mori, W. B., and Dawson, J. M., Computation of the electric and magnetic fields induced in a plasma created by ionization lasting a finite interval of time, *IEEE Trans. Plasma Sci.*, 21, 57, 1993

13. Kalluri, D. K., Goteti, V. R., and Sessler, A. M., WKB solution for wave propagation in a time-varying magnetoplasma medium: Longitudinal propagation, *IEEE Trans. Plasma Sci.*, 21, 70, 1993.

14. Lee, J. H., Wave propagation in a time-varying switched magnetoplasma medium: Transverse propagation, MS thesis, University of Massachusetts Lowell, Lowell, 1994.

15. Kalluri, D. K., Frequency shifting using magnetoplasma medium: Flash ionization, *IEEE Trans. Plasma Sci.*, 21, 77, 1993.

16. Kalluri, D. K., Conversion of a whistler wave into a controllable helical wiggler magnetic field, *J. Appl. Phys.*, 79, 6770, 1996.

17. Kalluri, D. K., Frequency upshifting with power intensification of a whistler wave by a collapsing plasma medium, *J. Appl. Phys.*, 79, 3895, 1996.

18. Heald, M. A. and Wharton, C. B., *Plasma Diagnostics with Microwaves*, Wiley, New York, 1965, 36.

19. Blachman, N., *Mathematica: A Practical Approach*, Prentice-Hall, Englewood Cliffs, NJ, 1992.

Part II

Numerical Simulation: FDTD for Time-Varying Medium

11

FDTD Method

Sections 11.1 through 11.10 introduce the FDTD method. Sections 11.3, 11.5, 11.8, 11.9, and 11.10 follow the notation and the subject development in Refs. [1,2]. Sections 11.4, 11.6, and 11.7 follow the subject development given in Ref. [3]. Sections 11.11 through 11.14 are an introduction to Appendices I, J, and K, which give the development of FDTD to a time-varying, dispersive, inhomogeneous magnetoplasma medium. Appendices also contain many more references.

Maxwell's equations in time domain are partial differential equations of hyperbolic type. Finite difference solutions of this type, if properly formulated and discretized, will lead to a step-by-step solution rather than requiring the solution of simultaneous equations. Step-by-step solutions are also called "marching" solutions in the sense that the solution at the present time step is expressed in terms of the solution at previous time steps, already obtained by a simple calculation. In that sense, when it works, the solution takes much less time than the solution of an equilibrium problem. The difficulty is that, if it is not properly done, numerical instability can set in and one can get absurd results. Let us explain the above-mentioned points through a simple example.

11.1 Air-Transmission Line

The simple wave equation for an air-transmission line is

$$\frac{\partial^2 V}{\partial x^2} - \frac{1}{c^2}\frac{\partial^2 V}{\partial t^2} = 0. \tag{11.1}$$

The boundary conditions for the short-circuited line shown in Figure 11.1 are

$$V(0,t) = 0, \tag{11.2}$$

$$V(\ell,t) = 0. \tag{11.3}$$

Let us assume that the line is charged and has an initial voltage of $V_0 \sin(\pi x/\ell)$: such a static voltage distribution can arise due to an overhead

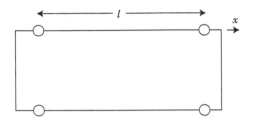

FIGURE 11.1   Air-transmission line.

cloud. The current I is zero. Suppose that the cloud moves away at $t = 0$ and we wish to find the voltage variation on the line as a function of x and t. Using the separation of variable technique, one can obtain an analytical solution for this simple problem. The initial conditions for the problem are

$$V(x,0) = V_0 \sin \frac{\pi x}{\ell}, \tag{11.4}$$

$$\frac{\partial V}{\partial t}(x,0) = 0. \tag{11.5}$$

Equation 11.5 is obtained from $I(x,0) = 0$, I being proportional to $\partial V/\partial t$. Equation 11.1 is a second-order differential equation, requiring two boundary conditions and two initial conditions. The analytical solution for this well-posed problem is given by

$$V(x,t) = V_0 \sin \frac{\pi x}{\ell} \cos \frac{\pi ct}{\ell}. \tag{11.6}$$

This solution will be useful in ascertaining the accuracy of the numerical solution, which we will obtain by using finite differences.

11.2   FDTD Solution

We shall obtain FDTD solution by using central difference approximations for the space and time second-order partial derivatives. Let

$$t = n\,\Delta t, \quad n = 0, 1, 2, \ldots, \tag{11.7}$$

$$x = i\,\Delta x, \quad i = 0, 1, 2, \ldots N. \tag{11.8}$$

We can now abbreviate (Figure 11.2)

$$V(x_i, t_n) = V_i^n. \tag{11.9}$$

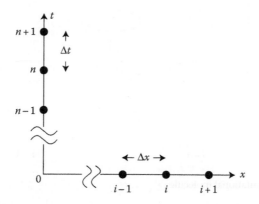

FIGURE 11.2   Notation for the grid points.

From central difference approximations

$$\left.\frac{\partial^2 V}{\partial x^2}\right|_{x_i,t_n} = \frac{V_{i+1}^n - 2V_i^n + V_{i-1}^n}{(\Delta x)^2} + 0(\Delta x)^2, \tag{11.10}$$

$$\left.\frac{\partial^2 V}{\partial t^2}\right|_{x_i,t_n} = \frac{V_i^{n+1} - 2V_i^n + V_i^{n-1}}{(\Delta t)^2} + 0(\Delta t)^2. \tag{11.11}$$

From Equations 11.1, 11.9, and 11.10 and treating V_i^{n+1} as the unknown to be calculated, we obtain the following:

$$\begin{aligned} V_i^{n+1} &= r^2 \left(V_{i+1}^n - 2V_i^n + V_{i-1}^n \right) + 2V_i^n - V_i^{n-1} \\ &= r^2 V_{i-1}^n + 2(1 - r^2)V_i^n + r^2 V_{i+1}^n - V_i^{n-1}, \end{aligned} \tag{11.12}$$

where

$$r = c\frac{\Delta t}{\Delta x}. \tag{11.13}$$

The computational molecule for the problem can be written as a sketch given in Figure 11.3.

The values in the circles are the weights of multiplication of the potentials at various grid points. The choice of $r = 1$ results in a simple computational molecule given in Figure 11.4 and Equation 11.14.

$$V_i^{n+1} = V_{i-1}^n + V_{i+1}^n - V_i^{n-1}. \tag{11.14}$$

We can apply this computational molecule and calculate the voltage at any time at any point on the transmission line. Let us illustrate this by hand computation to a simple example where the length of the line $\ell = 1$. Let us prepare a table where the values of the voltage will be recorded as shown in Figure 11.5.

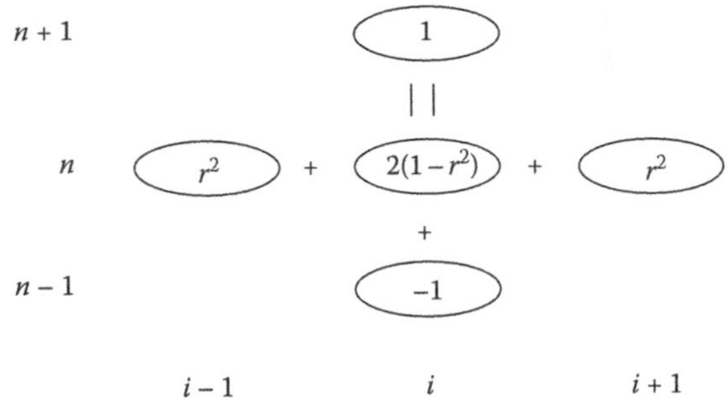

FIGURE 11.3 Computational molecule.

Note that the problem has a line of symmetry at $x = 0.5$ ($i = 5$). It then follows

$$V_4^n = V_6^n. \tag{11.15}$$

From the initial condition (Equation 11.4), $V_0^0 = 0$, $V_1^0 = \sin \pi(0.1) = 0.309$, and $V_2^0 = \sin \pi(0.2) = 0.58979$, and so on, are marked in Figure 11.5. The second initial condition given by Equation 11.5, when approximated by the central difference formula, gives

$$\left. \frac{\partial V}{\partial t} \right|_{t=0} = 0 = \frac{V_i^1 - V_i^{-1}}{2\Delta t} + O(\Delta t)^2 \tag{11.16}$$

$$V_i^1 = V_i^{-1}.$$

The first boundary condition (Equation 11.2) gives

$$V_0^n = 0. \tag{11.17}$$

From Equation 11.14, when applied for $n = 0, i = 1$

$$V_1^1 = V_0^0 + V_2^0 - V_1^{-1}. \tag{11.18}$$

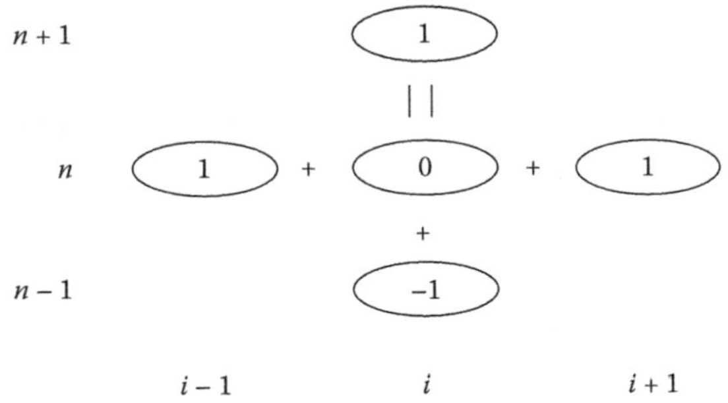

FIGURE 11.4 Computational molecule for $r = 1$.

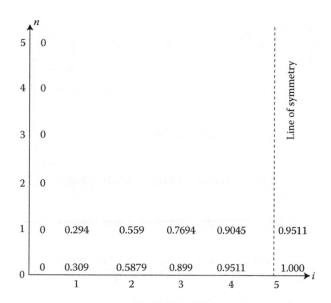

FIGURE 11.5 Values of the voltage variable at $n = 0$ and $n = 1$ time step levels.

From Equations 11.16 and 11.18,

$$2V_1^1 = 0 + 0.5879; \qquad V_1^1 = 0.294.$$

Applying Equation 11.14 at $n = 0$, $i = 2$ and using Equation 11.18,

$$V_2^1 = V_1^0 + V_3^0 - V_2^{-1}; \quad 2V_2^1 = V_1^0 + V_3^0; \quad V_2^1 = 0.559.$$

Similar computations at $i = 3, 4$ give

$$V_3^1 = 0.7694, \qquad V_4^1 = 0.9045.$$

For $n = 0$ and $i = 5$, Equation 11.14 gives

$$V_5^1 = V_4^0 + V_6^0 - V_5^{-1}.$$

From Equations 11.16, we have $V_5^1 = V_5^{-1}$, and from Equation 11.15, we have $V_4^0 = V_6^0$.

$$V_5^1 = V_4^0 = 0.9511.$$

Note that we obtained the voltage at the time level $t = \Delta t$ by a step-by-step calculation. Applying Equation 11.14 at interior points and applying Equation 11.15 for points on the line of symmetry, we can calculate the voltage at the next time step and then at the next time step, and so on, and march the

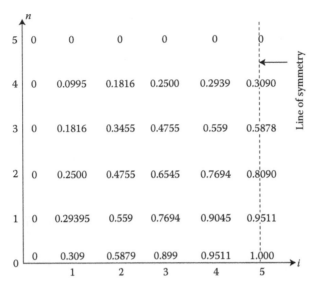

FIGURE 11.6 Values of the voltage variable up to $n = 5$.

solution to any time level. Figure 11.6 shows the table of values up to $n = 5$. Let us compare the exact solution with the value obtained by FDTD for V_3^1. The exact solution is given by substituting $x = 3\Delta x$ and $t = \Delta t$ in Equation 11.6 with $V_0 = 1$ and $\ell = 1$:

$$V_3^1 = \sin 3\pi\, \Delta x \, \cos \pi c\, \Delta t,$$

$$\Delta x = 0.1; \quad c\frac{\Delta t}{\Delta x} = 1; \quad \Delta t = \frac{\Delta x}{c} = \frac{0.1}{c},$$

$$V_3^1 = \sin 0.3\pi \, \cos 0.1\pi = 0.7694,$$

which is the value obtained by FDTD. The result is surprising since the used central difference approximation is only accurate up to second order, that is, of the order of $O\left[(\Delta x)^2, (\Delta t)^2\right]$. This result is not accidental but it can be shown that the FDTD solution is exact for $r = 1$. For example, if we have chosen $\Delta x = 0.25$, but we choose Δt such that $r = c(\Delta t/\Delta x) = 1$, we again obtain the exact answer.

11.3 Numerical Dispersion

For the simple one-dimensional wave equation given by Equation 11.1, the harmonic solution of a positive-going wave is

$$V = A\, e^{j(\omega t - kx)}. \tag{11.19}$$

A negative-going wave is also a solution and the solution (Equation 11.6) is a superposition of the above two waves satisfying the initial and boundary conditions. The relation between ω and k is

$$\omega = ck, \tag{11.20}$$

where c is the velocity of light. The $\omega = ck$ diagram is a straight line. The phase velocity $v_p = \omega/k$ and the group velocity $v_g = d\omega/dk$ are both equal to c. When we approximate Equation 11.1 by using central difference formulas, we obtained the computational molecule given by Equation 11.12. Since the central difference formula we used has an error of the order of $(\Delta t)^2$ and $(\Delta x)^2$, we expect the wave solutions to be approximate. The error can be studied by studying a monotonic wave of frequency ω and an approximate corresponding wave number $\tilde{k}$, and the solution can be written as

$$\tilde{V} \approx A\, e^{j(\omega t - \tilde{k}x)}. \tag{11.21}$$

The relation between ω and the numerical wave number $\tilde{k}$ is called numerical dispersion relation. We can obtain this relation by substituting Equation 11.21 into Equation 11.12. The constant A will cancel and hence we will concentrate on the exponential factors involved in Equation 11.12. The exponential factor ψ, $\exp[j(\omega t - kx)]$, in V will assume the form ψ_i^n in V_i^n:

$$\psi_i^n = e^{j\left(\omega n\,\Delta t - \tilde{k}i\,\Delta x\right)}. \tag{11.22}$$

Thus Equation 11.12 can be approximated as

$$e^{j\left[\omega(n+1)\,\Delta t - \tilde{k}i\,\Delta x\right]} = r^2\left\{e^{j\left[\omega n\,\Delta t - \tilde{k}(i+1)\,\Delta x\right]} - 2e^{j(\omega n\,\Delta t - \tilde{k}i\,\Delta x)} + e^{j\left[\omega n\,\Delta t - \tilde{k}(i-1)\,\Delta x\right]}\right\}$$
$$+ 2e^{j\left(\omega n\,\Delta t - \tilde{k}i\,\Delta x\right)} - e^{j\left[\omega(n-1)\,\Delta t - \tilde{k}i\,\Delta x\right]}.$$

Factoring out $e^{j\left(\omega n\,\Delta t - \tilde{k}i\,\Delta x\right)}$, one can obtain

$$e^{j\omega\Delta t} = r^2\left(e^{-j\tilde{k}\,\Delta x} - 2 + e^{j\tilde{k}\,\Delta x}\right) + 2 - e^{-j\omega\Delta t}. \tag{11.23}$$

Grouping the terms, Equation 11.23 can be written as

$$\cos(\omega\,\Delta t) = r^2\left[\cos\left(\tilde{k}\,\Delta x\right) - 1\right] + 1. \tag{11.24}$$

Let us now examine two special cases

Case 1: r = 1
Equation 11.24 gives

$$\cos \omega \Delta t = \cos \tilde{k} \Delta x, \quad \omega \Delta t = \tilde{k} \, \Delta x, \quad \tilde{v}_p = \frac{\omega}{\tilde{k}} = \frac{\Delta x}{\Delta t} = c. \qquad (11.25)$$

The last equality in Equation 11.25 is because

$$r = c \frac{\Delta t}{\Delta x} = 1.$$

For this case, the numerical wave has the same phase velocity as the physical wave; the numerical approximation has not introduced any dispersion. This explains the exact answer we obtained when we used $r = 1$.

Case 2: Very fine mesh $\Delta t \to 0$, $\Delta x \to 0$
The arguments of the cosine functions in Equation 11.24 are small, and we can use the approximations

$$\cos \theta = (1 - \sin^2 \theta)^{1/2} \approx 1 - \tfrac{1}{2} \sin^2 \theta \approx 1 - \tfrac{1}{2} \theta^2 \qquad (11.26)$$

and obtain

$$1 - \tfrac{1}{2}(\omega \, \Delta t)^2 = r^2 \left[1 - \tfrac{1}{2} \tilde{k}^2 (\Delta x)^2 - 1 \right] + 1,$$

$$-\tfrac{1}{2}(\omega \Delta t)^2 = c^2 \frac{(\Delta t)^2}{(\Delta x)^2} \left[-\frac{1}{2} \tilde{k}^2 (\Delta x)^2 \right], \qquad (11.27)$$

$$\omega^2 = c^2 \tilde{k}^2.$$

As expected, the numerical wave and the physical wave are identical.
General case:
Let us look at a general case of arbitrary r and R, where

$$R = \frac{\lambda_0}{\Delta x}. \qquad (11.28)$$

R is the number of cells per wavelength. Equation 11.24 can be written as

$$\tilde{k} = \frac{1}{\Delta x} \cos^{-1} \left\{ 1 + \frac{1}{r^2} [\cos(\omega \, \Delta t) - 1] \right\}, \qquad (11.29)$$

and the normalized numerical phase velocity is given by

$$\frac{\tilde{v}_p}{c} = \frac{\omega}{\tilde{k}c} = \frac{(\omega \, \Delta x/c)}{\cos^{-1} \left\{ 1 + ((\cos(\omega \, \Delta t) - 1)/r^2) \right\}}.$$

Noting that

$$\frac{\omega \, \Delta x}{c} = \frac{2\pi f \, \Delta x}{c} = \frac{2\pi \, \Delta x}{\lambda_0} = \frac{2\pi}{R},$$

$$\omega \, \Delta t = 2\pi f \, \Delta t = \frac{2\pi f}{c} \frac{c \, \Delta t}{\Delta x} \Delta x = \frac{2\pi}{\lambda_0} r \, \Delta x = 2\pi \frac{r}{R},$$

we obtain

$$\frac{\tilde{v}_p}{c} = \frac{2\pi}{R \cos^{-1}\left\{1 + (1/r^2)\left[\cos 2\pi(r/R) - 1\right]\right\}}. \tag{11.30}$$

Let us look at a couple of numbers obtained from Equation 11.30.

Case 1: $r = 0.5$, $R = 10$: Let $r = 0.5$ and the number of cells per wavelength $R = 10$; the normalized numerical phase velocity

$$\frac{\tilde{v}_p}{c} = 0.9873. \tag{11.31a}$$

This is an error of 1.27%.

Case 2: $r = 0.5$, $R = 20$:

$$\frac{\tilde{v}_p}{c} = 0.99689. \tag{11.31b}$$

By doubling R, the phase velocity error is reduced to 0.31%, 0.25 of the previous value. Figure 11.7 shows the normalized phase velocity versus R with r as a parameter. As we note from Figure 11.7, the decrease in accuracy with a decrease in r can be compensated for by increasing R, that is, by increasing the number of cells per wavelength.

11.4 Stability Limit and Courant Condition

In any one time step, a point on a plane wave propagating through a one-dimensional FDTD grid cannot pass through more than one cell. If the medium is free space and if the point exactly passes through one cell, then $c(\Delta t/\Delta x) = r = 1$. This condition is called the courant condition. If $r > 1$, then the point passes through a distance more than that of one cell, resulting in a velocity $\Delta x/\Delta t > c$, violating the principle that the wave velocity in free space cannot be greater than that of the light. Thus, the choice of $r > 1$ results in an instability of the algorithm, which cannot be corrected by choosing a large

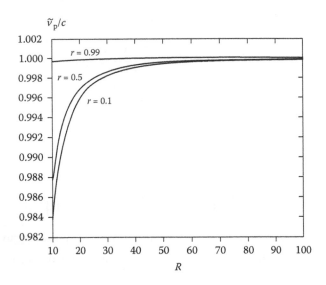

FIGURE 11.7   Normalized numerical phase velocity versus R with r as the parameter.

value for R. For two- and three-dimensional problems, the stability criterion can be stated as

$$r = \frac{c\Delta t}{\sqrt{N}\Delta} \leq 1,$$ (11.32)

where

$$\Delta = \Delta x = \Delta y = \Delta z,$$ (11.33)

and N is the dimensionality.

11.5   Open Boundaries

The problem solved in Section 11.1 has bounded space $0 < x < l$ and the problem is solved by using the boundary conditions at the end points $x = 0$ and $x = l$. The problem of finding the reflection and transmission coefficients involves open boundaries and the domain is $-\infty < x < \infty$. Unless we terminate the problem space we cannot solve the problem (Figure 11.8).

Suppose we decide to consider the problem space as $-L_1 < x < L_2$. At $x = -L_1$, the reflected wave is to be an outgoing wave and should not be reflected back into the problem space. This can be ensured if the wave is absorbed by the boundary. Such a boundary condition is called absorbing boundary condition. A simple way of implementing the absorbing boundary condition is illustrated below. Let the left boundary $x = -L_1$ have the space index $i = 0$

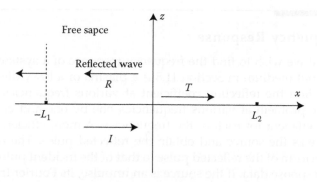

FIGURE 11.8 One-dimensional problem involving open boundaries.

and let us choose

$$r = c\frac{\Delta t}{\Delta x} = 0.5. \qquad (11.34)$$

The above equation tells us that it takes two time steps for a wavefront to cross one cell, that is,

$$E_0^n = E_1^{n-2}. \qquad (11.35)$$

As you approach the left boundary, store the value of E_1^{n-2} in E_0^n. A similar arrangement at $x = L_2$ terminates the right boundary.

11.6 Source Excitation

The problem solved in Section 11.1 has the initial value of the voltage of the charged line given. For a problem in Section 11.5, the incident wave is specified. If the incident wave is a monochromatic wave, one can simulate it by assigning a sinusoidal value at a fixed point. This is called a hard source. A propagating wave sees the hard source as a metal wall and when the pulse passes through it, it will be reflected. Suppose that the fixed point is indexed by $i = F$, then

$$E_F^n = \sin(2 * \text{pi} * \text{freq} * dt * n) \qquad (11.36)$$

generates a hard source at the space point $i = F$.

On the other hand,

$$\text{pulse} = \sin(2 * \text{pi} * \text{freq} * dt * n) \qquad (11.37)$$

$$E_F^n = E_F^n + \text{pulse} \qquad (11.38)$$

generates a soft source. With a soft source, the propagating pulse will pass through.

11.7 Frequency Response

Suppose that we wish to find the frequency response of a system. Suppose that the output medium in Section 11.5 is a plasma or a lossy dielectric and we wish to find the reflection coefficient at various frequencies. One way is to run the problem at various frequencies one by one and calculate the reflection coefficient for each of the frequencies. A more efficient way is to have a pulse as the source and obtain the reflected pulse. The ratio of the Fourier transform of the reflected pulse to that of the incident pulse gives the frequency response data. If the source is an impulse, its Fourier transform is 1, that is, the frequency spectrum has an amplitude of 1 at all frequencies. A more practical source is a Gaussian pulse given by

$$f(t) = e^{-1/2((t-t_0)/\sigma)^2}, \tag{11.39}$$

which is centered at t_0 and has a spread (standard deviation) of σ (Figure 11.9). As an example [3], suppose that $t_0 = 40\Delta t$, $t = n\Delta t$, and $\sigma = 12\Delta t$, then

$$f(n) = e^{-0.5((n-40)/12)^2}, \tag{11.40}$$

$$f(0) = f(80) = e^{-0.5(10/3)^2} = e^{-5.55} \approx 0. \tag{11.41}$$

The pulse can be made narrower by choosing a smaller value of σ. For example, if $\sigma = 5\sqrt{2}\Delta t$,

$$f(0) = f(80) = e^{-0.5\left(40/5\sqrt{2}\right)^2} = e^{-0.5*64*0.5} = e^{-16} = -140 \text{ dB}. \tag{11.42}$$

The amplitude of the Fourier transform of Equation 11.39 is sketched in Figure 11.10. The truncation of the pulse at $t = 0$ or $t = 40\Delta t$ did not introduce unwanted high frequencies.

From the fast Fourier transform (FFT), it is clear that the amplitude of the signal is quite adequate, say up to 3 GHz. However, we will be concerned

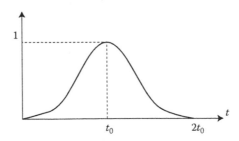

FIGURE 11.9 Gaussian pulse.

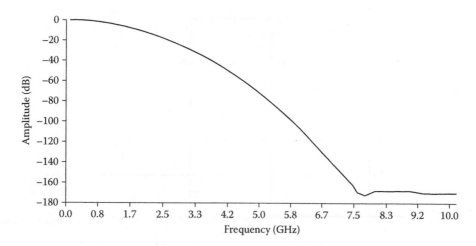

FIGURE 11.10 Fourier transform of the Gaussian pulse $\Delta t = 1.875 \times 10^{-11}$, $\sigma = 5\sqrt{2}\Delta t$, $t_0 = 40\Delta t$.

about noise if say 7 or 8 GHz is of interest, since the signal amplitude is 120 dB at this frequency.

After obtaining the reflected pulse, one can obtain the FFT of the reflected pulse and thus one can obtain the ratio of the reflected pulse amplitude to the incident pulse amplitude at various frequencies.

11.8 Waves in Inhomogeneous, Nondispersive Media: FDTD Solution

Let us apply the FDTD method to wave propagation in an inhomogeneous, nondispersive dielectric medium. Consider the one-dimensional case, where the fields as well as the permittivity vary in one spatial dimension:

$$\varepsilon = \varepsilon(x), \tag{11.43}$$

$$\bar{D} = \hat{z}D_z(x), \tag{11.44a}$$

$$\bar{E} = \hat{z}E_z(x), \tag{11.44b}$$

$$\bar{H} = \hat{y}H_y(x), \tag{11.45}$$

$$D_z = \varepsilon(x)E_z. \tag{11.46}$$

A simple example of Equation 11.46 is a dielectric slab of width d and permittivity ε_2. The input medium has permittivity ε_1 and the output medium has permittivity ε_3 as shown in Figure 11.11.

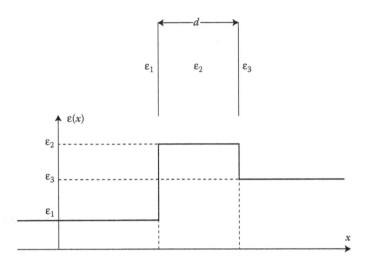

FIGURE 11.11   Example of an inhomogeneous medium.

The relevant Maxwell's equations for a nonmagnetic medium are

$$\bar{\nabla} \times \bar{E} = -\mu_0 \frac{\partial \bar{H}}{\partial t}, \tag{11.47}$$

$$\bar{\nabla} \times \bar{H} = \frac{\partial \bar{D}}{\partial t}. \tag{11.48}$$

From Equations 11.47, 11.44b, and 11.45,

$$\frac{\partial H_y}{\partial t} = \frac{1}{\mu_0} \frac{\partial E_z}{\partial x}. \tag{11.49}$$

From Equations 11.48, 11.45, and 11.44a,

$$\frac{\partial D_z}{\partial t} = \frac{\partial H_y}{\partial x}. \tag{11.50}$$

We can discretize Equation 11.49 by using the central difference formula for the first derivative. Let $t = n\Delta t$ and $x = i\Delta x$, and approximate Equation 11.49 at the spatial index $(i + (1/2))$ and time index n:

$$\left.\frac{\partial H_y}{\partial t}\right|_{i+(1/2)}^{n} = \frac{1}{\mu_0} \left.\frac{\partial E_z}{\partial x}\right|_{i+(1/2)}^{n}, \tag{11.51}$$

$$\frac{\left.H_y\right|_{i+(1/2)}^{n+(1/2)} - \left.H_y\right|_{i+(1/2)}^{n-(1/2)}}{\Delta t} = \frac{1}{\mu_0} \frac{\left.E_z\right|_{i+1}^{n} - \left.E_z\right|_{i}^{n}}{\Delta x}, \tag{11.52}$$

$$H_y\Big|_{i+(1/2)}^{n+(1/2)} = H_y\Big|_{i+(1/2)}^{n-(1/2)} + \frac{\Delta t}{\mu_0\,\Delta x}\left[E_z\Big|_{i+1}^{n} - E_z\Big|_{i}^{n}\right]. \tag{11.53}$$

We can discretize Equation 11.50 by using FDTD at the space index i and time index $(n + (1/2))$:

$$\frac{\partial D_z}{\partial t}\Big|_{i}^{n+(1/2)} = \frac{\partial H_Y}{\partial x}\Big|_{i}^{n+(1/2)}, \tag{11.54}$$

$$D_z\Big|_{i}^{n+1} = D_z\Big|_{i}^{n} + \frac{\Delta t}{\Delta x}\left[H_y\Big|_{i+(1/2)}^{n+(1/2)} - H_y\Big|_{i-(1/2)}^{n+(1/2)}\right]. \tag{11.55}$$

The algebraic Equation 11.46 can be discretized at the space index i and time index i:

$$E_z\Big|_{i}^{n+1} = \frac{1}{\varepsilon_i}D_z\Big|_{i}^{n+1}, \tag{11.56}$$

where

$$\varepsilon_i = \varepsilon(i\Delta x). \tag{11.57}$$

Equations 11.53, 11.55, and 11.56 give a leap-frog step-by-step computational algorithm to compute and march the solution in the time domain. Figure 11.12 helps us to understand the algorithm.

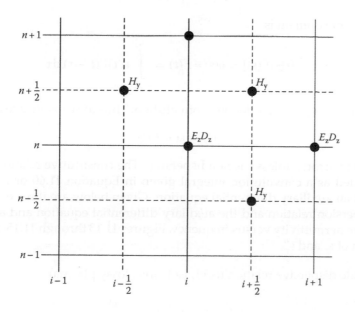

FIGURE 11.12 Leap-frog computational algorithm.

The left-hand side of Equation 11.53 can be computed from the values of $H_y|_{i+(1/2)}^{n-(1/2)}$ and $E_z|_i^n$ and $E_z|_{i+1}^n$ on the right-hand side of Equation 11.53, which are the values at previous time steps and have already been computed. Thus, H_y values at time step $[n + (1/2)]$ are calculated. The left-hand side of Equation 11.55 is now computed from D_z at time step n and H_y at time step $[n + (1/2)]$. Thus D_z values at time step $(n + 1)$ are computed. Using Equation 11.56, from D_z, E_z can be computed. Repeating these three computational steps, the values of the fields can be computed at the next level of time step and the solution is thus marched in time. To start the process, you have to supply the initial values: H_y at the time step $-1/2$ $(n = 0, t = -\Delta t/2)$ and E_z at the time step 0 $(n = 0, t = 0)$.

11.9 Waves in Inhomogeneous, Dispersive Media

For a dispersive medium, where ε is a function of the frequency

$$\varepsilon = \varepsilon(\omega), \tag{11.58}$$

the algebraic Equation 11.46 will be true only if a monochromatic wave propagation is considered. A pulse propagation problem requires the constitutive relation for the dielectric medium to be written as

$$D(s) = \varepsilon(s)E(s), \tag{11.59}$$

which in time domain is

$$D(t) = L^{-1}[\varepsilon(s)E(s)] = \varepsilon(s) * E(t) = \int_{-\infty}^{\infty} \varepsilon(\tau)E(t - \tau)\, d\tau. \tag{11.60}$$

In the above equation, $*$ denotes convolution. Thus the constitutive relation

$$D(t) = \varepsilon(\omega)E(t) \tag{11.61}$$

will lead to errors, unless ε is nondispersive. The constitutive relation needs to be stated as a convolution integral given in Equation 11.60 or it can be stated as an auxiliary differential equation. Given below are three examples of a dispersion relation and the auxiliary differential equation and a sketch of relative permittivity versus frequency. Figures 11.13 through 11.15 give the variation of ε_r' and ε_r''.

1. Drude dispersive relation (cold, isotropic, lossy plasma)

$$\varepsilon_r' - j\varepsilon_r'' = \varepsilon_p(\omega) = 1 - \frac{\omega_p^2}{\omega(\omega - j\nu)} \tag{11.62}$$

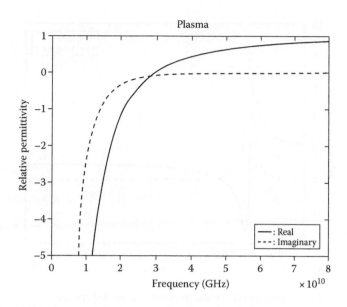

FIGURE 11.13 Relative permittivity versus frequency: Drude dispersion.

Constitutive relation as an auxiliary differential equation

$$\frac{d\bar{J}}{dt} + v\bar{J} = \varepsilon_0 \omega_p^2 \bar{E} \tag{11.63}$$

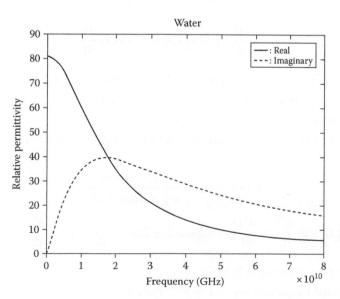

FIGURE 11.14 Relative permittivity versus frequency: Debye dispersion.

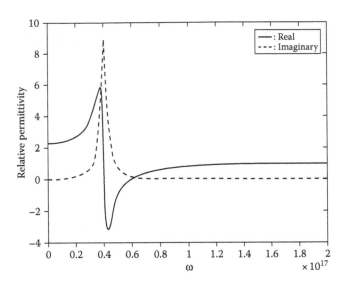

FIGURE 11.15   Relative permittivity versus frequency: Lorentz dispersion.

Example values: plasma frequency $f_p = (\omega_p/2\pi) = 30$ GHz Collision frequency $\nu = 2 \times 10^{10}$ rad/s

2. Debye dispersion relation (water)

$$\varepsilon_r(\omega) = \varepsilon_\infty + \frac{\varepsilon_s - \varepsilon_\infty}{1 + j\omega t_0} \tag{11.64}$$

$$t_0 \frac{d\bar{D}}{dt} + \bar{D} = \varepsilon_s \varepsilon_0 \bar{E} + t_0 \varepsilon_\infty \varepsilon_0 \frac{d\bar{E}}{dt} \tag{11.65}$$

Example values: low-frequency relative permittivity $\varepsilon_s = 81$
High-frequency relative permittivity $\varepsilon_\infty = 1.8$
Relaxation time $t_0 = 9.4 \times 10^{-12}$ s

3. Lorentz dispersive material (second-order, optical material)

$$\varepsilon_r(\omega) = \varepsilon_\infty + \frac{(\varepsilon_s - \varepsilon_\infty)\omega_R^2}{\omega_R^2 + 2j\omega\delta - \omega^2} \tag{11.66}$$

$$\omega_R^2 \bar{D} + 2\delta\frac{d\bar{D}}{dt} + \frac{d^2\bar{D}}{dt^2} = \omega_R^2 \varepsilon_s \varepsilon_0 \bar{E} + 2\delta\varepsilon_\infty\varepsilon_0\frac{d\bar{E}}{dt} + \varepsilon_\infty\varepsilon_0\frac{d^2\bar{E}}{dt^2} \tag{11.67}$$

Example values: $\varepsilon_s = 2.25$, $\varepsilon_\infty = 1$
Resonant frequency $\omega_R = 4 \times 10^{16}$ rad/s
Damping constant $\delta = 0.28 \times 10^{16}/$s

The technique of obtaining the auxiliary differential equation from the dispersion relation is illustrated for the Debye material

$$\bar{D}(\omega) = \varepsilon_0 \varepsilon_r(\omega)\bar{E}(\omega) = \varepsilon_0 \left[\varepsilon_\infty + \frac{\varepsilon_s - \varepsilon_\infty}{1 + j\omega t_0} \right] \bar{E}(\omega), \qquad (11.68)$$

$$\bar{D}(\omega)(1 + j\omega t_0) = \varepsilon_0 \varepsilon_\infty (1 + j\omega t_0)\bar{E}(\omega) + \varepsilon_0(\varepsilon_s - \varepsilon_\infty)\bar{E}(\omega), \qquad (11.69)$$

$$j\omega t_0 \bar{D}(j\omega) + \bar{D}(j\omega) = \varepsilon_0(\varepsilon_\infty + \varepsilon_s - \varepsilon_\infty)\bar{E}(j\omega) + \varepsilon_0 \varepsilon_\infty j\omega t_0 \bar{E}(j\omega). \qquad (11.70)$$

Using $\partial/\partial t = j\omega$,

$$t_0 \frac{\partial \bar{D}}{\partial t} + \bar{D} = \varepsilon_0 \varepsilon_\infty t_0 \frac{\partial \bar{E}}{\partial t} + \varepsilon_0 \varepsilon_s \bar{E}. \qquad (11.71)$$

Note, however, that in example (1), the constitutive relation in time domain relates the current density with the electric field. The medium is modeled as a conductor. The frequency domain constitutive relation relates $\bar{D}$ with $\bar{E}$.

11.10   Waves in Debye Material: FDTD Solution

The auxiliary differential Equation 11.65 is the constitutive relation for the Debye material. The computational algorithm consists of Equations 11.53 and 11.55 but Equation 11.56 needs to be replaced. The replacement can be obtained by discretizing Equation 11.65 through FDTD. Applying Equation 11.65 at the space step i and the time step $[n + (1/2)]$,

$$\varepsilon_0 \varepsilon_\infty t_0 \frac{dE_z}{dt}\bigg|_i^{n+(1/2)} + \varepsilon_0 \varepsilon_s E_z|_i^{n+(1/2)} = t_0 \frac{dD_z}{dt}\bigg|_i^{n+(1/2)} + D_z|_i^{n+(1/2)}. \qquad (11.72)$$

Using central difference approximation in Equation 11.68,

$$\varepsilon_0 \varepsilon_\infty t_0 \frac{E_z|_i^{n+1} - E_z|_i^n}{\Delta t} + \varepsilon_0 \varepsilon_s \ E_z|_i^{n+(1/2)} = t_0 \frac{D_z|_i^{n+1} - D_z|_i^n}{\Delta t} + D_z|_i^{n+(1/2)}.$$

$$\qquad (11.73)$$

Since E_z and D_z are available at integral time steps (n integer), we cannot directly get values for $E_z|_i^{n+(1/2)}$ and $D_z|_i^{n+(1/2)}$ in Equation 11.69. However, we can use averaging operator on these (averaging operator is accurate to second order in Δt):

$$E_z|_i^{n+(1/2)} = \frac{E_z|_i^{n+1} + E_z|_i^n}{2}, \qquad (11.74)$$

$$D_z|_i^{n+(1/2)} = \frac{D_z|_i^{n+1} + D_z|_i^n}{2}. \qquad (11.75)$$

Substituting Equations 11.74 and 11.75 into Equation 11.73 and rearranging the unknown to be on the left-hand side of the equation, we obtain

$$
E_z|_i^{n+1} = \frac{\Delta t + 2t_0}{2t_0\varepsilon_\infty + \varepsilon_s\,\Delta t}\,D_z|_i^{n+1} + \frac{\Delta t - 2t_0}{2t_0\varepsilon_\infty + \varepsilon_s\,\Delta t}\,D_z|_i^{n} + \frac{2t_0\varepsilon_\infty - \varepsilon_s\,\Delta t}{2t_0\varepsilon_\infty + \varepsilon_s\,\Delta t}\,E_z|_i^{n}.
$$

$$(11.76)$$

Equations 11.55, 11.56, and 11.76 give the computational algorithm for the Debye material.

11.11   Total Field/Scattered Field Formulation

The approach is based on dealing with total fields in region 1 that includes the interacting structure and only scattered fields in region 2, which is terminated by a domain where absorbing boundary conditions are implemented. Appendix I illustrates its use in studying the interaction of an electromagnetic pulse with a switched plasma slab.

11.12   Perfectly Matched Layer: Lattice Truncation

Section 11.8 introduced a very simple technique to terminate the problem space. A perfectly matched layer (artificial material) called PML absorbs an EMW that enters it. The artificial material properties are designed so that the impedance matching boundary condition is satisfied and no refection occurs. Appendix I illustrates its use in studying the interaction of an electromagnetic pulse with a switched plasma slab.

11.13   Exponential Time Stepping

For a highly lossy medium, the exponential decay of waves is so rapid that the use of central difference approximation of the first derivative used in the standard Yee algorithm cannot be easily used. An alternative time-stepping expression valid for highly lossy media is called exponential time stepping and is discussed in Refs. [1,2]. On a lossy plasma, the auxiliary differential equation is

$$
\frac{dJ}{dt} + vJ = \varepsilon_0\omega_p^2(z,t)E.
$$

$$(11.77)$$

The exponential time-stepping algorithm for it can be obtained:

$$J\big|_k^{n+1/2} = e^{-\nu|_k^n \Delta t} J\big|_k^{n-1/2} + \frac{\varepsilon_0}{\nu|_k^n} \left(1 - e^{-\nu|_k^n \Delta t}\right) \omega_p^2\big|_k^n E\big|_k^n. \tag{11.78}$$

Equation 11.78 is obtained by solving Equation 11.77, over one time step, assuming ν, ω_p, and E_k are constant over a time step and have values at the center of the time step. Let the beginning of the time step be indexed by $(n - 1/2)$ and the end of the time step be indexed by $(n + 1/2)$.

On the other hand, if we use central difference approximation for the derivative

$$\frac{dJ}{dt}\bigg|_k^n = \frac{J\big|_k^{n+1/2} - J\big|_k^{n-1/2}}{\Delta t} \tag{11.79}$$

and use the averaging operator for $J\big|_k^n$

$$J\big|_k^n = \frac{J\big|_k^{n+1/2} + J\big|_k^{n-1/2}}{2}, \tag{11.80}$$

Equation 11.77 then gives

$$J\big|_k^{n+1/2} = \frac{1 - ((\nu|_k^n \Delta t)/2)}{1 + ((\nu|_k^n \Delta t)/2)} J\big|_k^{n-1/2} + \varepsilon_0 \; \omega_p^2\big|_k^n E\big|_k^n,$$

$$J\big|_k^{n+1/2} = \frac{(1 - ((\nu|_k^n \Delta t)/2))^2}{\left(1 - ((\nu|_k^n \Delta t)/2)^2\right)} J\big|_k^{n-1/2} + \varepsilon_0 \; \omega_p^2\big|_k^n E\big|_k^n. \tag{11.81}$$

Equation 11.81 is approximately the same as Equation 11.78 provided $(\nu \Delta t/2)^2 \ll 1$ and is neglected compared with 1. Expanding the exponential terms in Equation 11.78 and neglecting terms of the order $(\nu \Delta t)^2$, we obtain

$$J\big|_k^{n+1/2} = \left(1 - \nu|_k^n \Delta t\right) J\big|_k^{n-1/2} + \varepsilon_0 \; \Delta t \omega_p^2\big|_k^n E\big|_k^n. \tag{11.82}$$

Equation 11.81 is the same as Equation 11.82 if we neglect $(\nu \Delta t/2)^2$. Moreover, in the limit $\nu \to 0$, Equations 11.78 and 11.81 are the same.

11.14 FDTD for a Magnetoplasma

Appendix J deals with the FDTD simulation of EMW transformation in a dynamic magnetized plasma for the case of wave propagation in the direction of static magnetic field. The source wave is assumed to be an R wave. The results obtained earlier using analytical approximations based on (a) the WKB method for slow switching of the plasma medium (Chapter 9) and (b) Green's function technique for rapid switching of the plasma medium (Chapter 8) are verified. The code is capable of obtaining results for the arbitrary rise time of the plasma profile.

Sudden switching as well as switching with a finite rise time of periodic magnetoplasma layers is studied through this code. Thus the results of Section 2.7 are extended for the case of magnetoplasma layers alternating with dielectric layers.

A novel successive reduction method has been developed and applied to obtain the amplitudes and frequencies of the new modes generated by the switching medium. Appendices M, N, O, P, and Q also deal with the one-dimensional FDTD technique for a switched medium in a cavity and are included in Part III of this book.

11.15   Three-Dimensional FDTD

In Appendix K, a three-dimensional FDTD algorithm is developed to study the transformation of an EMW by a dynamic (time-varying) inhomogeneous magnetized plasma medium. The current density vector is positioned at the center of the Yee cube [4] to accommodate the anisotropy of the plasma medium due to the presence of a static magnetic field. An appropriate time-stepping algorithm is used to obtain accurate solutions for arbitrary values of the collision frequency and the electron cyclotron frequency. The technique is illustrated by calculating the frequency shifts in a cavity due to a switched magnetoplasma medium with a time-varying and space-varying electron density profile.

References

1. Taflove, A., *Computational Electrodynamics: The Finite-Difference Time-Domain Method*, Artech House Inc., Norwood, MA, 1995.
2. Taflove, A. and Hagness, S., *Computational Electrodynamics: The Finite-Difference Time-Domain Method*, 2nd edition, Artech House Inc., Norwood, MA, 2000.
3. Sullivan, D. M., *Electromagnetic Simulation Using FDTD Method*, IEEE Press, New York, 2000.
4. Yee, K., Numerical solution of initial boundary value problems involving Maxwell equations in isotropic media, *IEEE Trans. Ant. Prop.*, 14(3), 302–307, 1966.

Appendix I

FDTD Simulation of Electromagnetic Pulse Interaction with a Switched Plasma Slab*

Dikshitulu K. Kalluri, Joo Hwa Lee, and Monzurul M. Ehsan

I.1 Introduction

The interaction of an EMW with a plasma slab is experimentally more realizable than an unbounded plasma medium. When an incident wave enters a preexisting plasma slab, the wave experiences space discontinuity. If the plasma frequency is lower than the incident wave frequency, then the incident wave will be partially reflected and transmitted. When the plasma frequency is higher than that of the incident wave, the wave is totally reflected because the dielectric constant in the plasma is less than zero. However, if the width of the plasma slab is sufficiently thin, the wave can be transmitted, which is known as a tunneling effect [1]. For this time-invariant plasma, the reflected and transmitted waves have the same frequency as the source wave frequency and we call these as A waves. The wave inside the plasma has a different wave number but the same frequency due to the requirement of the boundary conditions.

When a source wave is propagating in free space and suddenly a plasma slab is created, the wave inside will experience a time discontinuity in the properties of the medium. Hence, the switching action generates new waves whose frequencies are upshifted and then the waves propagate out of the slab. We call these waves as B waves. The phenomenon is illustrated in Figure I.1. In Figure I.1, the source wave of the frequency ω_0 is propagating in free space. At $t = 0$, a plasma slab of the plasma frequency ω_p is created. The A waves in Figure I.1 have the same frequency as that of the source wave. The B waves are newly created due to the creation of the plasma and upshifted to $\omega_1 = \sqrt{\omega_0^2 + \omega_p^2} = -\omega_2$. The negative value for the frequency of the second B wave shows that it is a backward-propagating wave. These waves, however, have the same wave number as that of the source wave as long as they remain

---

* Reprinted from *Int. J. Infrared Millim. Waves*, 24(3), 349–365, March 2003. With permission.

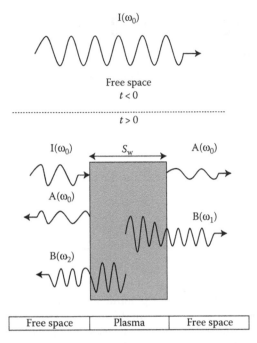

FIGURE I.1   Effect of switching an isotropic plasma slab. The A waves have the same frequency as the incident wave frequency (ω_0) and the B waves have upshifted frequency $\omega_1 = (\omega_0^2 + \omega_p^2)^{1/2} = -\omega_2$. (From Kalluri, D. K., *IEEE Trans. Plasma Sci.*, 16(1), 11–16, 1988. With permission.)

in the slab. As the B waves come out of the slab, the waves will encounter space discontinuity and therefore the wave number will change accordingly. The B waves are created only at the time of switching and, hence, exist for a finite time.

Analytical solution for a sudden switching of a plasma slab requires computationally expensive numerical inversion of Laplace transform. It is extensively studied as Kalluri has published a theoretical work on this problem using a Laplace transform method [2,3]. However, extension of such methods to the more realistic problems of plasma profiles of finite rise time and spatial inhomogeneities becomes very cumbersome. The FDTD method discussed in this paper handles such practical profiles with ease.

I.2   Development of FDTD Equations

Consider a continuous source wave of frequency ω_0 propagating in free space. At $t = 0$, a plasma slab is created with a spatial distribution. For this problem, the condition of an infinite unbounded space cannot be assumed since, after switching, the plasma medium is of finite extent with a defined spatial plasma density profile. Hence space and time domain considerations must be taken into account in FDTD simulation.

I.2.1 Total-Field and Scattered-Field Formulation

The consideration of the space formulation in the FDTD method brings out the realization of a continuous source wave. Unless the source is located at a very distant place, the reflected wave by the plasma slab will eventually arrive at the source and corrupt the source wave. Consider a sinusoidal source wave at a grid location s:

$$E\big|_s^n = \exp[j(\omega_0 n\,\Delta t - k_0 s\,\Delta z)]. \tag{I.1}$$

This wave propagates to the region of interest and is eventually reflected back to the source location unless the source wave is located at a very distant position. The source wave behaves as a *hard* source [3] and prevents the movement of the reflected wave from passing and propagating toward infinity. Consequently, the hard source causes nonphysical reflection of the scattered wave to the system of interest. In order to simulate the source condition properly, the source must be hidden to the reflected wave, yet continuously feed the system. This is realized by introducing the concept of total-field/scattered-field formulation [4]. This approach is based on the linearity of Maxwell's equations and decomposition of the electric and magnetic fields as

$$\mathbf{E}_{tot} = \mathbf{E}_{inc} + \mathbf{E}_{scat}, \tag{I.2}$$

$$\mathbf{H}_{tot} = \mathbf{H}_{inc} + \mathbf{H}_{scat}, \tag{I.3}$$

where $\mathbf{E}_{inc}$ and $\mathbf{H}_{inc}$ are incident fields and assumed to be known at all space points of the FDTD grid at all time steps. $\mathbf{E}_{scat}$ and $\mathbf{H}_{scat}$ are scattered fields (either reflected wave or transmitted wave) and unknown initially. To implement this idea, we divide the computation domain into two regions, Region 1 and Region 2, as shown in Figure I.2.

Region 1 is the total-field region and the plasma slab is embedded within this region. The incident wave is initially assigned to the total field. Region 2 is the scattered-field region and is located in free space.

The FDTD formulation at the interface of Region 1 and Region 2 requires various field components E and H from both regions to advance one time step. Let us assume that the interface belongs to Region 1 and the electric field component $E_{tot}\big|_L^{n+1}$ lies on the interface as indicated by E_{incL} in Figure I.2. The FDTD equation for this component is given by

$$E_{tot}\big|_L^{n+1} = E_{tot}\big|_L^n - \frac{\Delta t}{\varepsilon_0\,\Delta z}\left(H_{tot}\big|_{L+(1/2)}^{n+(1/2)} - H_{tot}\big|_{L-(1/2)}^{n+(1/2)}\right). \tag{I.4}$$

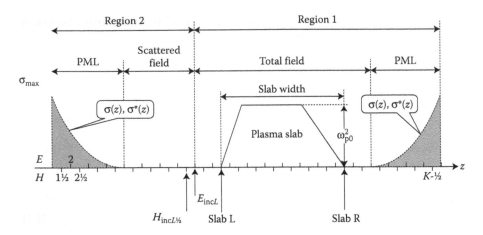

FIGURE I.2 Configuration of geometry.

However, $H_{\text{tot}}\big|_{L-(1/2)}^{n+(1/2)}$ is located in Region 2 and is not defined. At this location $(L - \frac{1}{2})$, the scattered field $H_{\text{scat}}\big|_{L-(1/2)}^{n+(1/2)}$ is available. Hence, Equation I.4 needs some modification in such a way that

$$E_{\text{tot}}\big|_L^{n+1} = E_{\text{tot}}\big|_L^n - \frac{\Delta t}{\varepsilon_0\,\Delta z}\left(H_{\text{tot}}\big|_{L+(1/2)}^{n+(1/2)} - H_{\text{scat}}\big|_{L-(1/2)}^{n+(1/2)}\right) + \frac{\Delta t}{\varepsilon_0\,\Delta z}H_{\text{inc}}\big|_{L-(1/2)}^{n+(1/2)}.$$

(I.5)

Note that

$$H_{\text{tot}}\big|_{L-(1/2)}^{n+(1/2)} = H_{\text{scat}}\big|_{L-(1/2)}^{n+(1/2)} + H_{\text{inc}}\big|_{L-(1/2)}^{n+(1/2)},$$ (I.6)

where $H_{\text{inc}}\big|_{L-(1/2)}^{n+(1/2)}$ is assumed to be known, as mentioned earlier.

In a similar way, the formulation for the H field component at $L - \frac{1}{2}$ needs some modification. At this space point, the H field component is the scattered field because it is in Region 2. The FDTD equation is written as

$$H_{\text{scat}}\big|_{L-(1/2)}^{n+(1/2)} = H_{\text{scat}}\big|_{L-(1/2)}^{n-(1/2)} - \frac{\Delta t}{\mu_0\,\Delta z}\left(E_{\text{scat}}\big|_L^n - E_{\text{scat}}\big|_{L-1}^n\right).$$ (I.7)

Again, $E_{\text{scat}}\big|_L^n$ is located in Region 1 and therefore it is not defined. The modification of Equation I.7 is made using known quantities by using

$$H_{\text{scat}}\big|_{L-(1/2)}^{n+(1/2)} = H_{\text{scat}}\big|_{L-(1/2)}^{n-(1/2)} - \frac{\Delta t}{\mu_0\,\Delta z}\left(E_{\text{tot}}\big|_L^n - E_{\text{scat}}\big|_{L-1}^n\right) + \frac{\Delta t}{\mu_0\,\Delta z}E_{\text{inc}}\big|_L^n.$$ (I.8)

Note that

$$E_{\text{scat}}\big|_L^n = E_{\text{tot}}\big|_L^n - E_{\text{inc}}\big|_L^n.$$ (I.9)

In short, the concept of splitting the computational domain into Region 1 (total-field region) and Region 2 (scattered-field region) is to provide the incident field (source wave) to Region 1 (where materials of interest are embedded) and to separate the scattered field out from the total field by canceling it with the incident field at the interface. In this simulation, Region 1 extends to the end of the PML on the right-hand side.

I.2.2 Lattice Truncation: PML

The reflected and transmitted waves will propagate to the ends of the computational boundaries. To prevent numerical reflection of the waves from the boundaries of the computational domain, an absorption boundary condition such as PML is needed. The concept of the absorption boundary condition is that when an EMW enters a lossy medium, no reflection by the lossy medium will occur if the impedance matching condition is satisfied. Consider the Maxwell's equations in a lossy medium for an R-wave propagation:

$$\frac{\partial H}{\partial t} + \sigma^* H = -\frac{1}{\mu_0} \frac{\partial E}{\partial z}, \tag{I.10}$$

$$\frac{\partial E}{\partial t} + \sigma E = -\frac{1}{\varepsilon_0} \frac{\partial H}{\partial z}, \tag{I.11}$$

where σ is the electric conductivity and σ^* is the magnetic conductivity. The magnetic conductivity σ^* is not a physical quantity but is introduced for the requirement of impedance matching. Since the PMLs are positioned in the boundaries that are in free space, the introduction of the magnetic conductivity will not change the physics of the region of interest. The impedance matching condition [5] is well known: $\sigma/\varepsilon_0 = \sigma^*/\mu_0$. To simplify computation we assume that $\varepsilon_0 = \mu_0 = 1$. This assumption implies normalization of the variables. For example, the velocities are all normalized with respect to the velocity of light in free space $c = 3 \times 10^8$ m/s. The impedance matching condition then simplifies to $\sigma = \sigma^*$.

As the wave propagates in the lossy medium, the wave will decay. However, if the PML is finite, then the wave will arrive at the boundary and reflect back to the system. This reflection is generally very weak and can be ignored. In numerical calculation, however, there will be numerical reflection at each space step since the unit space and time steps are finite and therefore the source wave will experience the sudden changes in the conductivity, resulting in numerical reflection. To reduce this reflection, the conductivity must vary smoothly from zero to a maximum value and numerical simulation shows that the following quadratic form gives very low reflection [5].

$$\sigma = \sigma^* = \sigma_{max} \left(\frac{z}{W} \right)^2, \tag{I.12}$$

where z is measured from the beginning of the PML from the inside direction and W is the width of the PML. The maximum value of the conductivity is determined to obtain the minimum reflection from PML.

I.2.3 FDTD Formulation for an R Wave in a Switched Plasma Slab

The Maxwell's equations containing a damping constant v for an R-wave [6] propagation in a magnetized plasma medium are written as

$$\frac{\partial H}{\partial t} + \sigma H = -\frac{1}{\mu_0}\frac{\partial E}{\partial z}, \tag{I.13}$$

$$\frac{\partial E}{\partial t} + \sigma E = -\frac{1}{\varepsilon_0}\frac{\partial H}{\partial z} - \frac{1}{\varepsilon_0}J, \tag{I.14}$$

$$\frac{dJ}{dt} + vJ = \varepsilon_0\omega_p^2(z,t)E. \tag{I.15}$$

In the above equation, ω_p is the plasma frequency and v is the collision frequency [1].

The FDTD equations of Equations I.13 through I.15 can be written as

$$H\big|_{k+(1/2)}^{n+(1/2)} = e^{-\sigma\Delta t}H\big|_{k+(1/2)}^{n-(1/2)} - \frac{1}{\mu_0\sigma\,\Delta z}\left(1 - e^{-\sigma\Delta t}\right)\left(E\big|_{k+1}^{n} - E\big|_{k}^{n}\right), \tag{I.16}$$

$$J\big|_{k}^{n+(1/2)} = e^{(-v)\Delta t}J\big|_{k}^{n-(1/2)} + \frac{\varepsilon_0}{v}\left(1 - e^{(-v)\Delta t}\right)\omega_p^2\big|_{k}^{n}E\big|_{k}^{n}, \tag{I.17}$$

$$E\big|_{k}^{n+1} = e^{-\sigma\Delta t}E\big|_{k}^{n} - \frac{1}{\varepsilon_0\sigma}\left(1 - e^{-\sigma\Delta t}\right)\left[\frac{1}{\Delta z}\left(H\big|_{k+(1/2)}^{n+(1/2)} - H\big|_{k-(1/2)}^{n+(1/2)}\right) + J\big|_{k}^{n+(1/2)}\right]. \tag{I.18}$$

In Equation I.17, $v = 0$ when the plasma is considered lossless. Numerical results obtained in the next two sections are based on such a consideration.

The algorithm given in Equations I.16 through I.18 is implemented in various media as follows:

Free space: $\sigma = 0$, $J = 0$, $\omega_p^2 = 0$
Plasma slab: $\sigma = 0$
PML: $J = 0$, $\omega_p^2 = 0$

It is noted that the following approximation can be made for small σ:

$$\lim_{\sigma \to 0}\frac{\left(1 - e^{-\sigma\Delta t}\right)}{\sigma} = \Delta t. \tag{I.19}$$

I.3 Interaction of a Continuous Wave with a Switched Plasma Slab

Figure I.3 shows profiles of plasma creation in space and time as (a) plasma frequency increases with a rise time T_r to the maximum values ω_{p0}^2, and (b) spatial distribution of the square of the plasma frequency, constant over a width $(S_W - S_L - S_R)$, a linear rising profile of width (S_L), and a linear declining profile of width (S_R).

A result of the transient wave at the right-hand side of the slab of $\omega_0 = 0.8$, $\omega_p = 1.0$, the slab width $S_W = 4.0\pi/\omega_p$, $S_L = 0$, $S_R = 0$, and the rise time $T_r = 0$ is shown in Figure I.4. The real part of the electric field component at

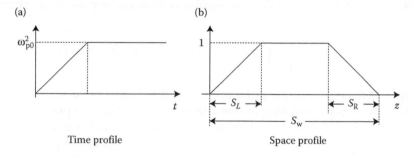

(a)   (b)

Time profile   Space profile

FIGURE I.3 Plasma profiles in space and time domain.

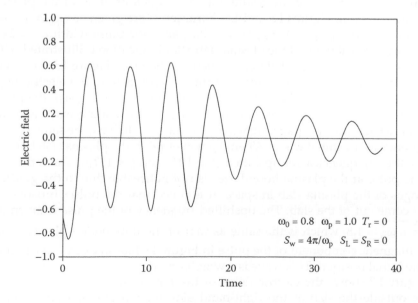

$\omega_0 = 0.8$ $\omega_p = 1.0$ $T_r = 0$

$S_W = 4\pi/\omega_p$ $S_L = S_R = 0$

FIGURE I.4 A transient wave due to creation of a plasma slab.

"Slab R" in Figure I.2 is shown and this wave will disappear quickly. Since the source wave frequency is less than the plasma frequency and the slab is sufficiently large, only the transmitted wave exists and soon propagates away. The result is in good agreement with Ref. [1]. The angular frequency of this wave is $\sqrt{\omega_0^2 + \omega_p^2} = 1.28$, as expected.

I.4   Interaction of a Pulsed Wave with a Switched Plasma Slab

For the case of pulsed signals, the interaction of a newly created plasma slab is easily shown. Consider a Gaussian pulse with frequency ω_0 in free space:

$$E(z, t) = \exp\left[-\left(\frac{\omega_0 t - k_0(z - z_0)}{k_0 z_w / 2} \right)^2 + j(\omega_0 t - k_0 z) \right], \quad (I.20)$$

where z_0 is the center of the pulse at $t = 0$ and z_w is the pulse width. When a pulsed wave is propagating in free space along the z-axis, a plasma slab is created, which encloses the whole pulse. The wave then will interact with the plasma slab. The interaction will split the pulse into a forward-propagating wave and a backward-propagating wave. The magnetic field component of the pulse will have an additional mode that is the wiggler mode whose frequency is zero. Some of the results for the sudden creation of the plasma slab are shown in Figures I.5 and I.6. Parameters are given as $\omega_{p0} = 1.5\,\omega_0$, $T_r = 0$, $S_w = 16\,\lambda_0$, $S_L = 0.2 S_w$, $S_R = 0.3 S_w$, and tshe pulse width $z_w = 2\lambda_0$. The spatial distribution of the plasma slab after its creation is illustrated with a dotted line. The electric field component is shown in Figure I.5. The electric field in space domain at different times is presented. The electric field at $t = 0^+$ is the initial pulse shape. At $t = 5T_0$ (here, $T_0 = 2\pi/\omega_0$), the pulse is split into two waves, where one is a forward-propagating wave and the other is a backward-propagating wave. They appear to have the same spatial frequency as the initial wave. When the waves reach the plasma slab boundaries, their frequencies get upshifted ($t = 10T_0$ and $t = 15T_0$). The reflection of the pulse at the plasma boundaries is not noticeable due to the gradual changes of the plasma slab in space. It is noted that the pulses broaden as they come out of the slab. The upshifted frequency of the pulse is given by $\sqrt{\omega_0^2 + \omega_p^2} = 1.8$, which is the same as that of the unbounded plasma case. The magnetic component of the pulse in Figure I.6 has an extra wave sitting at the initial position. This wave is a wiggler mode whose frequency is zero.

Figure I.7 shows the electric fields of the forward-propagating waves in time outside the slab at the right-hand side for various plasma frequencies, $\omega_{p0} = 0$, $\omega_{p0} = 0.5\omega_0$, $\omega_{p0} = 1\omega_0$, $\omega_{p0} = 2\omega_0$, and $\omega_{p0} = 4\omega_0$ from top

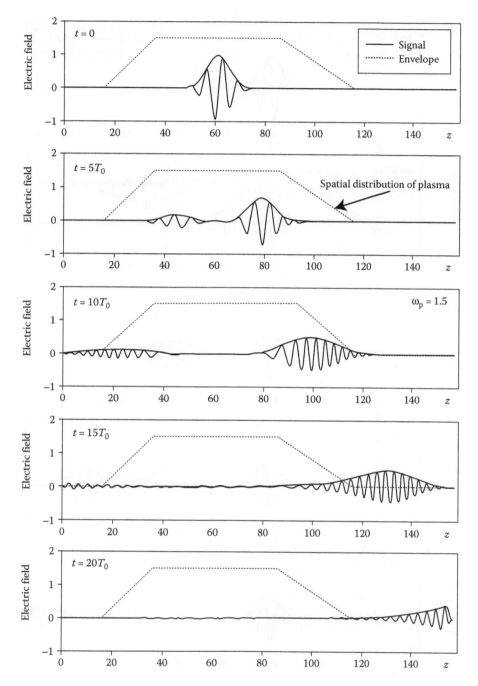

FIGURE I.5 The time evolution of the electric field after sudden creation of plasma. $T_r = 0$, $\omega_p = 1.5\omega_0$, $z_w = \lambda_0$, $S_W = 16\lambda_0$, $S_L = 0.2S_W$, and $S_R = 0.3S_W$.

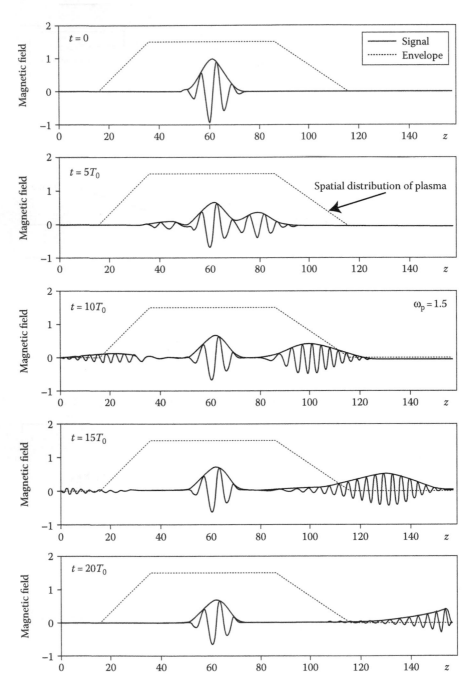

FIGURE 1.6   The time evolution of the magnetic field after sudden creation of plasma. $T_r = 0$, $\omega_p = 1.5\omega_0$, $z_w = \lambda_0$, $S_w = 16\lambda_0$, $S_L = 0.2S_w$, and $S_R = 0.3_w$.

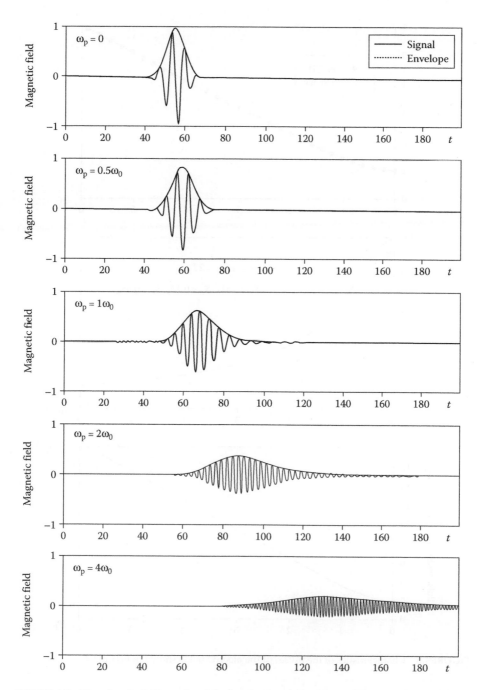

FIGURE I.7 The electric fields at the right-hand side of the plasma slab for various plasma frequencies. $T_r = 2T_0$, $z_w = \lambda_0$, $S_W = 16\lambda_0$, $S_L = 0.2S_W$, and $S_R = 0.3S_W$.

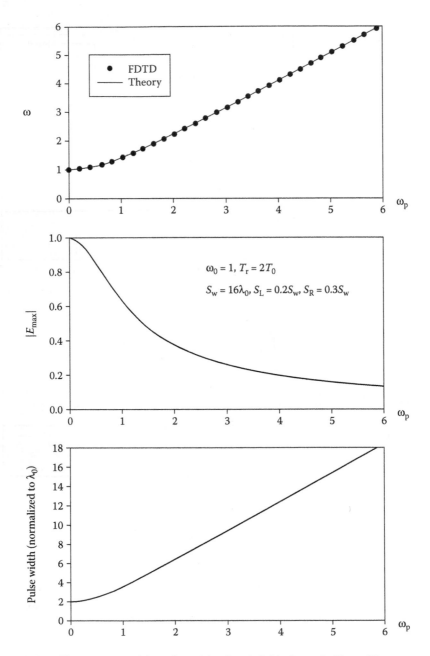

FIGURE I.8 Characteristics of the pulses of the electric fields shown in Figure I.7.

to bottom, respectively, and pulse width $z_w = 2\lambda_0$. The slab parameters are $S_w = 16\lambda_0$, $S_L = 0.2S_w$, and $S_R = 0.3S_w$, where the gradual changes in the boundaries of the slab reduce the reflection of the pulse. The plasma slab is created with rise time $T_r = 2T_0$ and this will suppress the generation of backward propagation of the pulse. With these parameters, the pulse will keep its shape but become a higher-frequency pulse as the plasma frequency increases. The widths of the pulse decrease as ω_{p0} increases and it can be understood as pulse broadening. The pulse broadening shows the clear nature of the dispersive characteristics of a plasma medium. Since the group velocity depends on the plasma frequency as $v_g = c\omega_0/\sqrt{\omega_0^2 + \omega_p^2}$, as the plasma frequency becomes higher, the arrival of the pulse is more delayed. The plasma creation does not provide any additional energy to the pulse and therefore the amplitude of the wave decreases as the pulse broadens. The frequency, the peak amplitude, and the pulse width of the electric field component of the pulse outside at the right edge of the plasma slab are shown in Figure I.8. The pulse duration is measured by locating points on the envelope function that are e^{-1} $(= 0.368)$ of the maximum value. The frequency of the pulse is the same as that of the unbounded plasma creation, that is, $\omega = \sqrt{\omega_0^2 + \omega_{p0}^2}$.

References

1. Kalluri, D. K., *Electromagnetics of Complex Media*, CRC Press, Boca Raton, FL, 1998.
2. Kalluri, D. K., On reflection from a suddenly created plasma half-space: Transient solution, *IEEE Trans. Plasma Sci.*, 16(1), 11–16, 1988.
3. Kalluri, D. K. and Goteti, V. R., Frequency shifting of electromagnetic radiation by sudden creation of a plasma slab, *J. Appl. Phys.*, 72(10), 4575–4580, 1992.
4. Taflove, A., *Computational Electrodynamics: The Finite Difference Time Domain Method*, Artech House, Boston, MA, 1995.
5. Berenger, J. P., A perfectly matched layer for the absorption of electromagnetic waves, *J. Comput. Phys.*, 114, 185–200, 1994.
6. Lee, J. H., Kalluri, D. K., and Nigg, G. C., FDTD simulation of electromagnetic transformation in a dynamic magnetized plasma, *Int. J. Infrared Millim. Waves*, 21(8), 1223–1253, 2000.

Appendix J

FDTD Simulation of EMW Transformation in a Dynamic Magnetized Plasma*

Joo Hwa Lee, Dikshitulu K. Kalluri, and Gary Nigg

J.1 Introduction

The propagation of an EMW in a dynamic magnetized plasma is a subject of interest. The interaction of the wave with the time-varying medium invokes frequency transformation of the original signal. Frequency transformation of a wave is of great significance in high-frequency electrical engineering. It is cost effective to construct high-frequency generators in certain frequency bands and very difficult and expensive to do the same in certain other frequency bands. A frequency transformer is thus a potentially useful device.

The interaction of the wave with a plasma is described by Maxwell's equations and the constitutive relation having a time-varying plasma frequency term. Most problems involving time-varying plasmas do not have exact solutions and are likely to be described by high-order differential equations. Hence, only very simple problems are solved exactly and many problems are left unsolved. Due to their high-order complexity, approximate solutions have been given in several forms. WKB methods are used for slowly varying plasma density profiles [1,2], and the perturbation methods based on the development of the time-domain Green's function [3,4] are used for fast-profile plasma problems.

In order to verify and explore the range of validity of our earlier results, we used the FDTD method [5], which is a very powerful method to analyze electromagnetic problems due to its simplicity and accuracy. The FDTD is a natural representation of Maxwell's equations in a discretized form. The FDTD method is well suited to study plasma problems in general and in particular the problems of interest here and in earlier work, namely the effect on a preexisting EMW of monotonically increasing the plasma density in time from zero to some upper value. The FDTD method generates, as most numerical techniques do, the total field solution and so it is not easy to analyze

---

* Reprinted from *Int. J. Infrared Millim. Waves*, 21(8), 1223–1252, August 2000. With permission.

the modes imbedded in the solution. We have developed a new technique of successive reduction to obtain the frequencies and amplitudes of the various modes buried in the time series generated by the FDTD method. Several illustrative examples of FDTD simulations of EMW transformations in dynamic magnetized plasmas are given.

J.2   One-Dimensional FDTD Algorithm

J.2.1   One-Dimensional Equations

Consider a cold magnetoplasma medium with independent prescriptions for an overall time variation and space variation of the plasma density in the presence of a uniform external magnetic field. Maxwell's equations and the constitutive relation for the magnetized plasma medium [6] are given by

$$\nabla \times \mathbf{E} = -\mu_0 \frac{\partial \mathbf{H}}{\partial t}, \tag{J.1}$$

$$\nabla \times \mathbf{H} = \varepsilon_0 \frac{\partial \mathbf{E}}{\partial t} + \mathbf{J}, \tag{J.2}$$

$$\frac{\partial \mathbf{J}}{\partial t} = \varepsilon_0 \omega_p^2(\mathbf{r}, t)\mathbf{E} + \boldsymbol{\omega}_b \times \mathbf{J}, \tag{J.3}$$

where ε_0 is the permittivity of free space, μ_0 the permeability of free space, ω_p^2 the square of plasma frequency, $\boldsymbol{\omega}_b = q\mathbf{B}_0/m$ the cyclotron frequency, $\mathbf{B}_0$ the external static magnetic field, and q and m are the electric charge and mass of an electron, respectively.

For the one-dimensional case, the EMW is assumed to be propagating in the z-direction, and the static magnetic field is also assumed to be in the z-direction, $\boldsymbol{\omega}_b = \omega_b \hat{z}$, and $\omega_p^2 = \omega_p^2(z, t)$. The natural propagating transverse modes for a uniform plasma are circularly polarized waves. For the time-varying uniform plasma beginning from zero plasma density with a preexisting circularly polarized wave, the spatially helical field structure for the fields is preserved although the rate of advance with time will change as the plasma density changes. For the right circularly polarized wave, that is, the one whose electric field rotates in time around the static magnetic field in the same way as electrons do, this spatial structure is left-handed helical and this is captured by writing the complex field and current variables in the following way:

$$\mathbf{E} = \left(\hat{x} - j\hat{y}\right) E(z, t), \tag{J.4}$$

$$\mathbf{H} = \left(j\hat{x} + \hat{y}\right) H(z, t), \tag{J.5}$$

$$\mathbf{J} = \left(\hat{x} - j\hat{y}\right) J(z, t), \tag{J.6}$$

where $j = \sqrt{-1}$.

From Equations J.1 through J.3 and Equations J.4 through J.6, we obtain the following one-dimensional equations:

$$\frac{\partial H}{\partial t} = -\frac{1}{\mu_0}\frac{\partial E}{\partial z}, \tag{J.7}$$

$$\frac{\partial E}{\partial t} = -\frac{1}{\varepsilon_0}\frac{\partial H}{\partial z} - \frac{1}{\varepsilon_0}J, \tag{J.8}$$

$$\frac{dJ}{dt} = \varepsilon_0\omega_p^2(z,t)E + j\omega_b J. \tag{J.9}$$

Equations J.7 through J.9 are also valid for the isotropic case when ω_b is made equal to zero.

J.2.2 FDTD Formulation

Let $z = k\Delta z$ and $t = n\Delta t$, where k and n are the indices of the grid and Δz and Δt are the space and time intervals. We can use the standard Yee grids to discretize Equations J.7 through J.9 [5]. As Equation J.9 implies, E and J lie on the same space coordinate but will be offset by $1/2$ time step [7]. Figure J.1 shows the positioning of **E**, **H**, and **J** in the space and time grids.

By using central difference formulas, we generate the following FDTD equations for Equations J.7 through J.9:

$$\frac{H\big|_{k+(1/2)}^{n+(1/2)} - H\big|_{k+(1/2)}^{n-(1/2)}}{\Delta t} = -\frac{1}{\mu_0}\frac{E\big|_{k+1}^{n} - E\big|_{k}^{n}}{\Delta z}, \tag{J.10}$$

$$\frac{E\big|_{k}^{n+1} - E\big|_{k}^{n}}{\Delta t} = -\frac{1}{\varepsilon_0}\frac{H\big|_{k+(1/2)}^{n+(1/2)} - H\big|_{k-(1/2)}^{n+(1/2)}}{\Delta z} - \frac{1}{\varepsilon_0}J\big|_{k}^{n+(1/2)}, \tag{J.11}$$

$$\frac{J\big|_{k}^{n+(1/2)} - J\big|_{k}^{n-(1/2)}}{\Delta t} = \varepsilon_0\,\omega_p^2\big|_{k}^{n}\,E\big|_{k}^{n} + \frac{j\omega_b}{2}\left(J\big|_{k}^{n+(1/2)} + J\big|_{k}^{n-(1/2)}\right). \tag{J.12}$$

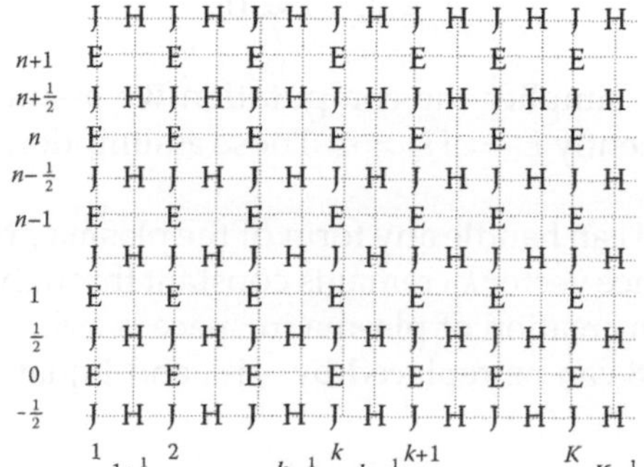

FIGURE J.1 Position of E, H, and J for the one-dimensional problem.

These equations, arranged in the order given below, lead to a step-by-step explicit computation of the fields in the time domain:

$$H\Big|_{k+(1/2)}^{n+(1/2)} = H\Big|_{k+(1/2)}^{n-(1/2)} - \frac{\Delta t}{\mu_0 \, \Delta z}\left(E\Big|_{k+1}^{n} - E\Big|_{k}^{n}\right), \tag{J.13}$$

$$J\Big|_{k}^{n+(1/2)} = \frac{1 + j\omega_b \, \Delta t/2}{1 - j\omega_b \, \Delta t/2}J\Big|_{k}^{n-(1/2)} + \frac{\Delta t \, \varepsilon_0}{1 - j\omega_b \, \Delta t/2} \, \omega_p^2\Big|_{k}^{n} E\Big|_{k}^{n}, \tag{J.14}$$

$$E\Big|_{k}^{n+1} = E\Big|_{k}^{n} - \frac{\Delta t}{\varepsilon_0 \, \Delta z}\left(H\Big|_{k+(1/2)}^{n+(1/2)} - H\Big|_{k-(1/2)}^{n+(1/2)}\right) - \frac{\Delta t}{\varepsilon_0}J\Big|_{k}^{n+(1/2)}. \tag{J.15}$$

Initial conditions are given at $n = -1/2$ for J and H and at $n = 0$ for E. For a large ω_b, the exponential time-stepping algorithm [8] can be applied to Equation J.14 and it becomes

$$J\Big|_{k}^{n+(1/2)} = e^{j\omega_b \, \Delta t}J\Big|_{k}^{n-(1/2)} + \Delta t \, \varepsilon_0 e^{j\omega_b \, \Delta t/2}\text{sinc}\left(\frac{\omega_b \, \Delta t}{2}\right)\omega_p^2\Big|_{k}^{n} E\Big|_{k}^{n}, \tag{J.16}$$

where $\text{sinc}\,x = \sin x/x$.

J.3 Illustrative Examples of FDTD Simulation

J.3.1 Time-Varying Unbounded Plasma with an Arbitrary Profile

As a first example, we consider the creation of a uniform unbounded isotropic plasma ($\omega_b = 0.0$). Initially, an incident R wave (source wave) of frequency ω_0 is propagating in free space in the positive z-direction:

$$E = E_0 e^{j(\omega_0 t - k_0 z)}, \quad t < 0, \tag{J.17}$$

$$H = H_0 e^{j(\omega_0 t - k_0 z)}, \quad t < 0, \tag{J.18}$$

$$J = 0, \quad t < 0, \tag{J.19}$$

where $k_0 = \omega_0/c$. To simplify the computation we assume $c = \varepsilon_0 = \mu_0 = 1$, $\omega_0 = 1$, and consequently $E_0 = H_0 = 1$. These assumptions imply normalization of the variables.

The FDTD method can handle any form of the plasma profile. In simulating this problem, the space vector k_0 remains constant throughout the conversion [9] since the uniform creation of plasma in space is assumed. Hence, $\partial/\partial z$ in Equations J.7 and J.8 can be replaced by $-jk_0$ and Equations J.7 through J.9 reduce to

$$\frac{\partial H}{\partial t} = \frac{jk_0}{\mu_0}E, \tag{J.20}$$

$$\frac{\partial E}{\partial t} = \frac{jk_0}{\varepsilon_0}H - \frac{1}{\varepsilon_0}J, \qquad (J.21)$$

$$\frac{dJ}{dt} = \varepsilon_0\omega_p^2(t)E + j\omega_b J. \qquad (J.22)$$

From these equations, the FDTD equations for explicit computation can be readily obtained as

$$H\big|^{n+(1/2)} = H\big|^{n-(1/2)} + \frac{jk_0\,\Delta t}{\mu_0}E\big|^n, \qquad (J.23)$$

$$J\big|^{n+(1/2)} = e^{j\omega_b\,\Delta t}J\big|^{n-(1/2)} + \Delta t\,\varepsilon_0 e^{j\omega_b\,\Delta t/2}\mathrm{sinc}\left(\frac{\omega_b\,\Delta t}{2}\right)\omega_p^2\big|^n E\big|^n, \qquad (J.24)$$

$$E\big|^{n+1} = E\big|^n + \frac{jk_0\,\Delta t}{\varepsilon_0}H\big|^{n+(1/2)} - \frac{\Delta t}{\varepsilon_0}J\big|^{n+(1/2)}. \qquad (J.25)$$

FDTD simulations give the total field solutions in the form of time series as many numerical methods do. After the plasma creation is completed, we take a time series of the fields and use it to obtain frequencies and amplitudes. To identify individual modes contained in the total fields, we may think of conventional methods such as FFT or the Prony method. However, these methods do not give the required accuracy for the problems under consideration. Hence, we have developed a new technique, a successive reduction method (SRM) discussed in Appendix A, to overcome such difficulties.

J.3.2 Validation of FDTD

Using the above FDTD equations together with the SRM, we considered several examples and compared results with WKB and Green's function methods. A linear ramp transition time density program of the plasma as shown in Figure J.2,

$$\omega_p^2(t) = \begin{cases} 0, & t < 0, \\ \omega_{p0}^2\dfrac{t}{T_r}, & 0 < t < T_r, \\ \omega_{p0}^2, & t > T_r \end{cases} \qquad (J.26)$$

is an example of a time behavior having an exact solution that can be used to test the validity of the three methods, that is, WKB, Green's function method, and FDTD. In the above equation, T_r is the rise time as shown in Figure J.2 and ω_{p0}^2 is the square of the saturation plasma frequency. The exact solution and Green's function method to this plasma profile can be found in Ref. [3]

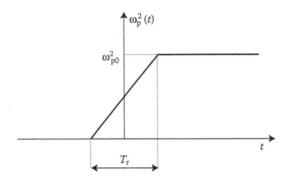

FIGURE J.2   Square of the plasma frequency for a linear time profile.

and the WKB solution is equivalent to those in Refs. [2,10]. The exact solution is repeated here for convenience.

$$
E(t) = \begin{cases}
e^{j\omega_0 t}, & t < 0, \\[2ex]
c_1 Ai\left[\left(\dfrac{T_r}{\omega_{p0}^2}\right)^{2/3}\left(\omega_0^2 + \dfrac{\omega_{p0}^2 t}{T_r}\right)\right] \\[2ex]
\quad + c_2 Bi\left[\left(\dfrac{T_r}{\omega_{p0}^2}\right)^{2/3}\left(\omega_0^2 + \dfrac{\omega_{p0}^2 t}{T_r}\right)\right], & 0 < t < T_r \\[2ex]
T e^{j\omega_1 t} + R e^{-j\omega_1 t}, & t > T_r,
\end{cases}
\qquad (\text{J.27})
$$

where $\omega_1 = (\omega_0^2 + \omega_{p0}^2)^{1/2}$, and Ai and Bi are the Airy functions. The amplitude of the incident field is normalized to one. The unknown coefficients c_1, c_2, T, and R can be determined by using the continuity conditions on E and dE/dt at the two temporal interfaces $t = 0$ and $t = T_r$. The coefficient T is the amplitude of the forward-propagating wave and R is the amplitude of the backward-propagating wave after the plasma is fully created. Figures J.3 and J.4 show such comparisons for the two coefficients $|T|$ and $|R|$, respectively. The frequency of the incident wave ω_0 is taken as 1 and the saturation plasma frequency ω_{p0}^2 is taken as $3\omega_0^2$ for computations ($\omega_{p0}^2 = 3\omega_0^2$). The FDTD results and the exact solution agree in a wide range of rise times of the plasma profile. It is seen that for the fast profile (T_r is small), the Green's function method is valid, whereas for the slow profile (T_r is large), the WKB method is valid, as expected.

J.3.3   Characteristics of Plasma Transition Time Behaviors

Different time behaviors with the same rise time T_r are now considered. Instead of using a simple linear profile in the plasma creation, the cases of

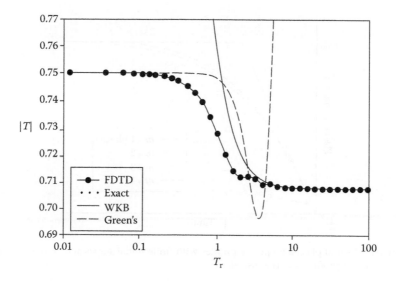

FIGURE J.3 The amplitude of the transmitted wave $|T|$ for a linear time profile.

the following time profiles of plasma creation (Figure J.5) are considered:

$$\omega_p^2(t) = \begin{cases} 0, & t < 0, \\ \omega_{p0}^2 \left[1 - \left(1 - \dfrac{t}{T_r} \right)^m \right], & 0 < t < T_r, \\ \omega_{p0}^2, & t > T_r. \end{cases} \quad (J.28)$$

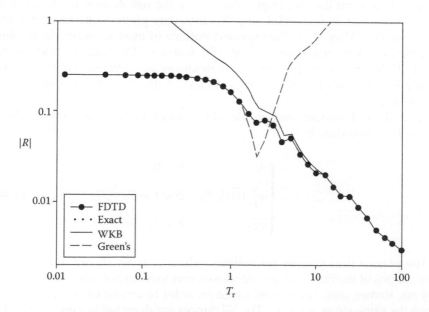

FIGURE J.4 The amplitude of the reflected wave $|R|$ for a linear time profile.

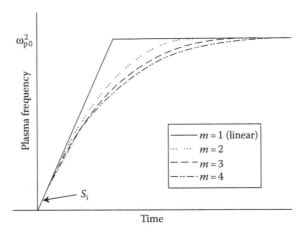

FIGURE J.5 A set of plasma frequency profiles with finite initial slopes and zero final slope at $t = T_r$. The linear profile ($m = 1$) is also shown.

Here m is the power factor. This profile has a finite initial slope but a zero slope at $t = T_r$ for $m > 1$. The rise time is defined as the time by which the plasma creation is completed. In this figure the time profiles with different power factors are plotted. The initial slope of the profile is expressed as $S_i = d\omega_p^2(t)/dt = \omega_{p0}^2 m/T_r$ at $t = 0$. This slope is kept constant by keeping m/T_r constant for various profiles. The $|R|$ curves are shown in Figure J.6a with respect to S_i. The WKB solution [1] is given by $|R| = S_i/8\omega_0^3$. For low values of the slopes, the $|R|$ values are identical regardless of the power factor of the plasma profiles. For large values of S_i, the results merge into those of sudden creation. In Figure J.6b, the $|R|$ curves are plotted with respect to the rise time (T_r). They show the expected pattern of monotonically decreasing functions and do not show any special features. The transmission coefficient $|T|$ is not shown because the differences for different ms are almost negligible and transmission coefficients are all like that shown with dots in Figure J.3.

Next, the polynomial time profile with zero initial slope but a finite slope at $t = T_r$ is considered:

$$\omega_p^2(t) = \begin{cases} 0, & t < 0, \\ \omega_{p0}^2(t/T_r)^m, & 0 < t < T_r \\ \omega_{p0}^2, & t > T_r. \end{cases} \qquad (J.29)$$

The plasma time profiles for various values of m are shown in Figure J.7. The origins of the time axis are not shown and they do not coincide for different ms. Rather, each curve is relocated in order to emphasize that all curves have the same slope at $t = T_r$. The $|R|$ curves are depicted in Figure J.8a. The

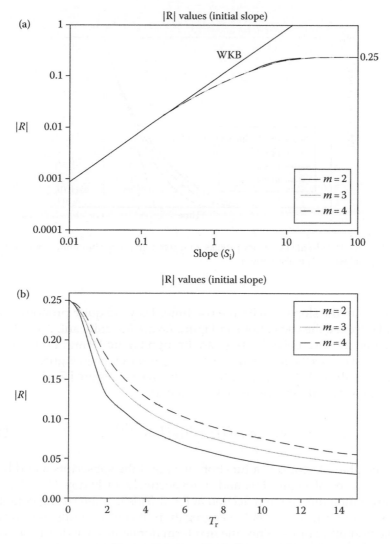

FIGURE J.6 Reflection coefficient ($|R|$ values) for finite initial slopes. (a) Initial slope S_i versus $|R|$ value and (b) rise time versus $|R|$ value.

finial slope is obtained as $S_f = d\omega_p^2(t)/dt = \omega_{p0}^2 m/T_r$ at $t = T_r$. The WKB solution given by $|R| = S_f/8\omega_f^{7/2}$ in Ref. [10] is also shown. For low values of the slope, the WKB solution is valid. Figure J.8b shows the same curves with rise time as the independent variable. From Figures J.6 and J.8, it is inferred that the amplitude of the backward-propagating wave is governed by the sharp changes in the time behaviors of plasma creation rather than the details of the smoother parts of the profiles. The curves in Figure J.8 show oscillatory

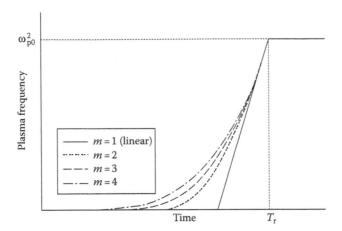

FIGURE J.7   A set of plasma frequency profiles with zero initial slope but finite slopes at $t = T_r$. The linear profile ($m = 1$) is also shown.

behavior with the slope or with the rise time. They are quite pronounced for $m = 2$. However, the oscillations in Figure J.6 are insignificant. A qualitative explanation for such behavior is given through Figure J.9 and Equation J.30. Figure J.9a sketches the reflected rays for the profile given in Figure J.5. Figure J.9b sketches the reflected rays for the profile given in Figure J.7. The $|R|$ value in either case may be approximately written as

$$|R| \approx A_{a,b}(T_r) + B_{a,b}(T_r)e^{j\phi(T_r)}, \tag{J.30}$$

where the phase factor ϕ is a function of T_r and the subscripts a and b represent the cases of Figure J.9a and b, respectively. In Figure J.9a, $A_a(T_r)$ is the contribution due to the reflection at $t = 0$. This is quite strong compared to the second term B_a that is the aggregation of contributions due to weak reflections at other times. Since the first term dominates, we do not see oscillation in the graphs of Figure J.7. In Figure J.9b, $A_b(T_r)$ is the aggregation of the contributions due to weak reflections at various times. The second term $B_b(T_r)e^{j\phi(T_r)}$ is the contribution due to the reflection at $t = T_r$ where there is a sharp change in the slope of the plasma density curve. For certain ranges of T_r, $A_b(T_r)$ and $B_b(T_r)$ are comparable. Moreover, the phase of the second term $e^{j\phi(T_r)}$ causes constructive and destructive interferences giving rise to the oscillations.

Next we consider the hyperbolic function for the plasma profile given by Equation J.31:

$$\omega_p^2(t) = \frac{\omega_{p0}^2}{1 + e^{-t/T_r}}. \tag{J.31}$$

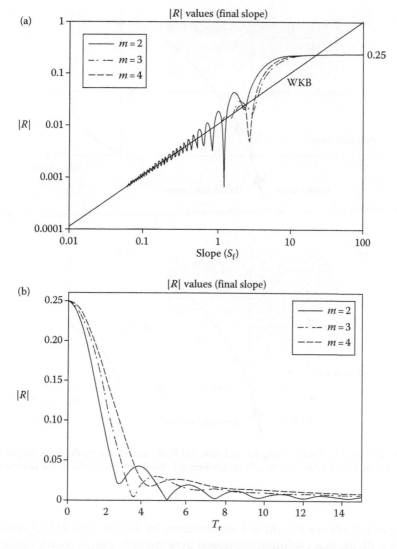

FIGURE J.8 Reflection coefficient ($|R|$ values) for finite final slopes. (a) Final slope S_f versus $|R|$ value and (b) rise time versus $|R|$ value.

This function is defined in the entire time region. It is smooth everywhere and its typical shape is shown in Figure J.10. The $|R|$ values are shown in Figure J.11. The slope of the profiles at $t = 0$ is given by

$$S_0 = \frac{\omega_{p0}^2}{4T_r}. \qquad (J.32)$$

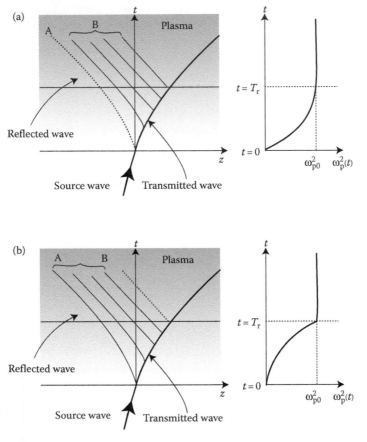

FIGURE J.9 Sketch of waves in space and time. (a) Reflected waves for the finite initial slope of the plasma time profile and (b) reflected waves for the finite final slope of the plasma time profile.

Figure J.11 shows the $|R|$ values obtained by FDTD. The FDTD solution agrees with the exact solution expressed in terms of hypergeometric functions [11]. The WKB method predicts $|R|$ values to be zero [10]. However, the FDTD method accurately calculates the $|R|$ values. It is noted that the $|R|$ values decrease quickly with the slope.

J.3.4   FDTD Analysis of a Magnetized Plasma Creation

In Section 3.2, we have considered isotropic cases where the cyclotron frequency ω_b is set to zero. When a plasma is created in the presence of a static magnetic field, where the direction of wave propagation is the same as the direction of the static magnetic field, the source wave splits into three new waves. The WKB solution has been considered in Refs. [1,12] for the following

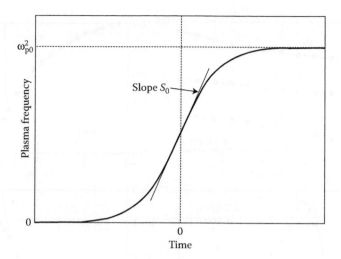

FIGURE J.10   Square of the plasma frequency profile for a hypergeometric time profile.

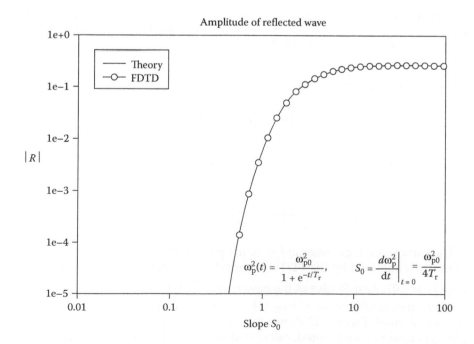

FIGURE J.11   Reflection coefficient ($|R|$ values) for a hypergeometric time profile. Slope S_0 versus $|R|$ value.

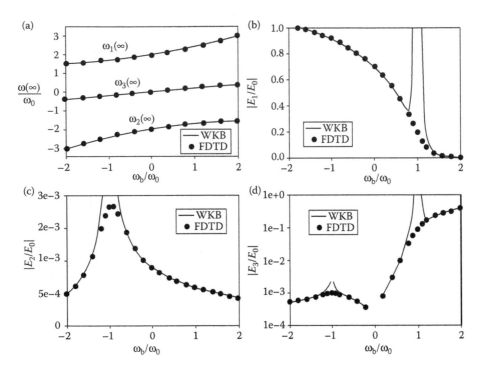

FIGURE J.12 Magnetized plasma creation. (a) Saturation frequency of new waves, (b) amplitude of the electric field with frequency $\omega_1(\infty)$, (c) amplitude of the electric field with frequency $\omega_2(\infty)$, and (d) amplitude of the electric field with frequency $\omega_3(\infty)$.

plasma creation profile:

$$\omega_p^2(t) = \omega_{p0}^2(1 - e^{-t/T_r}). \tag{J.33}$$

The frequencies ω_n of the new waves were found in the WKB method by solving the cubic equation [1,12]:

$$\omega_n^3 - \omega_b\omega_n^2 - (\omega_0^2 + \omega_p^2(t))\omega_n + \omega_b\omega_0^2 = 0. \tag{J.34}$$

The saturation frequencies at $t = \infty$ are plotted in Figure J.12 together with the results obtained by the FDTD where the parameters are $\omega_{p0}^2 = 3\omega_0^2$ and $\omega_{p0}^2/T_r = 0.01$. Amplitudes of the new waves by the WKB method, however, contain singularities at $\omega_b = \pm\omega_b$ [1] and are not valid in the regions around the singularities. Figure J.12 shows the amplitudes of the new waves, after the plasma creation is completed, calculated by FDTD. It is noted that the FDTD method with the SRM calculates successfully the amplitude at $\omega_b = \pm\omega_0$ for all the modes.

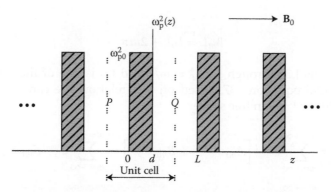

FIGURE J.13 Periodic layers of plasma in the presence of static magnetic field B_0. Amplitudes shown in the subsequent figures are computed in the middle of free space or in the middle of the plasma.

J.3.5 Periodic Magnetoplasma Layers: Sudden Switching

We next consider an infinite periodic-layer problem. Interaction of an EMW with a suddenly created spatially periodic plasma is of interest [13]. An EMW with right circular polarization (R wave) is assumed to be propagating in free space for $t < 0$. A static magnetic field of strength B_0 in the direction of propagation of the wave is also assumed to be present. At time $t = 0$, the plasma that is periodic in space is created as shown in Figure J.13. The profile is given by

$$\omega_p^2(z,t) = u(t)\, \omega_p^2(z) = u(t)\, \omega_{p0}^2 \sum_{m=-\infty}^{m=+\infty} [u(z - mL) - u(z - d - mL)]. \quad (J.35)$$

Here, $u(t)$ is the unit step function, L is the length of the spatial period, d is the width of the plasma layer, and ω_{p0} is the maximum plasma frequency. By applying (a) boundary conditions at the interfaces of free space and the plasma medium, and (b) the periodic boundary condition, that is, the Bloch condition,

$$E(z) = e^{j\beta L} E(z + L), \quad (J.36)$$

where β is the propagation constant, we obtain the dispersion relation [13] as

$$\cos \beta L = \cos k\eta d \, \cos 2kl - \frac{1}{2}\left(\eta + \frac{1}{\eta}\right) \sin k\eta d \, \sin 2kl, \quad (J.37)$$

where the refractive index η of the magnetoplasma [6] is given by

$$\eta = \sqrt{1 - \frac{\omega_{p0}^2}{\omega(\omega - \omega_b)}} \quad (J.38)$$

and

$$\beta_m L = k_0 L + 2m\pi. \tag{J.39}$$

In Equations J.37 through J.39, $k = \omega/c$ and β_m is one of the roots of the transcendental equation J.37. Based on this relation, we can construct the electric field for $t > 0$ in free space:

$$E(z,t) = \sum_{m=-\infty}^{m=+\infty} e^{j\omega_m t} \left[A_{mf} e^{-jk_m z} + B_{mf} e^{jk_m z} \right] = \sum_{m=-\infty}^{m=+\infty} e^{j\omega_m t} E_{mf}(z), \tag{J.40}$$

$$H(z,t) = \sum_{m=-\infty}^{m=+\infty} e^{j\omega_m t} \left[C_{mf} e^{-jk_m z} + D_{mf} e^{jk_m z} \right] = \sum_{m=-\infty}^{m=+\infty} e^{j\omega_m t} H_{mf}(z), \tag{J.41}$$

and in the plasma medium

$$E(z,t) = \sum_{m=-\infty}^{m=+\infty} e^{j\omega_m t} \left[A_{mp} e^{-jk_m \eta_m z} + B_{mp} e^{jk_m \eta_m z} \right] = \sum_{m=-\infty}^{m=+\infty} e^{j\omega_m t} E_{mp}(z), \tag{J.42}$$

$$H(z,t) = \sum_{m=-\infty}^{m=+\infty} e^{j\omega_m t} \left[C_{mp} e^{-jk_m \eta_m z} + D_{mp} e^{jk_m \eta_m z} \right] = \sum_{m=-\infty}^{m=+\infty} e^{j\omega_m t} H_{mp}(z), \tag{J.43}$$

where $k_m = \omega_m/c$.

It is noted from Equation J.39 that for any β we have the following condition:

$$e^{-j\beta_m L} = e^{-j(k_0 L + 2\pi m)} = e^{-jk_0 L} = \text{constant.} \tag{J.44}$$

This indicates that we can simulate the unbounded periodic media problem easily without having a large number of cells; only one unit cell with a free space layer and a plasma slab is needed. Hence Equations J.13 through J.15 can be used with the following conditions at the boundaries:

$$E|_{K+1}^n = E|_1^n e^{-jk_0 L}, \tag{J.45}$$

$$H|_{1/2}^{n+(1/2)} = H|_{K+(1/2)}^{n+(1/2)} e^{jk_0 L}, \tag{J.46}$$

where K is the last index in the computational domain of the unit cell. The computational domain used in the simulations is shown in Figure J.13 in which the unit cell is chosen as the space between the planes P and Q.

The coefficients A_{mf} and B_{mf} in Equation J.40 can be determined numerically. The frequency ω_m can be first found from Equation J.37 and then the amplitudes (E_{mf}) of the time series of the fields at two fixed space points in the free space are estimated. The unknown coefficients A_{mf} and B_{mf} can be found

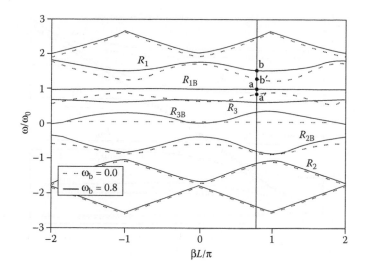

FIGURE J.14 Dispersion relation of a periodic magnetoplasma. Parameters are $\omega_{p0} = 1.2\omega_0$, $d = 0.2\,\lambda_0$, and $L = 0.6\,\lambda_0$.

by constructing a system of two equations from Equation J.40. Similarly, A_{mp} and B_{mp} can be obtained in the plasma region.

Figure J.14 shows the ω–β diagram of the infinite number of plasma layers for $d = L/3$ and $\omega_{p0}^2 = 1.2$. These results are obtained by solving Equation J.37 and are verified by the FDTD method. The solid lines are for the magneto-plasma ($\omega_b = 0.8\omega_0$) and the dotted lines are for the isotropic case ($\omega_b = 0$). For the isotropic case, the wiggler field ($\omega = 0$) is not found from the dispersion relation in the previous research [13]. However, this mode becomes apparent by imposing ω_b and taking the limit of ω_b approaching zero.

The solid vertical line represents the condition to be satisfied by the new modes for the given parameters [13]. Its intersection points with various branches of the ω–β diagram (like points a and b shown in the diagram) give the frequencies of the new waves generated by the switching action. From the diagram, it is obvious that as the strength of the gyrofrequency increases from 0 to 0.8, the frequencies of the modes labeled as R_1 and R_{1B} increase. The values obtained by Kuo et al. for the isotropic case [6] for R_1 and R_{1B} modes, 0.8379 and 1.2733, are shown by the points a' and b'. The logic behind the symbols used for labeling the modes is given next. The case of switching an unbounded magnetoplasma was considered earlier [3] and the modes generated were labeled as R_1, R_2, and R_3. R_1 and R_3 are forward-propagating waves with upshifted and downshifted frequencies, respectively. R_2 is a backward-propagating wave with upshifted frequency (ω/ω_0 is negative but the magnitude is greater than 1). This case is a particular case of $d = L$ of the problem under consideration in this section. For $d < L$, the switched medium is a periodic medium and hence gives rise to an infinite number of branches in the ω–β diagram.

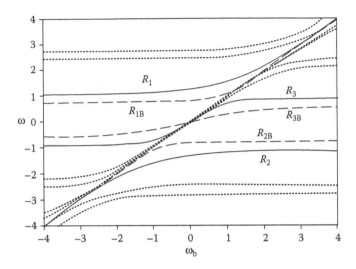

FIGURE J.15   Frequencies versus ω_b. Parameters are $\omega_{p0} = 1.2\omega_0$, $d = 0.2\lambda_0$, and $L = 0.6\lambda_0$.

The additional modes of significant amplitudes are labeled as R_{1B}, R_{2B}, and R_{3B}. The second subscript signifies that these modes are connected with the satisfaction of the Bloch condition (Equation J.36). Of course there are many more waves generated as shown in Figure J.14 but they are of small amplitudes. These additional small amplitude modes act as noise signals in the time series. This is the reason for the failure of the conventional signal processing technique [14].

In Figure J.15, the dependence of the new frequencies on ω_b is described. There are again an infinite number of modes generated and the solid and broken lines in the Figure J.15 show significant modes. In the presence of the external magnetic field, additional forward-propagating modes are generated below the line $\omega = \omega_b$. However, only two modes (R_3 and R_{3B}) are significant.

In Figures J.16 through J.18, the frequencies and amplitudes ($|E_{mf}|$) of the electric field of the significant modes in the middle of free space versus the thickness d of the plasma layer are shown. These results are obtained by the FDTD method. The waves whose frequencies are negative are backward-propagating waves. R_1 wave is always upshifted and the frequency increases with ω_b. R_{1B} is a downshifted wave for small ω_b; however, as ω_b increases it becomes an upshifted wave. It is noted that the amplitudes of $R_1(R_{1B})$ are low (high) for d around 0.5. R_3 is a downshifted wave for positive ω_b and its amplitude is significant and increases with ω_b. Another downshifted mode is the Bloch mode R_{3B}. ω_b has only a marginal effect on R_2 and R_{2B}.

In Figure J.19, the electric and magnetic field amplitudes of the individual modes are shown as a function of position in the computational domain for the isotropic case ($\omega_b = 0$). These results are obtained by the FDTD method. The electric field of the upshifted mode R_1 is weaker in free space and stronger in

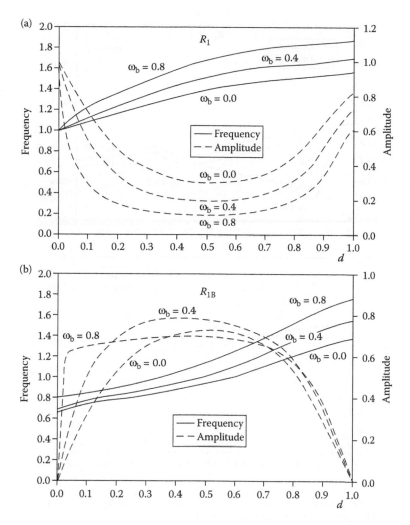

FIGURE J.16 Frequencies and amplitudes of (a) R_1 mode and (b) R_{1B} mode in the middle of free space. Other parameters are given in Figure J.6.

plasma. However, the electric field of the downshifted mode R_{1B} is stronger in free space. R_3 mode has interesting features. Its frequency and the electric field are zero. Its magnetic field is constant (dc magnetic field) in free space and space-varying (wiggler field) in plasma. Its amplitude decreases in plasma. The upshifted modes of the field have nearly constant fields in the plasma because this frequency is greater than the plasma frequency (ω_{p0}) and the plasma behaves like a dielectric. However, the downshifted frequency modes have minimum amplitudes in the middle of the plasma because this frequency ($\omega < \omega_{p0}$) is in the cutoff domain for the plasma. So the mode behavior strongly depends on the width of the plasma layer.

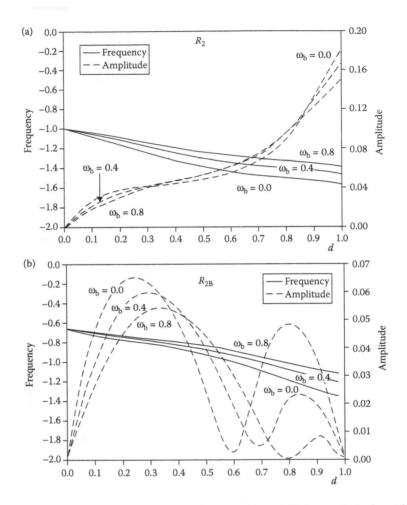

FIGURE J.17 Frequencies and amplitudes of (a) R_2 mode and (b) R_{2B} mode in the middle of free space. Other parameters are given in Figure J.6.

J.3.6 Periodic Plasma Layers: Finite Rise Time

The sudden creation of plasma layers is an ideal model, but in reality the plasma density increases from 0 to a maximum value with a finite rise time. The linear profile model of such a process is given by

$$
\omega_p^2(z,t) = \begin{cases}
0, & t < 0, \\[2mm]
\dfrac{t}{T_r}\omega_{p0}^2 \displaystyle\sum_{m=-\infty}^{m=+\infty} [u(z-mL) - u(z-d-mL)], & 0 < t < T_r, \\[4mm]
\omega_{p0}^2 \displaystyle\sum_{m=-\infty}^{m=+\infty} [u(z-mL) - u(z-d-mL)], & t > T_r.
\end{cases}
$$

$$\tag{J.47}$$

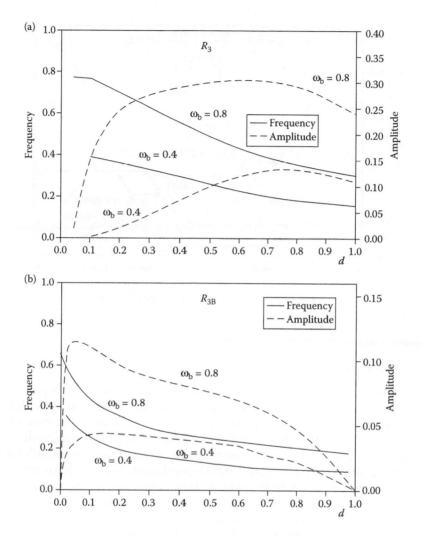

FIGURE J.18 Frequencies and amplitudes of (a) R_3 mode and (b) R_{3B} mode in the middle of free space. Other parameters are given in Figure J.6.

By changing T_r we can simulate an arbitrary time rate of plasma creation. In this simulation, we will concentrate on the fields after the plasma is fully created. The spectrum estimation is not appropriate during the plasma creation because their amplitudes and frequencies change with time. Figure J.20 describes the effect of the rise time of plasma creation on the amplitude of the electric fields at the center of free space and plasma. Upshifted mode R_1 is strong and its amplitude increases with the rise time. R_1 mode is the modified source wave. However, other modes become weaker with rise time.

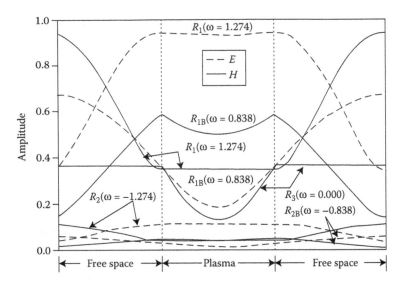

FIGURE J.19 Variation of amplitudes of several modes in a unit cell for the case of $\omega_b = 0$. Other parameters are given in Figure J.6.

J.4 Conclusions

We have seen that the FDTD method can easily simulate various problems and gives very accurate solutions. A new successive reduction algorithm is presented to overcome the conventional limitations of FFT. The new algorithm resolves several modes even if some of them have low amplitudes (up to 80 dB). Using this new method, comparisons are made among Green's function method, the WKB method, and FDTD simulation. It is seen that for the rise time approximately equal to the period of the source wave, the perturbation method by Green's function and the WKB method are not accurate.

We have also considered the case of the magnetoplasma creation. In the magnetoplasma problem the WKB method fails in the neighborhood of cyclotron resonance. However, the FDTD method performs successfully.

The FDTD method is also used to study the effect of creating periodic plasma layers in the presence of a static magnetic field. The characteristics of additional downshifted modes are examined.

J.A Appendix: Spectrum Estimation Using an SRM

J.A.1 Spectrum Analysis Using FFT

Spectrum analysis is a useful tool to extract information from the signals. Usually an FFT is used to analyze the signals in the frequency domain.

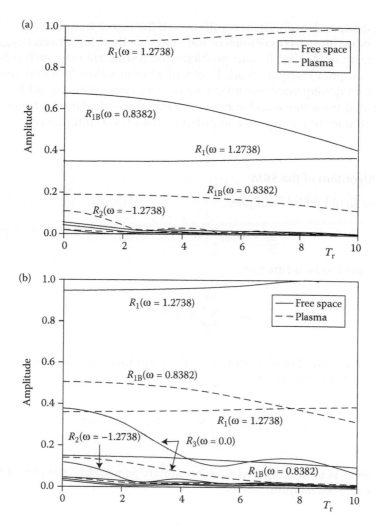

FIGURE J.20   Amplitudes versus rise time in the middle of free space and in the middle of the plasma: (a) electric fields and (b) magnetic fields.

However, the frequency accuracy is limited to the sampling frequency that is half the inverse of the sampling rate. The maximum error in the frequency estimation is half the frequency spacing. This frequency estimation error may not be serious for the modes whose amplitudes are large. However, small amplitude modes can be buried and undetectable due to the errors of poor frequency estimation. This will result in misinterpretation of the results.

The Prony method [15] is an alternative method to find complex frequencies and amplitudes from the signal. This method can be useful for a signal with a limited number of modes and very low noise. However, our extensive simulations show that as the number of modes increases this method produces very unstable and inconsistent results even if the signal-to-noise ratio is high. Several other improved versions of the Prony method have been suggested in the literature [14] but for our problems the results are not much different from the original Prony method. To obtain a better value than that afforded by FFT for any frequency that might not be very close to a natural FFT value, we searched for a stable and consistent method to calculate the frequencies and amplitudes of the signals computed by the FDTD simulations.

J.A.2   Algorithm of the SRM

Let s be a signal with N samples:

$$s_n = s(t_n), \quad n = 1, 2, \ldots, N. \tag{J.A.1}$$

s_n is assumed to be written as

$$s_n = \sum_{m=1}^{M} A_m e^{2\pi j f_m t_n}, \tag{J.A.2}$$

where M is the number of modes, A_m is the amplitude, and f_m is the frequency of the modes. Let FFT be an FFT operator given by

$$s_k = \text{FFT}(s) = \frac{1}{N} \sum_{n=1}^{N} s_n e^{-2\pi j (n-1)(k-1)/N}. \tag{J.A.3}$$

Our object is to find A_m and f_m as accurately as possible. We may think of the least squares method:

$$\text{SSE} = \sum_{n=1}^{N} \left| s_n - \sum_{m=1}^{M} A_m e^{2\pi j f_m t_n} \right|^2. \tag{J.A.4}$$

However, this equation is highly nonlinear in A_m and f_m and, therefore, it is not easy to handle. The authors have used the following procedures to find these parameters accurately:

Step 1: Let $A_m = 0$ for $m = 1, 2, \ldots, M$

Step 2: Repeat Steps 3–5 for $m = 1, 2, \ldots, M$

Step 3: $\hat{s}_n = s_n - \sum_{l=1, l \neq m}^{M} A_l e^{2\pi j f_l t_n}$ for $n = 1, 2, \ldots, N$

Step 4: Choose the frequency that gives the maximum amplitude by the $\text{FFT}(\hat{s}_n)$ and call it f_m

Step 5: Accurately find the frequency by maximizing $|A_m|$ given by Equation J.A.5 in the neighborhood of the frequency f_m using an optimization technique available in MATLAB® as *fmin*

$$A_m = \frac{1}{N} \sum_{n=1}^{N} \hat{s}_n e^{-2\pi j f_m t_n}. \tag{J.A.5}$$

Step 6: Repeat Steps 2 through 5 if the following condition is not satisfied:

$$\sum_{n=1}^{N} \left| s_n - \sum_{l=1}^{M} A_l e^{2\pi j f_l t_n} \right|^2 < \text{tolerance.} \tag{J.A.6}$$

This procedure eliminates the mutual interference between two adjacent components and accurately finds frequencies and amplitudes. Successive over relaxation (SOR) may be used in estimating the frequency to save computational time.

Figures J.A1 and J.A2 are examples of the signal analysis by the above procedure. The following model is considered to test the validity of the procedure:

$$s(t) = 0.45 \exp[2\pi j(-1.41)t] + 0.15 \exp[2\pi j(-0.83)t]$$

$$+ 0.00887 \exp[2\pi j(-1.13)t] + 0.95 \exp[2\pi j(1.02)t] \tag{J.A.7}$$

$$+ 0.01 \exp[2\pi j(1.18)t] + \text{noise}(t),$$

where the random noise power is -40 dB relative to the signal power. Figure J.A1 shows the real and imaginary parts of the sampled signal $s(t)$.

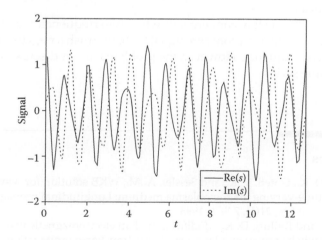

FIGURE J.A1 Time series of a signal s.

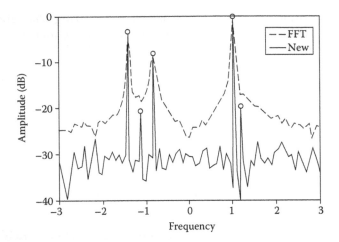

FIGURE J.A2 Comparison of the FFT and the SRM.

The number of samples is 128 and the sampling time interval is 0.1. Figure J.A2 shows the amplitudes of the spectra $s(f)$ in frequency domain. The broken line is obtained by an ordinary FFT method. This reveals three distinct peaks. The solid line is obtained by the new method and shows five distinct peaks. For the strong modes there exist small differences in frequencies and amplitudes between the two methods. The two weak modes are not shown in the ordinary FFT, while they are distinct in the new method. The residues of the new method are uniform in the entire frequency range as expected by the randomness of the noise. Our exhaustive simulations show that the Prony method and the modified Prony method do not work for a large number of modes. In addition, even for a small number of modes, they fail when the signal has noise.

This method has a slow convergence when two frequencies are very close. Note that we need to use a complex signal to distinguish a negative frequency from a positive frequency. However, the problems under discussion generate a complex time series.

References

1. Kalluri, D. K., Goteti, V. R., and Sessler, A. M., WKB solution for wave propagation in a time-varying magnetoplasma medium: Longitudinal propagation, *IEEE Trans. Plasma Sci.*, 21(1), 70–76, 1993.
2. Lee, J. H. and Kalluri, D. K., Modification of an electromagnetic wave by a time-varying switched magnetoplasma medium: Transverse propagation, *IEEE Trans. Plasma Sci.*, 26(1), 1–6, 1998.

3. Huang, T. T., Lee, J. H., Kalluri, D. K., and Groves, K. M., Wave propagation in a transient plasma—development of a Green's function, *IEEE Trans. Plasma Sci.*, 26(1), 19–25, 1998.
4. Kalluri, D. K. and Huang, T. T., Longitudinal propagation in a magnetized time-varying plasma: Development of Green's function, *IEEE Trans. Plasma Sci.*, 26(3), 1022–1030, 1998.
5. Yee, K. S., Numerical solution of initial boundary value problems involving Maxwell's equations in isotropic media, *IEEE Trans. Ant. Prop.*, AP-14, 302–307, 1966.
6. Kalluri, D. K., *Electromagnetics of Complex Media*, CRC Press, Boca Raton, FL, August 1998.
7. Young, J. L., A full finite difference time domain implementation for radio wave propagation in a plasma, *Radio Sci.*, 29, 1513–1522, 1994.
8. Taflove, A., *Computational Electrodynamics: The Finite-Difference Time-Domain Method*, Artech House, Boston, MA, 1995.
9. Jiang, C. L., Wave propagation and dipole radiation in a suddenly created plasma, *IEEE Trans. Ant. Prop.*, AP-23, 83–90, 1975.
10. Banos, Jr., A. Mori, W. B., and Dawson, J. M., Computation of the electric and magnetic fields induced in a plasma created by ionization lasting a finite interval of time, *IEEE Trans. Plasma Sci.*, 21, 57–69, 1993.
11. Lekner, J., *Theory of Reflection of Electromagnetic and Particle Waves*, Kluwer Academic Publishers, Boston, 1987.
12. Lee, J. H., Wave propagation in a time-varying switched magnetoplasma medium: Transverse propagation, MS thesis, University of Massachusetts Lowell, Lowell, 1994.
13. Kuo, S. P. and Faith, J., Interaction of an electromagnetic wave with a rapid created spatially periodic plasma, *Phys. Rev. E*, 56(2), 1–8, 1997.
14. Osborne, M. R. and Smyth, G. K., A modified Prony algorithm for fitting functions defined by difference equations, *SIAM J. Sci. Stat. Comput.*, 12, 362–382, 1991.
15. Kannan, N. and Kundu, D., On modified EVLP and ML methods for estimating superimposed exponential signals, *Signal Process.*, 39, 223–233, 1994.

Appendix K

*Three-Dimensional FDTD Simulation of EMW Transformation in a Dynamic Inhomogeneous Magnetized Plasma**

Joo Hwa Lee and Dikshitulu K. Kalluri

K.1 Introduction

Frequency shifting of an EMW in a time-varying plasma has been extensively studied [1–7] and some experiments [1–3] were carried out to demonstrate frequency shifting. Analytical studies make various assumptions and use simplified geometries including one-dimensional models, flash ionization, and slow or fast creation of the plasma medium [4–7].

A limited number of theoretical and numerical studies of three-dimensional models are reported. Buchsbaum et al. [8] examined the perturbation theory for various mode configurations of a cylindrical cavity coaxial with a plasma column and coaxial with the static magnetic field. Gupta [9] used a moment method to study cavities and waveguides containing anisotropic media and compared the results with those from the perturbation method. A transmission-line-matrix (TLM) method was developed to study the interaction of an EMW with a time-invariant and space-invariant magnetized plasma in three-dimensional space [10]. Mendonça [11] presented a mode coupling theory in a cavity for a space-varying and slowly created isotropic plasma.

Since the introduction of the FDTD method [12], it has been widely used in solving many electromagnetic problems including those concerned with plasma media [13,14]. For the anisotropic cases [15,16], the equations for the components of the current density vector become coupled and the implementation of the conventional FDTD scheme is difficult. We propose a new FDTD method to overcome this difficulty.

In this paper, we use the FDTD method to analyze the interaction of an EMW with a magnetoplasma medium created in a cavity. This paper is organized as follows. The FDTD algorithm is derived first and the implementation of

* IEEE. Reprinted from *IEEE Trans. Ant. Prop.*, 47 (7), 1146–1151, 1999. With permission.

perfect electric conductor (PEC) boundary conditions is investigated. The algorithm is verified by using a mode coupling theory and a perturbation technique. The application of the algorithm is illustrated by computing the new frequencies and amplitudes of the coupled modes generated due to a switched magnetized time-varying and space-varying plasma in a cavity with an initial TM mode excitation.

K.2 Three-Dimensional FDTD Algorithm

K.2.1 Maxwell's Equation

Consider a time-varying magnetoplasma medium with collisions. Maxwell's equations and constitutive relation for a cold magnetoplasma are given by

$$\nabla \times \mathbf{E} = -\mu_0 \frac{\partial \mathbf{H}}{\partial t}, \tag{K.1}$$

$$\nabla \times \mathbf{H} = \varepsilon_0 \frac{\partial \mathbf{E}}{\partial t} + \mathbf{J}, \tag{K.2}$$

$$\frac{d\mathbf{J}}{dt} + \nu \mathbf{J} = \varepsilon_0 \omega_p^2(\mathbf{r}, t)\mathbf{E} + \boldsymbol{\omega}_b(\mathbf{r}, t) \times \mathbf{J}, \tag{K.3}$$

where ε_0 is the permittivity of free space, μ_0 is the permeability of free space, ω_p^2 is the square of the plasma frequency, $\boldsymbol{\omega}_b = e\mathbf{B}_0/m_e$ is the electron gyrofrequency, $\mathbf{B}_0$ is the external static magnetic field, and e and m_e are the electric charge and mass of an electron, respectively. The field components in the Cartesian coordinate are expressed as

$$\mathbf{E} = E_x \hat{x} + E_y \hat{y} + E_z \hat{z}, \tag{K.4}$$

$$\mathbf{H} = H_x \hat{x} + H_y \hat{y} + H_z \hat{z}, \tag{K.5}$$

$$\mathbf{J} = J_x \hat{x} + J_y \hat{y} + J_z \hat{z}, \tag{K.6}$$

$$\boldsymbol{\omega}_b = \omega_{bx} \hat{x} + \omega_{by} \hat{y} + \omega_{bz} \hat{z}. \tag{K.7}$$

Substituting Equations K.4 through K.7 into Equations K.1 through K.3 gives the following equations:

$$\frac{\partial H_x}{\partial t} = -\frac{1}{\mu_0} \left(\frac{\partial E_z}{\partial y} - \frac{\partial E_y}{\partial z} \right), \tag{K.8}$$

$$\frac{\partial E_x}{\partial t} = \frac{1}{\varepsilon_0} \left(\frac{\partial H_z}{\partial y} - \frac{\partial H_y}{\partial z} - J_x \right), \tag{K.9}$$

$$\begin{bmatrix} \dfrac{dJ_x}{dt} \\[2mm] \dfrac{dJ_y}{dt} \\[2mm] \dfrac{dJ_z}{dt} \end{bmatrix} = \mathbf{\Omega} \begin{bmatrix} J_x \\ J_y \\ J_z \end{bmatrix} + \varepsilon_0 \omega_p^2(\mathbf{r}, t) \begin{bmatrix} E_x \\ E_y \\ E_z \end{bmatrix}, \tag{K.10}$$

where

$$\mathbf{\Omega} = \begin{bmatrix} -\nu & -\omega_{bz} & \omega_{by} \\ \omega_{bz} & -\nu & -\omega_{bx} \\ -\omega_{by} & \omega_{bx} & -\nu \end{bmatrix}. \tag{K.11}$$

The other components of **E** and **H** can be obtained in a similar manner.

K.2.2 FDTD Equation

The usual grid configuration for **J** is to place J_x, J_y, and J_z [17] at the locations of E_x, E_y, and E_z, respectively. This configuration works fine as long as the equations for the components of **J** are not coupled. When coupled as in Equation K.10, J_y and J_z are needed at the position of J_x to update J_x. However, the estimations of J_y and J_z at this position can be very complex; the maintenance of second-order accuracy requires averaging the values at four diagonal positions. Moreover, implementation of the averaging on the boundary is troublesome because some quantities outside the boundary are needed. We overcame these difficulties by placing **J** at the center of the Yee cube as shown in Figure K.1. The components of **J** are located at the same space point. It is now possible to solve Equation K.10 analytically since all variables are now available at the same space point. Treating **E**, ω_b, ν, and ω_p as constants, each having an average value observed at the center of the time step, the Laplace transform of Equation K.10 leads to

$$\mathbf{J}(s) = (s\mathbf{I} - \mathbf{\Omega})^{-1} \mathbf{J}_0 + \varepsilon_0 \omega_p^2 \frac{1}{s} (s\mathbf{I} - \mathbf{\Omega})^{-1} \mathbf{E}, \tag{K.12}$$

where **I** is the identity matrix. The inverse Laplace transform of Equation K.12 leads to the explicit expression for **J**(*t*) as

$$\mathbf{J}(t) = \mathbf{A}(t)\mathbf{J}_0 + \varepsilon_0 \omega_p^2 \mathbf{K}(t)\mathbf{E}, \tag{K.13}$$

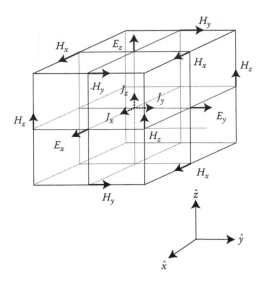

FIGURE K.1 Positioning of the electric, magnetic, and current density field vector components about a cubic unit cell.

where

$$\mathbf{A}(t) = \exp[\mathbf{\Omega} t]$$

$$= e^{-vt} \begin{bmatrix} C_1\omega_{bx}\omega_{bx} + \cos\omega_b t & C_1\omega_{bx}\omega_{by} - S_1\omega_{bz} & C_1\omega_{bx}\omega_{bz} + S_1\omega_{by} \\ C_1\omega_{by}\omega_{bx} + S_1\omega_{bz} & C_1\omega_{by}\omega_{by} + \cos\omega_b t & C_1\omega_{by}\omega_{bz} - S_1\omega_{bx} \\ C_1\omega_{bz}\omega_{bx} - S_1\omega_{by} & C_1\omega_{bz}\omega_{by} + S_1\omega_{bx} & C_1\omega_{bz}\omega_{bz} + \cos\omega_b t \end{bmatrix},$$

$$\tag{K.14}$$

$$\mathbf{K}(t) = \mathbf{\Omega}^{-1}(\exp[\mathbf{\Omega} t] - \mathbf{I}) = \frac{e^{-vt}}{\omega_b^2 + v^2}$$

$$\times \begin{bmatrix} C_2\omega_{bx}\omega_{bx} + C_3 & C_2\omega_{bx}\omega_{by} - C_4\omega_{bz} & C_2\omega_{bx}\omega_{bz} + C_4\omega_{by} \\ C_2\omega_{by}\omega_{bx} + C_4\omega_{bz} & C_2\omega_{by}\omega_{by} + C_3 & C_2\omega_{by}\omega_{bz} - C_4\omega_{bx} \\ C_2\omega_{bz}\omega_{bx} - C_4\omega_{by} & C_2\omega_{bz}\omega_{by} + C_4\omega_{bx} & C_2\omega_{bz}\omega_{bz} + C_3 \end{bmatrix},$$

$$\tag{K.15}$$

$$S_1 = \sin\frac{\omega_b t}{\omega_b}, \tag{K.16}$$

$$C_1 = \frac{(1 - \cos\omega_b t)}{\omega_b^2}, \tag{K.17}$$

$$C_2 = \frac{(e^{vt} - 1)}{v - vC_1 - S_1}, \tag{K.18}$$

$$C_3 = v(e^{vt} - \cos\omega_b t) + \omega_b \sin\omega_b t, \tag{K.19}$$

$$C_4 = e^{vt} - \cos\omega_b t - vS_1, \tag{K.20}$$

and $\omega_b^2 = \omega_{bx}^2 + \omega_{by}^2 + \omega_{bz}^2$. The FDTD equations for **J** are now expressed in terms of their values at a previous time step without having to solve simultaneous equations at each step:

$$
\begin{bmatrix} J_x|_{i,j,k}^{n+(1/2)} \\ J_y|_{i,j,k}^{n+(1/2)} \\ J_z|_{i,j,k}^{n+(1/2)} \end{bmatrix} = \mathbf{A}(\Delta t) \begin{bmatrix} J_x|_{i,j,k}^{n-(1/2)} \\ J_y|_{i,j,k}^{n-(1/2)} \\ J_z|_{i,j,k}^{n-(1/2)} \end{bmatrix}
$$

$$
+ \frac{\varepsilon_0}{2} \left. \omega_p^2 \right|_{i,j,k}^n \mathbf{K}(\Delta t) \begin{bmatrix} E_x|_{i+(1/2),j,k}^n + E_x|_{i-(1/2),j,k}^n \\ E_y|_{i,j+(1/2),k}^n + E_y|_{i,j-(1/2),k}^n \\ E_z|_{i,j,k+(1/2)}^n + E_z|_{i,j,k-(1/2)}^n \end{bmatrix}. \qquad \text{(K.21)}
$$

In the above equation, the time step begins at $n - (1/2)$ and ends at $n + (1/2)$. This formulation is valid for arbitrary values of v and ω_b; the idea is similar to the exponential time stepping [18] for a high conductivity case. Also, if v and ω_b are functions of time and space, Equation K.21 can be used by replacing those by $v|_{i,j,k}^n$ and $\omega_b|_{i,j,k}^n$ in Equations K.14 through K.20.

Using the grid in Figure K.1, the following FDTD equations for the x-components of **E** and **H** can be written as

$$
E_x|_{i+(1/2),j,k}^{n+1} = E_x|_{i+(1/2),j,k}^n
$$

$$
+ \frac{1}{\varepsilon_0} \left[\frac{\Delta t}{\Delta y} \left(H_z|_{i+(1/2),j+(1/2),k}^{n+(1/2)} - H_z|_{i+(1/2),j-(1/2),k}^{n+(1/2)} \right) \right.
$$

$$
\left. - \frac{\Delta t}{\Delta z} \left(H_y|_{i+(1/2),j,k+(1/2)}^{n+(1/2)} - H_y|_{i+(1/2),j,k-(1/2)}^{n+(1/2)} \right) \right]
$$

$$
- \frac{\Delta t}{2\varepsilon_0} \left(J_x|_{i+1,j,k}^{n+(1/2)} + J_x|_{i,j,k}^{n+(1/2)} \right), \qquad \text{(K.22)}
$$

$$
H_x|_{i,j+(1/2),k+(1/2)}^{n+(1/2)} = H_x|_{i,j+(1/2),k+(1/2)}^{n-(1/2)}
$$

$$
- \frac{1}{\mu_0} \left[\frac{\Delta t}{\Delta y} \left(E_z|_{i,j+1,k+(1/2)}^n - E_z|_{i,j,k+(1/2)}^n \right) \right.
$$

$$
\left. - \frac{\Delta t}{\Delta z} \left(E_y|_{i,j+(1/2),k+1}^n - E_y|_{i,j+(1/2),k}^n \right) \right]. \qquad \text{(K.23)}
$$

The other components can be written in a similar way.

The stability condition of this method is not easily expressed in a simple form due to the complexity of the algorithm. Nevertheless, it can be said that the standard stability criterion can be applied, that is, the stability is governed by the mode whose phase velocity is fastest in the medium [19,20].

K.2.3 PEC Boundary Conditions

Figure K.2 shows the locations of the fields on the $x-y$ plane. The indices k and $k + (1/2)$ of the planes correspond to those in the FDTD equations. J_x, J_y, J_z, E_x, E_y, and H_z are located on the k plane, whereas E_z, H_x, and H_y are located on the $k + (1/2)$ plane. For a perfect conductor, the tangential components of **E** field and the normal component of **H** field vanish on the boundaries. Hence, it is natural to choose PEC boundaries to correspond to a k rather than a $k + (1/2)$ plane. For example, the plane $k = 1$ may be used for the bottom PEC of a cavity. The boundary conditions are satisfied by assigning zero value to E_x, E_y, J_x, J_y, and H_z on the $k = 1$ plane. Since J_z is not zero on the $k = 1$ plane, we need to update its value with time. From Equation K.21, the values

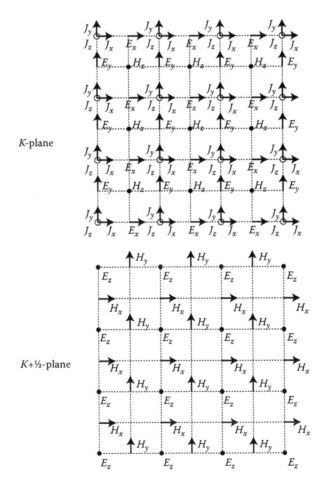

FIGURE K.2 Construction of planes for a simple implementation of the PEC boundary conditions.

$E_z|_{i,j,3/2}^n$ and $E_z|_{i,j,1/2}^n$ are needed to obtain the updated value $J_z|_{i,j,1}^{n+(1/2)}$. We note that the PEC boundary condition $\partial E_n/\partial n = 0$ will permit us to modify Equation K.21 for $k = 1$ as given below:

$$\begin{bmatrix} J_x|_{i,j,1}^{n+(1/2)} \\ J_y|_{i,j,1}^{n+(1/2)} \\ J_z|_{i,j,1}^{n+(1/2)} \end{bmatrix} = \mathbf{A}(\Delta t) \begin{bmatrix} J_x|_{i,j,1}^{n-(1/2)} \\ J_y|_{i,j,1}^{n-(1/2)} \\ J_z|_{i,j,1}^{n-(1/2)} \end{bmatrix}$$

$$+ \frac{\varepsilon_0}{2} \omega_P^2|_{i,j,1}^n \mathbf{K}(\Delta t) \begin{bmatrix} E_x|_{i+(1/2),j,1}^n + E_x|_{i-(1/2),j,1}^n \\ E_y|_{i,j+(1/2),1}^n + E_y|_{i,j-(1/2),1}^n \\ 2E_z|_{i,j,(3/2)}^n \end{bmatrix}. \tag{K.24}$$

The algorithm given above leads to explicit computation while maintaining second-order accuracy.

K.3 Validation of the Algorithm

The algorithm is validated by considering two test cases for which results are obtained by other techniques. The first test case involves switching of a plasma medium in a rectangular microwave cavity with dimensions L_x, L_y, and L_z. Several cavity modes are excited due to the interaction of the incident mode and the plasma medium. Hence, the new fields may be written in terms of cavity modes as

$$\mathbf{E}(\mathbf{r}, t) = \sum a_l(t) \mathbf{E}_l(\mathbf{r}), \tag{K.25}$$

where $\mathbf{E}_l(\mathbf{r})$ is a normalized cavity mode and

$$\int \mathbf{E}_l(\mathbf{r}) \mathbf{E}_{l'}^*(\mathbf{r}) \, d^3r = \delta_{ll'}. \tag{K.26}$$

In the above equation, $\mathbf{E}^*$ is the complex conjugate of $\mathbf{E}$. For the space-varying but isotropic plasma creation, the differential equations for a are obtained by the mode coupling theory as [11]

$$\frac{\partial^2 a_l(t)}{\partial t^2} + \omega_l^2 a_l + \sum_{l'} C_{ll'} a_{l'} = 0, \tag{K.27}$$

where

$$\omega_l^2 = c^2 k^2 = c^2 \left[\left(\frac{m\pi}{L_x} \right)^2 + \left(\frac{p\pi}{L_y} \right)^2 + \left(\frac{q\pi}{L_z} \right)^2 \right], \tag{K.28}$$

$$C_{ll'} = \int \omega_p^2(\mathbf{r}, t) \mathbf{E}_l(\mathbf{r}) \mathbf{E}_{l'}^*(\mathbf{r}) \, d^3 r, \tag{K.29}$$

and m, p, and q are the mode numbers.

Consider the creation of a time-varying and space-varying plasma medium in the cavity with a plasma frequency profile $\omega_p^2(\mathbf{r}, t)$:

$$\omega_p^2(\mathbf{r}, t) = \omega_{p0}^2 (1 - e^{-t/T_r}) \exp\left[-\left(\alpha_x \frac{x - x_0}{L_x} \right)^2 - \left(\alpha_y \frac{y - y_0}{L_y} \right)^2 - \left(\alpha_z \frac{z - z_0}{L_x} \right)^2 \right],$$

$$t > 0, \tag{K.30}$$

where ω_{p0}^2 is the saturation plasma frequency, T_r is a rise time, and the spatial variation is Gaussian. If the spatial variation of the plasma is only in the x-direction, that is, $\alpha_y = \alpha_z = 0$, newly excited modes should have the same mode numbers for p and q. Therefore, we need to consider the changes in mode number m only to describe the fields in the cavity.

Figure K.3 shows the comparison of the results from FDTD with those from the mode coupling theory for the initial TM_{111} mode excitation. Isotropic and inhomogeneous plasma distribution is considered for comparison. The following parameters are used: $L_x = 4L_y = 4L_z$, $T_r = 0$, $x_0 = L_x/2$, $\omega_{p0} = \omega_0$, $v = 0$, $\alpha_x = 1$, $\alpha_y = \alpha_z = 0$, and ω_0 is the frequency of the initial mode. The results for the mode coupling theory are obtained by numerically solving the differential Equations K.27 with the initial conditions obtained from the initial excitation. The results for the FDTD are obtained first by calculating $\mathbf{E}(\mathbf{r}, n\Delta t)$ using our algorithm and then by computing $a_l(t)$:

$$a_l(t) = a_l(n \, \Delta t) = \int \mathbf{E}(\mathbf{r}, n \, \Delta t) \mathbf{E}_l^*(\mathbf{r}) \, d^3 r. \tag{K.31}$$

Figure K.3 shows very good agreement.

The second validation is for the case of a magnetized plasma. Results based on a perturbation method are available for a magnetized low-density plasma-filled cavity in Ref. [21]. In the perturbation theory, the frequency shift due to homogeneous magnetized plasma can be obtained as

$$\frac{\Delta\omega}{\omega_0} \approx -\frac{1}{2} \int \varepsilon_0 \mathbf{E}(\mathbf{r}) \Delta\varepsilon \mathbf{E}^*(\mathbf{r}) \, d^3 r, \tag{K.32}$$

where $\Delta\varepsilon = \varepsilon - \mathbf{I}$ and ε is the dielectric tensor of the magnetized plasma medium. For the magnetized plasma problem, a differential equation similar to Equation K.27 is not available. Nevertheless, the new fields can be

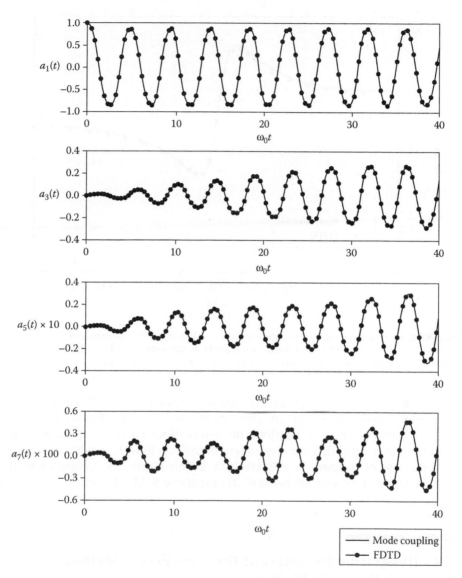

FIGURE K.3 Mode amplitudes for the case of a switched isotropic plasma in a cavity. Comparisons of the FDTD and the mode coupling theory are shown. The coefficients are normalized with respect to $a_1(0)$. $a_m(t)$ indicates $a_{m11}(t)$.

approximated by Equation K.25 and the empty cavity modes can be used to extract the modes from FDTD simulation for a weak plasma density since the coupling is not strong. During the FDTD computations, $a_l(t)$ is computed using Equation K.31 and the frequency change is obtained from the time series of $a_l(t)$ by the Prony method [22]. Figure K.4 shows the comparison of the perturbation and FDTD methods for an initial TM$_{111}$ mode excitation.

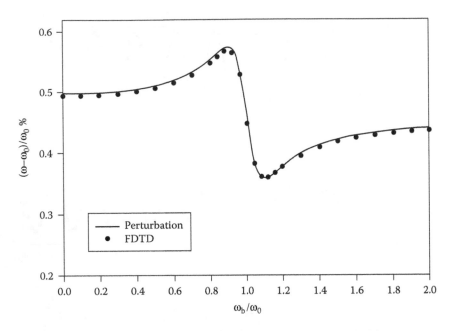

FIGURE K.4 Frequency shift due to the creation of a low-density homogeneous magnetized plasma in a cavity. The FDTD and perturbation methods are compared. The ratio of frequency shift (%) versus the cyclotron frequency ω_b/ω_0 is plotted. $\omega_p/\omega_0 = 0.1$ and $\nu/\omega_0 = 0.1$.

A lossy ($\nu/\omega_0 = 0.1$) homogeneous plasma medium ($\omega_p/\omega_0 = 0.1$) is used and the dimensions of the cavity are given by $9L_x = 4L_y = 3L_z$. The background magnetic field $\omega_b(\mathbf{r}) = \omega_b\hat{z}$ is along the z-axis. Since we have computed the final frequency shift that is independent of the rise time T_r, we have considered the case of sudden ($T_r = 0$) creation. In this figure, the frequency-shift ratio is depicted for various background magnetic field intensities.

K.4 Illustrative Examples of the New FDTD Method for a Dynamic Medium

Figure K.5 shows the frequency shifting due to the interaction of the EMW with a time-varying and space-varying magnetized plasma. A rectangular cavity with an initial excitation of TM_{111} mode is considered with the following parameters: $6L_x = 3L_y = 2L_z$, $T_r = 100/\omega_0$, $x_0 = L_x/2$, $y_0 = L_y/2$, $z_0 = L_z/2$, $\omega_{p0} = \omega_0$, and $\nu = 0$. $\alpha_x = \alpha_y = \alpha_z = 0$ is used for homogeneous plasma distribution [(a) and (b)] and $\alpha_x = \alpha_y = \alpha_z = 2$ is used for inhomogeneous plasma distribution [(c) and (d)]. The electron cyclotron frequency ω_b is chosen to be zero for (a) and (c), and $\omega_b = \omega_0\hat{z}$ for (b) and (d). The results are obtained from the spectrum of the time series of E_x at $x = \frac{1}{15}L_x$,

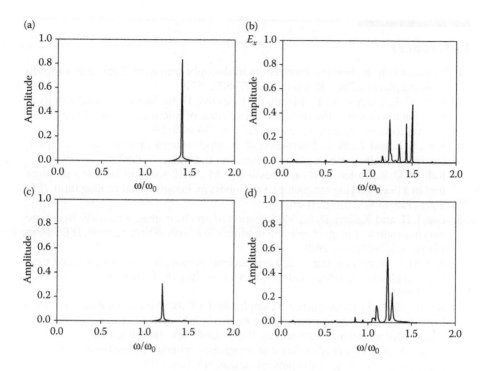

FIGURE K.5 Amplitude spectra of the electric field E_x at $x = \frac{1}{15}L_x$, $y = \frac{7}{15}L_y$, $z = \frac{7}{15}L_z$. See the text for the values of the parameters. (a) Homogeneous-isotropic, (b) homogeneous-anisotropic, (c) inhomogeneous-isotropic, and (d) inhomogeneous-anisotropic.

$y = \frac{7}{15}L_y$, $z = \frac{7}{15}L_z$ taken after the plasma is almost fully created, that is, after $t = 6.4\,T_r$. For a homogeneous and isotropic creation of plasma (Figure K.5a) the frequency shift due to the interaction of the EMW with the plasma is the same as that of an unbounded plasma. For the cavity with an anisotropic uniform plasma (Figure K.5b), several new modes appear. It is noted that some of the modes have higher frequencies than that of the unbounded isotropic plasma. A comparison of Figure K.5a with Figure K.5c and of Figure K.5b with Figure K.5d shows that a nonuniformly filled plasma cavity displays lower values of frequency shifts for the same maximum values of ω_{p0} as predicted in Ref. [11].

K.5 Conclusion

Three-dimensional FDTD formulations are constructed for time-varying inhomogeneous magnetoplasma medium. An explicit formulation is accomplished by locating **J** at the center of the Yee cube and by using an appropriate time-stepping algorithm.

References

1. Yablonovitch, E., Spectral broadening in the light transmitted through a rapidly growing plasma, *Phys. Rev. Lett.*, 31, 877–879, 1973.
2. Joshi, C. J., Clayton, C. E., Marsh, K., Hopkins, D. B., Sessler, A., and Whittum, D., Demonstration of the frequency upshifting of microwave radiation by rapid plasma creation, *IEEE Trans. Plasma Sci.*, 18, 814–818, 1990.
3. Kuo, S. P. and Faith, J., Interaction of an electromagnetic wave with a rapidly created spatially periodic plasma, *Phys. Rev. E*, 56(2), 1–8, 1997.
4. Kalluri, D. K., Goteti, V. R., and Sessler, A. M., WKB solution for wave propagation in a time-varying magnetoplasma medium: Longitudinal propagation, *IEEE Trans. Plasma Sci.*, 21(1), 70–76, 1993.
5. Lee, J. H. and Kalluri, D. K., Modification of an electromagnetic wave by a time-varying switched magnetoplasma medium: Transverse propagation, *IEEE Trans. Plasma Sci.*, 26(1), 1–6, 1998.
6. Kalluri, D. K. and Huang, T. T., Longitudinal propagation in a magnetized time-varying plasma: development of Green's function, *IEEE Trans. Plasma Sci.*, 26(3), 1–9, 1998.
7. Kalluri, D. K., *Electromagnetics of Complex Media*, CRC Press: Boca Raton, FL, 1998.
8. Buchsbaum, S. J., Mower, L., and Brown, S. C., Interaction between cold plasmas and guided electromagnetic waves, *Phys. Fluids*, 3(5), 806–819, 1960.
9. Gupta, R. R., *A study of cavities and waveguides containing anisotropic media*, PhD dissertation, Syracuse University, Syracuse, NY, July 1965.
10. Kashiwa, T., Yoshida, N., and Fukai, I., Transient analysis of a magnetized plasma in three-dimensional space, *IEEE Trans. Ant. Prop.*, AP-36, 1096–1105, 1988.
11. Mendonça, J. T. and Oliveira e Silva, L., Mode coupling theory of flash ionization in a cavity, *IEEE Trans. Plasma Sci.*, 24(1), 147–151, 1996.
12. Yee, K. S., Numerical solution of initial boundary value problems involving Maxwell's equations in isotropic media, *IEEE Trans. Ant. Prop.*, AP-14, 302–307, 1966.
13. Young, J. L., A full finite difference time domain implementation for radio wave propagation in a plasma, *Radio Sci.*, 29, 1513–1522, 1994.
14. Cummer, S. A., An analysis of new and existing FDTD methods for isotropic cold plasma and a method for improving their accuracy, *IEEE Trans. Ant. Prop.*, AP-45, 392–400, 1997.
15. García, S. G., Hung-Bao, T. M., Martín, R. G., and Olmedo, B. G., On the application of finite methods in time domain to anisotropic dielectric waveguides, *IEEE Trans. Microwave Theory*, 44, 2195–2205, 1996.
16. Schneider, J. and Hudson, S., The finite-difference time-domain method applied to anisotropic material, *IEEE Trans. Ant. Prop.*, AP-41, 994–999, 1993.
17. Sano, E. and Shibata, T., Fullwave analysis of picosecond photoconductive switches, *IEEE J. Quant. Electron.* 26(2), 372–377, 1990.
18. Taflove, A., *Computational Electrodynamics: The Finite-Difference Time-Domain Method*, Artech House, Boston, MA, 1995.
19. Taflove, A. and Brodwin, M. E., Numerical solution of steady-state electromagnetic scattering problems using the time-dependent Maxwell's equations, *IEEE Trans. Microwave Theory Tech.*, MTT-23(8), 623–630, 1975.

20. Young, J. L. and Brueckner, F. P., A time domain numerical model of a warm plasma, *Radio Sci.*, 29(2), 451–463, 1994.
21. Harrington, R. F., *Field Computation by Moment Methods*, Robert E. Krieger Publishing Company, Malabar, FL, 1968.
22. Kannan, N. and Kundu, D., On modified EVLP and ML methods for estimating superimposed exponential signals, *Signal Process.*, 39, 223–233, 1994.

Part III

Application: Frequency and Polarization Transformer—Switched Medium in a Cavity

12

Time-Varying Medium in a Cavity and the Effect of the Switching Angle

12.1 Introduction

In Part I of this book, the theory of the transformation of an EMW by a time-varying plasma medium is discussed. The medium is assumed to be unbounded (Chapters 3, 7, 8, and 9) or semi-infinite (Chapter 4) or a slab in free space (Chapter 5). In all these cases, unless the plasma is created very fast, the newly created traveling waves in the time-varying medium leave the boundaries before the interaction with the time-varying medium is completed. In a cavity the interaction continues since the waves are bounded by the boundaries.

12.2 Sudden Creation in a Cavity and Switching Angle

A one-dimensional cavity (Figure 12.1) is approximately simulated by placing two PEC metal plates parallel to each other such that the distance they are apart (in the z-direction) is very small compared to their length and width. This way, the only dimension that matters is the z-direction. Thus one could easily verify the simulation results in the laboratory, as the setup is relatively simple. A source of electric field is placed between the two plates such that there is a standing wave inside the cavity. In this section, equations will be set up for the case where we start with a standing wave that is elliptically polarized. At time $t = 0$, this cavity is suddenly and uniformly filled with a plasma. Then the corresponding field equations are derived after the plasma has been created.

12.2.1 Fields before Switching

Let E_0 be the amplitude of the electric field, d the distance between the plates, ω_0 the angular frequency of the standing wave, ϕ_0 the switching angle, and z the distance along the cavity.

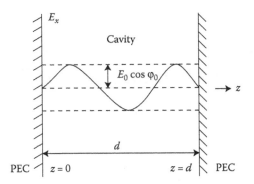

FIGURE 12.1 One-dimensional cavity used for the theory and simulation. It also sketches E_x for $m = 3$, $\varphi_0 = 45°$, and $t = 0$.

The source frequency ω_0 and d need to be chosen so that the tangential electric field at the two PEC plates is zero to satisfy the boundary conditions. Thus the equation between ω_0 and d is given by

$$\omega_0 = \left(\frac{m\pi c}{d}\right). \tag{12.1}$$

In Equation 12.1, $m = 1, 2, 3, \ldots, \infty$ and c is the speed of light in free space.

Let E denote electric field intensity, E^- denote E before plasma is created, E^+ denote E after plasma is created, E_x^- denote the x-component of E^-, and E_x^+ denote the x-component of E^+.

A similar notation to the one above is used for the y- and z-component of E by replacing the x subscript with y or z. The notation for the corresponding magnetic and current density fields is obtained by replacing E by H and J, respectively. Let the standing electric field intensity in the cavity before the plasma is created have the following properties:

$$E_z^- = 0, \tag{12.2}$$

$$E_x^- = E_0 \sin\left(\frac{m\pi z}{d}\right) \cos(\omega_0 t + \phi_0), \tag{12.3}$$

$$E_y^- = (\tan \gamma) E_0 \sin\left(\frac{m\pi z}{d}\right) \cos(\omega_0 t + \phi_0 + \delta). \tag{12.4}$$

In Equation 12.3, ϕ_0 is called the switching angle, which is the phase of E_x^- at the instant the plasma is created. In Equation 12.4, γ and δ are the polarization parameters of the elliptically polarized source wave [1].

By substituting the electric field components in Equations 12.2 through 12.4 in the following Maxwell's equation:

$$\nabla \times E = -\mu_0 \frac{\partial H}{\partial t}, \tag{12.5}$$

the magnetic field intensity components come out to be

$$H_x^- = -(\tan \gamma) H_0 \cos\left(\frac{m\pi z}{d}\right) \sin(\omega_0 t + \phi_0 + \delta), \tag{12.6}$$

$$H_y^- = H_0 \cos\left(\frac{m\pi z}{d}\right) \sin(\omega_0 t + \phi_0), \tag{12.7}$$

$$H_z^- = 0, \tag{12.8}$$

where

$$H_0 = -E_0 \frac{m\pi}{d\mu_0 \omega_0}. \tag{12.9}$$

12.2.2 Sudden Creation and Results

A sudden creation of the plasma is equivalent to a step profile for the plasma density and Figure 12.2 shows such a profile. The formulation, solution, and results are presented in Appendix L.

It is shown that by choosing appropriate values of the source wave parameters (E_0, γ, ϕ_0, and δ), and plasma parameter (ω_p), one can get any desired wiggler magnetic field (both magnitude and direction). This wiggler magnetic field varies as a sinusoid in the cavity and has a wavelength that is of the same value as the wavelength of the source wave.

It is also shown that by suitable choice of the above parameters, one can transform an elliptically polarized standing wave to a circularly polarized wave, and vice versa. Also the frequency of the source wave can be transformed to a desired upshifted value by choosing a plasma of appropriate angular frequency ω_p.

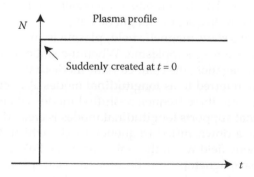

FIGURE 12.2   Electron density profile in time.

12.3   FDTD Method for a Lossy Plasma with Arbitrary Space and Time Profiles for the Plasma Density

The theory developed in Appendix L is based on a rather ideal model, where the plasma is created uniformly and instantaneously. However, for more realistic models where the plasma profile has arbitrary space and time plasma profiles, FDTD numerical solutions can be obtained. The formulation and the results are given in Appendix M. The effects of the rise time of the profile (parameter K), space profile (plasma created in a width d_1 of the cavity of width d), and the effect of collision frequency (parameter k_ν) are presented after computing them through FDTD algorithm.

The time-varying profiles show that the strength of the wiggler field significantly depends on the rate of growth of the plasma. To obtain the largest possible value of the wiggler field, we would need to create the plasma at the fastest rate.

The space-varying profiles show that new modes are excited. The introduction of collisions shows that the standing wave and the wiggler field decay differently, depending on the relative frequency of the plasma with respect to the original signal frequency. Thus one can select a desired plasma frequency so that the desired wave in the output is the dominant signal.

The robustness of the FDTD equations has been proved by the simulation of the total magnetic field in a lossy plasma having both a temporal and a spatial profile. Indeed, one should be able to apply this FDTD model and study in great detail the transformation of an electromagnetic signal under any arbitrary space and time plasma profile.

12.4   Switching a Magnetoplasma: Longitudinal Modes

A source wave splits into three modes when an unbounded plasma medium is created (see Chapter 3). Two of the modes are propagating modes with upshifted frequencies. The third mode is a wiggler magnetic field, which is a space-varying dc (zero frequency) magnetic field.

In the presence of a static magnetic field, plasma is an anisotropic medium and is referred to as a magnetoplasma. When the static magnetic field is in the direction of propagation, the characteristic modes are circularly polarized and the modes are referred to as longitudinal modes. A circularly polarized source wave splits into three frequency-shifted modes when an unbounded magnetoplasma that supports longitudinal modes is created (see Chapter 7). The third mode is a downshifted frequency mode, which degenerates into the wiggler magnetic field when the static magnetic field is reduced to zero and the anisotropic magnetoplasma becomes an isotropic plasma.

Sudden creation of an unbounded magnetoplasma is an idealization to simplify the problem but is not easily obtainable even in an approximate

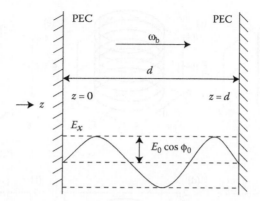

FIGURE 12.3 One-dimensional cavity used for the theory and simulation. It also sketches E_x for $m = 3$, $\varphi_0 = 45°$, and $t = 0$. The static magnetic field in the z-direction creates longitudinal modes.

sense. Since it is easier to create experimentally a switched medium in a bounded space, we have taken up the study of the switched medium in a *cavity*. The switched medium is a magnetoplasma that supports longitudinal modes (see Figure 12.3).

In Appendix N, analytical expressions for the three transformed waves are obtained by using the Laplace transform technique. Then the numerical FDTD model is developed to compute the transformed waves for an arbitrary space and time profile of the magnetoplasma. One illustrative result for the three modes (see Figures 12.4 and 12.5 and Table 12.1) is presented as graphs to illustrate the effects of the source and system parameters.

We note that an elliptically polarized R source wave produces two R waves one with frequency higher than the source frequency (upshifted) and another with downshifted frequency. The additional mode is an upshifted L wave.

12.5 Switching a Magnetoplasma Medium: X Wave

When the static magnetic field is perpendicular to the direction of propagation, the characteristic modes are ordinary (O) and extraordinary modes (X)

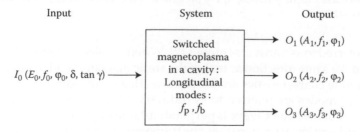

FIGURE 12.4 Frequency transformer effect.

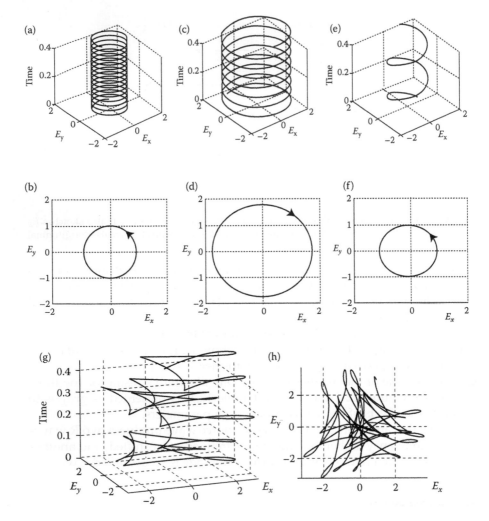

FIGURE 12.5   Variation of the output electric field with time. The units for the x- and y-axes are V/m. The unit for the vertical axis is ns. Parameters: $f_0 = 10$ GHz, $f_p = 17.63$ GHz, $f_b = 20$ GHz, $\tan \gamma = 4$, $\delta = -\pi/4$, $\phi_0 = 0$, $E_0 = 1$ V/m, $z = d/2$, and $m = 1$. (a) Highest frequency mode (mode 1), (b) x–y projection of (a), (c) intermediate frequency mode (mode 2), (d) x–y projection of (c), (e) lowest frequency mode (mode 3), (f) x–y projection of (e), (g) total electric field, and (h) x–y projection of (g).

and are referred to as transverse modes [2]. When the electric field of the source wave is in the same direction as that of the static magnetic field, it has no effect and the mode characteristics are the same as that of an isotropic plasma and hence these modes are referred to as ordinary. On the other hand, when the electric field of the source wave is perpendicular to the static magnetic field as well as the direction of propagation, the transverse modes are called extraordinary and have a component of the electric field in the direction of propagation.

TABLE 12.1

E_x and E_y for the Three Modes

Quantity	E_x	E_y	Polarization
A_1 (V/m)	0.94	0.94	R
f_1 (GHz)	30.88	30.88	
ϕ_1 (degrees)	55.86	−34.14	
A_2 (V/m)	1.76	1.76	L
f_2 (GHz)	15.14	15.14	
ϕ_2 (degrees)	252.95	−17.05	
A_3 (V/m)	1.06	1.06	R
f_3 (GHz)	4.28	4.28	
ϕ_3 (degrees)	20.41	−69.59	

Parameters: $f_0 = 10$ GHz, $f_p = 17.63$ GHz, $f_b = 20$ GHz, $\tan\gamma = 4$, $\delta = -\pi/4$, $\phi_0 = 0$, $E_0 = 1$ V/m, $z = d/2$, and $n = 1$.

A linearly polarized source wave splits into four frequency-shifted modes when an *unbounded* magnetoplasma that supports the X modes is created (Appendix C). Two of them are forward-propagating modes of different frequencies. The other two are the corresponding backward-propagating modes. Moreover, wiggler (zero frequency but space varying) electric as well as magnetic fields are also created and they degenerate into a wiggler magnetic field only when the static magnetic field is reduced to zero and the anisotropic magnetoplasma becomes isotropic (Appendix C).

Appendix O discusses a frequency transformer that converts a linearly polarized standing wave into frequency-shifted extraordinary standing waves. The transformer is a one-dimensional *cavity* in which a magneto-plasma that supports "transverse modes" is created. Theoretical derivation for the case of sudden and uniform creation of the magnetoplasma is given. It is shown that the switching would result in the transformation of the original source wave into three new waves, each having a unique frequency, amplitude, and phase. One of these modes is a dc component of zero frequency called the wiggler mode having a wiggler electric as well as a wiggler magnetic component. The wiggler electric field is the unique aspect of the switching problem studied in this appendix. A few illustrative results are presented as graphs and tables to illustrate the effects of source and system parameters. They include a case of conversion from linear to circular polarization.

Appendix P discusses the same problem but allows a finite rise time T_r in the creation of the plasma. The FDTD technique is used to study the effect of the rise time on the various modes. A bounded medium, a cavity shown in Figure 12.6, was considered with arbitrary space and time profiles for the electron density.

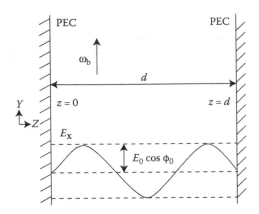

FIGURE 12.6 The one-dimensional cavity. Inside the cavity, a sketch of E_x for $m = 3$, $\varphi_0 = 45°$, and $t = 0$ is shown. The static magnetic field in the y-direction creates transverse modes when the plasma medium is switched on.

12.6 Switching Off the Magnetoplasma by Collapse of the Ionization: Whistler Source Wave

The relative permittivity ε_p for R wave propagation is given by

$$\varepsilon_p = 1 - \frac{\omega_p^2}{\omega_0(\omega_0 - \omega_b)}. \tag{12.10}$$

When $\omega_0 \ll \omega_b$ and $(\omega_p^2/\omega_0\omega_b) \gg 1$, ε_p can be quite large and is approximately given by

$$\varepsilon_p \approx \frac{\omega_p^2}{\omega_0\omega_b}, \quad \omega_0 \ll \omega_b \quad \text{and} \quad \frac{\omega_p^2}{\omega_0\omega_b} \gg 1. \tag{12.11}$$

Such an R wave is called a whistler wave, since the whistlers in radio reception were explained in terms of the propagation of electromagnetic signals in lightning. The whistler mode is called helicon mode in the literature on solid-state plasmas.

A big change in ε_p can be obtained by collapsing the electron density (by switching off the source of ionization), thereby converting the magnetoplasma medium into free space ($\varepsilon_p = 1$).

In Appendix E, the author discussed the conversion of a whistler wave by a collapsing ionization. It was shown that the collapse of the ionization converted the whistler wave to a much higher frequency wave with power intensification. An ideal model of an unbounded medium was used to simplify the problem and establish the physical basis of the processes. Two extreme cases of collapse, (i) instantaneous collapse and (ii) very slow decay, were considered. In all these cases of large values for ε_p given by Equation

12.11, it was shown that the two additional modes of different frequencies created by switching off the magnetoplasma were of insignificant amplitude.

We are motivated to study this problem since it appears to lead to a practical frequency and a polarization transformer, with frequency transformation ratios of orders of magnitude. In particular, it appears that one can transform the frequency from 2.45 to 300 GHz with the output having significantly higher electric field amplitude as compared to the input electric field. Details are given in Appendix Q.

12.7 Switching Off the Magnetoplasma by Collapse of the Background Magnetic Field: Whistler Source Wave

A big change in ε_p can also be obtained by collapsing the background static magnetic field. In this case the medium is converted from a magnetoplasma with $\varepsilon_p \gg 1$ to an isotropic plasma with $\varepsilon_p < 1$. In Appendix F, an ideal model of an unbounded medium was used to simplify the problem and establish the physical basis of the processes. Two extreme cases of collapse, (i) instantaneous collapse and (ii) very slow decay, were considered. It was shown that the output frequency will be reduced to zero giving an output of a wiggler magnetic field.

In particular, we are interested in investigating the possibility of producing an mm-wavelength wiggler magnetic field. A wiggler magnetic field is used in an FEL, which uses a wiggler magnetic field obtained by using physical magnets. Construction of an mm-wavelength physical wiggler magnet is a difficult task.

In this section, it is shown [3,4] that such a wiggler can be obtained by switching off the background magnetic field.

The set of Figures 12.7 and 12.8 are for the case of decaying quasistatic magnetic field in the presence of a uniform constant electron density. The results are obtained by using the FDTD technique.

For $t < 0$, the parameters are given by

$$\omega_p^2(t) = \omega_{p0}^2, \tag{12.12a}$$

$$\omega_b(t) = \omega_{b0}. \tag{12.12b}$$

For $t > 0$,

$$\omega_b(t) = \omega_{b0} f(t), \tag{12.13a}$$

$$\omega_p^2 = \omega_{p0}^2, \tag{12.13b}$$

where $f(t)$ is given by

$$f(t) = e^{-bt/T}. \tag{12.14}$$

In Equation 12.14, T is the period of the source whistler wave.

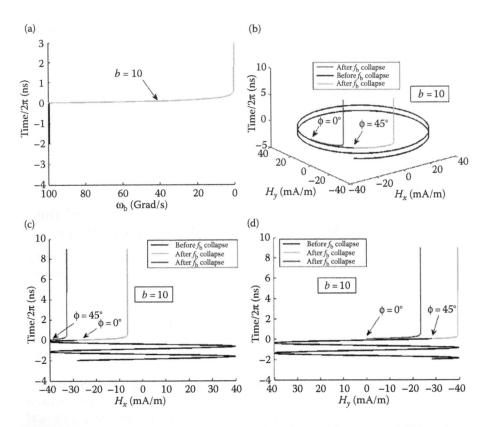

FIGURE 12.7   (See color insert following page 202.) Evolution of the magnetic field intensity as f_b collapses under constant plasma density. $b = 10$, $\omega_0 = 150$ Grad/s, $\omega_{b0} = 150$ Grad/s, $m = 1$, and $z \approx 0$. (a) w_b profile, (b) H_x–H_y variation in time, (c) H_x-variation in time, and (d) H_y-variation in time.

The effect of this temporal change is the collapse of the magnetoplasma medium to the limit of the isotropic plasma medium. Figure 12.7 shows that the source whistler wave of frequency ω_0 is converted to a wiggler (zero frequency but spatially varying) magnetic field. The wiggler wavelength is the same as the wavelength of the source whistler standing wave, as is to be expected since the wavelength is conserved over a temporal discontinuity in the properties of the medium. However, the strength and the direction of the wiggler field are controlled by the switching angle ϕ_0.

Figure 12.8 shows the evolution of the x-component of the whistler wave into a wiggler magnetic field of wiggler wavelength of the order of millimeters. In Figure 12.8a, the parameters are $\omega_0 = 1$ Grad/s, $\omega_{b0} = 10$ Grad/s ($B_0 = 0.0568$ T), and $\omega_{p0} = 1000$ Grad/s ($N_0 = 3.13 \times 10^{20}/\text{m}^3$). The refractive index $n_R \approx 1000/3$ and the wavelength of the whistler $\lambda_w = 5.66$ mm. For $m = 10$, the plate separation $d = m\lambda_w/2 = 2.83$ cm. When the static magnetic field is switched off, the wiggler of wavelength λ_w is produced. For the

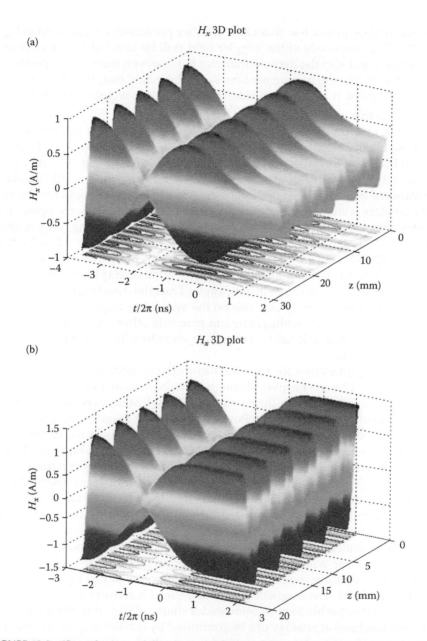

FIGURE 12.8 **(See color insert following page 202.)** Evolution of the x-component of the magnetic field intensity as f_b collapses under constant plasma density. $b = 10$, $\phi = 45°$, $m = 10$, (a) $\omega_0 = 1$ Grad/s, $\omega_{p0} = 1000/s$, $\omega_{b0} = 10$ Grad/s, $d = 2.83$ cm, (b) $\omega_0 = 1$ Grad/s, $\omega_{p0} = 150$ Grad/s, $\omega_{b0} = 1.1$ Grad/s, and $d = 2$ cm.

purpose of illustration, $b = 10$ is chosen. Other parameters are marked in Figure 12.8. The amplitude of the wiggler field will be affected by the collision frequency ν and also the decay parameter b. However, the wiggler period is unaffected. While the requirement on B_0 can be easily met, the electron density requirement is on the high side. The millimeter wiggler wavelength is achieved in this case because $n_R = \varepsilon_p^{1/2}$ is large.

One can also produce a high value for n_R for another set of parameters, where ω_0 is a little lower than ω_{b0}. These are $\omega_0 = 1\,\mathrm{Grad/s}$, $\omega_{b0} = 1.1\,\mathrm{Grad/s}$ ($B_0 = 0.0063\,\mathrm{T} = 63\,\mathrm{Gauss}$), and $\omega_{p0} = 150\,\mathrm{Grad/s}$ ($N_0 = 7 \times 10^{18}/\mathrm{m}^3$) giving $n_R \approx 475$. The whistler wavelength $\lambda_w \approx 4\,\mathrm{mm}$ and for $m = 10$, the plate separation $d = 2\,\mathrm{cm}$. The evolution of the wiggler magnetic field for this set of parameters is shown in Figure 12.8b. Since ω_0 is close to ω_{b0}, the operating point is close to resonance and hence even if ν/ω_p is of the same value as that of Figure 12.8a, the attenuation of the amplitude of the wiggler will be higher in Figure 12.8b had the collisions been taken into account. The remarks made above apply to the y-component as well. The relative amplitudes of the two components depend on the switching angle. Thus the direction of the wiggler magnetic field in the cavity depends on the switching angle.

By using appropriate scaling, one can generate other sets of data for producing wiggler magnetic fields of short wavelength with the desired number of spatial loops m.

A physical explanation for the two major effects discussed in this section and in Section 12.6 can be given through a mechanical analogy.

A physical explanation for the conversion of a whistler wave into a wiggler magnetic field can be given as follows. At $t = 0$, the magnetoplasma, in the presence of a standing wave of $1\,\mathrm{Grad/s}$, behaves like a dielectric of high permittivity. When the imposed magnetic field is reduced, from Equation 12.11, it can be seen that the plasma becomes a dielectric of even higher permittivity and hence higher refractive index n. Since the wave number is conserved in a time-varying medium, a conservation law that the product $n\,\omega$ is a constant [3, p. 209] applies. This results in a decrease of the angular frequency ω of the standing wave. When the imposed magnetic field is ultimately reduced to zero, the frequency of the standing wave reduces to zero and the whistler wave becomes the wiggler magnetic field. Figure 12.9 sketches the electric and magnetic fields in the cavity at $t = 0$, assuming $m = 1$ in Equations 12.3, 12.4, 12.7, and 12.9. The magnetic field at any time t is zero at $z = d/2$. The field line is like a flexible blade in a cosinusoidal shape, clamped to the center of a rod. This mechanical analogy can be continued by considering the boundary condition on the PEC plates at $z = 0$ and $z = d$, which is the condition that the normal derivative of the tangential component of the magnetic field is zero. This Neuman boundary condition is analogous to the requirement that the blade can only make a sliding contact on the end plates. For $t < 0$, the magnetic field line of the R standing wave rotates describing a circle in the end plates as shown in Figure 12.9. The angular velocity of this rotating magnetic field is ω_0. As the imposed magnetic field is reduced, the speed of rotation

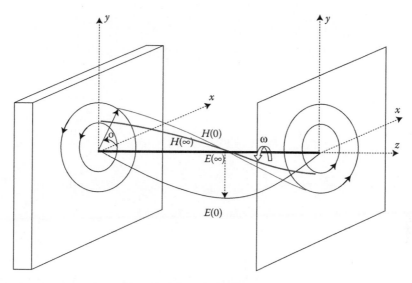

FIGURE 12.9 (See color insert following page 202.) Sketch of a mechanical analog of the conversion process, shown for $m = 1$ and E_0 being negative.

decreases and, finally, when the imposed magnetic field becomes zero, the magnetic field becomes stationary. $H(\infty)$ at the ends, however, is not zero. Its values, for any $t > 0$ and $t \to \infty$, are given in Ref. [3]. It can be shown that the difference between the radii of the two circles in Figure 12.9 is less than $1/(n_R^2)$. The electric field line is analogous to a skipping rope [2] clamped at the ends $z = 0$ and $z = d$. For $t < 0$, the skipping rope is rotating at the angular speed of ω_0. When the imposed magnetic field is reduced to 0, $E(\infty)$ reduces to 0.

The same analogy can be used to explain the conversion of the whistler wave to the electric field-intensified higher-frequency wave (see Section 12.6) when the ionization source is switched off. In this case, from Equation 12.11, we see that as ω_p is reduced the refractive index decreases; hence, the angular speed will increase. The amplitudes of the electric and magnetic fields can be obtained from Ref. [3]. A complete account on this section can be seen in a publication in preparation.

References

1. Balanis, C. A., *Advances Engineering Electromagnetics*, Wiley, New York, 1989.
2. Booker, H. G., *Cold Plasma Waves*, Kluwer, Hingham, MA, 1984.
3. Kalluri, D. K., *Electromagnetics of Complex Media*, CRC Press, Boca Raton, Florida, 1999.
4. Eker, S., Effect of switching-off a magnetoplasma medium on a whistler wave in a cavity: FDTD simulation, Master's thesis, University of Massachusetts Lowell, Lowell, 2006.

Appendix L

Plasma-Induced Wiggler Magnetic Field in a Cavity*

Monzurul M. Ehsan and Dikshitulu K. Kalluri

L.1 Introduction

A source wave splits into three modes when an unbounded plasma medium is suddenly created [1]. Two of the modes are propagating modes with upshifted frequencies. The third mode is a wiggler magnetic field. It is a space-varying dc (zero frequency) magnetic field. Such a mode arises due to the current in the plasma medium. Several investigators studied this mode from various viewpoints [2–6].

Sudden creation of an *unbounded* plasma is an idealization to simplify the mathematics but is not easily obtainable even in an approximate sense. Creation of a plasma in a cavity is experimentally feasible. A preliminary theoretical study of effects of such a creation has been undertaken recently [7].

This article gives a comprehensive account of such a study regarding the wiggler magnetic field. The emphasis is on the creation of the wiggler magnetic field of any desired magnitude and direction by a suitable choice of the switching angle and other parameters of the source wave as well as the electron density of the switched plasma medium. The wiggler magnetic field is used in an FEL to generate coherent radiation [8].

L.2 Sudden Creation in a Cavity

A one-dimensional cavity (Figure L.1) is approximately simulated by placing two PEC metal plates parallel to each other such that their distance apart (in the z-direction) is very small compared to their length and width. Thus, the only dimension that matters is the z-direction. Thus one could easily verify the simulation results in the laboratory, as the setup is relatively simple. A

---

* Reprinted from *Int. J. Infrared Millim. Waves*, 24(8), 1215–1234, August 2003. With permission.

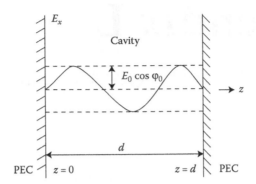

FIGURE L.1 Shows the one-dimensional cavity used for the theory and simulation. It also sketches E_x for $n = 3$, $\varphi_0 = 45°$, and $t = 0$.

source of electric field is placed between the two plates such that there is a standing wave inside the cavity. In this paper, equations will be set up for the case where we start with a standing wave that is elliptically polarized. At time $t = 0$, this cavity is suddenly and uniformly filled with a plasma. Then the corresponding field equations are derived after the plasma has been created.

L.2.1 Fields before Switching

Let E_0 be the amplitude of the electric field, d the distance between the plates, ω_0 the angular frequency of the standing wave, ϕ_0 the switching angle, and z the distance along the cavity.

The source frequency ω_0 and d need to be chosen so that the tangential electric field at the two PEC plates is zero to satisfy the boundary conditions. Thus the equation between ω_0 and d is given by

$$\omega_0 = \left(\frac{n\pi c}{d}\right). \tag{L.1}$$

In Equation L.1, $n = 1, 2, 3, \ldots, \infty$ and c is the speed of light in free space.

Let E denote electric field intensity, E^- denote E before plasma is created, E^+ denote E after plasma is created, E_x^- denote the x-component of E^-, and E_x^+ denote the x-component of E^+.

A similar notation to the one above is used for the y- and z-component of E by replacing subscript x with y or z. The notation for the corresponding magnetic and current density fields is obtained by replacing E by H and J, respectively. Let the standing electric field intensity in the cavity before the plasma is created have the following properties:

$$E_z^- = 0, \tag{L.2}$$

$$E_x^- = E_0 \sin\left(\frac{n\pi z}{d}\right) \cos(\omega_0 t + \phi_0), \tag{L.3}$$

$$E_y^- = (\tan \gamma) E_0 \sin\left(\frac{n\pi z}{d}\right) \cos(\omega_0 t + \phi_0 + \delta). \tag{L.4}$$

In Equation L.3, ϕ_0 is called the switching angle, which is the phase of E_x^- at the instant the plasma is created. In Equation L.4, γ and δ are the polarization parameters of the elliptically polarized source wave [9].

By substituting the electric field components in Equations L.2 through L.4 in the following Maxwell's equation:

$$\nabla \times \mathbf{E} = -\mu_0 \frac{\partial \mathbf{H}}{\partial t}, \tag{L.5}$$

the magnetic field intensity components come out to be

$$H_x^- = -(\tan \gamma) H_0 \cos\left(\frac{n\pi z}{d}\right) \sin(\omega_0 t + \phi + \delta), \tag{L.6}$$

$$H_y^- = H_0 \cos\left(\frac{n\pi z}{d}\right) \sin(\omega_0 t + \phi), \tag{L.7}$$

$$H_z^- = 0, \tag{L.8}$$

where

$$H_0 = -E_0 \frac{n\pi}{d\mu_0 \omega_0}. \tag{L.9}$$

One can express E_x^- as the product of a space component, $f_1^-(z)$, and a time component, $f_2^-(t)$, as follows:

$$E_x^- = f_1^-(z) f_2^-(t). \tag{L.10}$$

In Equation L.10, $f_1^-(z) = E_0 \sin(n\pi z/d)$, while $f_2^-(t) = \cos(\omega_0 t + \phi)$.

Similarly, H_y^- can be decomposed into a space component, $f_3^-(z)$, and a time component, $f_4^-(t)$, as follows:

$$H_y^- = f_3^-(z) f_4^-(t). \tag{L.11}$$

In Equation L.11, $f_3^-(z) = H_0 \cos(n\pi z/d)$, while $f_4^-(t) = \sin(\omega_0 t + \phi)$.

Such decompositions are useful in obtaining the fields after the sudden creation of a plasma in the cavity.

L.2.2   Step Plasma Profile

Now let the cavity between the plates be suddenly filled at time $t = 0$ (Figure L.2) with a uniform plasma of angular frequency ω_p given by $\omega_p^2 = (2\pi f_p)^2 = (q^2 N/m\varepsilon_0)$, where q is the electron charge, m is the electron mass, ε_0 is the permittivity of free space, and N is the electron density. Another way of describing the sudden creation is to state that the medium in the cavity is switched from free space to plasma at $t = 0$.

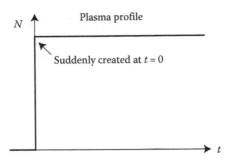

FIGURE L.2   Electron density profile in time.

L.2.3   Fields after Switching

The fields after switching the plasma are obtained from the Maxwell equations given in Equations L.5 and L.12, and the constitutive relation for a cold plasma given in Equation L.13.

$$\nabla \times \mathbf{H} = \varepsilon_0 \frac{\partial \mathbf{E}}{\partial t} + \mathbf{J}, \tag{L.12}$$

$$\frac{d\mathbf{J}}{dt} = \varepsilon_0 \omega_p^2 \mathbf{E}. \tag{L.13}$$

L.2.3.1   *Solution for the Electric Field*

Similar to the previous section, one can break E_x^+ into the product of a space component, $f_1^+(z)$, and a time component, $f_2^+(t)$, as follows:

$$E_x^+ = f_1^+(z) f_2^+(t). \tag{L.14}$$

The z-component E_z^+ is clearly zero since it is zero before the plasma is created. As the plasma is suddenly and uniformly created, the components of the fields that depend on space do not change with the plasma creation [1]. Then

$$f_1^+(z) = f_1^-(z) = E_0 \sin\left(\frac{\pi z}{d}\right). \tag{L.15}$$

From Equations L.5, L.12, and L.13, Kalluri [1] has shown that the time component of the electric field after the plasma is created depends on the plasma frequency through the following relation:

$$\frac{d^2 f_2^+}{dt^2} + (\omega_0^2 + \omega_p^2) f_2^+ = 0 \quad \text{for} \quad 0 \le z \le d, \ t > 0, \tag{L.16}$$

$$s^2 F_2^+(s) - s f_2^+(0) - \left(\frac{df_2^+}{dt}\right)_{t=0} + \left(\omega_0^2 + \omega_p^2\right) F_2^+(s) = 0. \tag{L.17}$$

Equation L.17 is obtained by taking the Laplace transform of Equation L.16, and using the notation that $F_2^+(s)$ is the Laplace transform of $f_2^+(t)$.

From the constraints at $t = 0$ of continuity of the electric field and the magnetic field (which is proportional to the derivative of the electric field),

$$f_2^+(0) = f_2^-(0) = [\cos(\omega_0 t + \phi_0)]_{t=0} = \cos\phi_0, \tag{L.18}$$

$$\left(\frac{df_2^+}{dt}\right)_{t=0} = \left(\frac{df_2^-}{dt}\right)_{t=0} = [-\omega_0 \sin(\omega_0 t + \phi_0)]_{t=0} = -\omega_0 \sin\phi_0. \tag{L.19}$$

Combining Equations L.17 through L.19, we obtain

$$F_2^+\left(s^2 + \omega_0^2 + \omega_p^2\right) = s\cos\phi_0 - \omega_0\sin\phi_0. \tag{L.20}$$

Taking the Laplace inverse of F_2^+ in Equation L.20,

$$f_2^+(t) = \cos\phi_0 \cos\omega_2 t - \frac{\omega_0}{\omega_2}\sin\phi_0 \sin\omega_2 t, \tag{L.21}$$

where

$$\omega_2^2 = \omega_0^2 + \omega_p^2. \tag{L.22}$$

Thus from Equations L.15 and L.22,

$$E_x^+ = E_0 \sin\left(\frac{n\pi z}{d}\right)\left[\cos\phi_0 \cos(\omega_2 t) - \frac{\omega_0}{\omega_2}\sin\phi_0 \sin(\omega_2 t)\right], \tag{L.23}$$

which can be written as

$$E_x^+ = E_0 \sin\frac{n\pi z}{d}\left[A_1 \cos(\omega_2 t + A_2)\right], \tag{L.24}$$

$$A_1 = \sqrt{\left(\frac{\omega_0}{\omega_2}\right)^2 \sin^2\phi_0 + \cos^2\phi_0}, \tag{L.25}$$

$$A_2 = \tan^{-1}\left[\frac{\omega_0 \tan\phi_0}{\omega_2}\right]. \tag{L.26}$$

Comparing Equation L.24 with Equation L.3, we note that the switching action changes the x-component from E_0 to $E_0 A_1$ in amplitude, from ϕ_0 to $\tan^{-1}((\omega_0 \tan\phi_0)/\omega_2)$ in phase, and from ω_0 to $\omega_2 = \sqrt{\omega_0^2 + \omega_p^2}$ in frequency.

Since E_y^- can be obtained from E_x^- by replacing ϕ_0 with $\phi_0 + \delta$, and multiplying by the amplitude ratio, one can deduce that E_y^+ can be obtained from E_x^+

in Equation L.24 by replacing ϕ_0 with $\phi_0 + \delta$, and premultiplying the results by $\tan \gamma$.

$$E_y^+ = (\tan \gamma)E_0 \sin \frac{n\pi z}{d} [A_3 \cos(\omega_2 t + A_4)], \qquad (\text{L.27})$$

$$A_3 = \sqrt{\left(\frac{\omega_0}{\omega_2}\right)^2 \sin^2(\phi_0 + \delta) + \cos^2(\phi_0 + \delta)}, \qquad (\text{L.28})$$

$$A_4 = \tan^{-1}\left[\frac{\omega_0 \tan(\phi_0 + \delta)}{\omega_2}\right]. \qquad (\text{L.29})$$

L.2.3.2   Solution for the Magnetic Field

The z-component H_z^+ is clearly zero since this component is zero before the plasma is created. As shown in the previous section, one can break H_y^+ into a space component, $f_3^+(z)$, and a time component, $f_4^+(t)$, as follows:

$$H_y^+ = f_3^+(z)f_4^+(t). \qquad (\text{L.30})$$

Since the component of the fields that depends on space does not change with the plasma creation (as was explained when stating Equation L.15),

$$f_3^+(z) = f_3^-(z) = H_0 \cos\left(\frac{n\pi z}{d}\right). \qquad (\text{L.31})$$

From Equations L.5, L.12, and L.13, Kalluri [1] has shown that the time component of the magnetic field after the plasma is created depends on the plasma frequency through the following relation:

$$\frac{d^3 f_4^+}{dt^3} + \left(\omega_0^2 + \omega_p^2\right)\frac{df_4^+}{dt} = 0. \qquad (\text{L.32})$$

Let $f_5^+(t) = df_4^+/dt$; then Equation L.32 becomes

$$\frac{d^2 f_5^+}{dt^2} + \left(\omega_0^2 + \omega_p^2\right)f_5^+ = 0. \qquad (\text{L.33})$$

Taking the Laplace transform of Equation L.33,

$$s^2 F_5^+(s) - sf_5^+(0) - \frac{df_5^+}{dt}(0) + \left(\omega_0^2 + \omega_p^2\right) F_5^+(s) = 0. \qquad (\text{L.34})$$

From the constraints for continuity of the first and second derivatives of the magnetic field at $t = 0$, one obtains

$$f_5^+(0) = \left(\frac{df_4^+}{dt}\right)_{t=0} = \left(\frac{df_4^-}{dt}\right)_{t=0} = \{\omega_0 \cos(\omega_0 t + \phi_0)\}_{t=0} = \omega_0 \cos\phi_0,$$

(L.35)

$$\left(\frac{df_5^+}{dt}\right)_{t=0} = \left(\frac{d^2f_4^+}{dt^2}\right)_{t=0} = \left(\frac{d^2f_4^-}{dt^2}\right)_{t=0}$$

$$= \left\{-\omega_0^2 \sin(\omega_0 t + \phi_0)\right\}_{t=0} = -\omega_0^2 \sin\phi_0.$$

(L.36)

From Equations L.34 through L.36,

$$F_5^+\left(s^2 + \omega_0^2 + \omega_p^2\right) = s\omega_0 \cos\phi_0 - \omega_0^2 \sin\phi_0.$$

(L.37)

Getting the inverse Laplace transform of Equation L.37 and using Equation L.22,

$$f_5^+(t) = \omega_0 \cos\phi_0 \cos\omega_2 t - \frac{\omega_0^2}{\omega_2} \sin\phi_0 \sin\omega_2 t,$$

(L.38)

$$f_4^+(t) = \int f_5^+(t)\, dt$$

$$= \left(\frac{\omega_0}{\omega_2}\right) \cos\phi_0 \sin\omega_2 t + \left(\frac{\omega_0}{\omega_2}\right)^2 \sin\phi_0 \cos\omega_2 t + k_1.$$

(L.39)

The constant k_1 can be obtained by considering the values at $t = 0$:

$$f_4^+(0) = f_4^-(0) = \sin\phi_0.$$

(L.40)

Then from Equations L.39 and L.40,

$$\sin\phi_0 = \left(\frac{\omega_0}{\omega_2}\right)^2 \sin\phi_0 + k_1.$$

(L.41)

Then from Equations L.39 and L.41,

$$f_4^+(t) = \left\{\left(\frac{\omega_0}{\omega_2}\right) \cos\phi_0 \sin\omega_2 t + \left(\frac{\omega_0}{\omega_2}\right)^2 \sin\phi_0 \cos\omega_2 t\right.$$

$$\left. + \sin\phi_0 - \left(\frac{\omega_0}{\omega_2}\right)^2 \sin\phi_0\right\}.$$

(L.42)

Then from Equations L.30, L.31, and L.42,

$$H_y^+ = H_0 \cos\left(\frac{n\pi z}{d}\right)\left[\left(\frac{\omega_0}{\omega_2}\right)\cos\phi_0 \sin\omega_2 t + \left(\frac{\omega_0}{\omega_2}\right)^2 \sin\phi_0 \cos\omega_2 t\right.$$

$$\left. + \sin\phi_0 - \left(\frac{\omega_0}{\omega_2}\right)^2 \sin\phi_0\right]. \tag{L.43}$$

Expressing H_y^+ as a wiggler magnetic field, H_{y_wigg} (that depends only on z), and a standing magnetic field, H_{y_sw} (that depends both on z and t),

$$H_y^+ = (H_{y\_sw} + H_{y\_wigg}), \tag{L.44}$$

$$H_{y\_wigg} = H_0 \left(\cos\frac{n\pi z}{d}\right)\left(\frac{\omega_p^2}{\omega_2^2}\right)\sin\phi_0, \tag{L.45}$$

$$H_{y\_sw} = H_0 \cos\frac{n\pi z}{d}[A_5 \cos(\omega_2 t + A_6)], \tag{L.46}$$

$$A_5 = \sqrt{\left(\frac{\omega_0}{\omega_2}\right)^4 \sin^2\phi_0 + \left(\frac{\omega_0}{\omega_2}\right)^2 \cos^2\phi_0} == \frac{\omega_0}{\omega_2}A_1, \tag{L.47}$$

$$A_6 = \tan^{-1}\left[-\frac{\omega_2}{\omega_0}\cot\phi_0\right]. \tag{L.48}$$

Since H_x^- can be obtained from H_y^- by replacing ϕ_0 with $\phi_0 + \delta$ and multiplying by $(-\tan\gamma)$, one can deduce that H_x^+ can be obtained from H_y^+ in Equation L.53 by substituting ϕ_0 with $\phi_0 + \delta$ and premultiplying the results by $(-\tan\gamma)$. Thus

$$H_x^+ = (H_{x\_sw} + H_{x\_wigg}), \tag{L.49}$$

$$H_{x\_wigg} = -(\tan\gamma)H_0\left(\cos\frac{n\pi z}{d}\right)\left(\frac{\omega_p^2}{\omega_2^2}\right)\sin(\phi_0 + \delta), \tag{L.50}$$

$$H_{x\_sw} = -(\tan\gamma)H_0 \cos\frac{n\pi z}{d}[A_7 \cos(\omega_2 t + A_8)], \tag{L.51}$$

$$A_7 = \sqrt{\left(\frac{\omega_0}{\omega_2}\right)^4 \sin^2(\phi_0 + \delta) + \left(\frac{\omega_0}{\omega_2}\right)^2 \cos^2(\phi_0 + \delta)} = \frac{\omega_0}{\omega_2}A_3, \tag{L.52}$$

$$A_8 = \tan^{-1}\left[-\frac{\omega_2}{\omega_0}\cot(\phi_0 + \delta)\right]. \tag{L.53}$$

Expressing the total wiggler magnetic field by adding Equations L.45 and L.50 and expressing as a vector,

$$\mathbf{H_{wigg}} = \mathbf{H_n}\cos\left(\frac{n\pi z}{d}\right), \tag{L.54}$$

where $\mathbf{H_n}$ is the normalized wiggler magnetic field and is given by

$$\mathbf{H_n} = \frac{H_0 \omega_p^2}{\omega_2^2} \left\{ [-\tan \gamma \cdot \sin(\phi_0 + \delta)] \hat{x} + (\sin \phi_0) \hat{y} \right\}. \qquad (L.55)$$

From Equation L.54, the wiggler wavelength is computed as

$$\lambda_{wigg} = \frac{2d}{n}. \qquad (L.56)$$

Clearly, λ_{wigg} has the same value as the wavelength of the initial source wave.

L.2.4 Solution for Special Cases

By substituting appropriate values of δ and $\tan \gamma$ in the equations derived in the previous section, the solutions to several special and important cases can be obtained. Table L.1 lists these special cases.

L.2.5 Results and Observations

It is possible to completely eliminate either the x- or the y-component of the wiggler magnetic field. If $\phi_0 = 0$ is substituted into Equation L.45, we obtain $H_{y\_wigg} = 0$. This means that for an arbitrary polarized source wave (since polarization depends on γ and δ) and a plasma of any angular frequency ω_p, one can completely eliminate H_{y_wigg} simply by appropriately choosing the time instant when the plasma is created. Similarly, from Equation L.50, if $\phi_0 = -\delta$, H_{x_wigg} is zero.

From Equation L.55, the amplitude of the wiggler magnetic field is proportional to ω_p^2/ω_2^2. Thus, as the concentration of the plasma increases, so does the amplitude. However, from Equation L.22, as ω_p increases even further, the value of $\omega_p^2/\omega_2^2 \rightarrow 1$, meaning that the wiggler field magnitude finally

TABLE L.1

Type of Source Wave for Given δ and $\tan \gamma$

δ	$\tan \gamma$	Type of Source Wave
$-\dfrac{\pi}{2}$	1	Right circularly polarized
$\dfrac{\pi}{2}$	1	Left circularly polarized
0	α	Linearly polarized with angle $\tan^{-1} \alpha$ with the x-axis
—	0	Linearly polarized in the x-direction
—	∞	Linearly polarized in the y-direction

saturates. In fact this saturated value can be computed from Equation L.55. The maximum magnitude of H_{y_wigg} is H_0, while the maximum magnitude of H_{x_wigg} is $H_0 \tan \gamma$.

For a better appreciation of the results given for the electric fields in Equations L.24 through L.29 and for the magnetic fields in Equations L.43 through L.55, Figures L.3 through L.6 are presented. The parameters common to all figures are $n = 1$, $f_0 = 10\,\text{GHz}$, and the amplitude of the x-component of the electric field $E_0 = 1\,\text{V/m}$. For an actual electric field of arbitrary strength E_0, all field quantities need to be multiplied by E_0. The electric field is plotted at the center of the cavity ($z = d/2$), where it is maximum. The magnetic field is plotted at the left edge of the cavity ($z = 0$), where the magnetic field is maximum. For other values of z, the electric field values have to be multiplied by $\sin(n\pi z/d)$ and the magnetic field by $\cos(n\pi z/d)$, respectively.

Figure L.3 shows a 3D view of the evolution of the electric field as the plasma is created. The standing source wave before the plasma is created is circularly polarized ($\gamma = \pi/4$ and $\delta = -\pi/2$), with a frequency of $f_0 = 10\,\text{GHz}$. A plasma of frequency $f_p = 17.32\,\text{GHz}$ is suddenly created at $t = 0$, when the switching angle of the source wave, $\phi_0 = 5\pi/4$. After the plasma is created, the frequency of the standing wave increases to 20 GHz (from Equation L.29), while the amplitude becomes smaller as can be determined from Equations L.30 and L.35. Note that the waves are more closely packed after the plasma is created. From the x–y projection of the plot shown on the right-hand side of Figure L.3 (the solid ellipse), it is clear that the polarization of the wave changes after the plasma is created. In fact, in this case, the source wave polarization transforms from circular to elliptical. For comparison, two additional projections (in dotted lines) are also plotted for $\phi_0 = 0$, labeled as A, and $\phi_0 = \pi/2$, labeled as B. These projections indicate that as the switching angle,

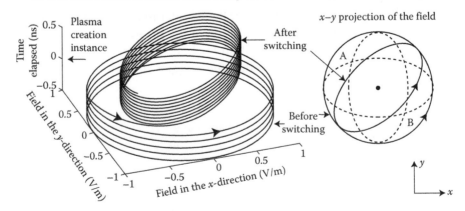

FIGURE L.3 Demonstration of the change in the electric field as the plasma is created. Note the increase in frequency and decrease in amplitude after the plasma is created. Parameters: $f_0 = 10\,\text{GHz}$, $f_p = 17.32\,\text{GHz}$, $E_0 = 1\,\text{V/m}$, $z = d/2$, $\delta = -\pi/2$, $\varphi_0 = 5\pi/4$, and $\tan \gamma = 1$.

ϕ_0, varies from $0 \rightarrow 2\pi$, the major axis of the projected ellipse rotates such that the angle between it and the negative x-axis is also ϕ_0. This means that a circularly polarized wave can be transformed into an elliptically polarized wave having the major axis in any desired direction, merely by suitably choosing the appropriate moment when the plasma is created. Also the dimensions of the ellipse can be controlled by appropriately selecting ω_p.

Figure L.4 depicts the 3D view of the changes in the total magnetic field as the plasma is switched on at $t = 0$. The standing source wave before the plasma is created has the same properties as already described for Figure L.3. As in the previous figure, the frequency is upshifted to 20 GHz, while the amplitude of the standing wave diminishes as computed from Equations L.46 and L.51. The x–y projection of the plot is shown on the right-hand side of Figure L.4 (the solid ellipse). The center of the ellipse and the center of the circle are clearly separated by an offset. The value of the offset for this specific case is $H_{x\_wigg} \cong 2.0\,\text{mA/m}$ and $H_{y\_wigg} \cong 2.0\,\text{mA/m}$. This offset is actually the wiggler magnetic field that was derived in Equations L.45 and L.50. Note from Figure L.3 that there is no corresponding offset in the electric field. Thus there is no wiggler electric field. To understand the effect of ϕ_0, two additional projections (in dotted lines) are also plotted for $\phi_0 = 0$, labeled as A, and for $\phi_0 = \pi/2$, labeled as B. These projections demonstrate that as the switching angle, ϕ_0, varies from $0 \rightarrow 2\pi$, the ellipse rotates on the inside boundary of the outer circle. The directed line, H_n, joining the center of the circle with the center of the ellipse, makes an angle of ϕ_0 with the negative x-axis, and represents the vector wiggler magnetic field. This angle can easily be deduced from Equations L.9 and L.55.

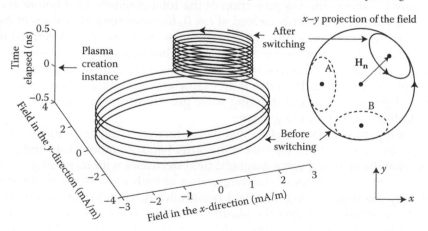

FIGURE L.4 Demonstration of the change in the magnetic field as the plasma is created. Note the increase in frequency and decrease in amplitude after the plasma is created. Parameters: $f_0 = 10\,\text{GHz}, f_p = 17.32\,\text{GHz}, E_0 = 1\,\text{V/m}, z = 0, \delta = -\pi/2, \varphi_0 = 5\pi/4$, and $\tan \gamma = 1$.

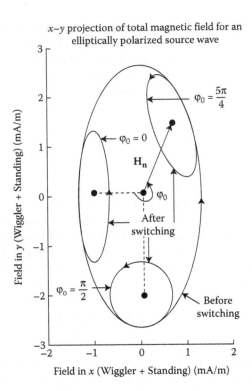

x–y projection of total magnetic field for an
elliptically polarized source wave

Field in x (Wiggler + Standing) (mA/m)

FIGURE L.5 The displacement of the postplasma signal projection from the center of the figure represents the value of the wiggler magnetic field. Parameters: $f_0 = 10\,\text{GHz}$, $f_p = 17.32\,\text{GHz}$, $z = 0, n = 1, E_0 = 1\,\text{V/m}$, $\varphi_0 = 5\pi/4$, $\delta = -\pi/2$, and $\tan\gamma = 0.5$.

Figure L.5 shows the x–y projection of the total magnetic field before and after the plasma is suddenly created at $t = 0$, for three separate values of ϕ_0. The parameters for the source wave in this case are the same as described in Figure L.4, with the exception that $\tan\gamma = 0.5$. The interesting result here is the projection for the case when $\phi_0 = \pi/2$, where the transformed standing wave is circularly polarized. Thus this is an example of a scenario where an originally elliptically polarized standing wave gets transformed to an upshifted circularly polarized standing wave.

From Equation L.55 it is clear that as ω_p increases, the amplitude of the wiggler magnetic field increases. On the other hand, Equations L.46 and L.51 show that the standing wave magnetic field decreases with increasing ω_p. If $\omega_p \gg \omega_0$, the amplitudes of standing waves of both E and H are negligible, and only the wiggler magnetic field is significant. Thus to obtain only the wiggler magnetic field after the plasma is created, one can create a plasma of very high concentration. This is demonstrated in Figure L.6. The source wave in Figure L.6 has the same parameters as that of Figure L.4, with the exception of the plasma frequency, which in this case is $f_p = 87\,\text{GHz}$. From the x–y projection of the plot shown on the right-hand side of the figure, it is clear that the amplitude of the standing wave after the plasma is created is very small compared to the amplitude of the wiggler field. Comparison between Figures L.4 and L.6 shows that increasing the electron density of the plasma causes the wiggler field to become larger ($2.0\,\text{mA/m} \rightarrow 2.6\,\text{mA/m}$), while the standing wave amplitude diminishes ($0.7\,\text{mA/m} \rightarrow 0.1\,\text{mA/m}$).

L.3 Conclusion and Future Work

By choosing appropriate values of the source wave parameters (E_0, γ, ϕ_0, and δ) and plasma parameter (ω_p), one can obtain any desired wiggler magnetic field (both magnitude and direction). This wiggler magnetic field varies as a sinusoid in the cavity and has a wavelength that is of the same value as the wavelength of the source wave.

Again by suitable choice of the above parameters, one can transform an elliptically polarized standing wave to a circularly polarized wave, and vice versa. Also the frequency of the source wave can be transformed to a desired upshifted value by choosing a plasma of appropriate angular frequency ω_p.

The theory developed in this paper is based on a rather ideal model, where the plasma is created uniformly and instantaneously. However, for more realistic models where the plasma profile is close but not exact to this ideal case, the modified solutions can be obtained by perturbation techniques [1], based

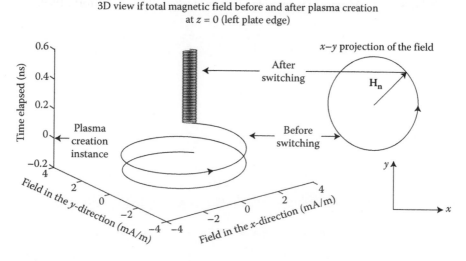

FIGURE L.6 Demonstration of the change in the magnetic field as the plasma is created for a dense plasma. Note the reduction in size of the standing wave, and the dominance of the wiggler magnetic field. Parameters: $f_0 = 10\,\text{GHz}$, $\varphi_0 = 5\pi/4$, $f_p = 87\,\text{GHz}$, $E_0 = 1\,\text{V/m}$, $z = 0$, $\delta = -\pi/2$, and $\tan\gamma = 1$.

on the solutions given in this paper. For arbitrary space and time plasma profiles, FDTD numerical solutions can be obtained. Work in this regard is in progress and will be published soon.

References

1. Kalluri, D. K., *Electromagnetics of Complex Media*, CRC Press, Boca Raton, FL, 1999.
2. Kalluri, D. K., Lee, J. H., and Ehsan, M. M., FDTD simulation of electromagnetic pulse interaction with a switched plasma slab, *Int. J. Infrared Millim. Waves*, 24, 349–365, 2003.
3. Jiang, C. L., Wave propagation and dipole radiation in a suddenly created plasma, *IEEE Trans. Ant. Prop.*, 23, 83–90, 1975.
4. Wilks, S. C., Dawson, J. M., and Mori, W. B., Frequency up-conversion of electromagnetic radiation with use of an overdense plasma, *Phys. Rev. Lett.*, 61, 337–340, 1988.
5. Kalluri, D. K., Effect of switching a magnetoplasma medium on a traveling wave: Longitudinal propagation, *IEEE Trans. Ant. Prop.*, 37, 1638–1642, 1989.
6. Mendonca, J. T. and Oliveira, L., Mode coupling theory of flash ionization in a cavity, *IEEE Trans. Plasma Sci.*, 24, 147–151, 1996.
7. Vemuri, K., Theory and computer simulation of electromagnetic wave interaction with a dynamic plasma in a cavity, MS thesis, Department of Electrical and Computer Engineering, University of Massachusetts, Lowell, MA, 2001.
8. Granastein, V. L. and Alexeff, I., *High-Power Microwave Sources*, Artech House, Boston, MA, 1987.
9. Balanis, C. A., *Advanced Engineering Electromagnetics*, Wiley, New York, 1989.

Appendix M

Plasma-Induced Wiggler Magnetic Field in a Cavity: II—The FDTD Method for a Switched Lossy Plasma*

Monzurul M. Ehsan and Dikshitulu K. Kalluri

M.1 Introduction

In a previous paper [1], Appendix L, we theoretically derived the various equations for the electric and magnetic fields in a cavity excited by an elliptically polarized standing source wave, after a plasma is instantaneously created in it. Also we assumed that the plasma was lossless and uniformly distributed in the cavity.

However, in real life, a plasma cannot be instantaneously created nor will the spatial distribution be uniform. Unfortunately, analytical solutions are not available for such arbitrary space and time profiles. To get to such practical solutions of a more realistic plasma profile, a numerical approach is presented in this paper. The FDTD method is used [2–8].

This article will derive an FDTD model for the cavity problem and then use it to solve a wide range of practical real-life scenarios. The authors have also previously published a related FDTD simulation in a plasma slab [2].

M.2 Algorithm

The case under study is a one-dimensional cavity. The spatial variations are in the z-direction. Thus all fields are functions of the spatial coordinate z and time t.

---

* Reprinted from *Int. J. Infrared Millim. Waves*, 24(10), 1655–1676, October 2003. With permission.

M.2.1   Initial Conditions

The setup used in this paper is the same as the one used in Part I of the present work [1]. However, sufficient outline will now be described to make the material self-contained. However, readers are advised to read Part I of the present work [1] for some theory and additional details.

This is a one-dimensional cavity that has been created by placing two PEC metal plates parallel to each other such that their distance apart (in the z-direction) is very small compared to their length and width. A signal source is placed between the plates such that initially there is an elliptically polarized standing source wave in the cavity.

Let E_0 be the amplitude of the electric field, d the distance between the plates, ω_0 the angular frequency of the standing wave, and z the distance along the cavity. Let E denote electric field intensity and E_x denote the x-component of E. A similar notation is used for the y- and z-component of E by replacing subscript x with y or z. The symbols for the corresponding magnetic field and current density fields are obtained by replacing E by H and J, respectively. Let the standing wave electric field in the cavity before the plasma is created be defined by the following equations:

$$E_x = E_0 \sin\left(\frac{n\pi z}{d}\right)\cos(\omega_0 t + \phi_0), \tag{M.1}$$

$$E_y = (\tan\gamma)E_0 \sin\left(\frac{n\pi z}{d}\right)\cos(\omega_0 t + \phi_0 + \delta), \tag{M.2}$$

$$E_z = 0. \tag{M.3}$$

In Equation M.2, ϕ_0 is called the switching angle, which is the phase of E_x at the instant the plasma is created, while γ and δ are the polarization parameters of the elliptically polarized source wave [3]. By substituting the electric field components in Equations M.1 through M.3 into Equation M.4, we obtain Equations M.5 through M.7:

$$\nabla \times \mathbf{E} = -\mu_0 \frac{\partial \mathbf{H}}{\partial t}, \tag{M.4}$$

$$H_x = -(\tan\gamma)H_0 \cos\left(\frac{n\pi z}{d}\right)\sin(\omega_0 t + \phi + \delta), \tag{M.5}$$

$$H_y = H_0\cos\left(\frac{n\pi z}{d}\right)\sin(\omega_0 t + \phi), \tag{M.6}$$

$$H_z = 0, \tag{M.7}$$

where

$$H_0 = -E_0\frac{n\pi}{d\mu_0\omega_0}. \tag{M.8}$$

The source frequency ω_0 and d need to be chosen so that the tangential electric field at the two PEC plates is zero to satisfy the boundary conditions.

Then the equation relating ω_0 and d is given by

$$\omega_0 = \left(\frac{n\pi c}{d}\right). \tag{M.9}$$

In Equation M.9, $n = 1, 2, 3, \ldots, \infty$ and c is the speed of light in free space. Since initially there is no plasma, the current density is

$$J_x = J_y = J_z = 0. \tag{M.10}$$

M.2.2   FDTD Equations

By expanding the Maxwell equation M.11, and expressing the time derivatives in terms of space derivatives when there are no field variations with x or y, one obtains Equations M.12 and M.13:

$$\nabla \times \mathbf{H} = \varepsilon_0 \frac{\partial \mathbf{E}}{\partial t} + \mathbf{J}, \tag{M.11}$$

$$\frac{\partial E_x}{\partial t} = -\frac{1}{\varepsilon_0} \frac{\partial H_y}{\partial z} - \frac{1}{\varepsilon_0} J_x, \tag{M.12}$$

$$\frac{\partial E_y}{\partial t} = \frac{1}{\varepsilon_0} \frac{\partial H_x}{\partial z} - \frac{1}{\varepsilon_0} J_y. \tag{M.13}$$

The cavity space is divided into an FDTD grid and E, H, and J field components are sampled as shown in Figure M.1. In this figure, the vertical axis represents time, while the horizontal axis represents space. By sampling the fields as indicated, one is able to systematically march forward in time and explicitly calculate future field values from their values in the past.

Discretizing Equations M.12 through M.13 using central difference formulas, one obtains the following FDTD equations for E_x and E_y:

$$E_{x_k}^{n+1} = E_{x_k}^{n} - \frac{\Delta t}{\varepsilon_0 \Delta z} \left(H_{y_{k+(1/2)}}^{n+(1/2)} - H_{y_{k-(1/2)}}^{n+(1/2)}\right) - \frac{\Delta t}{\varepsilon_0} J_{x_k}^{n+(1/2)}, \tag{M.14}$$

$$E_{y_k}^{n+1} = E_{y_k}^{n} + \frac{\Delta t}{\varepsilon_0 \Delta z} \left(H_{x_{k+(1/2)}}^{n+(1/2)} - H_{x_{k-(1/2)}}^{n+(1/2)}\right) - \frac{\Delta t}{\varepsilon_0} J_{y_k}^{n+(1/2)}. \tag{M.15}$$

Similarly, by expanding the Maxwell equation M.4 and expressing time derivatives in terms of space derivatives when there are no field variations with x or y, one obtains

$$\frac{\partial H_x}{\partial t} = \frac{1}{\mu_0} \frac{\partial E_y}{\partial z}, \qquad \frac{\partial H_y}{\partial t} = -\frac{1}{\mu_0} \frac{\partial E_x}{\partial z}. \tag{M.16}$$

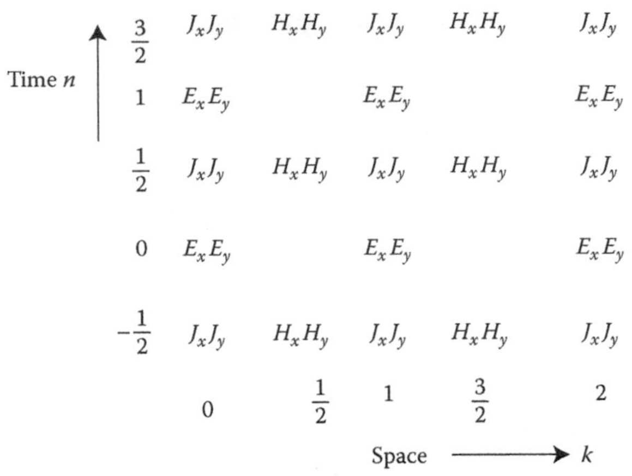

FIGURE M.1   FDTD layout, FDTD grid for the computation.

Discretizing Equation M.16 using central difference formulas, one yields the following FDTD equations for H_x and H_y:

$$(H_x)_{k+(1/2)}^{n+(1/2)} = (H_x)_{k+(1/2)}^{n-(1/2)} + \frac{\Delta t}{\mu_0 \Delta z} \left[(E_y)_{k+1}^n - (E_y)_k^n \right], \tag{M.17}$$

$$(H_y)_{k+(1/2)}^{n+(1/2)} = (H_y)_{k+(1/2)}^{n-(1/2)} - \frac{\Delta t}{\mu_0 \Delta z} \left[(E_x)_{k+1}^n - (E_x)_k^n \right]. \tag{M.18}$$

The equation that connects current density with other field variables [4] is given by the constitutive relation:

$$\frac{d\mathbf{J}}{dt} + v\mathbf{J} = \varepsilon_0 \omega_p^2 \mathbf{E}, \tag{M.19}$$

where v is the collision frequency of the plasma. Expressing Equation M.19 as components and taking the Laplace transform over one time step, we have

$$v \begin{bmatrix} L(J_x) \\ L(J_y) \end{bmatrix} + \begin{bmatrix} sL(J_x) - J_x(0) \\ sL(J_y) - J_y(0) \end{bmatrix} = \frac{\varepsilon_0 \omega_p^2}{s} \begin{bmatrix} E_x \\ E_y \end{bmatrix}, \tag{M.20}$$

where $L(\alpha)$ is the Laplace transform of α. In Equation M.20, ω_p, v, E_x, and E_y are assumed to be constant over the time step. Equation M.20 is based on "exponential time-stepping" [5] instead of central differencing, thus permitting a large value for v to be used in the simulation. It is important to very carefully sample the right value at the correct space and time coordinate. If any of these quantities vary with time, we will take the value at the midpoint of the time step. Equation M.20 can be written as

$$[sI - \Omega] \begin{bmatrix} L(J_x) \\ L(J_y) \end{bmatrix} = \begin{bmatrix} J_x(0) \\ J_y(0) \end{bmatrix} + \frac{\varepsilon_0 \omega_p^2}{s} \begin{bmatrix} E_x \\ E_y \end{bmatrix}, \tag{M.21}$$

where

$$\Omega = \begin{bmatrix} -v & 0 \\ 0 & -v \end{bmatrix}. \tag{M.22}$$

From Equation M.21, we obtain

$$\begin{bmatrix} L(J_x) \\ L(J_y) \end{bmatrix} = \frac{1}{[sI - \Omega]} \begin{bmatrix} J_x(0) \\ J_y(0) \end{bmatrix} + \frac{\varepsilon_0 \omega_p^2}{s[sI - \Omega]} \begin{bmatrix} E_x \\ E_y \end{bmatrix}. \tag{M.23}$$

We obtain Equation M.24 by getting the inverse Laplace transform of Equation M.23:

$$\begin{bmatrix} J_x \\ J_y \end{bmatrix} = e^{\Omega t} \begin{bmatrix} J_x(0) \\ J_y(0) \end{bmatrix} + \frac{e^{\Omega t} - I}{\Omega} \varepsilon_0 \omega_p^2 \begin{bmatrix} E_x \\ E_y \end{bmatrix}. \tag{M.24}$$

By applying Taylor's expansion to $e^{\Omega t}$, Equation M.24 can be written as

$$\begin{bmatrix} J_x \\ J_y \end{bmatrix} = \begin{bmatrix} e^{-vt} & 0 \\ 0 & e^{-vt} \end{bmatrix} \begin{bmatrix} J_x(0) \\ J_y(0) \end{bmatrix} + \frac{\varepsilon_0 \omega_p^2}{v} \begin{bmatrix} 1 - e^{-vt} & 0 \\ 0 & 1 - e^{-vt} \end{bmatrix} \begin{bmatrix} E_x \\ E_y \end{bmatrix}. \tag{M.25}$$

Equation M.25 states that the present values of the current density (J_x, J_y) are obtained from the previous values of the current density $(J_x(0), J_y(0))$ and the present values of the electric field components (E_x, E_y). Converting this into an FDTD equation, we will calculate the current density values at time step $(n+(1/2))$ from the current density value at time step $(n-(1/2))$ and the value of electric field intensity at time step n. Hence, the FDTD equation for the current densities is given by Equation M.26. In Equation M.26, note that the values of ω_p, v, E_x, and E_y are taken at the midpoint (at time step n) between time steps where the current densities are sampled: $(n - (1/2))$ and $(n + (1/2))$. This is in accordance with the careful sampling required for these variables for the "exponential stepping" algorithm as noted in the text following Equation M.20.

$$\begin{bmatrix} J_{x_k}^{n+(1/2)} \\ J_{y_k}^{n+(1/2)} \end{bmatrix} = e^{-(v\Delta t)} \begin{bmatrix} 1 & 0 \\ 0 & 1 \end{bmatrix} \begin{bmatrix} J_{x_k}^{n-(1/2)} \\ J_{y_k}^{n-(1/2)} \end{bmatrix}$$

$$+ \frac{\varepsilon_0 \omega_p^2 |_n^k e^{-(v\Delta t)}}{v^2} \begin{bmatrix} v\left(e^{(v\Delta t)} - 1\right) & 0 \\ 0 & v\left(e^{(v\Delta t)} - 1\right) \end{bmatrix} \begin{bmatrix} E_{x_k}^n \\ E_{y_k}^n \end{bmatrix}. \tag{M.26}$$

Thus in Equation M.26, ω_p, v, E_x, and E_y are the values at space step k and time step n.

M.2.3   The FDTD Computation Process

The data points for H_x, H_y, J_x, and J_y are evaluated at time step $n = -1/2$ using Equations M.5, M.6, and M.10. Then E_x and E_y are calculated at time step $n = 0$ using Equations M.1 and M.2. At this point, all the necessary data points have been initialized, and we are ready for FDTD computation.

Now data points for H_x, H_y, J_x, and J_y at time step $n = 1/2$ are evaluated from data points of time step $n = 0$ and $n = -1/2$, using the FDTD equations M.17, M.18, and M.26. Then data points for E_x and E_y at time step $n = 1$ are computed using the FDTD equations M.14 and M.15, using data points from time step $n = 1/2$ and $n = 0$. This process can continue indefinitely for subsequent time steps, until the entire region of interest has been solved.

It is important to choose the time increment (Δt) and the space increment (Δz), so that instabilities are avoided [5,6]. In the simulation, $\Delta z/\lambda = 1/50$ and $c\Delta t/\Delta z = 0.5$ were used, where λ is the wavelength of the source wave. However, Δz could have been chosen to be as large as $1/10\ \lambda$ without having any instability (at the expense of a slight degradation in the image quality but faster computation).

M.3   Results and Observations

Several cases have been simulated by the FDTD equations derived in the previous section. The results for these cases will now be presented.

M.3.1   Effect of Time-Varying Profile

In a previous paper [1], we derived the theoretical results for the scenario when the plasma is suddenly (instantaneously) created in the cavity. We showed that the magnetic field of the output wave consists of a wiggler and a standing wave with upshifted frequency. In this section, we will study the behavior of these two fields when the plasma is created at various rates. Suppose the plasma is built up exponentially in time as follows:

$$f_p^2(t) = \begin{cases} 0, & t < 0, \\ f_{p\_max}^2 \left[1 - \exp\left(\frac{-Kt}{T}\right) \right], & t > 0. \end{cases} \qquad \text{(M.27)}$$

where f_{p_max} is the maximum value of the plasma frequency, T is the period of the source signal, and K is a constant (that determines the growth rate). Note that in Equation M.27, there is no spatial variation in the plasma profile. This constraint will enable us to attribute any observed effect to the time profile. Also, the larger the value of K, the faster the plasma growth rate. This is demonstrated in Figure M.2.

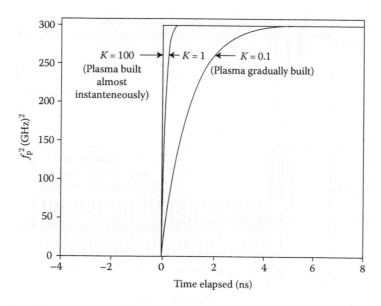

FIGURE M.2 Plasma concentration—time variation. Parameters: $f_0 = 10$ GHz and $f_{p\_max} = 17.32$ GHz.

For a linearly polarized source wave (tan $\gamma = 0$ in Equation M.5), Figure M.3 shows the value of the total magnetic field for two different values of K. In the top graph ($K = 100$), the plasma is almost instantaneously created, while in the bottom curve ($K = 0.1$), the plasma is created gradually. Clearly, the value of the wiggler magnetic field (measured as displacement from the origin) is larger for $K = 100$. However, note that the final output frequency of the standing wave does not depend on K, but only depends on f_{p_max}. On the other hand, the final steady amplitude of the standing wave is different for the two cases.

Figure M.4 demonstrates the transformation in the total magnetic field for a circularly polarized source wave (tan $\gamma = 1$, $\delta = -\pi/2$ in Equation M.5). Note how the polarization of the postplasma standing wave magnetic field changes from elliptical to "circularish" as K decreases from 100 to 1. Also the direction and magnitude of the wiggler magnetic field change as depicted in Figure M.4.

Figure M.5 shows the $x-y$ projection of the time evolution of the total magnetic field for a circularly polarized source wave for various K values. From Figure M.5, it is clear that, starting with a preplasma circularly polarized standing wave magnetic field, the corresponding postplasma field is circularly polarized when K is small, but elliptically polarized when K is large. Thus one could get an output wave at a desired polarization by precisely controlling the rate at which the plasma is created.

The magnitude of this field also depends on K. Let L_1, L_2, and L_3 be the dimensions defined in Figure M.5. The relation $L_1 < L_3 < L_2$ is evident from inspection of the top and bottom graphs of Figure M.5.

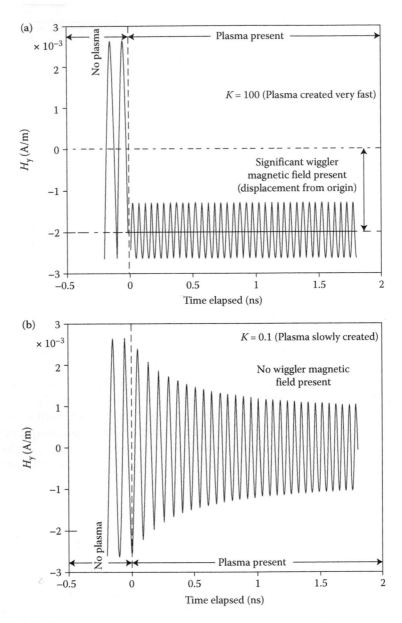

FIGURE M.3 H_y—time variation before and after plasma creation for different rates of plasma growth. Parameters: $f_0 = 10\,GHz$, $f_{p\_max} = 17.32\,GHz$, $E_0 = 1\,V/m$, $z = 0$, $\phi_0 = \pi/2$, and $\tan\gamma = 0$.

M.3.2 Effect of Space-Varying Profile

In a previous paper [1], we showed that when the plasma is suddenly created with a space-invariant profile, the original signal frequency is upshifted. In this section, we will investigate the scenario when the plasma is suddenly

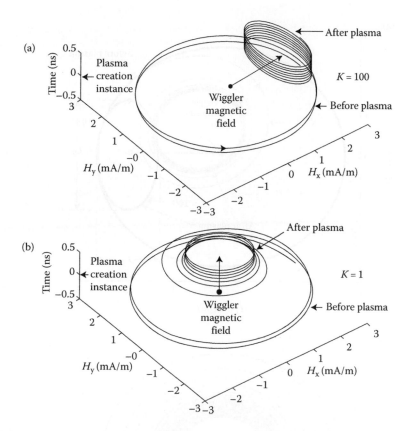

FIGURE M.4 3D view of total magnetic field for different rates of plasma growth. Demonstration of the change in the magnetic field as the plasma is created for two values of K. Parameters: $f_0 = 10\ \text{GHz}, f_{p\_max} = 17.32\ \text{GHz}, E_0 = 1\ \text{V/m}, z = 0, \delta = -\pi/2, \phi_0 = \pi$, and $\tan \gamma = 1$.

created, but with a space-variant profile. We will show that, in fact, several new modes are generated in the process.

Let the plasma profile be distributed as follows in space:

$$f_p^2(z,t) = \begin{cases} f_{p\_max}^2 & \dfrac{d - d_1}{2} < z < \dfrac{d + d_1}{2}, \quad t > 0, \\ 0 & \text{otherwise}, \end{cases} \tag{M.28}$$

where d is the distance between the plates, and d_1 and f_{p_max} are constants. This spatial profile is depicted in Figure M.6. Note that when $d_1/d = 1$, there is no spatial variation, and we have a uniform plasma profile throughout the cavity. For a linearly x-polarized source wave ($\tan \gamma = 0$ in Equation M.5), Figure M.7 shows the variation of the total magnetic field for two spatial profiles. Note that when the plasma is uniformly distributed ($d_1/d = 1.0$) as depicted in Figure M.7a, there is just a single standing wave mode in the

(a)

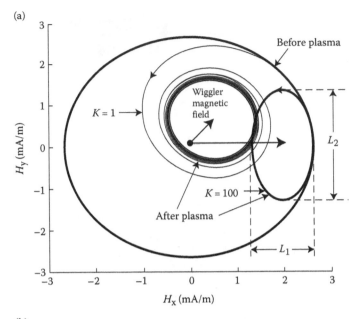

(b)

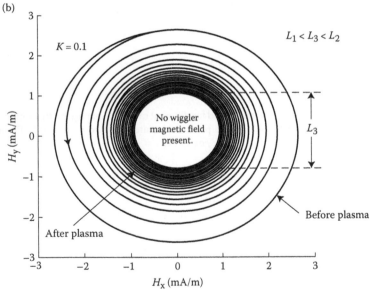

FIGURE M.5 $x-y$ projection of total magnetic field for various K. Demonstration of the change in the magnetic field as the plasma is created for three values of K. Parameters: $f_0 = 10$ GHz, $f_{p\_max} = 17.32$ GHz, $E_0 = 1$ V/m, $z = 0$, $\delta = -\pi/2$, $\phi_0 = \pi$, and $\tan \gamma = 1$.

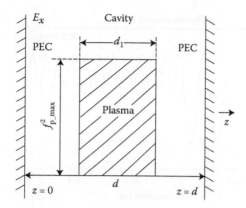

FIGURE M.6 Plasma concentration—space variation.

output. However, several modes are obtained when the plasma is created in only part of the cavity as demonstrated in Figure M.7b and c for $d_1/d = 0.3$.

The presence of several frequencies becomes more evident when the FFT is taken as demonstrated in Figure M.8. In Figure M.8a, for $d_1/d = 1.0$, there is just a single frequency at 20 GHz. In Figure M.8b, for $d_1/d = 0.3$ and sample point being in air, several modes can be seen, of which the following five frequencies are most significant: 8, 13, 18, 20, and 22 GHz. In Figure M.8c for $d_1/d = 0.3$ (sample point is in plasma), the three significant modes are at 18, 20, and 22 GHz.

M.3.3 Effect of Space-Varying Profile: Comparison of Signal Spectrum at a Cavity Point in Free Space with the One in Plasma

In this section [8], the difference in frequency content of different points in a non-homogeneous medium is discussed. Assume that the cavity is 30% filled with plasma using the profile discussed in Equation M.28. This is demonstrated in the right part of Figure M.7b and c.

Point A is at $z = d/6$ and point B is at $z = d/2$. The locations are carefully chosen so as to ensure that we have a maximum value for the magnetic field at each point for the source wave. This can easily be verified from the parameters listed for the figure. Figure M.7b demonstrates the variation of the magnetic field at the point in air and Figure M.7c demonstrates the same variation at the point in plasma. Clearly the transformed signal is different at each point. Figure M.8b shows that the transformed signal has significant modes at the point in air at 8, 13, 18, 20, and 22 GHz. However, of these five modes, the modes at 8 and 13 GHz are insignificant in the transformed signal in the cavity. This is expected as these modes are evanescent since the plasma frequency $f_p = 17.32$ GHz.

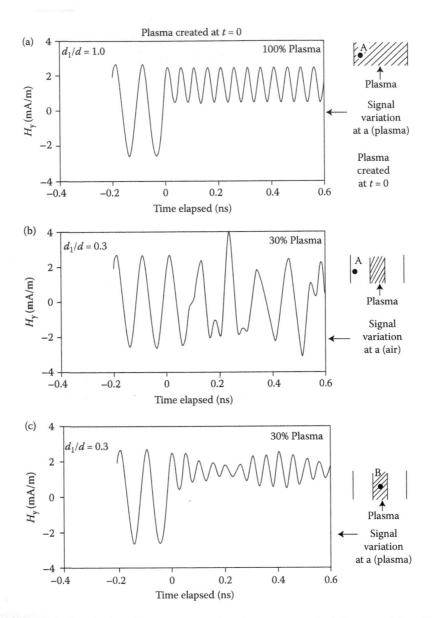

FIGURE M.7 H_y—time variation before and after plasma creation for different spatial profiles. Parameters: $f_0 = 10$ GHz, $f_{p\_max} = 17.32$ GHz, $E_0 = 1$ V/m, $\phi_0 = \pi/4$, $n = 6$, and $\tan \gamma = 0$.

M.3.4   Effect of Collisions

In the previous two sections, the computations were done assuming that the collision frequency for the plasma is zero. In reality, the plasma is indeed lossy and thus the collision frequency is not zero. We have shown [1] that after the

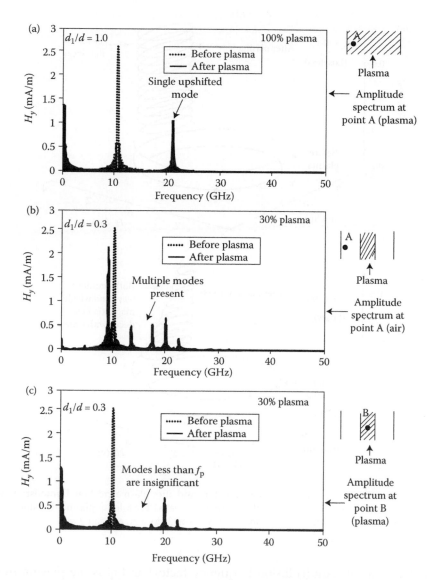

FIGURE M.8 Amplitude spectrum of H_y before and after plasma creation for different spatial profiles. Parameters: $f_0 = 10$ GHz, $f_{p\_max} = 17.32$ GHz, $E_0 = 1$ V/m, $\phi_0 = \pi/4$, $n = 6$, and $\tan \gamma = 0$.

plasma is created in the plasma, the magnetic field will consist of a standing wave and a wiggler wave. In this section, we will study how these two waves decay when we have a lossy plasma. The extent of this loss will be determined by the collision frequency, ν, of the plasma defined as follows:

$$\nu = k_\nu \cdot 2\pi f_p, \qquad (M.29)$$

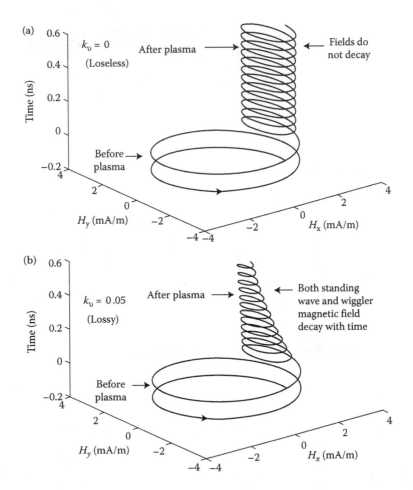

FIGURE M.9 3D view of total magnetic field before and after plasma creation. Demonstration of the change in the magnetic field as the plasma is created for a lossy plasma. Parameters: $f_0 = 10\,\text{GHz}, f_p = 17.32\,\text{GHz}, E_0 = 1\,\text{V/m}, z = 0, \delta = -\pi/2, \phi_0 = \pi$, and $\tan \gamma = 1$.

where k_v is a constant (collision frequency factor) and f_p is the plasma frequency. Figure M.9 shows the effect of a lossy plasma on the magnetic field. Clearly, both the standing wave and the wiggler field decay when collisions are present. According to theory [6], the time taken, t_{pn}, for a signal with frequency ω when placed in a lossy plasma of angular frequency ω_p and collision frequency v to decay to e^{-1} of its original value is

$$t_{pn} = \frac{1}{v}\left[1 + \frac{\omega_p^2}{(\omega_n^2 - \omega_0^2)} + \frac{2\omega_p^2\omega_0^2}{(\omega_n^2 - \omega_0^2)^2}\right], \quad n = 1, 2, 3, \dots. \tag{M.30}$$

Let t_{wigg} be the time taken by the wiggler magnetic field to decay to e^{-1} of its original value and let t_{sw} be the corresponding time for the standing wave. The equations for t_{wigg} and t_{sw} can be obtained by substituting $\omega_n = 0$ and $\omega_n = \sqrt{\omega_0^2 + \omega_p^2}$, respectively, in Equation M.30:

$$t_{wigg} = \frac{1}{v}\left[1 + \frac{\omega_p^2}{\omega_0^2}\right], \qquad t_{sw} = \frac{2}{v}\left[1 + \frac{\omega_0^2}{\omega_p^2}\right]. \qquad \text{(M.31)}$$

From Equation M.31, it can be inferred that for a given collision frequency (v) and signal frequency (ω_0), the larger the value of the plasma frequency (ω_p), the larger is the value of t_{wigg} and the smaller is the value of t_{sw}. This will be demonstrated in Figure M.10. In Figure M.10a, the rate of decay of the wiggler magnetic field is greater than the rate of decay of the standing wave field. The plasma frequency is 0.9 times the original signal frequency. From this we can deduce that if we are only interested in the standing wave magnetic field, we could use a plasma of low frequency relative to the source frequency, as the wiggler field would decay quickly, leaving a dominant standing wave. In Figure M.10b, the rate of decay of the wiggler magnetic field is less than the rate of decay of the standing wave field. The plasma frequency is 4.5 times he original signal frequency. Thus if we need just the wiggler magnetic field, we could create a plasma with a high frequency relative to the source frequency, as the standing wave would decay soon, leaving a dominant wiggler field.

M.3.5 Simultaneous Effect of Space Profile, Time Profile, and Collision

In Sections 3.1 through 3.3, we have independently investigated the effect of a time profile, space profile, and collisions. In this section, we will present the results due to the simultaneous combination of all these three effects.

Suppose that the time profile is given by Equation M.27 of Section 3.1, space profile by Equation M.28 of Section 3.2, and collision frequency by Equation M.29 of Section 3.3.

Figure M.11 shows the total magnetic field before and after the plasma creation with all three effects characterized by parameters $K = 10$ (in Equation M.27), $d_1/d = 0.5$ (in Equation M.28), and $k_v = 0.03$ (in Equation M.29), for a circularly polarized source wave ($\tan \gamma = 1$, $\delta = -\pi/2$ in Equation M.5). From the 3D view of the evolution of the total magnetic field as depicted in Figure M.11a, the effect of frequency upshift and presence of multiple frequencies are clearly visible, while the decay of the fields and shift in the wiggler magnetic field are evident from the x–y projection in Figure M.11b.

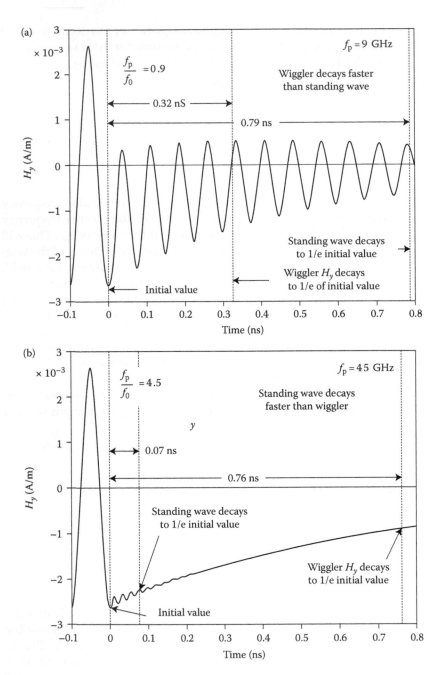

FIGURE M.10 Demonstration of different rates between standing and wiggler magnetic fields for two plasma concentrations: Parameters: $k_v = 0.05$, $f_0 = 10\,\text{GHz}$, $E_0 = 1\,\text{V/m}$, $z = 0$, $\phi_0 = \pi/2$, and $\tan \gamma = 0$.

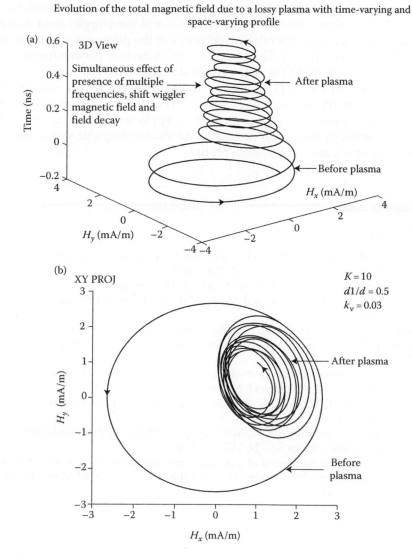

Evolution of the total magnetic field due to a lossy plasma with time-varying and space-varying profile

FIGURE M.11 Evolution of the total magnetic field. Parameters: $K = 10$, $d_1/d = 0.5$, $k_v = 0.03$, $f_0 = 10\,\text{GHz}$, $z = 0$, $f_{\text{p\_max}} = 17.32\,\text{GHz}$, $E_0 = 1\,\text{V/m}$, $\delta = -\pi/2$, $\phi_0 = \pi$, and $\tan\gamma = 1$.

M.4 Conclusion

The time-varying profiles show that the strength of the wiggler field significantly depends on the rate of growth of the plasma. To obtain the largest possible value of the wiggler field, we would need to create the plasma at the highest rate.

The space-varying profiles show that new modes are excited. The introduction of collisions shows that the standing wave and the wiggler field decay differently, depending on the relative frequency of the plasma with respect to the original signal frequency. Thus one can select a desired plasma frequency so that the desired wave in the output is the dominant signal.

The robustness of the FDTD equations has been proved by the simulation of the total magnetic field in a lossy plasma having both a temporal and a spatial profile. Indeed one should be able to apply this FDTD model and study in great detail the transformation of an electromagnetic signal under any arbitrary space and time plasma profile.

References

1. Ehsan, M. M. and Kalluri, D. K., Plasma induced wiggler magnetic field in a cavity, *Int. J. Infrared Millim. Waves*, 24, 1215–1234, 2003.
2. Kalluri, D. K., Lee, J. H., and Ehsan, M. M., FDTD Simulation of electromagnetic pulse interaction with a switched plasma slab, *Int. J. Infrared Millim. Waves*, 24, 349–365, 2003.
3. Balanis, C. A., *Advanced Engineering Electromagnetics*, Wiley, New York, 1989.
4. Kalluri, D. K., *Electromagnetics of Complex Media*, CRC Press LLC, Boca Raton, FL, 1999.
5. Taflove, A., *Computational Electrodynamics: The Finite-Difference Time-Domain Method*, Artech House, Norwood, MA, 1995.
6. Sullivan, D. M., *Electromagnetic Simulation Using the FDTD Method*, IEEE Press, New York, 2000.
7. Mendonca, J. T. and Oliveira, L., Mode coupling theory of flash ionization in a cavity, *IEEE Trans. Plasma Sci.*, 24, 147–151, 1996.
8. Ehsan, M. M., Transformation of a source wave in a cavity by a switched magnetoplasma: Longitudinal modes, Doctoral thesis, University of Massachusetts Lowell, Lowell, 2004.

Appendix N

Frequency and Polarization Transformer: Longitudinal Modes*

Dikshitulu K. Kalluri and Monzurul M. Ehsan

N.1 Introduction

A source wave splits into three modes when an unbounded plasma medium is created [1]. Two of the modes are propagating modes with upshifted frequencies. The third mode is a wiggler magnetic field, which is a space varying-DC (zero frequency) magnetic field.

In the presence of a static magnetic field, plasma is an anisotropic medium and is referred to as a magnetoplasma. When the static magnetic field is in the direction of propagation, the characteristic modes are circularly polarized and the modes are referred to as longitudinal modes [2]. A circularly polarized source wave splits into three frequency-shifted modes when an unbounded magnetoplasma that supports longitudinal modes is created [1]. The third mode is a downshifted-frequency mode, which degenerates into the wiggler magnetic field when the static magnetic field is reduced to zero and the anisotropic magnetoplasma becomes an isotropic plasma [1].

Sudden creation of an unbounded magnetoplasma is an idealization to simplify the problem but is not easily obtainable even in an approximate sense. Since it is easier to create experimentally a switched medium in a bounded space, we have taken up the study of the switched medium in a *cavity*. The switched medium is a magnetoplasma that supports longitudinal modes.

In this paper, analytical expressions for the three transformed waves are obtained by using the Laplace transform technique. Then a numerical FDTD model is developed to compute the transformed waves for an arbitrary space and time profile of the magnetoplasma. A few instructive results are presented as graphs to illustrate the effects of the source and system parameters.

The authors have recently published similar work [3], where there was no external magnetic field. Additional related FDTD work by the authors includes a simulation in a switched plasma [4]. Some details of the work omitted in this paper can be obtained from Refs. [3–5].

* Reprinted from *Int. J. Infrared Millim. Waves*, 25(2), 327–353, February 2004. With permission.

N.2   Theoretical Derivation

The source wave is initially in a one-dimensional cavity as shown in Figure N.1. This is made up of two PEC plates that are placed parallel to each other with a separation of d in the z-direction. The dimension of d is much smaller than the plate dimensions. The initial source signal is a standing sinusoidal wave with z dependence inside the cavity. Then a plasma is suddenly and uniformly created in this cavity in the presence of an external static magnetic field in the z-direction. The characteristic waves in such a magnetoplasma are circularly polarized [1].

N.2.1   Fields before Switching

Let E_0 be the amplitude of the electric field, d the distance between the plates, ω_0 the angular frequency of the standing wave, and z the distance along the cavity from the left plate edge. Let E denote electric field intensity, E^- denote E before a plasma is created, E^+ denote E after the plasma is created, E_x^- denote the x-component of E^- and E_x^+ denote the x-component of E^+. A similar notation to the above is used for the y- and z-component of E by replacing subscript x with y or z. The notation for the corresponding magnetic and current density fields is obtained by replacing E by H and J, respectively. Let the electric field intensity of the elliptically polarized source wave in the cavity before the plasma is created have the following field components:

$$\left\{ \begin{aligned} &E_x^-(z,t) = E_0 \sin\left(\frac{n\pi z}{d}\right) \cos(\omega_0 t + \phi_0), \\ &E_y^-(z,t) = (\tan\gamma)E_0 \sin\left(\frac{n\pi z}{d}\right) \cos(\omega_0 t + \phi_0 + \delta), \\ &E_z^-(z,t) = 0. \end{aligned} \right\} \tag{N.1}$$

In Equation N.1, ϕ_0 is called the switching angle, which is the phase of E_x^- at the instant the plasma is created. The notation of the definition of the

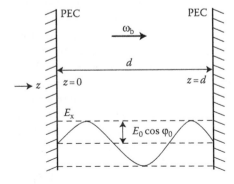

FIGURE N.1   One-dimensional cavity used for the theory and simulation. It also sketches E_x for $n = 3$, $\varphi_0 = 45°$, and $t = 0$.

elliptical wave is in accordance with Ref. [6], where tan γ is the amplitude ratio, while δ is the phase difference between the x- and y-component. By substituting the electric field components of Equation N.1 into the Maxwell's equation N.2 given below, one obtains the field equations for the magnetic field (Equation N.3):

$$\nabla \times \mathbf{E} = -\mu_0 \frac{\partial \mathbf{H}}{\partial t}, \tag{N.2}$$

$$\left\{ \begin{aligned} &H_x^-(z,t) = -(\tan \gamma) H_0 \cos\left(\frac{n\pi z}{d}\right) \sin(\omega_0 t + \phi_0 + \delta), \\ &H_y^-(z,t) = H_0 \cos\left(\frac{n\pi z}{d}\right) \sin(\omega_0 t + \phi_0), \\ &H_z^-(z,t) = 0, \\ &H_0 = -E_0 \frac{n\pi}{d\mu_0\omega_0} = \frac{-E_0}{\mu_0 c} = \frac{-E_0}{\eta_0}. \end{aligned} \right\} \tag{N.3}$$

The source frequency ω_0 and d need to be chosen so that the tangential electric field at the two PEC plates is zero to satisfy the boundary conditions. Thus the relation between ω_0 and d is given by Equation N.4, where $n = 1, 2, 3, \ldots, \infty$ and c is the speed of light in free space:

$$\omega_0 = \left(\frac{n\pi c}{d}\right). \tag{N.4}$$

One can express E_x^- as a product of a space component, $f_1^-(z)$, and a time component, $f_2^-(t)$, as follows:

$$\left\{ \begin{aligned} &E_x^-(z,t) = f_{Ex1}^-(z) f_{Ex2}^-(t), \\ &f_{Ex1}^-(z) = E_0 \sin\left(\frac{n\pi z}{d}\right), \\ &f_{Ex2}^-(t) = \cos(\omega_0 t + \phi_0). \end{aligned} \right\} \tag{N.5}$$

Such decompositions are useful in obtaining the fields after the sudden creation of plasma in the cavity. Similarly, let

$$\left\{ \begin{aligned} &E_y^-(z,t) = f_{Ey1}^-(z) f_{Ey2}^-(t), \\ &H_x^-(z,t) = f_{Hx1}^-(z) f_{Hx2}^-(t), \\ &H_y^-(z,t) = f_{Hy1}^-(z) f_{Hy2}^-(t), \\ &f_{Ey1}^-(z) = E_0 \sin\left(\frac{n\pi z}{d}\right), \\ &f_{Ey2}^-(t) = (\tan \gamma) \cos(\omega_0 t + \phi_0 + \delta) \\ &f_{Hx1}^-(z) = H_0 \cos\left(\frac{n\pi z}{d}\right), \\ &f_{Hx2}^-(t) = (-\tan \gamma) \sin(\omega_0 t + \phi_0 + \delta), \\ &f_{Hy1}^-(z) = H_0 \cos\left(\frac{n\pi z}{d}\right), \\ &f_{Hy2}^-(t) = \sin(\omega_0 t + \phi_0). \end{aligned} \right\} \tag{N.6}$$

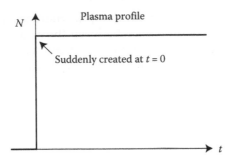

FIGURE N.2   Electron density profile in time.

N.2.2   Creation of Magnetoplasma

As depicted by the profile in Figure N.2, the cavity is uniformly and suddenly filled with a plasma of plasma frequency ω_p, where $\omega_p^2 = (2\pi f_p)^2 = q^2 N/m\varepsilon_0$, q is the electron charge, m is the electron mass, ε_0 is the permittivity of free space, and N is the electron density. The cavity is also subjected to an external magnetic field of gyrofrequency ω_b in the z-direction, where $\omega_b = q\mathbf{B_0}/m$ and $\mathbf{B_0}$ is the strength and direction of the external magnetic field.

N.2.3   Fields after Switching

Similar to the previous section, one can break $E_x^+(z,t)$ into a product of a space component, $f_{Ex1}^+(z)$, and a time component, $f_{Ex2}^+(t)$, as follows:

$$E_x^+(z,t) = f_{Ex1}^+(z)f_{Ex2}^+(t) = f_{Ex1}^-(z)f_{Ex2}^+(t). \tag{N.7}$$

Since our plasma profile does not vary across the cavity, the space-dependent field component does not change when the plasma is created. Thus, in Equation N.7, $f_{Ex1}^-(z) = f_{Ex2}^+(z)$. In the analysis that we are about to present, we will determine $f_{Ex2}^+(t)$ and then multiply that by $f_{Ex1}^-(z)$ to get $E_x^+(z,t)$. Kalluri [1] has shown that the general relation between $\mathbf{E}$, $\mathbf{J}$, ω_b, and ω_p is given by

$$\nabla^2\mathbf{E} - \nabla(\nabla.\mathbf{E}) - \frac{1}{c^2}\frac{\partial^2\mathbf{E}}{\partial t^2} - \frac{1}{c^2}\omega_p^2\mathbf{E} + \mu_0\mathbf{J}\times\omega_b = 0. \tag{N.8}$$

The fields after switching the plasma can be obtained from Equation N.2, Equation N.8, and the following Maxwell equation:

$$\nabla\times\mathbf{H} = \varepsilon_0\frac{\partial\mathbf{E}}{\partial t} + \mathbf{J}. \tag{N.9}$$

After expanding Equations N.2, N.8, and N.9 into the individual components and after some algebra and Laplace transformations, one obtains the six s-domain equations specified in Equation N.10:

$$\begin{cases} s^2 F_{Ex2}^+(s) \left[s^2 + \omega_0^2 + \omega_p^2\right] - F_{Jy2}^+(s) \left[\mu_0\omega_b c^2\right] = s\cos\phi_0 - \omega_0\sin\phi_0, \\ s^2 F_{Ey2}^+(s) \left[s^2 + \omega_0^2 + \omega_p^2\right] + F_{Jx2}^+(s) \left[\mu_0\omega_b c^2\right] \\ \qquad = s\cos(\phi_0 + \delta)(\tan\gamma) - \omega_0\sin(\phi_0 + \delta)(\tan\gamma), \\ F_{Ey2}^+(s) = a_1\left[sF_{Hx2}^+(s) - \sin(\phi_0 + \delta)(-\tan\gamma)\right], \\ F_{Ex2}^+(s) = -a_1\left[sF_{Hy2}^+(s) - \sin\phi_0\right], \\ F_{Hy2}^+(s) = a_2\varepsilon_0 E_0\left[sF_{Ex2}^+(s) - \cos\phi_0\right] + a_2 F_{Jx2}^+(s), \\ F_{Hx2}^+(s) = -a_2\varepsilon_0 E_0\left[sF_{Ey2}^+(s) - \cos(\phi_0 + \delta)(\tan\gamma)\right] - a_2 F_{Jy2}^+(s), \end{cases}$$

$$\text{(N.10)}$$

$$\begin{cases} a_1 = \mu_0 d H_0/n\pi E_0 = -1/\omega_0, \\ a_2 = d/n\pi H_0 = -1/\varepsilon_0 E_0\omega_0, \\ a_3 = \mu_0\omega_b\varepsilon_0 E_0 c^2, \\ a_4 = -\mu_0\omega_b c^2/a_1 a_2, \\ a_5 = -\omega_0\sin\phi_0 + c^2\mu_0\omega_b\varepsilon_0 E_0\cos(\phi_0 + \delta)(\tan\gamma), \\ a_6 = -\omega_0\sin(\phi_0 + \delta)(\tan\gamma) - c^2\mu_0\omega_b\varepsilon_0 E_0\cos\phi_0, \\ a_7 = -c^2\mu_0\omega_b\sin(\phi_0 + \delta)(\tan\gamma)/a_2, \\ a_8 = -c^2\mu_0\omega_b\sin\phi_0/a_2, \\ a_9 = \omega_0^2 + \omega_p^2, \\ a_{10} = \cos\phi_0, \\ a_{11} = a_5 - a_3\cos(\phi_0 + \delta)(\tan\gamma), \\ a_{12} = a_9\cos\phi_0 - a_7 - a_3 a_6, \\ a_{13} = a_5 a_9 - a_3 a_8 + a_4\cos(\phi_0 + \delta)(\tan\gamma), \\ a_{14} = -a_7 a_9 + a_4 a_6, \\ a_{15} = a_4 a_8, \\ a_{16} = 2a_9 + a_3^2, \\ a_{17} = a_9^2 - 2a_3 a_4, \\ a_{18} = a_4^2. \end{cases} \qquad \text{(N.11)}$$

In Equation N.10, $F(s) = L[f(t)] = \int_0^\infty e^{-st} f(t)\, dt$ and a_1 and a_2 are as defined in Equation N.11. In this equation set, there are six s-domain equations in the six unknown functions: $F_{Ex2}^+(s)$, $F_{Ey2}^+(s)$, $F_{Hx2}^+(s)$, $F_{Hy2}^+(s)$, $F_{Jx2}^+(s)$, and $F_{Jy2}^+(s)$. This set can be simultaneously solved for the six desired fields. Thus, solving for $F_{Ex2}^+(s)^+$ yields

$$F_{Ex2}^+(s) = \frac{a_{10}s^5 + a_{11}s^4 + a_{12}s^3 + a_{13}s^2 + a_{14}s + a_{15}}{s^6 + a_{16}s^4 + a_{17}s^2 + a_{18}}. \qquad \text{(N.12)}$$

By factoring the denominator of Equation N.12, one obtains three pairs of complex conjugate poles. These poles constitute the frequencies ω_1, ω_2, and ω_3 of the components of the transformed wave. Thus, Equation N.12 can be written as

$$F_{Ex2}^{+}(s) = \frac{B_1 s + C_1}{s^2 + \omega_1^2} + \frac{B_2 s + C_2}{s^2 + \omega_2^2} + \frac{B_3 s + C_3}{s^2 + \omega_3^2}. \tag{N.13}$$

Since $a_{10}-a_{15}$ and $\omega_1-\omega_3$ are known, the values of B_1, B_2, B_3, C_1, C_2, and C_3 can be computed. We know from inverse Laplace transformations that

$$L^{-1}\left(\frac{Bs + C}{s^2 + \omega^2}\right) = L^{-1}\left(\frac{(A\cos\phi)\,s + (-A\omega\sin\phi)}{s^2 + \omega^2}\right) = A\cos(\omega t + \phi), \tag{N.14}$$

where

$$B = A\cos\phi, \qquad C = -A\omega\sin\phi. \tag{N.15}$$

From Equation N.15,

$$A = \left(\frac{\sqrt{C^2 + \omega^2 B^2}}{\omega}\right). \tag{N.16}$$

Finally, ϕ is solved from Equations N.15 and N.16.

Thus, taking the inverse Laplace transform of Equation N.13 and combining with Equations N.5, N.14, N.15 and N.16, one obtains the final expression for $E_x^{+}(z,t)$:

$$E_x^{+}(z,t) = E_0 \sin\left(\frac{n\pi z}{d}\right) \sum_{i=1}^{i=3} A_i \cos[\omega_i t + \phi_i]. \tag{N.17}$$

N.2.4 Reduction of Field Equation for $E_x^{+}(z, t)$ When the External Magnetic Field is Zero (Isotropic Case)

Substituting $f_b = 0$ in the expression for $a_1 - a_{18}$ in Equation N.11 and substituting into Equation N.12 lead to

$$F_{Ex2}^{+}(s) = \frac{a_{10}s^5 + a_{11}s^4 + a_{12}s^3 + a_{13}s^2}{s^6 + a_{16}s^4 + a_{17}s^2} = \frac{a_{10}s^3 + a_{11}s^2 + a_{12}s + a_{13}}{s^4 + a_{16}s^2 + a_{17}}. \tag{N.18}$$

On closer examination, Equation N.18 reduces to

$$F_{Ex2}^{+}(s) = \frac{a_{10}s^3 + a_{11}s^2 + a_{12}s + a_{13}}{\left(s^2 + \omega_{up}^2\right)^2} = \frac{\left(s^2 + \omega_{up}^2\right)(A_1 s + B_1)}{\left(s^2 + \omega_{up}^2\right)^2}, \tag{N.19}$$

where

$$\omega_{up}^2 = \omega_0^2 + \omega_p^2; \quad A_1 = \cos\phi_0; \quad B_1 = -\omega_0 \sin\phi_0. \quad \text{(N.20)}$$

Combining Equations N.19 and N.20,

$$F_{Ex2}^+(s) = \frac{\cos\phi_0 s - \omega_0 \sin\phi_0}{s^2 + \omega_{up}^2}. \quad \text{(N.21)}$$

Taking the inverse Laplace transform of Equation N.21 and combining with Equation N.5, we obtain

$$E_x^+(z,t) = E_0 \sin\left(\frac{n\pi z}{d}\right)\left[\cos\phi_0 \cos(\omega_{up}t) - \frac{\omega_0}{\omega_{up}} \sin\phi_0 \sin(\omega_{up}t)\right]. \quad \text{(N.22)}$$

This derivation of the field is in agreement with a previous work [5].

N.2.5 Derivation of Other Fields

Similar to the derivation of $E_x^+(z,t)$ discussed above, one can obtain the final expressions of the other five fields: $E_y^+(z,t)$, $H_x^+(z,t)$, $H_y^+(z,t)$, $J_x^+(z,t)$, and $J_y^+(z,t)$.

N.3 Results

In a previous paper [5], we have shown that when the one-dimensional cavity in Section N.2 is suddenly and uniformly filled with a plasma of frequency f_p, the original source wave of frequency f_0 gets upshifted to a value of $f_{up} = \sqrt{f_0^2 + f_p^2}$. However, when there is an external magnetic field of strength f_b as described in Section N.2, the original single frequency source wave gets transformed into three waves having amplitudes and frequencies as described by Equation N.17. This is shown as a block diagram in Figure N.3, where the three output signals $O_1(A_1, f_1, \phi_1)$, $O_2(A_2, f_2, \phi_2)$, and $O_3(A_3, f_3, \phi_3)$ depend on the source wave input parameters E_0, f_0, ϕ_0, δ, and $\tan\gamma$ as well as the system parameters f_p, f_b.

N.3.1 Analysis of a Typical Transformation

In this section, we will transform an originally elliptically polarized source wave into a complex waveform by subjecting it to a magnetoplasma as

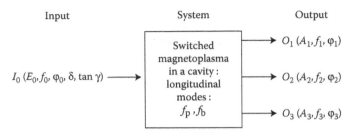

FIGURE N.3   Frequency transformer effect.

described in Section N.2. The input, system, and output parameters as computed by the theory of Section N.2 are given in Table N.1. Note from Table N.1 that the mode amplitudes and frequencies (unlike the mode phases) are identical for E_x and E_y. Moreover, the phases of E_x and E_y differ by 90°. Figure N.4a is a plot of mode 1, which has the highest frequency (f_1), and Figure N.4b shows its x–y projection. Mode 1 is an R-circularly polarized standing wave, hereafter abbreviated as "R wave." Mode 2 is an L-circularly polarized standing wave, abbreviated as "L wave." Its frequency f_2 is intermediate to the frequencies of modes 1 and 3. The smallest frequency mode is an R wave and its frequency f_3 is less than the source wave frequency f_0. The source wave is an elliptically polarized standing wave, yet the output waves are all circularly polarized. This may be understood by realizing that the source wave can be decomposed into an R wave and an L wave. Since the system is set up for longitudinal modes, each of the source waves give rise to the frequency-shifted circularly polarized standing waves. The net result is the three modes described above. It can be shown that if the static magnetic field is in the negative z-direction (sign of f_b is changed) and the sign of δ is changed, the results will be identical except that $\phi_1, \phi_2,$ and ϕ_3 change sign. Thus the results

TABLE N.1

E_x and E_y for the Three Modes

Quantity	E_x	E_y	Polarization
A_1 (V/m)	0.94	0.94	R
f_1 (GHz)	30.88	30.88	
ϕ_1 (degrees)	55.86	−34.14	
A_2 (V/m)	1.76	1.76	L
f_2 (GHz)	15.14	15.14	
ϕ_2 (degrees)	252.95	−17.05	
A_3 (V/m)	1.06	1.06	R
f_3 (GHz)	4.28	4.28	
ϕ_3 (degrees)	20.41	−69.59	

Parameters: $f_0 = 10$ GHz, $f_p = 17.63$ GHz, $f_b = 20$ GHz, $\tan \gamma = 4$, $\delta = -\pi/4$, $\phi_0 = 0$, $E_0 = 1$ V/m, $z = d/2$, and $n = 1$.

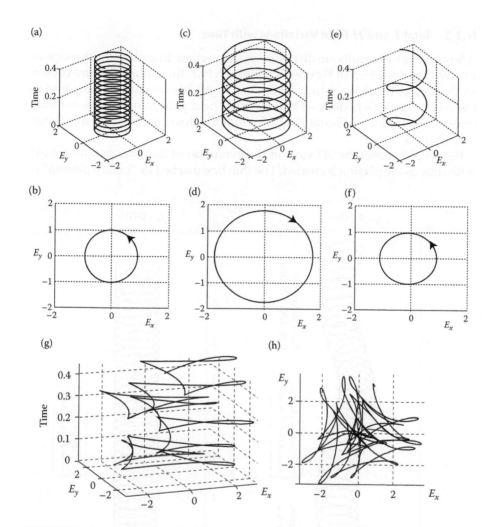

FIGURE N.4 Variation of the output electric field with time. The units for the x- and y-axes are V/m. The unit for the vertical axis is ns. Parameters: $f_0 = 10$ GHz, $f_p = 17.63$ GHz, $f_b = 20$ GHz, $\tan\gamma = 4$, $\delta = -\pi/4$, $\phi_0 = 0$, $E_0 = 1$ V/m, $z = d/2$, and $n = 1$. (a) Highest frequency mode (mode 1), (b) x–y projection of (a), (c) intermediate frequency mode (mode 2), (d) x–y projection of (c), (e) lowest frequency mode (mode 3), (f) x–y projection of (e), (g) total electric field, and (h) x–y projection of (g).

will be the same as in Table N.1, except that modes 1 and 3 will be L waves, whereas mode 2 will be an R wave.

Figure N.4g shows the total electric field and Figure N.4h shows its projection in the x–y plane. It is difficult to make sense of these two figures without decomposing them into the three modes. The results shown in Figure N.4g are verified by obtaining them through FDTD simulation discussed in Section N.4.

N.3.2   Total *E* and *H* Field Variations with Time

A few instructive results are discussed in this section. In each of the three cases that will be described in Figures N.5 through N.7, the parameters are chosen so that one of the three output modes has negligible amplitude compared to the other two. This is done so that the total field pattern can be more easily understood (unlike in Figure N.4g where all the three modes have significant amplitudes).

Figure N.5 shows the 3D view of the evolution of the total magnetic field with time as the plasma is created. The thin line marked as "before plasma" is

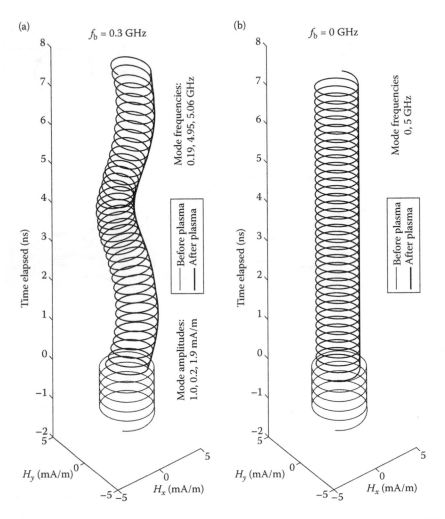

FIGURE N.5   Variation of total magnetic field with time. Parameters: $f_0 = 4$ GHz, $f_p = 3$ GHz, $\phi_0 = \pi/4$, $\delta = -\pi/2$, $\tan \gamma = 1$, $E_0 = 1$ V/m, $z = 0$, and $n = 1$.

the source wave. The thick line marked as "after plasma" is the total magnetic field, which is the sum of the magnetic fields of the three modes. From theory, we computed the frequencies of the three modes as 5.06, 4.95, and 0.19 GHz. The corresponding mode amplitudes are 1.9, 0.2, and 1.0 mA/m, respectively. The apparent "bending" of the wave in Figure N.5a is like the modulation of the mode at 0.19 GHz with the high-frequency carrier mode at 5.06 GHz. Note that in this scenario the mode at 4.95 GHz has negligible amplitude. In Figure N.5b, for the isotropic case ($f_b = 0$), the output consists of only two modes at 0 and 5 GHz. The component at 0 GHz is the wiggler magnetic field, while the single upshifted mode is the mode at 5 GHz. The figure demonstrates that the wiggler magnetic field of the isotropic case becomes the low-frequency R wave ($f_3 = 0.19\,\text{GHz}$) of significant amplitude for the anisotropic case under consideration ($f_b = 0.3\,\text{GHz}$).

Figure N.6a shows the 3D view of the evolution of the total electric field in the cavity for the given parameters. For the parameters in Figure N.6, the three mode frequencies are 56.07, 46.44, and 0.38 GHz. The corresponding mode amplitudes are 0.53, 0.43 and 0.04 V/m. Note that in this scenario, mode 3 at 0.38 GHz has relatively negligible amplitude. These components add up to form the interesting pattern that the x–y projection depicts in Figure N.6b. Figure N.6c shows the x–y projection for the isotropic case ($f_b = 0$), where there is just a single elliptically polarized standing wave mode at 50.99 GHz.

Figure N.7a shows the 3D view of the evolution of the total magnetic field for the given parameters. We observe that the components add to form the spiral-like structure. Note the transformation of the right circularly polarized source wave to the spiral-like structure. For the parameters given in Figure N.7, the three mode frequencies are 232.76, 35.03, and 2.45 GHz. The corresponding amplitudes are 0.01, 0.44, and 3.08 mA/m. It can be noted that the time difference between two consecutive loops in Figure N.7a, between point A (corresponding to $t_A = 0.013\,\text{ns}$) and point B (corresponding to $t_B = 0.040\,\text{ns}$), is about 0.027 ns. This time interval is approximately equal to the periodic time of the mode at 35.03 GHz. Thus the shape of the total waveform is like the modulation of the low-frequency component (at 2.45 GHz) with a high-frequency carrier of 35.03 GHz. Note that the effect of the component at 232.76 GHz is minimal since the amplitude of that mode is negligible compared to the amplitudes of the other two modes. The spiral-like structure is more easily appreciated when we take the x–y projection of Figure N.7a. This is depicted in Figure N.7b. Figure N.7c shows the total magnetic field in the cavity for the isotropic case ($f_b = 0$) where there are only two modes (at 87.57 and 0 GHz) after the plasma creation. The upshifted mode at 87.57 GHz is the dense helix seen at the center of the figure. The wiggler magnetic field is represented by the displacement of the mode at 87.57 GHz from the center (point x). We have recently published more detailed work for such a case [5].

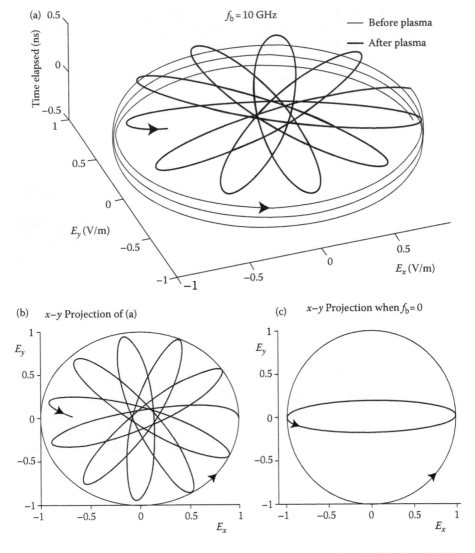

FIGURE N.6 Variation of total electric field with time. Parameters: $f_0 = 10$ GHz, $f_p = 50$ GHz, $\phi_0 = 0$, $\delta = -\pi/2$, $\tan \gamma = 1$, $E_0 = 1$ V/m, $z = d/2$, and $n = 1$.

N.3.3 Variation of the Modes with the Strength of the External Magnetic Field

The dependence of these three modes on the strength of the external magnetic field is depicted in Figure N.8. The plot for the electric field is taken in the middle of the cavity, while the magnetic field is taken at a plate edge. By so doing, we sample the fields at the points where they have the highest value.

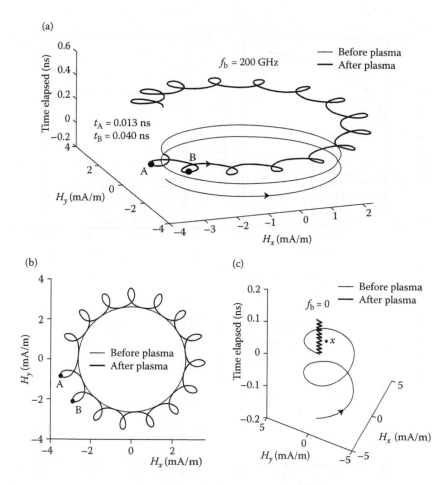

FIGURE N.7 (a) 3D view of the evolution of the total magnetic field for the given parameters. (b) This figure depicts that the spiral-like structure is more easily appreciated when we take the *x–y* projection of (a). (c) The total magnetic field in the cavity for the isotropic case (f_b=0) where there are only two modes (at 87.57 and 0 GHz) after the plasma creation. Parameters: $f_0 = 10$ GHz, $f_p = 87$ GHz, $\phi_0 = 0$, $\delta = -\pi/2$, $\tan\gamma = 1$, $E_0 = 1$ V/m, $z = 0$, and $n = 1$.

In Figure N.8a, as f_b is increased from 0 to 500 GHz, the frequency of the "lowest frequency" mode ("mode 3") increases from 0 to f_0, the frequency of the "intermediate frequency" mode ("mode 2") decreases from $\sqrt{f_0^2 + f_p^2}$ to f_0, and finally the frequency of the "highest frequency" mode ("mode 1") increases from $\sqrt{f_0^2 + f_p^2}$ to a large value. Figure N.8b depicts the variation of the phase for each of the three components of E_x. It can be seen from this figure that the phases reach a steady value as f_b increases. From Figure N.8c, when $f_b = 0$, the amplitude of the lowest frequency mode of E_x is zero,

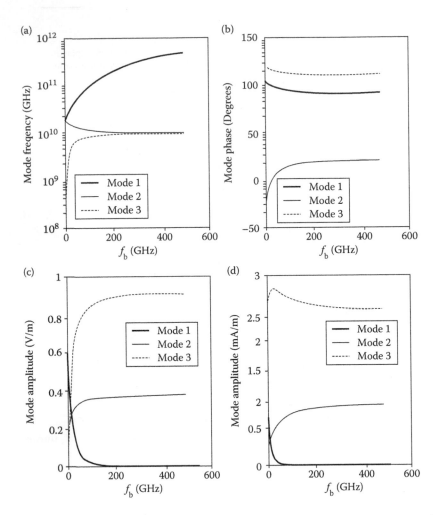

FIGURE N.8 Frequency, amplitude, and phase variation of modes with f_b. Parameters: $f_0 = 10\,\text{GHz}, f_p = 17.63\,\text{GHz}, \phi_0 = \pi/2, \tan\gamma = 1, \delta = -\pi/4, E_0 = 1\,\text{V/m}, z = d/2$ (for E_x), $z = 0$ (for H_y), and $n = 1$. (a) Mode freqencies (same for E_x and H_x), (b) E_x mode phases, (c) E_x mode amplitudes, and (d) H_x mode amplitudes.

while the other two modes (both at $\sqrt{f_0^2 + f_p^2}$) each have a nonzero value. This is consistent with the theory we developed in Ref. [5] for an isotropic case ($f_b = 0$), where it was shown that the output consists of a single upshifted mode at $\sqrt{f_0^2 + f_p^2}$. From Figure N.8d, when $f_b = 0$, the amplitude of the lowest frequency mode of H_y is nonzero. Also the other two modes (both at $\sqrt{f_0^2 + f_p^2}$) have a nonzero value. This is again consistent with the theory we developed in Ref. [5] for an isotropic case, where we have shown that the transformed H_y consists of a zero frequency mode known as the wiggler magnetic field and an additional mode at an upshifted frequency $\sqrt{f_0^2 + f_p^2}$. The E_y and H_x of the modes are not shown; however, they can be inferred since the modes are circularly polarized.

N.4 Finite-Difference Time-Domain

The results in Section N.3 were obtained from the theory developed in Section N.2, where it was assumed that the magnetoplasma was instantly built and that it was uniformly distributed in the cavity. However, these are very ideal requirements. In reality, the plasma is likely to take some time to build up. Also there is a likelihood of the plasma concentration to have some variation in space. Analytical computation of these scenarios is difficult to perform. However, we can numerically determine the field values under these circumstances using the FDTD technique. In this section, we will determine an FDTD model to compute the output field values when the plasma is created with arbitrary time and space profiles.

N.4.1 FDTD Equations

The one-dimensional cavity is converted into an FDTD grid as shown in Figure N.9. The field components E, H, and J are sampled at the space–time coordinates depicted in the figure. The vertical axis represents time, while the horizontal axis represents space. By sampling the fields in this way, we can march forward in time and compute future field values from their past values.

By expanding the Maxwell equation N.9 and expressing the time derivatives in terms of space derivatives when there are no field variations with x or y, and then discretizing using central difference formulas, one obtains the FDTD equations for E_x and E_y:

$$
\left.
\begin{aligned}
E_{x_k}^{n+1} &= E_{x_k}^n - \frac{\Delta t}{\varepsilon_0 \Delta z}\left(H_{y_{k+(1/2)}}^{n+(1/2)} - H_{y_{k-(1/2)}}^{n+(1/2)}\right) - \frac{\Delta t}{\varepsilon_0}J_{x_k}^{n+(1/2)}, \\
E_{y_k}^{n+1} &= E_{y_k}^n + \frac{\Delta t}{\varepsilon_0 \Delta z}\left(H_{x_{k+(1/2)}}^{n+(1/2)} - H_{x_{k-(1/2)}}^{n+(1/2)}\right) - \frac{\Delta t}{\varepsilon_0}J_{y_k}^{n+(1/2)}.
\end{aligned}
\right\}
\tag{N.23}
$$

Similarly by expanding the Maxwell equation N.2, one obtains the FDTD equations for H_x and H_y:

$$
\left.
\begin{aligned}
(H_x)_{k+(1/2)}^{n+(1/2)} &= (H_x)_{k+(1/2)}^{n-(1/2)} + \frac{\Delta t}{\mu_0 \Delta z}\left[(E_y)_{k+1}^n - (E_y)_k^n\right], \\
(H_y)_{k+(1/2)}^{n+(1/2)} &= (H_y)_{k+(1/2)}^{n-(1/2)} - \frac{\Delta t}{\mu_0 \Delta z}\left[(E_x)_{k+1}^n - (E_x)_k^n\right].
\end{aligned}
\right\}
\tag{N.24}
$$

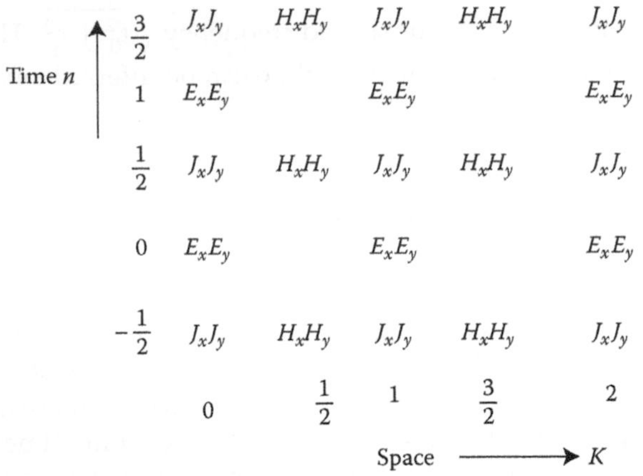

FIGURE N.9 FDTD grid for the computation.

The equation that connects current density with other field variables [1] is given by the constitutive relation

$$\frac{d\mathbf{J}}{dt} + \nu\mathbf{J} + (\mathbf{J} \times \boldsymbol{\omega_b}) = \varepsilon_0\omega_p^2\mathbf{E},$$ (N.25)

where ν is the collision frequency of the plasma and $\boldsymbol{\omega_b} = \omega_b\hat{z}$ is the external magnetic field. Using the exponential time-stepping technique [7], and after some algebra, one obtains the FDTD equations for J_x and J_y:

$$
\begin{bmatrix} J_{x_k}^{n+(1/2)} \\ J_{y_k}^{n+(1/2)} \end{bmatrix} = e^{-(\nu\Delta t)} \begin{bmatrix} \cos(\omega_b\Delta t) & -\sin(\omega_b\Delta t) \\ \sin(\omega_b\Delta t) & \cos(\omega_b\Delta t) \end{bmatrix} \begin{bmatrix} J_{x_k}^{n-(1/2)} \\ J_{y_k}^{n-(1/2)} \end{bmatrix}
$$
$$
+ \frac{\varepsilon_0\omega_p^2|_n^k E^{-(\nu\Delta t)}}{\nu^2 + \omega_b^2} \begin{bmatrix} k_{11} & k_{12} \\ k_{21} & k_{22} \end{bmatrix} \begin{bmatrix} E_{x_k}^n \\ E_{y_k}^n \end{bmatrix},
$$ (N.26)

where

$$
\begin{cases}
k_{11} = \nu\left(e^{\nu\Delta t} - \cos\omega_b\Delta t\right) + \omega_b\sin\omega_b\Delta t, \\
k_{12} = -e^{\nu\Delta t}\omega_b + \omega_b\cos\omega_b\Delta t + \nu\sin\omega_b\Delta t, \\
k_{21} = e^{\nu\Delta t}\omega_b - \omega_b\cos\omega_b\Delta t - \nu\sin\omega_b\Delta t, \\
k_{22} = \nu\left(e^{\nu\Delta t} - \cos\omega_b\Delta t\right) + \omega_b\sin\omega_b\Delta t.
\end{cases}
$$ (N.27)

For more details of the above derivation, one can refer to our recent publication [3], where we have shown in detail the steps for a derivation similar to Equation N.26 for the isotropic case ($\omega_b = 0$). A good reference for related FDTD modeling has been published by Sullivan [8].

N.4.2 FDTD Results

The field values in the FDTD simulation give the sum total of the three modes. The frequencies and amplitudes of the various modes can be obtained by taking the FFT of the total signal. This is in contrast to the theory developed in Section N.2 where we derived the fields of each mode (Equation N.17).

N.4.3 Effect of Time-Varying Profile

In this section, the effect of a time profile of the plasma creation will be studied. Suppose that the plasma profile varies in time as follows:

$$f_p^2(t) = \begin{cases} 0, & t < 0, \\ f_{p\_max}^2 \left[1 - \exp\left(\frac{-Kt}{T}\right) \right], & t > 0. \end{cases} \tag{N.28}$$

In Equation N.28, the larger the value of K, the higher the rate of plasma buildup in the cavity. Figure N.10 depicts the amplitudes of the various components as the plasma is built up at different rates for an R wave. From a comparison of Figure N.10a and b, it is clear that the lower the rate of growth of the plasma, the smaller the component amplitudes. Also the reduction in amplitudes is different for each of the three modes. This reduction appears to be highest for the intermediate frequency mode. The equivalent isotropic scenario ($f_b = 0$) is shown in Figure N.10c and d. Similar to the previous case, the amplitude of the upshifted mode is smaller when the plasma profile is built up at a lower rate. From Figure N.10b and d, it is evident that both demonstrate diminished amplitudes for the output due to the lower buildup of the plasma profile (as $K = 3$). It can also be noted that the amplitude of each of the three output modes in Figure N.10b is less than the single output mode in Figure N.10d. Figure N.11 shows similar results for an L source wave. Comparing Figure N.11a and b, it is clear that the output modes diminish when the rate of plasma growth is smaller. It is also apparent that the rate of reduction in amplitude is different for each of the modes. The figures also show that the rate of diminishing is smallest for the intermediate frequency mode. This is in contrast to the scenario for the R wave (where the rate of diminishing is highest for that mode).

N.4.4 Effect of Space-Varying Profile

In all the scenarios considered so far, the plasma profile was assumed to be uniform across the entire cavity. In this section, the effect of a space profile will be studied. Suppose that the plasma is distributed in the cavity as follows:

$$f_p^2(z, t) = \begin{cases} f_{p\_max}^2 & \frac{d - d_1}{2} < z < \frac{d + d_1}{2}, \quad t > 0, \\ 0 & \text{otherwise.} \end{cases} \tag{N.29}$$

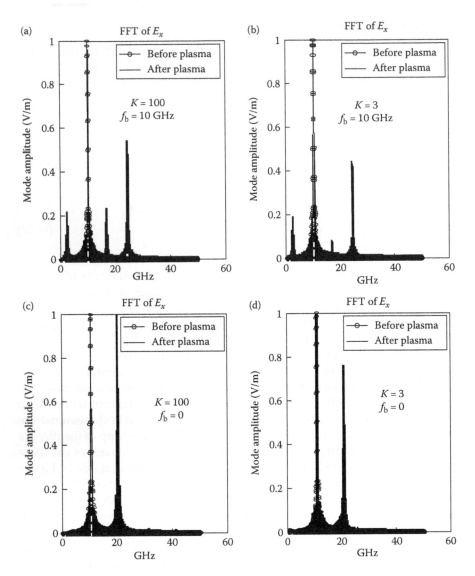

FIGURE N.10 Effect of plasma time profile (R wave). Mode frequencies of E_x for various plasma growth rates. (a) Fast-switching of magnetoplasma, (b) slow-switching of magneto-plasma, (c) fast-switching of isotropic plasma, and (d) slow-switching of isotropic plasma. Parameters: $f_0 = 10$ GHz, $f_p = 17.32$ GHz, $\tan \gamma = 1$, $\delta = -\pi/2$, $\phi_0 = 0$, $E_0 = 1$ V/m, $z = d/2$, and $n = 1$.

In Equation N.29, d is the distance between the plates, d_1 is the segment in the cavity where the plasma is created, and $f^2_{\text{P\_max}}$ is the square of the plasma frequency in the region where it exists. By appropriately selecting d_1, the fraction of the cavity containing the plasma is uniquely defined. Thus when d_1/d is 0.3, the middle 30% of the cavity is filled with plasma (while

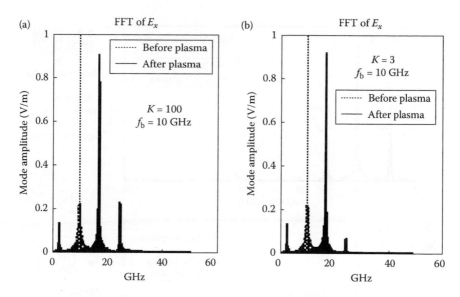

FIGURE N.11 Effect of plasma time profile (L wave). Mode frequencies of E_x for various plasma growth rates. (a) Fast-switching and (b) slow-switching. Parameters: $f_0 = 10$ GHz, $f_p = 17.32$ GHz, $\tan \gamma = 1$, $\delta = +\pi/2$, $\phi_0 = 0$, $E_0 = 1$ V/m, $z = d/2$, and $n = 1$.

the remaining 70% will have air). In Figure N.12a, the ratio d_1/d is 1 and thus the plasma is uniformly distributed in the entire cavity. Here three modes are observed. In Figure N.12b, the ratio d_1/d is 0.3 and thus only 30% of the cavity has plasma. Here numerous modes are observed. An explanation is that additional modes are excited when the cavity is partially filled. A qualitative explanation of this phenomenon is given in one of our recent publications [3] and in [9]. Comparing Figure N.12a and b, one can say that the ratio d_1/d will determine both the amplitudes and frequencies of the various modes. Thus, by appropriately filling the cavity with a plasma, one can achieve the desired output.

N.4.5 Effect of Space-Varying Profile: Comparison of a Signal Spectrum at a Cavity Point in Free Space with the One in Magnetoplasma

In this section [10], the difference in frequency content of different points in a nonhomogeneous medium is discussed. Assume that the cavity is 30% filled with magnetoplasma using the profile discussed in Equation N.29. This is demonstrated in the right part of Figure N.12.

Point A is at $z = d/6$ and point B is at $z = d/2$. The locations are carefully chosen so as to ensure that we have a maximum value for the magnetic field at each point for the source wave. This can easily be verified from the parameters listed for the figure. Figure N.12b demonstrates the amplitude spectrum of the magnetic field at the point in air and Figure N.12c

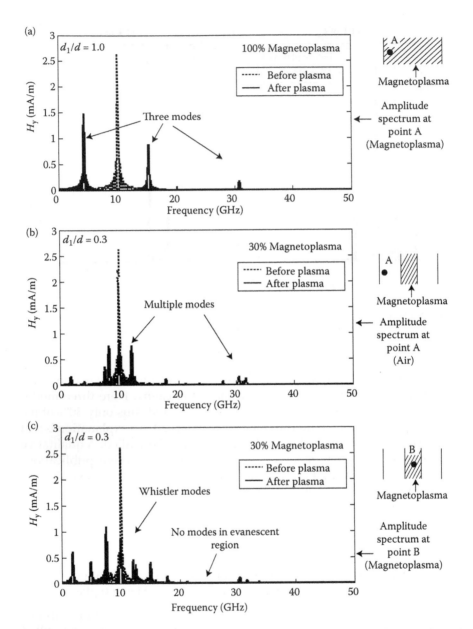

FIGURE N.12 Amplitude spectrum of H_y before and after plasma creation with spatial profile for two sample points. Parameters: $f_0 = 10$ GHz, $f_p = 17.32$ GHz, $f_b = 20$ GHz, $E_0 = 1$ V/m, $\phi_0 = \pi/4$, $n = 6$, and $\tan \gamma = 0$.

demonstrates the same variation at the point in the magnetoplasma. Clearly, the transformed signal is different at each point. Figure N.12b shows that the transformed signal has several significant modes. However, of these modes, the modes between 20 and 30 GHz are absent in the amplitude spectrum at point B (which is a point in the magnetoplasma), as depicted in Figure N.12c. From theory [1], the modes between $f_b = 20$ GHz and cutoff frequency $f_{c2} = f_b/2 + \sqrt{(f_b/2)^2 + f_p^2} = 30$ GHz are evanescent in the magnetoplasma. Thus the observation of the missing modes in Figure N.12c is in agreement with the theory.

In Figure N.12c, several whistler modes are clearly visible. Comparing Figure N.12c with Figure M.8c demonstrates the effect of the external magnetic field. These two figures have the same value of all parameters except for f_b. The comparison shows that the presence of the external magnetic field results in several low-frequency whistler modes being present in the spectrum in Figure N.12c (as opposed to the scenario in Figure M.8c).

N.4.6 Lossy Plasma

In all the cases considered so far, the plasma was assumed to be lossless. In the real-life scenario, the plasma is lossy due to collisions. The rate of this loss is defined by the collision frequency, v, of the plasma:

$$v = k_v \cdot 2\pi f_p, \tag{N.30}$$

where k_v is a constant (collision frequency factor) and f_p is the plasma frequency. Figure N.13 depicts the FFT of E_x with and without loss. Comparing Figure N.13a and b, it is clear that all the three components diminish in time

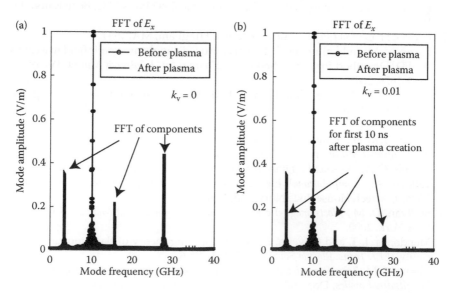

FIGURE N.13 Effect of a lossy plasma. FFT of E_x for a lossy plasma. (a) Loss-free plasma and (b) lossy plasma. Parameters: $f_0 = 10$ GHz, $f_p = 17.32$ GHz, $f_b = 15.8$ GHz, $\tan \gamma = 1$, $\delta = -\pi/2$, $\phi_0 = 0$, $E_0 = 1$ V/m, $z = d/2$, and $n = 1$.

due to the loss. However, the rate of diminishing is different for the various modes. For the parameters chosen in Figure N.13, the highest frequency mode diminishes at the highest rate.

N.5 Conclusion

A source wave transforms into three different frequency modes for the case of switching on of a magnetoplasma setup for longitudinal modes. These components have unique variations in both amplitude and frequency for varying strengths of the external magnetic field and plasma density. If the plasma is created gradually, then the amplitudes of the different modes change; however, the final frequency remains the same. If there is spatial variation in the plasma density, many new modes are introduced. Thus a source wave uniquely transforms into several modes based on the spatial variation in the plasma profile, the rate of plasma buildup, the plasma density, and the strength of the external magnetic field.

References

1. Kalluri, D. K., *Electromagnetics of Complex Media*, CRC Press LLC, Boca Raton, FL, 1999.
2. Booker, H. G., *Cold Plasma Waves*, Kluwer, Hingham, MA, 1984.
3. Ehsan, M. M. and Kalluri, D. K., Plasma induced wiggler magnetic field in a cavity II—FDTD method for a switched lossy plasma, *Int. J. Infrared Millim. Waves*, 24, 1655–1676, 2003.
4. Kalluri, D. K., Lee, J. H., and Ehsan, M. M., FDTD simulation of electromagnetic pulse interaction with a switched plasma slab, *Int. J. Infrared Millim. Waves*, 24, 349–365, 2003.
5. Ehsan, M. M. and Kalluri, D. K., Plasma induced wiggler magnetic field in a cavity, *Int. J. Infrared Millim. Waves*, 24, 1215–1234, 2003.
6. Balanis, C. A., *Advanced Engineering Electromagnetics*, Wiley, New York, 1989.
7. Taflove, A., *Computational Electrodynamics: The Finite-Difference Time-Domain Method*, Artech House, Inc., Norwood, MA, 1995.
8. Sullivan, D. M., *Electromagnetic Simulation Using the FDTD Method*, IEEE Press, New York, 2000.
9. Mendonca, J. T. and Oliveira, L., Mode coupling theory of flash ionization in a cavity, *IEEE Trans. Plasma Sci.*, 24, 147–151, 1996.
10. Ehsan, M. M., *Transformation of a source wave in a cavity by a switched magnetoplasma: Longitudinal modes*, Doctoral thesis, University of Massachusetts Lowell, Lowell, 2004.

Appendix O

Frequency and Polarization Transformer:
*Transverse Modes—I Zero Rise Time**

Dikshitulu K. Kalluri, Ahmad F. Khalifeh, and Sebahattin Eker

O.1   Introduction

A source wave splits into three modes when an *unbounded* isotropic plasma medium is created [1–17]. One is a forward-propagating mode with upshifted frequency and another is a backward-propagating mode of the same frequency. The third mode is a wiggler magnetic field, which is a space-varying dc (zero frequency) magnetic field.

In the presence of an external static magnetic field, plasma is an anisotropic medium and is referred to as a magnetoplasma. When the static magnetic field is perpendicular to the direction of propagation, the characteristic modes are ordinary (O) and extraordinary modes (X) and are referred to as transverse modes [18]. When the electric field of the source wave is in the same direction as that of the static magnetic field, it has no effect and the mode characteristics are the same as that of an isotropic plasma and hence these modes are referred to as O. On the other hand, when the electric field of the source wave is perpendicular to the static magnetic field as well as the direction of propagation, the transverse modes are called X and have a component of the electric field in the direction of propagation. A linearly polarized source wave splits into four frequency-shifted modes when an *unbounded* magnetoplasma that supports the X modes is created [1]. Two of them are forward-propagating modes of different frequencies. The other two are the corresponding backward-propagating modes. Moreover, wiggler (zero frequency but space varying) electric as well as magnetic fields are also created and they degenerate into a wiggler magnetic field only when the static magnetic field is reduced to zero and the anisotropic magnetoplasma becomes isotropic [1]. The wiggler fields have applications in an FEL [19].

Sudden creation of an unbounded plasma is an idealization to simplify the problem but is not easily obtainable even in an approximate sense. Since it

---

* Reprinted from *IEEE Trans. Ant. Prop.*, 55(6), 1789–1796, June 2007. With permission.

is easier to create experimentally a switched medium in a *bounded* space, we have taken up the study of the switched medium in a *cavity*. The switched medium is a magnetoplasma that supports X modes.

In this paper, analytical expressions for the transformed waves are obtained by using the Laplace transform technique. A few instructive results are presented as graphs to illustrate the effects of the source and system parameters. Some work has been published recently [20], where there was no external magnetic field. The case of longitudinal modes is discussed in [21]. A comparison of the results of the longitudinal modes with the results of this study is given in Section O.4.

O.2 Theoretical Derivation

The source wave is initially in a one-dimensional cavity as shown in Figure O.1. The cavity is made up of two PEC plates that are placed parallel to each other with a separation distance d in the z-direction. The dimension of d is much smaller than the plate dimensions. The initial source signal inside the cavity is an x-polarized standing sinusoidal wave with z dependence. Then plasma is suddenly and uniformly created in this cavity in the presence of an external static magnetic field in the y-direction. The characteristic wave in such a magnetoplasma is an X wave.

O.2.1 Fields before Switching

Let E_0 be the amplitude of the electric field, ω_0 the angular frequency of the standing wave, and z the distance in the coordinate direction along the cavity. Let E denote electric field intensity, E^- the electric field before plasma is created, E^+ the electric field after plasma is created, E_x^- the x-component of

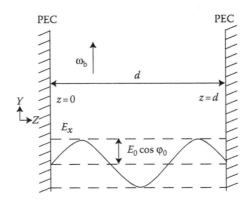

FIGURE O.1 The one-dimensional cavity. Inside the cavity, a sketch of E_x for $n = 3$, $\varphi_0 = 45°$, and $t = 0$ is shown.

E^-, and E_x^+ the x-component of E^+. A similar notation to the one above is used by replacing subscript x with y or z for, respectively, the y- or z-component of E. The notation for the corresponding magnetic and current density fields is obtained by replacing E by H and J, respectively. Let the electric field intensity of the linearly polarized source wave in the cavity before the plasma is created have the following field components:

$$\left.\begin{array}{l} E_x^-(z,t) = E_0 \sin\left(\dfrac{n\pi}{d}z\right)\cos(\omega_0 t + \phi_0), \\[2mm] E_y^-(z,t) = 0, \\[2mm] E_z^-(z,t) = 0. \end{array}\right\} \tag{O.1}$$

In Equation O.1, φ_0 is the phase of E_x^- at the instant the plasma is created, and it is called the switching angle. The magnetic field components of Equation O.3 can be obtained by substituting the electric field components of Equation O.1 into the Maxwell's Equation O.2 given below:

$$\nabla \times \mathbf{E} = -\mu_0 \frac{\partial \mathbf{H}}{\partial t}, \tag{O.2}$$

$$\left.\begin{array}{l} H_x^-(z,t) = 0, \\[2mm] H_y^-(z,t) = -H_0 \cos\left(\dfrac{n\pi}{d}z\right)\sin(\omega_0 t + \phi_0), \\[2mm] H_z^-(z,t) = 0, \\[2mm] H_0 = E_0 \dfrac{n\pi}{d\mu_0\omega_0} = \dfrac{E_0}{\mu_0 c} = \dfrac{E_0}{\eta_0}. \end{array}\right\} \tag{O.3}$$

The source frequency ω_0 and the distance d between the plates need to be chosen so that the tangential component of the electric field at the two PEC plates is zero, in order to satisfy the boundary conditions. Thus the relation between ω_0 and d is given by Equation O.4, where $n = 1,2,3,\ldots,\infty$ and c is the speed of light in free space:

$$\omega_0 = \left(\frac{n\pi}{d}c\right). \tag{O.4}$$

One can express E_x^- as a product of a space component $f_{Ex1}^-(z)$ and a time component $f_{Ex2}^-(t)$ as follows:

$$\left.\begin{array}{l} E_x^-(z,t) = f_{Ex1}^-(z)f_{Ex2}^-(t), \\[2mm] f_{Ex1}^-(z) = E_0 \sin\left(\dfrac{n\pi}{d}z\right), \\[2mm] f_{Ex2}^-(t) = \cos(\omega_0 t + \phi_0). \end{array}\right\} \tag{O.5}$$

Such decompositions are useful in obtaining the fields after the sudden creation of plasma in the cavity. Similarly, let

$$\left.\begin{array}{l} H_y^-(z,t) = f_{Hy1}^-(z)f_{Hy2}^-(t), \\[2mm] f_{Hy1}^-(z) = -H_0 \cos\left(\dfrac{n\pi}{d}z\right), \\[2mm] f_{Hy2}^-(t) = \sin(\omega_0 t + \phi_0). \end{array}\right\} \tag{O.6}$$

Since the cavity medium is initially free space, the current density inside the cavity must be zero. Thus

$$\begin{cases} J_x^-(z,t) = 0, \\ J_y^-(z,t) = 0, \\ J_z^-(z,t) = 0. \end{cases} \tag{O.7}$$

O.2.2 Creation of Magnetoplasma

As depicted by the profile in Figure O.2, the cavity is uniformly and suddenly filled with a plasma of plasma frequency ω_p, where

$$\omega_p^2 = (2\pi f_p)^2 = \frac{q^2 N}{m\varepsilon_0}. \tag{O.8}$$

In Equation O.8, q is the electron charge, m is the electron mass, ε_0 is the permittivity of free space, and N is the electron density. The cavity is also subjected to an external magnetic field of gyrofrequency ω_b in the y-direction, where

$$\omega_b = \hat{y}\frac{qB_0}{m}, \tag{O.9}$$

and B_0 is the strength of the external magnetic field.

O.2.3 Fields after Switching

The constitutive relation between the current density **J** and the electric field **E** in a magnetized plasma "neglecting collisions" is given in Ref. [1]:

$$\frac{\partial \mathbf{J}}{\partial t} = \varepsilon_0 \omega_p^2 \mathbf{E} - \mathbf{J} \times \boldsymbol{\omega}_b. \tag{O.10}$$

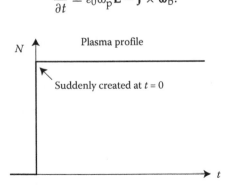

FIGURE O.2 Electron density profile in time.

In Equation O.10, the time derivative on the left-hand side should be a convective [22] rather than a partial derivative. The last term on the right-hand side of Equation O.10 should include the wave magnetic field. Trifkovic and Stanic [22] studied the effect of keeping the additional terms in Equation O.10 but assumed that the source wave amplitude was small and modeled the system as weakly nonlinear. They considered longitudinal propagation in an unbounded medium and showed that in addition to the main frequency-shifted TEM modes, stationary and electron wave modes of secondary importance are created. A similar analysis can be carried out for transverse modes to obtain the additional secondary modes.

The fields after switching the plasma can be obtained from Equation O.2, Equation O.10, and the following Maxwell equation:

$$\nabla \times \mathbf{H} = \varepsilon_0 \frac{\partial \mathbf{E}}{\partial t} + \mathbf{J}. \tag{O.11}$$

After expanding Equations O.2, O.10, and O.11 into the individual field components, the following set of equations relating the fields in the cavity are obtained:

$$\left. \begin{cases} \dfrac{\partial J_x^+}{\partial t} = \varepsilon_0 \omega_p^2 E_x^+ + J_z^+ \omega_b, \\[2mm] \dfrac{\partial J_z^+}{\partial t} = \varepsilon_0 \omega_p^2 E_z^+ - J_x^+ \omega_b, \\[2mm] \dfrac{\partial H_y^+}{\partial t} = -\dfrac{1}{\mu_0} \dfrac{\partial E_x^+}{\partial z}, \\[2mm] \dfrac{\partial E_x^+}{\partial t} = -\dfrac{1}{\varepsilon_0} \dfrac{\partial H_y^+}{\partial z} - \dfrac{1}{\varepsilon_0} J_x^+, \\[2mm] \dfrac{\partial E_z^+}{\partial t} = -\dfrac{1}{\varepsilon_0} J_z^+, \\[2mm] E_y^+(z,t) = H_x^+(z,t) = H_z^+(z,t) = J_y^+(z,t) = 0. \end{cases} \right\} \tag{O.12}$$

In the following analysis, we will determine the time-domain fields, and then multiply each component by the corresponding space-varying field component to find the total field. The fields are continuous over the time discontinuity for all z [1].

Therefore, the following equations apply:

$$E_x^+(z,t) = f_{Ex1}^+(z) f_{Ex2}^+(t) = f_{Ex1}^-(z) f_{Ex2}^+(t), \tag{O.13}$$

$$E_z^+(z,t) = f_{Ez1}^+(z) f_{Ez2}^+(t) = f_{Ez1}^-(z) f_{Ez2}^+(t), \tag{O.14}$$

$$H_y^+(z,t) = f_{Hy1}^+(z) f_{Hy2}^+(t) = f_{Hy1}^-(z) f_{Hy2}^+(t), \tag{O.15}$$

$$J_x^+(z,t) = f_{Jx1}^+(z) f_{Jx2}^+(t) = f_{Jx1}^-(z) f_{Jx2}^+(t), \tag{O.16}$$

$$J_z^+(z,t) = f_{Jz1}^+(z) f_{Jz2}^+(t) = f_{Jz1}^-(z) f_{Jz2}^+(t). \tag{O.17}$$

According to Equations O.5, O.6, and O.12, the space component of the electric fields in Equations O.13 and O.14 is $E_0 \sin((n\pi/d)z)$, the space component of the magnetic field in Equation O.15 is $-H_0 \cos((n\pi/d)z)$, and the space component of the current density in Equations O.16 and O.17 is $J_0 \sin((n\pi/d)z)$. H_0 is given in Equation O.3 and J_0 is the amplitude of the current density and can be found from Maxwell's Equation O.11. By making the appropriate substitutions, one can relate E_0, H_0, and J_0 by the following relation:

$$J_0 = -\omega_0 \varepsilon_0 E_0 = -\frac{n\pi}{d} H_0. \tag{O.18}$$

Substituting Equations O.13 through O.17 into Equation O.12 and with some mathematical simplifications, the following differential equations are obtained:

$$
\begin{cases}
J_0 \dfrac{\partial}{\partial t}\left(f_{Jx2}^+(t)\right) = \varepsilon_0 \omega_p^2 E_0 f_{Ex2}^+(t) + \omega_b J_0 f_{Jz2}^+(t), \\[2mm]
J_0 \dfrac{\partial}{\partial t}\left(f_{Jz2}^+(t)\right) = \varepsilon_0 \omega_p^2 E_0 f_{Ez2}^+(t) - \omega_b J_0 f_{Jx2}^+(t), \\[2mm]
\dfrac{\partial}{\partial t}\left(f_{Hy2}^+(t)\right) = \left(\dfrac{n\pi}{d\mu_0 H_0}\right) E_0 f_{Ex2}^+(t), \\[2mm]
\left(-\dfrac{n\pi}{d}\right) H_0 f_{Hy2}^+(t) = J_0 f_{Jx2}^+(t) + \varepsilon_0 E_0 \dfrac{\partial}{\partial t}\left(f_{Ex2}^+(t)\right), \\[2mm]
J_0 f_{Jz2}^+(t) = -\varepsilon_0 E_0 \dfrac{\partial}{\partial t}\left(f_{Ez2}^+(t)\right).
\end{cases}
\tag{O.19}
$$

In Equation O.19, there are five differential equations for the fields' time-domain components $f_{Ex2}^+(t)$, $f_{Ez2}^+(t)$, $f_{Hy2}^+(t)$, $f_{Jx2}^+(t)$, and $f_{Jz2}^+(t)$. This set of equations can be simultaneously solved using the Laplace transformation technique. The initial conditions used to solve this set are shown in Equation O.20:

$$
\begin{cases}
f_{Ex2}^+(0) = f_{Ex2}^-(0) = \cos(\phi_0), \\[1mm]
f_{Ez2}^+(0) = f_{Ez2}^-(0) = 0, \\[1mm]
f_{Hy2}^+(0) = f_{Hy2}^-(0) = \sin(\phi_0), \\[1mm]
f_{Jx2}^+(0) = f_{Jx2}^-(0) = 0, \\[1mm]
f_{Jz2}^+(0) = f_{Jz2}^-(0) = 0.
\end{cases}
\tag{O.20}
$$

The solution of the electric and magnetic fields of Equation set O.19 is given in the appendix and details of the solution are given in Ref. [21].

O.3 Illustrative Results

In a previous paper [20], it was shown that when the one-dimensional cavity in Section O.2 was suddenly and uniformly filled with plasma of frequency

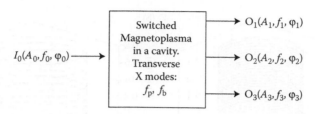

FIGURE O.3   Frequency transformer effect.

f_p, the original source wave of frequency f_0 got upshifted to a wave of frequency $f_{up} = \sqrt{f_0^2 + f_p^2}$, and a wiggler magnetic field of zero frequency was created. However, when there is an external magnetic field of strength f_b as described in Section O.2, the original single frequency source wave gets transformed into three waves, two of which have the frequencies given in the appendix and one dc mode of zero frequency (wiggler mode). The amplitudes and phases of the fields are also given in the appendix. These are shown as a block diagram in Figure O.3, where the three output waves $O_1(A_1, f_1, \varphi_1)$, $O_2(A_2, f_2, \varphi_2)$, and $O_3(A_3, f_3, \varphi_3)$ depend on the source wave input parameters A_0, f_0, and φ_0 as well as the system parameters f_p and f_b. The O_3 wave has a frequency $f_3 = 0$ and thus it is a wiggler mode.

The results for one set of input and system parameters are given in Table O.1. The source wave is assumed to be a fundamental ($n = 1$) x-polarized mode in the cavity of Figure O.1 with $E_0 = 1$ V/m, an input frequency $f_0 = 10$ GHz and a switching angle $\varphi_0 = 0°$. From Equation O.4, $d = 1.5$ cm. The cavity size can be changed by changing n and f_0.

TABLE O.1

Electric and Magnetic Fields of the Three Modes for the Following Parameters: $f_p = 17.63$ GHz, $f_b = 20$ GHz, $n = 1$, and $\varphi_0 = 0°$

Wave	Quantity	E_x	E_z	H_y
I_0 (source)	A_0	1 V/m	0 V/m	2.7 mA/m
	f_0 (GHz)	10	10	10
	φ_0 (degrees)	0	—	−90
O_1 (mode 1)	A_1	0.30 V/m	0.26 V/m	−0.262 mA/m
	f_1 (GHz)	30.72	30.72	30.72
	φ_1 (degrees)	0	90	270
O_2 (mode 2)	A_2	0.70 V/m	0.61 V/m	−1.38 mA/m
	f_2 (GHz)	13.33	13.33	13.33
	φ_2 (degrees)	0	−90	−90
O_3 (mode 3)	A_3	0	0	0
	f_3 (GHz)	0	0	0
	φ_3 (degrees)	—	—	—

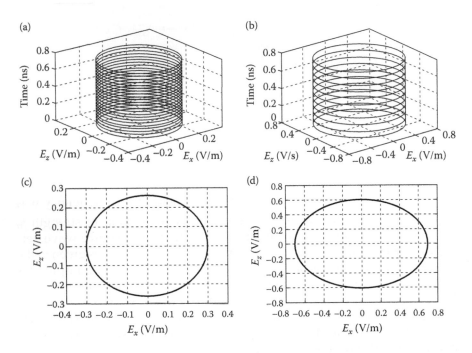

FIGURE O.4 Electric field of the modes after switching: Parameters: $f_0 = 10\,\text{GHz}, f_b = 20\,\text{GHz}$, $\varphi_0 = 0°, E_0 = 1\,\text{V/m}, z = d/2$, and $n = 1$. See Table O.1 for more information. (a) Mode 1: Higher upshifted frequency $f_1 = 30.72\,\text{GHz}$; (b) Mode 2: Lower upshifted frequency $f_2 = 13.33\,\text{GHz}$; (c) x–z projection of (a); and (d) x–z projection of (b).

The system parameters are the plasma frequency $f_p = 17.63$ GHz and the electron gyrofrequency $f_b = 20$ GHz. We note that O_1 mode has an output frequency $f_1 = 30.72$ GHz and O_2 mode has an output frequency $f_2 = 13.33$ GHz. Both frequencies are higher than the source frequency f_0 and hence these are upshifted modes. The amplitude of the wiggler mode O_3 varies as $\sin \varphi_0$ according to Equations O.A17 and O.A24 of the appendix, and hence its value is zero since $\varphi_0 = 0$. Figure O.4 shows modes 1 and 2. They are elliptical in the x–z plane.

Table O.2 parameters are the same as those of Table O.1; however, the switching angle is changed to $90°$. The O_3 mode has significant electric and magnetic wiggler fields. The strength of the wiggler fields is directly proportional to E_0 and hence they can be strong for practical values of E_0, although the results are given for a nominal value of $E_0 = 1$ V/m.

Figure O.5 shows the plot of the electric and magnetic fields' components including the wiggler fields for the parameters of Table O.2. Figure O.5a shows E_x at $z = d/2$. The dotted line represents E_x before switching as an x-polarized wave with $\varphi = 90°$. E_x after switching is represented by the bold line. Mode 2 has much higher amplitude than mode 1 and thus the lower

TABLE O.2

Electric and Magnetic Fields of the Three Modes for the Following Parameters: $f_p = 17.63$ GHz, $f_b = 20$ GHz, $n = 1$, and $\varphi_0 = 90°$

Wave	Quantity	E_x	E_z	H_y
I_0 (source)	A_0	1 V/m	0 V/m	2.7 mA/m
	f_0 (GHz)	10	10	10
	φ_0 (degrees)	90	—	0
O_1 (mode 1)	A_1	0.12 V/m	0.09 V/m	−0.0854 mA/m
	f_1 (GHz)	30.72	30.72	30.72
	φ_1 (degrees)	90	180	0
O_2 (mode 2)	A_2	0.72 V/m	0.62 V/m	−1.0389 mA/m
	f_2 (GHz)	13.33	13.33	13.33
	φ_2 (degrees)	90	0	0
O_3 (mode 3)	A_3	0	−0.37 V/m	−1.528 mA/m
	f_3 (GHz)	0	0	0
	φ_3 (degrees)	—	—	—

frequency mode is seen as the dominating mode in the figure. Figure O.5b shows E_z at $z = d/2$. This component is zero before switching since free space does not support a field component in the direction of propagation. After switching, the two upshifted modes, as well as a wiggler electric field, are generated. The amplitude of the wiggler electric field is seen as the vertical difference between the horizontal axes of the before-switching field and the horizontal axes of the after-switching field. Figure O.5c shows H_y at $z = 0$.

The amplitude of the wiggler magnetic field is also seen as the vertical difference between the two horizontal axes of the before-and-after-switching fields. The wiggler electric and magnetic fields constitute mode 3 of frequency $f_3 = 0$.

Table O.3 gives the case of a source wave of input frequency $f_0 = 10$ GHz and system parameters $f_p = f_b = 10$ GHz. We note that the standing wave of mode 2 is circularly polarized in the x–z plane since $A_2(E_x) = A_2(E_z) = 0.5$ V/m and the phase difference between E_x and E_z is 90°. Thus the system transforms the linearly polarized mode to a circularly polarized one. Figure O.6 shows the bandwidth of this structure for circularly polarized mode 2, by considering the effect of changing the values of f_p and f_b around f_0. Figure O.6a shows a change of 5% in f_2, when f_b or f_p is changed by about 10%. Figure O.6b shows the variation of $(1 - E_x/E_z)$ with $(1 - f_b/f_0)$ and $(1 - f_p/f_0)$. When the latter two are zero, $f_b = f_p = f_0$ and $(1 - E_x/E_z)$ is zero and the mode is circularly polarized. For a variation of 10% in f_p and/or f_b, we note that E_x/E_z is different from 1 by about 5%.

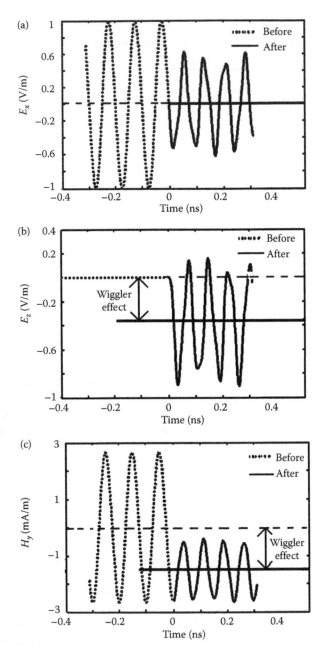

FIGURE O.5 (See color insert following page 202.) Total electric and magnetic fields before and after switching the medium: Parameters: $f_0 = 10$ GHz, $\varphi_0 = 90°$, $f_p = 17.63$ GHz, $f_b = 20$ GHz, $n = 1$, $E_0 = 1$ V/m, $z = d/2$ for (a) and (b), and $z = 0$ for (c). See Table O.2 for more information. (a) Total electric field of x-component; (b) Total electric field of z-component (including wiggler field); and (c) Total magnetic field of y-component (including wiggler magnetic).

TABLE O.3

Electric and Magnetic Fields of the Three Modes for the Following Parameters:
$f_p = 10$ GHz, $f_b = 10$ GHz, $n = 1$, Special Case: $f_0 = f_p = f_b = 10$ GHz

Wave	Quantity	E_x	E_z	H_y
I_0 (source)	A_0	1 V/m	0 V/m	2.7 mA/m
	f_0 (GHz)	10	10	10
	φ_0 (degrees)	90	—	0
O_1 (mode 1)	A_1	0.29 V/m	0.17 V/m	−0.4421 mA/m
	f_1 (GHz)	17.32	17.32	17.32
	φ_1 (degrees)	90	180	0
O_2 (mode 2)	A_2	0.50 V/m	0.50 V/m	−1.326 mA/m
	f_2 (GHz)	10	10	10
	φ_2 (degrees)	90	0	0
O_3 (mode 3)	A_3	0	−0.33 V/m	−0.8841 mA/m
	f_3 (GHz)	0	0	0
	φ_3 (degrees)	—	—	—

Table O.4 gives the results for the case when f_p is less than f_0. We note that while mode 1 is still an upshifted mode, mode 2 becomes downshifted. Thus, frequency downshifting is also possible.

Figure O.7 shows the effect of the electron gyrofrequency on the output frequencies and amplitudes of the fields of the various modes. In Figure O.7a,

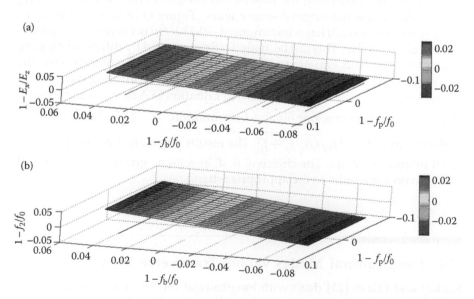

FIGURE O.6 (See color insert following page 202.) Effect of small changes in f_b and f_p on circular polarization of mode 2. The parameters are as specified in Table O.3.

TABLE O.4

Electric and Magnetic Fields of the Three Modes for the Following Parameters: $f_p = 10$ GHz, $f_b = 10$ GHz, $n = 1$, Special Case: $f_p < f_0$

Wave	Quantity	E_x	E_z	H_y
I_0 (source)	A_0	1 V/m	0 V/m	2.7 mA/m
	f_0 (GHz)	15	15	15
	φ_0 (degrees)	20	—	−70
O_1 (mode 1)	A_1	0.74 V/m	0.212 V/m	−1.522 mA/m
	f_1 (GHz)	19.50	19.50	19.50
	φ_1 (degrees)	−15.637	254.3623	74.3623
O_2 (mode 2)	A_2	0.24 V/m	0.363 V/m	−0.8025 mA/m
	f_2 (GHz)	12.02	12.02	12.02
	φ_2 (degrees)	−24.4207	65.5792	65.5792
O_3 (mode 3)	A_3	0	−0.09 V/m	−0.1649 mA/m
	f_3 (GHz)	0	0	0
	φ_3 (degrees)	—	—	—

it can be seen that the frequency of mode 1 tends to f_b when f_b increases, whereas mode 2 frequency decays to f_0. From Figure O.7b, it can be seen that while mode 1 amplitude decays to zero when f_b increases, mode 2 amplitude increases to the original source wave amplitude $E_0 = 1$ V/m. This shows that for certain input parameters where the electron gyrofrequency is much larger than the plasma frequency, the plasma has no effect and the only standing wave remaining is the original source wave. Figure O.7c shows that the E_z wiggler mode (mode 3) has a minimum of −0.43 V/m occurring at a gyrofrequency of 35.73 GHz. This can be analytically expressed by differentiating the wiggler electric field Equation with respect to ω_b and equating to zero. The minimum wiggler electric field is then obtained by substituting the value of ω_b in E_3 of the $E_z^+(z, t)$ equation shown in Equation O.A24. After some mathematical simplifications, one obtains $E_3(\text{minimum}) = -\left(\left(f_p/2\sqrt{f_0^2 + f_p^2} \right) E_0 \right)$ and it occurs at $f_b = (f_p/f_0)\sqrt{f_0^2 + f_p^2}$. The results given are directly applicable to all frequency bands. The choice of n, d, and ω_0 as given in Equation O.4 will convert the results to the appropriate band.

O.4 Longitudinal Modes versus Transverse Modes

Kalluri and Ehsan [21] dealt with longitudinal modes. The static magnetic field $\bar{B}_0$ was considered to be in the z-direction. The characteristic modes for this case were left and right circularly polarized modes, called L wave

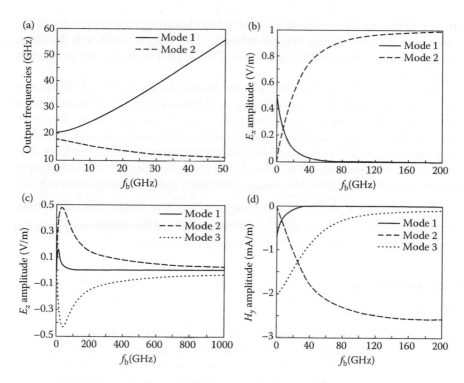

FIGURE O.7 Variation of the output frequencies and amplitudes of various modes with the external magnetic field strength. Parameters: $f_0 = 10$ GHz, $f_p = 17.63$ GHz, $\varphi_0 = 90°$, $E_0 = 1$ V/m, $n = 1$, $z = d/2$ for (b) and (c) and $z=0$ for (d). (a) Variation of output frequencies with f_b; (b) Variation of E_x amplitude with f_b; (c) Variation of E_z amplitude with f_b; and (d) Variation of H_y amplitude with f_b.

and R wave. Even if the source wave was elliptically polarized, the output waves were all circularly polarized modes. The parameters of the source wave (tan $\gamma = E_y/E_x$, δ the phase difference between the x- and y-component, and the switching angle ϕ_0) and the parameters of the magnetoplasma (f_p and f_b) were chosen to obtain one or more of the circularly polarized waves (L wave or R wave) with upshifted or downshifted frequencies of significant amplitudes. Thus it was shown that the setup acted as a frequency transformer with L or R waves as the outputs. However, this setup did not produce wiggler modes. Also, the output waves were TEM ($E_z = H_z = 0$). The paper also discussed at length the effect of finite rise time and nonuniform spatial plasma density profile.

In this paper, we studied the case where the source wave is x-polarized and the static magnetic field is in the y-direction. When the plasma is switched on suddenly, the output waves are all X waves. An X wave has an x-component as well as a z-component for the electric field. Thus we get elliptic polarization in the x–z plane. Moreover, it gives rise to a wiggler magnetic field in the y-direction and a wiggler electric field in the z-direction. These wiggler fields are due to the transient processes involved in changing vacuum modes to the transverse X modes in the magnetoplasma. The wiggler field will diminish in importance as the rise time increases. This aspect is dealt with in another article [23].

If the vacuum mode also has a y-component electric field and elliptically polarized, the output modes will have additionally a y-component for the electric field. The source wave y-component produces a so-called "O wave" since the static magnetic field in the y-direction will have no effect. The electrons traveling in the y-direction will not experience the magnetic force due to the static magnetic field. The case of the O wave is the same as the isotropic case discussed before [20]. The modes generated are a standing wave with E_y and H_x of frequency $\sqrt{\omega_0^2 + \omega_p^2}$ and a wiggler magnetic field in the x-direction. The superposition of these results with the results of this article will give the final results for a source wave with elliptic polarization.

O.5 Conclusion

A linearly polarized wave transforms into three different frequency modes for the case of switching on of a magnetoplasma in a cavity set up for transverse X modes. These components have unique variation in amplitude, frequency, and polarization. By proper choice of the source and plasma parameters, upshifted as well as downshifted standing waves can be obtained. Moreover, the polarization of the output waves can be controlled. The third mode is a wiggler mode (zero frequency) of electric and magnetic fields. The generation of the wiggler electric field is the unique aspect of the frequency and polarization transformer discussed in this paper.

The theory developed in this paper is based on a rather ideal model, where the plasma is created uniformly and instantaneously. In practice, plasma is created over a finite time rise T_r. The ideal case considered here has zero rise time; that is why the title of this article ends with "I zero rise time."

The case of finite rise time T_r brings into focus the transient phase in the mode conversion process. Moreover, the created plasma could have a nonuniform spatial plasma density. Thus the nonideal case involves the creation of a plasma whose temporal and spatial plasma density profiles are nonstep functions. The FDTD technique can be used to obtain a numerical solution for the case of arbitrary time and space profiles of the plasma density. The authors have completed work on this aspect and the results will be submitted in a companion paper [23].

The results show that the broad conclusion reached in this article is valid even when the plasma temporal profile is not a step profile. The final frequencies of various modes are unaffected by the rise time but the final amplitudes are dependent on the normalized value T_r/T, where T is the period of the source wave.

O.A Appendix

O.A.1 Fields' Equations

One can obtain the final expressions of the fields by simultaneously solving the Equation set O.19. The results are given in Equations O.A1 through O.A24.

$$E_x^+(z,t) = E_0 \sin\left(\frac{n\pi}{d}z\right)$$

$$\times \left(\sum_{i=1}^{2} \sqrt{A_i^2 + \frac{B_i^2}{\omega_i^2}} \cos(\omega_i t + \varphi_i)\right), \tag{O.A1}$$

where

$$\tan(\varphi_i) = -\frac{\omega_0}{\omega_i} \tan(\varphi_0), \tag{O.A2}$$

And ω_1 and ω_2 are given by Equations O.A3 through O.A6:

$$\omega_1 = \left(X + \sqrt{X^2 - Y}\right)^{1/2}, \tag{O.A3}$$

$$\omega_2 = \left(X - \sqrt{X^2 - Y}\right)^{1/2}, \tag{O.A4}$$

where

$$X = \left(\frac{\omega_0^2 + 2\omega_p^2 + \omega_b^2}{2}\right), \tag{O.A5}$$

$$Y = \omega_p^4 + \omega_0^2\left(\omega_p^2 + \omega_b^2\right). \tag{O.A6}$$

Also,

$$A_1 = \frac{\left(\omega_p^2 + \omega_b^2 - \omega_1^2\right)\cos(\varphi_0)}{\omega_2^2 - \omega_1^2}, \tag{O.A7}$$

$$B_1 = \frac{-\omega_0\left(\omega_p^2 + \omega_b^2 - \omega_1^2\right)\sin(\varphi_0)}{\omega_2^2 - \omega_1^2}, \tag{O.A8}$$

$$A_2 = \frac{\left(\omega_2^2 - \omega_p^2 - \omega_b^2\right)\cos(\varphi_0)}{\omega_2^2 - \omega_1^2}, \tag{O.A9}$$

$$B_2 = \frac{\omega_0\left(\omega_p^2 + \omega_b^2 - \omega_2^2\right)\sin(\varphi_0)}{\omega_2^2 - \omega_1^2}. \tag{O.A10}$$

The other fields are

$$H_y^+(z,t) = -H_0 \cos\left(\frac{n\pi}{d}z\right)$$

$$\times \left(\sum_{i=1}^{2} \sqrt{C_i^2 + \frac{D_i^2}{\omega_i^2}} \cos(\omega_i t + \phi_i) + C_3\right), \quad (O.A11)$$

where

$$\tan(\varphi_i) = -\frac{D_i}{\omega_i C_i}, \quad (O.A12)$$

$$C_1 = \frac{\left[\begin{array}{l}\omega_b^2\omega_1^2\sin(\varphi_0) + 2\omega_1^2\omega_p^2\sin(\varphi_0)\\ -\omega_p^4\sin(\varphi_0) - \omega_1^4\sin(\varphi_0)\end{array}\right]}{\omega_1^2\left(\omega_2^2 - \omega_1^2\right)}, \quad (O.A13)$$

$$D_1 = \frac{\left(\omega_p^2 + \omega_b^2 - \omega_1^2\right)\omega_0\cos(\varphi_0)}{\omega_2^2 - \omega_1^2}, \quad (O.A14)$$

$$C_2 = \frac{\left[\begin{array}{l}\omega_2^4\sin(\varphi_0) - \omega_b^2\omega_2^2\sin(\varphi_0)\\ -2\omega_2^2\omega_p^2\sin(\varphi_0) + \omega_p^4\sin(\varphi_0)\end{array}\right]}{\omega_2^2\left(\omega_2^2 - \omega_1^2\right)}, \quad (O.A15)$$

$$D_2 = \frac{\left(\omega_2^2 - \omega_p^2 - \omega_b^2\right)\omega_0\cos(\varphi_0)}{\omega_2^2 - \omega_1^2}, \quad (O.A16)$$

$$C_3 = \frac{\omega_p^4\sin(\varphi_0)}{\omega_1^2\omega_2^2}, \quad (O.A17)$$

$$E_z^+(z,t) = E_0 \sin\left(\frac{n\pi}{d}z\right)$$

$$\times \left(\sum_{i=1}^{2} \sqrt{E_i^2 + \frac{F_i^2}{\omega_i^2}} \cos(\omega_i t + \varphi_i) + E_3\right), \quad (O.A18)$$

where

$$\tan(\varphi_i) = -\frac{F_i}{\omega_i E_i}, \quad (O.A19)$$

$$E_1 = \frac{\omega_b\omega_p^2\omega_0\sin(\varphi_0)}{\omega_1^2\left(\omega_2^2 - \omega_1^2\right)}, \quad (O.A20)$$

$$F_1 = \frac{\omega_b \omega_p^2 \cos(\varphi_0)}{\omega_2^2 - \omega_1^2}, \tag{O.A21}$$

$$E_2 = -\frac{\omega_b \omega_p^2 \omega_0 \sin(\varphi_0)}{\omega_2^2 \left(\omega_2^2 - \omega_1^2\right)}, \tag{O.A22}$$

$$F_2 = -\frac{\omega_b \omega_p^2 \cos(\varphi_0)}{\omega_2^2 - \omega_1^2}, \tag{O.A23}$$

$$E_3 = -\frac{\omega_b \omega_p^2 \omega_0 \sin(\varphi_0)}{\omega_1^2 \omega_2^2}. \tag{O.A24}$$

References

1. Kalluri, D. K., *Electromagnetics of Complex Media*, CRC Press LLC, Boca Raton, FL, 1999.
2. Auld, B. A., Collins, J. H., and Zapp, H. R., Signal processing in a nonperiodically time-varying magnetoelastic medium, *Proc. IEEE*, 56, 258–272, 1968.
3. Jiang, C. L., Wave propagation and dipole radiation in a suddenly created plasma, *IEEE Trans. Ant. Prop.*, 23, 83–90, 1975.
4. Felsen, B. L. and Whitman, G. M., Wave propagation in time-varying media, *IEEE Trans. Ant. Prop.*, 18, 242–253, 1970.
5. Fante, R. L., Transmission of electromagnetic waves into time-varying media, *IEEE Trans. Ant. Prop.*, 19, 417–424, 1971.
6. Joshi, C. J., Clayton, C. E., Marsh, K., Hopkins, D. B., Sessler, A., and Whittum, D., Demonstration of the frequency upshifting of microwave radiation by rapid plasma creation, *IEEE Trans. Plasma Sci.*, 18, 814–818, 1990.
7. Kuo, S. P., Frequency up-conversion of microwave pulse in a rapidly growing plasma, *Phys. Rev. Lett.*, 65, 1000–1003, 1990.
8. Kuo, S. P. and Ren, A., Experimental study of wave propagation through a rapidly created plasma, *IEEE Trans. Plasma Sci.*, 21, 53–56, 1993.
9. Rader, M., Dyer, F., Matas, A., and Alexeff, I., Plasma-induced frequency shifts in microwave beams, In: *Conf. Rec. Abstracts, IEEE Int. Conf. Plasma Sci.*, Oakland, CA, p. 171, 1990.
10. Yablonovitch, E., Spectral broadening in the light transmitted through a rapidly growing plasma, *Phys. Rev. Lett.*, 31, 877–879, 1973.
11. Savage, Jr., R. L., Joshi, C. J., and Mori, W. B., Frequency up-conversion of electromagnetic radiation upon transmission into an ionization front, *Phys. Rev. Lett.*, 68, 946–949, 1992.
12. Lampe, M. and Walker, J. H., Interaction of electromagnetic waves with a moving ionization front, *Phys. Fluids*, 21, 42–54, 1978.
13. Banos, Jr., A., Mori, W. B., and Dawson, J. M., Computation of the electric and magnetic fields induced in a plasma created by ionization lasting a finite interval of time, *IEEE Trans. Plasma Sci.*, 21, 57–69, 1993.

14. Wilks, S. C., Dawson, J. M., and Mori, W. B., Frequency up-conversion of electromagnetic radiation with use of an overdense plasma, *Phys. Rev. Lett.*, 61, 337–340, 1988.
15. Mori, W. B., (Ed), Special issue on generation of coherent radiation using plasmas, *IEEE Trans. Plasma Sci.*, 21(2), 1993.
16. Kalluri, D. K., Effect of switching a magnetoplasma medium on a travelling wave: Longitudinal propagation, *IEEE Trans. Ant. Prop.*, 37, 1638–1642, 1989.
17. Kuo, S. P., Bivolaru, D., Orlick, L., Alexeff, I. and Kalluri, D. K., A transmission line filled with fast switched periodic plasma as a wideband frequency transformer, *IEEE Trans. Plasma Sci.*, 29, 365–370, 2001.
18. Booker, H. G., *Cold Plasma Waves*, Kluwer, Hingham, MA, 1984.
19. Granastein, V. L. and Alexeff, I., *High-power microwave sources*, Artech House, Boston, MA, 1987.
20. Ehsan, M. M. and Kalluri, D. K., Plasma induced wiggler magnetic field in a cavity, *Int. J. Infrared Millim. Waves*, 24, 1215–1234, 2003.
21. Kalluri, D. K. and Ehsan, M. M., Frequency and polarization transformer: Longitudinal modes, *Int. J. Infrared Millim. Waves*, 25, 327–353, 2004.
22. Trifkovic, Z. M. and Stanic, B. V., Nonlinear transformation of electromagnetic waves in suddenly created cold magnetized plasma: Longitudinal propagation, *J. Appl. Phys.*, 92(7), 3472–3479, 2002.
23. Khalifeh, A. F., Kalluri, D. K., and Eker, S., Frequency and polarization transformer: Transverse modes II FDTD for finite rise time, *IEEE Trans. Ant. Prop.*, 55, 2761–2765, 2007.

Appendix P

Frequency and Polarization Transformer: Transverse Modes—II Finite Rise Time*

Ahmad F. Khalifeh, Sebahattin Eker, and Dikshitulu K. Kalluri

P.1 Introduction

A standing source wave in a cavity transforms in a remarkable way when a medium surrounding the source wave changes in *time*. New modes are generated each with a distinct frequency, polarization, and amplitude. Plasma is a suitable time-varying electromagnetic medium whose relative permittivity can be altered by changing the ionization level of the medium. For example, a neutral gas can be converted into a plasma by a laser pulse or a strong electromagnetic pulse. A fast pulse can create the plasma with a small rise time T_r.

If the medium switching takes place in the presence of a static magnetic field, the switched on medium is a magnetoplasma medium, and frequency upshifting as well as downshifting is possible. It is even possible to create a standing wave of zero frequency, called a wiggler field. Such fields are used in FELs. Thus a switched magnetoplasma in a cavity acts as a generic frequency and polarization transformer. Such a transformer is important because radiation of the desired frequency and polarization is not always easily generated by conventional means. In general, a frequency and polarization transformer can convert off-the-shelf sources to the desired frequency and polarization source, for a given application. A brief account of the research work carried out by the authors and other investigators on plasma-induced frequency shifts is given next to put the present paper in context.

A source wave splits into three modes when an *unbounded* isotropic plasma medium is created [1–17]. One is a forward-propagating mode with upshifted frequency and another is a backward-propagating mode of the same frequency. The third mode is a wiggler magnetic field, which is a space-varying dc (zero frequency) magnetic field.

In the presence of an external static magnetic field, plasma is an anisotropic medium. When the static magnetic field is perpendicular to the direction of

---

* Reprinted from *IEEE Trans. Ant. Prop.*, 55(10), 2761–2766, October 2007. With permission.

wave propagation, the characteristic modes are O and X and are referred to as transverse modes [18]. When the electric field of the source wave is in the same direction as that of the static magnetic field, it has no effect and the mode characteristics are the same as that of an isotropic plasma and hence these modes are referred to as O. On the other hand, when the electric field of the source wave is perpendicular to the static magnetic field as well as the direction of wave propagation, the transverse modes are called X and have a component of the electric field in the direction of wave propagation. A linearly polarized source wave splits into four frequency-shifted modes when an *unbounded* magnetoplasma that supports the X modes is created [1]. Two of them are forward-propagating modes of different frequencies. The other two are the corresponding backward-propagating modes. Moreover, wiggler (zero frequency but space varying) electric as well as magnetic fields are also created and they degenerate into a wiggler magnetic field only when the static magnetic field is reduced to zero and the anisotropic magnetoplasma becomes isotropic [1]. The wiggler fields have applications in an FEL [19].

Sudden creation of an unbounded plasma is an idealization to simplify the problem but is not easily obtainable even in an approximate sense. Plasma in a cavity with a small distance between the plates is easier to create than the creation of an unbounded plasma medium assumed in Ref. [1]. In the latter case, one has to create a large volume of plasma with very low loss. If the plasma is bounded in the form of a slab in free space, the wave may escape the medium before the interaction of the wave with the time-varying medium is completed and the frequency change process will be incomplete. This will be the case if the width of the slab is the same as the cavity dimension. For this reason we have taken up the study of the switched medium in a cavity. The switched medium is a magnetoplasma that supports X modes.

In a preceding paper [20], the authors discussed the concept of a frequency and polarization transformer. The transformer was a one-dimensional cavity (see Figure 1 of Ref. [1]) in which a magnetoplasma that supported transverse modes was created. Theoretical derivation for the case of sudden and uniform creation of the magnetoplasma was given. In the conclusion section of [20], it was pointed out that realistic models where the plasma density profile was not a step profile need to be considered. Figure P.1 shows such a profile with a finite rise time T_r. The step profile is a particular case of $T_r = 0$.

A powerful numerical technique based on the FDTD method is used to study the effects of a finite rise time. The method is general enough to study the effect of a spatially nonuniform profile of plasma density as well and also to take into account other realistic situations rather than the idealized situation used in Ref. [20] to obtain an analytical solution. However, in this article, we consider only profiles with a finite but small value for the rise time T_r, to show that the broad conclusions reached in Ref. [20] are still valid; however, the amplitudes of the various modes will be affected.

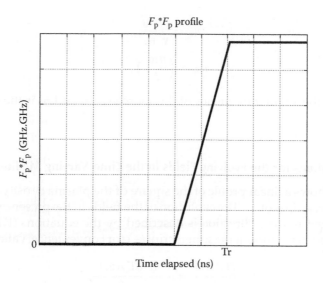

FIGURE P.1 Electron density profile with finite rise time T_r. The vertical axis is the square of the plasma frequency f_p in GHz. See Equation P.15 for the relation between the electron density and the plasma frequency in rad/s.

P.2 Formulation of the FDTD Solution

P.2.1 Fields before Switching

An x-polarized standing wave is assumed to be present in a one-dimensional cavity with free space as the medium. For a detailed description of the problem, see Ref. [20]. Let $t = 0$ mark the instant of time after which the medium in the cavity changes to a magnetoplasma.

The superscript $(-)$ indicates the steady-state fields for $t < 0$:

$$E_x^-(z,t) = E_0 \sin\left(\frac{n\pi}{d}z\right)\cos(\omega_0 t + \varphi_0), \qquad (\text{P.1})$$

$$E_z^-(z,t) = 0, \qquad (\text{P.2})$$

$$J_x^-(z,t) = 0, \qquad (\text{P.3})$$

$$J_z^-(z,t) = 0, \qquad (\text{P.4})$$

$$H_y^-(z,t) = -H_0 \cos\left(\frac{n\pi}{d}z\right)\sin(\omega_0 t + \varphi_0). \qquad (\text{P.5})$$

where

$$H_0 = \frac{E_0}{\eta_0}, \qquad (\text{P.6})$$

$$\eta_0 = \sqrt{\frac{\mu_0}{\varepsilon_0}}, \tag{P.7}$$

$$\omega_0 = \left(\frac{n\pi}{d}c\right), \tag{P.8}$$

where ω_0 is the angular frequency of the source wave and ϕ_0 is the switching angle.

P.2.2 Equations for the Evolving Fields in the Time-Varying Medium

Figure P.1 shows a linear profile of the square of the plasma density created at $t = 0$ with a rise time of T_r. The evolution of the fields in the presence of a static magnetic field in the y-direction is described by the equations (P.12) of Ref. [20], repeated here without the superscript (+) for convenience, valid for $t > 0$:

$$\frac{\partial H_y(z,t)}{\partial t} = -\frac{1}{\mu_0}\frac{\partial E_x(z,t)}{\partial z}, \tag{P.9}$$

$$\frac{\partial E_x(z,t)}{\partial t} = -\frac{1}{\varepsilon_0}\frac{\partial H_y(z,t)}{\partial z} - \frac{1}{\varepsilon_0}J_x(z,t), \tag{P.10}$$

$$\frac{\partial E_z(z,t)}{\partial t} = -\frac{1}{\varepsilon_0}J_z(z,t), \tag{P.11}$$

$$\frac{\partial J_x(z,t)}{\partial t} + \nu J_x(z,t) = \varepsilon_0\omega_p^2(z,t)E_x(z,t) + \omega_b(z,t)J_z(z,t), \tag{P.12}$$

$$\frac{\partial J_z(z,t)}{\partial t} + \nu J_z(z,t) = \varepsilon_0\omega_p^2(z,t)E_z(z,t) - J_x(z,t)\omega_b(z,t), \tag{P.13}$$

$$E_y(z,t) = H_x(z,t) = H_z(z,t) = J_y(z,t) = 0. \tag{P.14}$$

In the above equation,

$$\omega_p^2(z,t) = \frac{N_0(z,t)q^2}{m\varepsilon_0}, \tag{P.15}$$

$$\omega_b(z,t) = \frac{qB_0(z,t)}{m}, \tag{P.16}$$

where ω_p is the electron plasma frequency, N_0 is the electron density, ω_b is the electron gyrofrequency, B_0 is the imposed quasistatic magnetic flux density, ν is the collision frequency, q is the absolute value of the charge of an electron, and m is the mass of the electron.

The FDTD technique will eliminate the need for solving these equations simultaneously. It is a step-by-step solution of the fields, where the first field values are computed from initial source conditions, and then the rest are computed in a leap-frog fashion where future field values are calculated from the present and the past values.

P.2.3 FDTD Computation

The cavity will be represented by a two-dimensional FDTD grid, where E, H, and J field components are sampled in space and time. The grid is shown in Figure P.2.

Discretizing Equations P.9 through P.11 using central difference approximations, one obtains

$$(H_y)_{k+(1/2)}^{n+(1/2)} = (H_y)_{k+(1/2)}^{n-(1/2)} - \frac{\Delta t}{\mu_0 \Delta z}\left[(E_x)_{k+1}^n - (E_x)_k^n\right], \tag{P.17}$$

$$(E_x)_k^{n+1} = (E_x)_k^n - \frac{\Delta t}{\varepsilon_0 \Delta z}\left((H_y)_{k+(1/2)}^{n+(1/2)} - (H_y)_{k-(1/2)}^{n+(1/2)}\right) - \frac{\Delta t}{\varepsilon_0}(J_x)_k^{n+(1/2)}, \tag{P.18}$$

$$(E_z)_k^{n+1} = (E_z)_k^n - \frac{\Delta t}{\varepsilon_0}(J_z)_k^{n+(1/2)}. \tag{P.19}$$

Equations P.12 and P.13 for the current densities in a lossy plasma medium are written in matrix format as

$$\begin{bmatrix} \frac{dJ_x}{dt} \\ \frac{dJ_z}{dt} \end{bmatrix} = \begin{bmatrix} -\nu & \omega_b \\ -\omega_b & -\nu \end{bmatrix}\begin{bmatrix} J_x \\ J_z \end{bmatrix} + \varepsilon_0 \omega_p^2 \begin{bmatrix} E_x \\ E_z \end{bmatrix}, \tag{P.20}$$

where ν represents the collision frequency. One can use central difference approximations to discretize Equation P.20 but the algorithm becomes less accurate for large values of ν and/or ω_b [21].

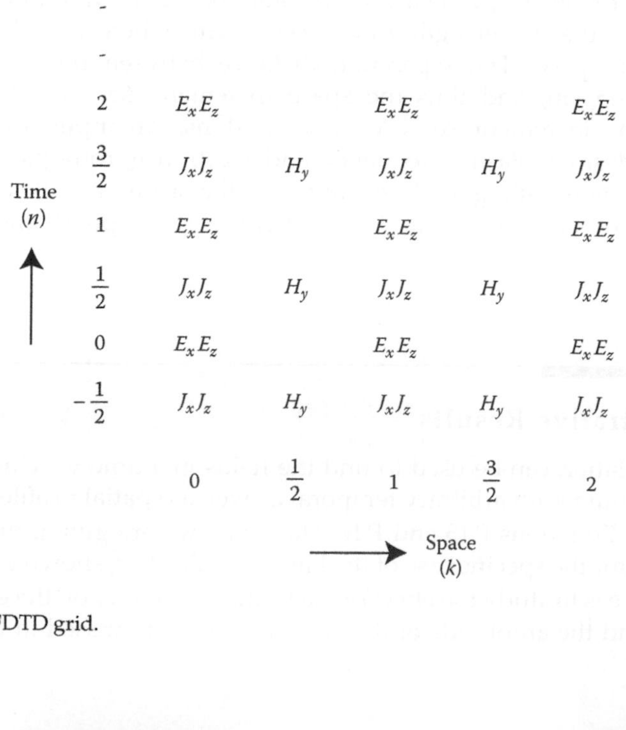

Time (n)						
2	$E_x E_z$		$E_x E_z$		$E_x E_z$	
$\frac{3}{2}$	$J_x J_z$	H_y	$J_x J_z$	H_y	$J_x J_z$	
1	$E_x E_z$		$E_x E_z$		$E_x E_z$	
$\frac{1}{2}$	$J_x J_z$	H_y	$J_x J_z$	H_y	$J_x J_z$	
0	$E_x E_z$		$E_x E_z$		$E_x E_z$	
$-\frac{1}{2}$	$J_x J_z$	H_y	$J_x J_z$	H_y	$J_x J_z$	

| 0 | $\frac{1}{2}$ | 1 | $\frac{3}{2}$ | 2 |

Space (k)

FIGURE P.2 FDTD grid.

Equation P.20 can be solved over one time step, assuming v, ω_b, and ω_p constant over the time step. The technique is similar to the one used in Ref. [21] and the results are

$$
\begin{bmatrix} (J_x)_k^{n+(1/2)} \\ (J_z)_k^{n+(1/2)} \end{bmatrix} = e^{-(v\Delta t)} \begin{bmatrix} \cos[(\omega_b)_k^n \Delta t] & \sin[(\omega_b)_k^n \Delta t] \\ -\sin[(\omega_b)_k^n \Delta t] & \cos[(\omega_b)_k^n \Delta t] \end{bmatrix} \begin{bmatrix} (J_x)_k^{n-(1/2)} \\ (J_z)_k^{n-(1/2)} \end{bmatrix}
$$
$$
+ \frac{\varepsilon_0 (\omega_p^2)_k^n e^{-(v\Delta t)}}{v^2 + (\omega_b^2)_k^n} \begin{bmatrix} k_{11} & k_{12} \\ k_{21} & k_{22} \end{bmatrix} \begin{bmatrix} (E_x)_k^n \\ (E_z)_k^n \end{bmatrix}, \qquad \text{(P.21)}
$$

where

$$
\left. \begin{cases} k_{11} = v \left(e^{v\Delta t} - \cos[(\omega_b)_k^n \Delta t] \right) + (\omega_b)_k^n \sin[(\omega_b)_k^n \Delta t], \\ k_{12} = e^{v\Delta t}(\omega_b)_k^n - (\omega_b)_k^n \cos[(\omega_b)_k^n \Delta t] - v \sin[(\omega_b)_k^n \Delta t], \\ k_{21} = -k_{12}, \\ k_{22} = k_{11}. \end{cases} \right\} \qquad \text{(P.22)}
$$

Equations P.17 through P.19 and P.21 in that order constitute a step-by-step leap-frog [22] FDTD algorithm.

The FDTD method computes future field values from the past and present values; thus errors accumulate as we move further in the computation process. The errors are minimized by a proper choice of the space and time steps [22]. For ensuring stability of the algorithm and to minimize discretization and numerical dispersion errors, Δz and $c\Delta t/\Delta z$ are chosen to be, respectively, $\lambda_0/50$ and 0.5. For an input frequency $f_0 = 10$ GHz, the wavelength $\lambda_0 = c/f_0 = 3$ cm, where c is the speed of light in free space. The separation distance between the two plates is $d = \lambda_0/2 = 1.5$ cm, and thus the space increment $\Delta z = \lambda_0/50 = 0.06$ cm and the time increment $\Delta t = \Delta z/2c = 0.001$ ns. An input array is used to get the desired plasma frequency and electron gyrofrequency profiles by appropriately filling each element of the array by its corresponding values of ω_p and ω_b at the particular space point and the time instant.

P.3   Illustrative Results

FDTD simulation can be used to find the fields in a time-varying magneto-plasma medium with arbitrary temporal as well as spatial profiles of ω_p^2 and ω_b given in Equations P.15 and P.16. However, we are giving the results in this section for the specific case of the linear profile of ω_p^2 shown in Figure P.1. The objective is to study the effect of the finite rise time T_r on the evolution of the fields and the amplitude of the various modes discussed in Ref. [20] for zero rise time. As an illustration of the effect of rise time, we have considered the following specific case: $f_0 = 10$ GHz, $f_b = 20$ GHz, $v = 0$, $\phi_0 = 90°$, and standing wave harmonic $n = 1$. The results for $T_r = 0$ are given in Table II and Figure 4 of Ref. [20]. The important aspects of these results are: the fields can be considered as composed of three modes: modes 1 and 2 are X standing waves of upshifted frequencies of 30.72 and 13.33 GHz. The amplitude of the second mode is approximately 6 times that of mode 1 and hence the standing wave pattern of the total field is dominated by mode 2. Mode 3 has a y-component magnetic wiggler field and a z-component electric wiggler field.

Figures P.3 and P.4 show the evolution of the electric and magnetic fields. The electric fields are shown at the spatial point $z = d/2$, where they are maximum. The magnetic field H_y is shown at the grid point closest to the plate at $z = 0$, where it is maximum. The final plasma frequency f_p at $t = T_r$ is 17.63 GHz for both the figures, but the rise time T_r is 0.1 ns for Figure P.3, whereas T_r is 1.0 ns for Figure P.4. The final frequencies of various modes

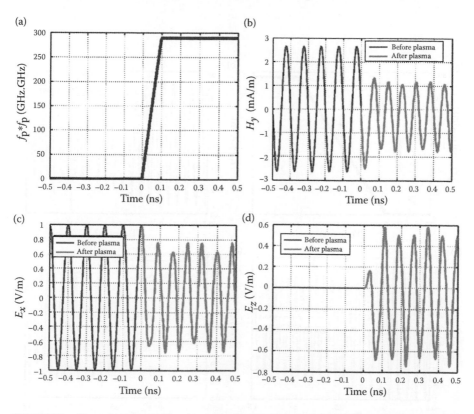

FIGURE P.3 **(See color insert following page 202.)** Evolution of the electric and magnetic fields for the case of $T_r = 0.1$ ns. (a) Electron density profile, (b) H_y, (c) E_x, and (d) E_z. The parameters are $f_0 = 10$ GHz, $f_p = 17.63$ GHz, $f_b = 20$ GHz, $\phi_0 = 90°$, $v = 0$, $n = 1$, $z = d/2$ for E, and $z \approx 0$ for H. See Equation P.16 for the relation between the imposed magnetic flux density B_0 and the electron gyrofrequency ω_b in rad/s.

are unaffected by the rise time T_r. However, the amplitudes of the various modes will be affected by the rise time. In particular, the magnitudes of the zero frequency wiggler fields (of H_y as well as E_z) become more and more weak as the rise time T_r increases. From these figures it is clear that whereas for $T_r = 0.1$ ns the wiggler components are still significant, for $T_r = 1$ ns the wiggler components are much less significant. A quantitative illustration of this point is made in Figure P.5 by an FFT plot of the amplitude spectrum of the three modes of E_z for, respectively, $T_r = 0.1$ ns (Figure P.5a) and $T_r = 1$ ns (Figure P.5b).

Figure P.6 is obtained from FFT plots as T_r is varied from 0 to 1 ns in steps of 0.05. It shows the effect of finite rise time T_r on the amplitude of the fields of each of the modes. Figure P.6a shows the amplitude of the H_y of each of the modes versus T_r, Figure P.6b that of the E_z component, and Figure P.6c that of the E_x component. For the data under consideration, the mode 2 component remains significant as expected.

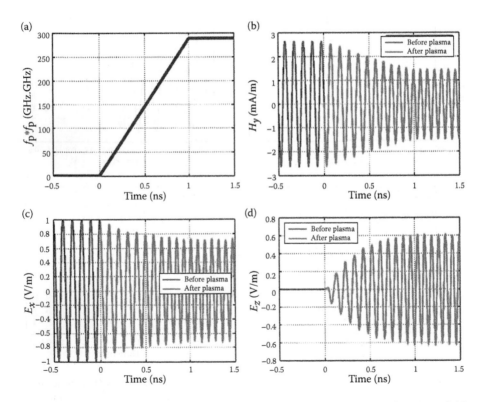

FIGURE P.4 (See color insert following page 202.) Evolution of the electric and magnetic fields for the case of $T_r = 1$ ns. (a) Electron density profile, (b) H_y, (c) E_x, and (d) E_z. The other parameters are the same as in Figure P.3.

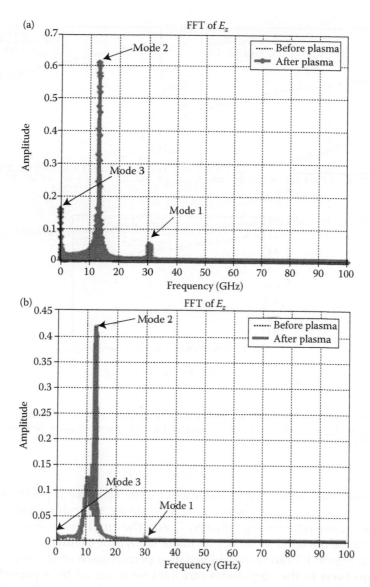

FIGURE P.5 FFT of E_z for (a) $T_r = 0.1$ ns and (b) $T_r = 1$ ns. The other parameters are the same as in Figure P.3.

P.4 Conclusion

Although the amplitudes are affected by the finite rise time, the general conclusion of [20] is valid for finite T_r, namely that the switching would result in the transformation of the original source wave into three new waves, each

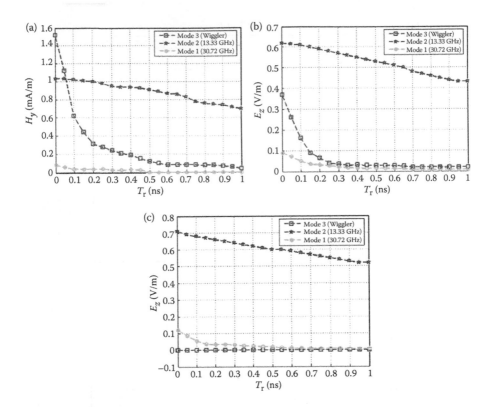

FIGURE P.6   **(See color insert following page 202.)** Amplitudes of the three modes versus rise time. (a) H_y, (b) E_z, and (c) E_x. Mode 3 is the wiggler. The other parameters are the same as those in Figure P.3.

having a unique frequency, amplitude and phase. The plasma parameters ω_p and ω_b and the source parameters ω_0 and ϕ_0 are the controlling parameters.

A quantitative relationship between T_r and the amplitudes of the various modes can be obtained by an approximate analytical technique. The profile shown in Figure P.1 can be considered as a perturbation of the step profile. The corrections to the amplitudes can then be obtained by a perturbation technique. The results of [20] serve as a zero order solution. By deriving and making use of a Green's function for the problem, the first-order correction term can be obtained.

Such a technique was used earlier for the isotropic case [1] and for the longitudinal modes [1]; however, in both the cases the medium was assumed to be unbounded. More recently, the method was applied to the transverse modes in a cavity and the technique was verified using FDTD [23].

We are in the process of obtaining an approximate analytical expression for the first-order correction terms for the amplitudes of the various modes as a function of the rise time T_r and the results will be published soon.

References

1. Kalluri, D. K., *Electromagnetics of Complex Media*, CRC Press LLC, Boca Raton, FL, 1999.
2. Auld, B. A., Collins, J. H., and Zapp, H. R., Signal processing in a nonperiodically time-varying magnetoelastic medium, *Proc. IEEE*, 56, 258–272, 1968.
3. Jiang, C. L., Wave propagation and dipole radiation in a suddenly created plasma, *IEEE Trans. Ant. Prop.*, 23, 83–90, 1975.
4. Felsen, B. L. and Whitman, G. M., Wave propagation in time-varying media, *IEEE Trans. Ant. Prop.*, 18, 242–253, 1970.
5. Fante, R. L., Transmission of electromagnetic waves into time-varying media, *IEEE Trans. Ant. Prop.*, 19, 417–424, 1971.
6. Joshi, C. J., Clayton, C. E., Marsh, K., Hopkins, D. B., Sessler, A., and Whittum, D., Demonstration of the frequency upshifting of microwave radiation by rapid plasma creation, *IEEE Trans. Plasma Sci.*, 18, 814–818, 1990.
7. Kuo, S. P., Frequency up-conversion of microwave pulse in a rapidly growing plasma, *Phys. Rev. Lett.*, 65, 1000–1003, 1990.
8. Kuo, S. P. and Ren, A., Experimental study of wave propagation through a rapidly created plasma, *IEEE Trans. Plasma Sci.*, 21, 53–56, 1993.
9. Rader, M., Dyer, F., Matas, A., and Alexeff, I., Plasma-induced frequency shifts in microwave beams, In: *Conf. Rec. Abstracts, IEEE Int Conf Plasma Sci*, Oakland, CA, p. 171, 1990.
10. E. Yablonovitch, E., Spectral broadening in the light transmitted through a rapidly growing plasma, *Phys. Rev. Lett.*, 31, 877–879, 1973.
11. Savage, Jr., R. L., Joshi, C. J., and Mori, W. B., Frequency up-conversion of electromagnetic radiation upon transmission into an ionization front, *Phys. Rev. Lett.*, 68, 946–949, 1992.
12. Lampe, M. and Walker, J. H., Interaction of electromagnetic waves with a moving ionization front, *Phys. Fluids*, 21, 42–54, 1978.
13. Banos, Jr., A., Mori, W. B., and Dawson, J. M., Computation of the electric and magnetic fields induced in a plasma created by ionization lasting a finite interval of time, *IEEE Trans. Plasma Sci.*, 21, 57–69, 1993.
14. Wilks, S. C., Dawson, J. M., and Mori, W. B., Frequency up-conversion of electromagnetic radiation with use of an overdense plasma, *Phys. Rev. Lett.*, 61, 337–340, 1988.
15. Mori, W. B., (Ed), Special issue on generation of coherent radiation using plasmas, *IEEE Trans. Plasma Sci.*, 21(2), 1993.
16. Kalluri, D. K., Effect of switching a magnetoplasma medium on a travelling wave: Longitudinal propagation, *IEEE Trans. Ant. Prop.*, 37, 1638–1642, 1989.
17. Kuo, S. P., Bivolaru, D., Orlick, L., Alexeff, I., and Kalluri, D. K., A transmission line filled with fast switched periodic plasma as a wideband frequency transformer, *IEEE Trans. Plasma Sci.*, 29, 365–370, 2001.
18. Booker, H. G., *Cold Plasma Waves*, Kluwer, Hingham, MA, 1984.
19. Granastein, V. L. and Alexeff, I., *High-Power Microwave Sources*, Artech House, Boston, MA, 1987.
20. Kalluri, D. K., Khalifeh, A. F., and Eker, S., Frequency and polarization transformer: Transverse modes: I zero rise time, *IEEE Trans. Ant. Prop.*, 6, 1789–1796, 2007.

21. Kalluri, D. K. and Ehsan, M. M., Frequency and polarization transformer: Longitudinal modes, *Int. J. Infrared Millim. Waves*, 25, 327–353, 2004.
22. Taflove, A., *Computational Electrodynamics: The Finite-Difference Time-Domain Method*, Artech House Inc., Norwood, MA, 1995.
23. Khalifeh, A. F., Perturbation techniques as applied to time-varying and space-varying electromagnetic systems, Doctoral dissertation, Department of Electrical Engineering, University of Massachusetts, Lowell, MA, 2006.

Appendix Q

Frequency Transformation of a Whistler Wave by a Collapsing Plasma Medium in a Cavity: FDTD Solution*

Dikshitulu K. Kalluri, Sebahattin Eker, and Monzurul Ehsan

Q.1 Introduction

A time-varying medium acts as a frequency transformer [1–19]. It splits a source wave into new waves whose frequencies are different from the frequency ω_0 of the source wave. The transformation can be represented by a black box (Figure Q.1) with the input terminals excited by an EMW of frequency ω_0 (rad/s) and amplitude A_0 and the nth frequency output terminal pair having an EMW of frequency ω_n and amplitude A_n. Equation for ω_n is easily obtained by noting that in a time-varying but space-invariant medium the wave number k is conserved [1–5], leading to the equation

$$\omega_n n(t) = \omega_0 n_0, \qquad (Q.1)$$

where $n(t)$ is the refractive index of the time-varying medium in the black box and n_0 is the refractive index of the input medium. Computation of A_n is more involved and discussed in Ref. [1].

A plasma medium is a suitable medium for affecting temporal changes in the dielectric properties of the medium. Many aspects of the frequency shifting by a transient magnetoplasma have been discussed by Kalluri [1].

Plasma in the presence of a static magnetic field B_0 is an anisotropic medium, whose dielectric properties are affected by ω_0, the plasma frequency ω_p,

$$\omega_p = \left(\frac{N_0 q^2}{m_e \varepsilon_0} \right)^{1/2}, \qquad (Q.2)$$

---

* Reprinted from *IEEE Trans. Ant. Prop.*, 57(7), 1921–1930, July 2009. With permission.

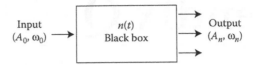

FIGURE Q.1   Black box description of frequency transformation.

and the electron gyrofrequency ω_b,

$$\omega_b = \left(\frac{qB_0}{m_e}\right). \tag{Q.3}$$

In the above equation, N_0 is the electron density (plasma density), q and m_e are the absolute values of the charge and mass of the electron, respectively, and ε_0 is the permittivity of free space.

Plane waves with left (L wave) or right (R wave) circular polarization are the normal modes in the case of wave propagation along the static magnetic field (longitudinal propagation). The relative permittivity ε_p for R wave propagation is given by the following [1]:

$$\varepsilon_p = 1 - \frac{\omega_p^2}{\omega_0(\omega_0 - \omega_b)}. \tag{Q.4}$$

When $\omega_0 \ll \omega_b$ and $\omega_p^2 \gg \omega_0\omega_b$, ε_p can be quite large and is approximately given by

$$\varepsilon_p \approx \frac{\omega_p^2}{\omega_0\omega_b}, \quad \omega_0 \ll \omega_b, \quad \omega_p^2 \gg \omega_0\omega_b. \tag{Q.5}$$

Such an R wave is called a whistler wave, since the whistlers in radio reception were explained [1,20] in terms of the propagation of electromagnetic signals in lightning in the earth's magnetosphere. The whistler mode is called the helicon mode in the literature on solid-state plasmas [20]. At the frequency of the whistler wave, the L mode is evanescent.

A big change in ε_p can be obtained by collapsing the electron density (by switching off the source of ionization), thereby converting the magnetoplasma medium into free space ($\varepsilon_p = 1$).

One of the authors investigated earlier [21] the conversion of a whistler wave by a collapsing magnetoplasma medium. It was shown that the collapse of the ionization converted the whistler wave to a much higher frequency wave with power intensification [21]. An ideal model of an unbounded medium was used to simplify the problem and establish the physical basis of the processes. Two extreme cases of collapse, (i) instantaneous collapse and (ii) very slow decay, were considered. In all these cases of large value for ε_p given by Equation Q.4, it was shown that the two additional modes of different frequencies created by switching off the magnetoplasma were of insignificant amplitude. In the rest of the paper, we will concentrate only on the significant mode.

In this paper, we consider the black box to be a bounded medium, in a cavity, with arbitrary space and time profiles for the electron density. Since the medium is bounded in a cavity, it is important to consider the state of the standing source wave at the beginning of the process of the change of the medium. We introduce the parameter, called switching angle ϕ_0, to consider this aspect of the problem.

Plasma in a cavity with a small distance between the plates is easier to create than the creation of an unbounded plasma medium assumed in Ref. [21]. In the latter case, one has to create a large volume of plasma with very low loss. If the plasma is bounded in the form of a slab in free space, the wave may escape through the boundary of the time-varying medium before the interaction of the wave with the medium is completed and the frequency change process may be incomplete [6]. The cavity walls prevent the leakage through the boundary, thus improving the conversion efficiency of the device.

We have modified in our laboratory the well-known FDTD technique [22–29] of simulating the wave interactions in a *time-varying* magnetized plasma medium [30–32]. We use this technique to study the conversion of a whistler wave by switching off the plasma medium in a cavity.

We are motivated to study this problem since it appears to lead to a frequency transformer, with frequency transformation ratios of orders of magnitude. The principle can be used to convert a readily available microwave source (say 2.45 GHz) to a terahertz radiation (of say 300 GHz).

The paper is organized as follows: Section Q.2 is used to formulate the problem using an ideal one-dimensional model. In Section Q.3, the FDTD technique used by us is explained. The significant modifications, as compared to the standard FDTD [22,29] when the magnetoplasma is itself *time-varying*, however, are discussed in Section Q.6.

In Section Q.4, the results for a one-dimensional model, with some notional parameter values for the source wave and the magnetoplasma, to illustrate various aspects of the transformation, are given. Actual values can be used when it becomes clear which of the competing and emerging scenarios for the black box can be done and are cost effective (see Section Q.6). In Section Q.5, a possible set of parameters to transform a (10 Grad/s) microwave source radiation to a (500 Grad/s) millimeter wave radiation is calculated and illustrated through the FDTD solution. In Section Q.6, the difficulties in obtaining experimentally the ideal model and the consequences of the departure from the ideal model are discussed. Black box descriptions involving new types of plasma sources are also discussed. The last section presents concluding remarks.

Q.2   Formulation of the Problem

A standing R wave is assumed to be present in a one-dimensional cavity containing a magnetoplasma medium (Figure Q.2).

Let $t = 0$ mark the instant of time after which the medium in the cavity changes. The superscript ($-$) indicates the steady-state fields for $t < 0$. Let the x-component of the electric field be denoted by

$$E_x^- = E_0 \sin\left(\frac{m\pi z}{d}\right) \cos(\omega_0 t + \phi_0). \tag{Q.6}$$

The y-component then is given by

$$E_y^- = E_0 \sin\left(\frac{m\pi z}{d}\right) \sin(\omega_0 t + \phi_0). \tag{Q.7}$$

Here ϕ_0 is the switching angle, which is the phase of E_x^- at $t = 0$. The number m is the index of the standing wave harmonic present in the cavity. (In Figure Q.2, $m = 3$.) The plate separation d and the source frequency ω_0 have to be chosen to satisfy the Maxwell's equations:

$$\omega_0 = \frac{m}{n_R} \frac{\pi c}{d}, \tag{Q.8}$$

where the refractive index n_R of the magnetoplasma medium is given by

$$n_R = \sqrt{\varepsilon_p} = \left[1 - \frac{\omega_p^2}{\omega_0(\omega_0 - \omega_b)}\right]^{1/2}. \tag{Q.9}$$

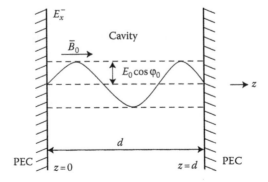

FIGURE Q.2   Sketch of E_x^- at $t = 0$ for $m = 3$ in a one-dimensional cavity.

The magnetic fields in the cavity can be obtained from Maxwell's equations:

$$H_x^-(z, t) = -H_0 \, \cos \frac{m\pi z}{d} \, \sin \left(\omega_0 t + \phi_0 - \frac{\pi}{2} \right),$$ (Q.10)

$$H_y^-(z, t) = H_0 \, \cos \frac{m\pi z}{d} \, \sin(\omega_0 t + \phi_0),$$ (Q.11)

where

$$H_0 = -\frac{E_0}{\eta_p}.$$ (Q.12)

In the above equation, η_p is the impedance given by

$$\eta_p = \sqrt{\frac{\mu_0}{\varepsilon_0 \varepsilon_p}} = \frac{\eta_0}{n_R},$$ (Q.13)

where η_0 is the impedance of free space. The currents in the plasma are obtained from the constitutive relation given by the following [1]:

$$\frac{d\bar{J}}{dt} + \nu \bar{J} = \varepsilon_0 \omega_p^2 \bar{E} + \bar{\omega}_b \times \bar{J}.$$ (Q.14)

For the steady-state case, Equation Q.14 gives

$$J_x^- = \frac{\varepsilon_0 \omega_p^2}{\omega_0 - \omega_b} E_0 \, \sin \left(\frac{m\pi z}{d} \right) \sin(\omega_0 t + \phi_0),$$ (Q.15)

$$J_y^- = \frac{-\varepsilon_0 \omega_p^2}{\omega_0 - \omega_b} E_0 \, \sin \left(\frac{m\pi z}{d} \right) \cos(\omega_0 t + \phi_0).$$ (Q.16)

In obtaining Equations Q.15 and Q.16, the collisions are neglected ($\nu = 0$). For small values of ν, Equations Q.15 and Q.16 can be easily modified by replacing E_0 by $E_0 e^{-\alpha_1 t}$, where α_1 is the attenuation constant, which will be small if ν is small.

Q.3 FDTD Solution

From the steady-state fields given in the previous section valid for $t < 0$, we will develop the FDTD technique for obtaining the fields for $t > 0$. We assume that the profiles of $\omega_p^2(z, t)$ and $\omega_b(z, t)$ are prescribed.

A detailed description of the FDTD technique for the case of switching on a magnetoplasma medium is given earlier [32]. Its adaptation for switching off

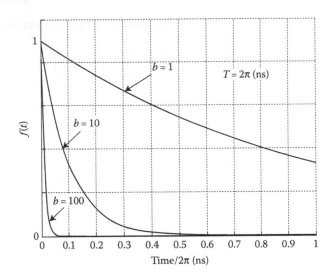

FIGURE Q.3   Plasma density profile function $f(t)$.

the medium is briefly discussed in the appendix. The time and space step sizes and Δt and Δz are chosen to ensure stability and adequate accuracy of the algorithm [22,24]. The courant number $r = c(\Delta t / \Delta z)$ is chosen to be 0.5 with the space increment $\Delta z = 0.012$ cm and the time increment $\Delta t = 0.0002$ ns. Thus the stability is ensured since the velocity of the input wave as well as the output waves is less than or equal to c. The number of spatial cells per wavelength $R = \lambda / \Delta z$ is maintained to be greater than 30, for the simulations shown in the figures (Figure Q.3).

Q.4   Results

The results obtained for various space–time variations of the plasma parameter, ω_p^2, are presented. The input wave electric field amplitude E_0 is taken to be 1 V/m. The output field amplitudes are thus the normalized values. For $t < 0$, ω_p^2 and ω_b are invariant in space and time and are given by

$$\omega_p^2 = \omega_{p0}^2, \quad t < 0, \tag{Q.17a}$$

$$\omega_b = \omega_{b0}, \quad t < 0. \tag{Q.17b}$$

The first set of results (given in Figures Q.4 through Q.9) is for temporally decaying, but spatially uniform, plasma under a constant static magnetic field. For $t > 0$,

$$\omega_p^2(t) = \omega_{p0}^2 f(t), \tag{Q.18a}$$

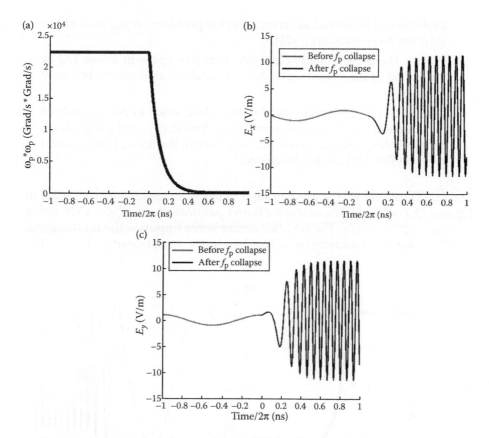

FIGURE Q.4 (See color insert following page 202.) Evolution of the electric field intensity as plasma collapses under constant external magnetic field. (a) Plasma profile, (b) E_x variation in time, and (c) E_y variation in time. $b = 10$, $\phi_0 = 90°$, $\omega_0 = 1\,\text{Grad/s}$, $\omega_{p0} = 150\,\text{Grad/s}$, $\omega_{b0} = 100\,\text{Grad/s}$, $m = 1$, $z = d/2$, $v/\omega_{p0} = 0$, and $E_0 = 1\,\text{V/m}$.

$$f(t) = e^{-bt/T}, \tag{Q.18b}$$

$$\omega_b = \omega_{b0}. \tag{Q.18c}$$

In the above equation, T is the period of the source wave: $T = 1/f_0 = 2\pi/\omega_0$. Throughout the paper, ω_0 is taken to be $1\,\text{Grad/s}$ (except in Figure Q.9) and $T = 2$, $\pi \approx 6.28\,\text{ns}$.

Figure Q.3 shows a sketch of $f(t)$ for three values of b (100, 10, and 1). Assuming $e^{-5} \approx 0$, the three values of b represent the following typical cases

i. $b = 100$: The plasma decays to zero density in about $1/20$ of the period of the source wave. A theoretical solution can be obtained by idealizing the problem as that of suddenly switching off the ionization. The problem can be solved as an initial value problem on the lines of the solution given by Kalluri [21].

ii. $b = 10$: The plasma decays and becomes free space in about $1/2$ of the period of the source wave. A theoretical solution cannot be easily obtained.

iii. $b = 1$: The plasma decays and becomes free space in five periods of the source wave. This corresponds to a slow decay, and the medium change being slow, an "adiabatic" solution on the lines of the solution given in Ref. [21] can be developed.

The FDTD technique is applied to solve (ii), and the results are presented in Figures Q.4 and Q.5. The magnetoplasma parameters are $\omega_{p0} = 150\,\text{Grad/s}$ and $\omega_{b0} = 100\,\text{Grad/s}$. The whistler source wave (input) is the fundamental standing wave in the cavity ($m = 1$) of frequency $\omega_0 = 1\,\text{Grad/s}$. The electric

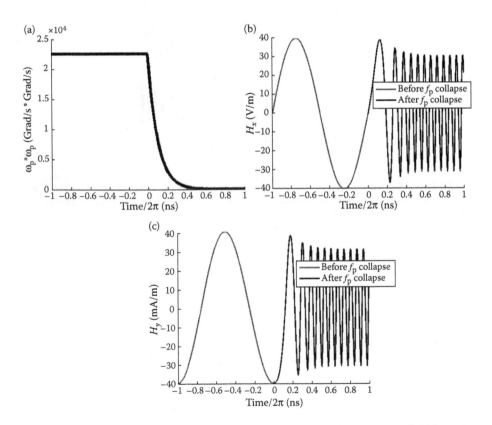

FIGURE Q.5 **(See color insert following page 202.)** Evolution of the magnetic field intensity as plasma collapses under constant external magnetic field. (a) Plasma profile, (b) H_x variation in time, and (c) H_y variation in time. $b = 10$, $\phi_0 = 90°$, $\omega_0 = 1\,\text{Grad/s}$, $\omega_{p0} = 150\,\text{Grad/s}$, $\omega_{b0} = 100\,\text{Grad/s}$, $m = 1$, $z \approx 0$, and $\nu/\omega_{p0} = 0$.

field is sampled at $z = d/2$ and the magnetic field at $z \approx 0$ (the nearest grid point to $z = 0$). After the collapse of the plasma, the frequency of the wave (output wave) is upshifted to 15 Grad/s. The upshift ratio is the same as the refractive index $n_R \approx n_w = \omega_{p0}/(\omega_0\omega_{b0})^{1/2}$ as shown by Kalluri [21]. The amplitude of the electric field is amplified approximately 11.5 times. The magnetic field is reduced to about 70% of the source wave magnetic field. The cavity acts like a frequency transformer upshifting the frequency by about 15 times. Moreover, the electric field is intensified by an order of magnitude. In Figure Q.6, the effect of collision frequency (lossy plasma) is illustrated. The parameter $k_\nu = \nu/\omega_p$ is taken to be 0.01. Although the amplitudes of the electric and

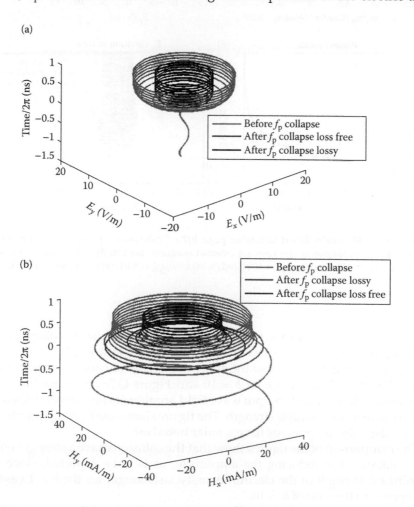

FIGURE Q.6 **(See color insert following page 202.)** Evolution of the electric and magnetic field intensities as lossy plasma collapses under constant external magnetic field in the cavity. (a) E_x–E_y variation in time and (b) H_x–H_y variation in time. $b = 10$, $\phi_0 = 45°$, $\omega_0 = 1\,\mathrm{Grad/s}$, $\omega_{p0} = 150\,\mathrm{Grad/s}$, $\omega_{b0} = 100\,\mathrm{Grad/s}$, and $k_\nu = \nu/\omega_{p0} = 0.01$.

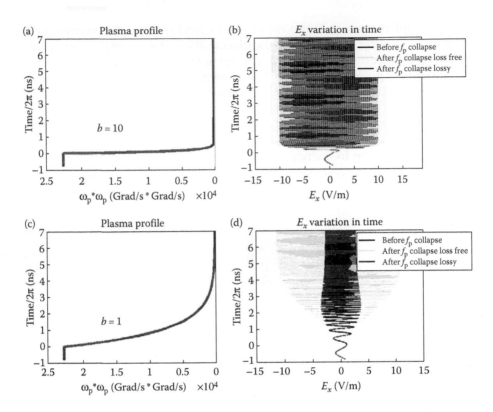

FIGURE Q.7 **(See color insert following page 202.)** Evolution of the electric field intensity as lossy plasma collapses under constant external magnetic field in the cavity. $\phi_0 = 45°$, $\omega_0 = 1\,\text{Grad/s}$, $\omega_{p0} = 150\,\text{Grad/s}$, $\omega_{b0} = 100\,\text{Grad/s}$, and $\nu/\omega_{p0} = 0.01$. (a) $b = 10$ and (c) $b = 1$.

magnetic fields are reduced, the output wave after the plasma collapses and the medium becomes free space is still of considerable strength.

Figure Q.7 is drawn to show the effect of b on the attenuation due to colli-sions. Figure Q.7a and b are for $b = 10$ and Figure Q.7c and d are for $b = 1$. The figures show that the output wave field amplitudes are further decreased but are still of considerable strength. The figure shows the E_x component only but all the field components have similar behavior.

The comparison of the figures show that the collisions have a stronger effect for a smaller b in reducing the amplitude, as is to be expected, since the significant strength of the electron density lasts longer for the $b = 1$ case as compared to the case of $b = 10$.

The switching angle has no effect on the amplitude of the newly created upshifted R wave since the new mode is fed by the energy of the source wave at the instant of switching. For the R source wave, the magnitudes of the electric and magnetic field vectors are unaffected by the switching angle

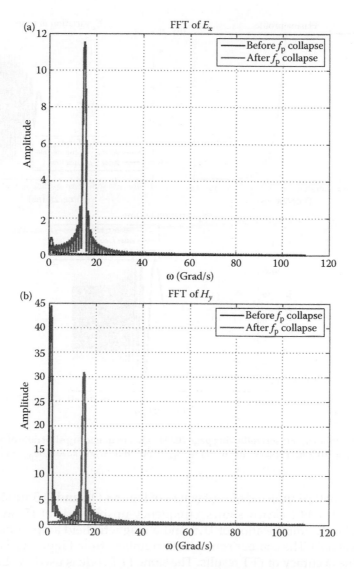

FIGURE Q.8 **(See color insert following page 202.)** FFT of the (a) electric and (b) magnetic fields before and after the collapse of the plasma under a steady magnetic field. $b = 10$, $\phi_0 = 45°$, $\omega_0 = 1\,\mathrm{Grad/s}$, $\omega_{p0} = 150\,\mathrm{Grad/s}$, $\omega_{b0} = 100\,\mathrm{Grad/s}$, $m = 1$, $z = d/2$ (for E), and $z \approx 0$ (for H).

and are given, respectively, by $E_0 \sin m\pi z/d$ and $H_0 \cos m\pi z/d$. Hence the electric and magnetic energies at the instant of switching are unaffected by the switching angle. However, the phase of the upshifted R wave will be affected by the switching angle.

In preparation for a discussion of the effects of switching from a uniform plasma to an inhomogeneous plasma in the cavity, the FFTs [25] of the

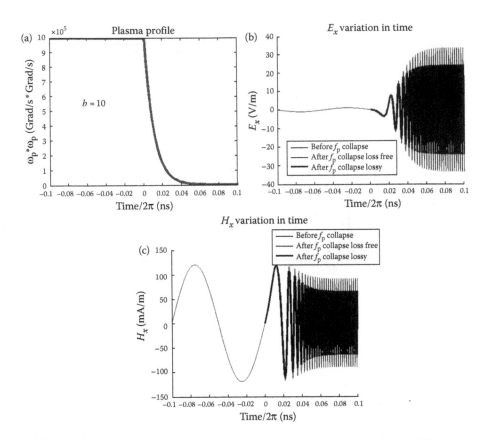

FIGURE Q.9   (See color insert following page 202.) Higher frequency-upshift ratio of 50: $b = 10$, $\phi_0 = 90°$, $\omega_0 = 10\,\mathrm{Grad/s}$, $\omega_{p0} = 1000\,\mathrm{Grad/s}$, $\omega_{b0} = 50\,\mathrm{Grad/s}$, and $k_v = v/\omega_{p0} = 0.01$.

amplitudes of the input and output waves of Figures Q.4 and Q.5 are shown in Figure Q.8. The FFT clearly shows a frequency-upshift ratio of 15, an electric field amplitude amplification of 11.5, and a magnetic field amplitude reduction ratio of 0.70. The comparison of these numbers from Figures Q.4 and Q.5 confirm the accuracy of FFT results. The same FFT code is used to obtain the results of Figure Q.10.

Q.5   Higher Frequency-Upshift Ratio

In this section we illustrate that in principle it is possible to transform a standard microwave source (say $\omega_0 = 10\,\mathrm{Grad/s}$) to a millimeter wave radiation (say 500 Grad/s) with significant strength of the fields of the output standing wave. The higher upshift ratio is obtained by increasing ω_{p0} and

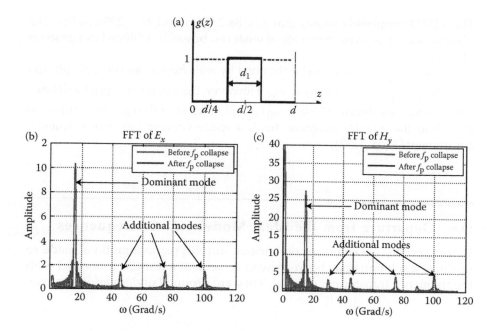

FIGURE Q.10 **(See color insert following page 202.)** FFT of the electric and magnetic fields before and after the collapse of the plasma space profile under constant magnetic field. $b = 10$, $\phi_0 = 45°$, $\omega_0 = 1\,\text{Grad/s}$, $\omega_{p0} = 150\,\text{Grad/s}$, $\omega_{b0} = 100\,\text{Grad/s}$, $m = 1$, $z = d/4$ (for E), $z \approx d/4$ (for H), and $d_1/d = 0.3$.

decreasing ω_{b0}, keeping $\omega_{p0}^2 \gg \omega_0\omega_{b0}$, but relaxing the second inequality as $\omega_0 < \omega_{b0}$. We choose $\omega_{b0} = 50\,\text{Grad/s}$ ($B_0 = 0.284\,\text{T}$) and $\omega_{p0} = 1000\,\text{Grad/s}$ ($N_0 = 3.13 \times 100\,N^{14}/\text{cm}^3$). From Equation Q.9, the refractive index $n_R \approx 50$ and the frequency of the output wave $\omega \approx 500\,\text{Grad/s}$. The results shown in Figure Q.9 are for a nominal value of $E_0 = 1\,\text{V/m}$ and $H_0 = 118\,\text{mA/m}$. The electric field of the output wave $E \approx 33\,\text{V/m}$ and the magnetic field of the output wave $H = 87\,\text{mA/m}$. If we assume that the plasma is lossy and take $k_\nu = 0.01$, the output wave is somewhat damped but still retains a significant value of $E \approx 23\,\text{V/m}$ and $H \approx 64\,\text{mA/m}$. If the decay is slower (b lower than 10), there will be more damping. The rate of collapse of the plasma depends on the recombination and other mechanisms of deionization.

The frequency of the output radiation can be further increased by reducing the ratio ω_{b0}/ω_0. For example, if $\omega_{b0}/\omega_0 = 2$, $\omega = 1000\,\text{Grad/s}$. Further decrease in the ratio, bringing the input frequency closer to the resonance, increases the damping of the waves in a lossy plasma [20, p. 95]. For example, if $\omega_{b0}/\omega_0 = 1.1$, $\omega = 1000\sqrt{10}\,\text{Grad/s}$, but it will be difficult to maintain the input wave so close to the electron cyclotron resonance.

Let us choose $\omega_{b0}/\omega_0 = 2$ to upshift the magnetron frequency of 2.45 GHz to terahertz radiation of 300 GHz. The plasma parameters are $N_0 = 1.11 \times 100^{16}/\text{cm}^3$ and $B_0 = 0.11\,\text{T}$. From Equation Q.9, $n_R \approx \omega_{p0}/\omega_0$ and $\omega = \omega_{p0}$.

The FDTD simulation shows that $E = 88.2\,\text{V/m}$ and $H = 235\,\text{mA/m}$. The plasma density is on the high side of what can be easily achieved in a gaseous plasma.

The same upshift ratio can be obtained by *switching on* an isotropic plasma $\omega = \sqrt{\omega_0^2 + \omega_{p0}^2} \approx \omega_{p0}$, for $\omega_0 \ll \omega_{p0}$ However, the upshifted signal will have a very small amplitude, $H/H_0 \approx \omega_0/\omega_{p0}$ [33]. Most of the energy in this case goes into the wiggler magnetic field, a space-varying but zero-frequency component [33, p. 1233].

Q.6   Departure from the Ideal Model and Consequences

In this section, we comment on various simplifying assumptions made to facilitate a solution and obtain answers to several questions connected with the frequency transformer concept. They are grouped under various items:

Q.6.1   Three Waves

The frequencies of the output waves are obtained from Equation Q.1 by substituting Equation Q.9 into Equation Q.1:

$$\omega^2 \left(1 - \frac{\omega_p^2(t)}{\omega(\omega_0 - \omega_b)}\right) = \omega_0^2 n_0^2. \tag{Q.19}$$

Solving for ω, we get the cubic equation for ω [3, pp. 150–152] in the uniform unbounded time-varying magnetoplasma medium (longitudinal propagation):

$$\omega^3 - \omega_b\omega^2 - \left(n_0^2\omega_0^2 + \omega_p^2(t)\right)\omega + n_0^2\omega_0^2\omega_b = 0. \tag{Q.20}$$

Thus there will be three output waves. While all the three waves will be present during the period of evolution of the fields with varying amplitudes and frequencies, the significant mode that emerges is the wave with frequency

$$\omega \approx \omega_0 \frac{\omega_{p0}}{\sqrt{\omega_0(\omega_b - \omega_0)}}, \tag{Q.21}$$

where $\omega_0 < \omega_b$ and $\omega_{p0} \gg \omega_0\omega_b$. In this paper, we verified that the above equation is still true for a time-varying medium in a cavity by using the FDTD technique to solve the coupled differential equations.

Q.6.2   Static Magnetic Field in Other Directions

If the static magnetic field B_0 is not along the z-axis but makes a small angle θ_{B_0} with the z-axis, many of the results are qualitatively still correct, since the refractive index of the quasi-longitudinal whistler mode is given by the following [20]:

$$n \approx \frac{\omega_p}{\sqrt{\omega\omega_b \cos \theta_{B_0}}} \approx \frac{\omega_p}{\sqrt{\omega\omega_p}} \left[1 + \frac{1}{2}\theta_{B_0}^2 \right], \quad \theta_{B_0} \ll 1. \qquad (Q.22)$$

For a larger value of θ_{B_0}, the frequencies are obtained by using for n the refractive index from the Appleton–Hartree equation [20]. The maximum possible value for θ_{B_0} is 90°, in which case the modes are O and X modes. The X mode also has a high refractive index for $\omega_0 \approx \sqrt{\omega_b^2 + \omega_p^2} = \omega_{uh}$. In such a case, the source frequency is near the upper hybrid resonance ω_{uh} and in the presence of even a low-loss mechanism, the wave is damped heavily. Thus we chose the longitudinal whistler as the source wave.

Q.6.3   Cavity, Plasma Decay Model, and Collision Frequency

In this paper, we arbitrarily chose an exponential decay model for the collapsing plasma and avoided specifying at this time the specific decay processes. The purpose of choosing $e^{-bt/T}$ decay is to classify the problem as rapid decay, very slow decay, or decay in a period comparable to the source period. The approximate analytical solution of the first two compared well with our FDTD solution, thus validating the FDTD results. The last decay-value problem is not easily amenable to even approximate analytical techniques even for an exponential profile. However, it was important to study this profile to ensure that the output wave has significant field strength.

If the physical decay process is considered to be due to recombination of a plasma, an approximate analytical expression for the plasma density may be represented as [34]

$$\omega_p^2(t) = \frac{\omega_{p0}^2}{1 + \omega_{p0}^2 \alpha t}, \qquad (Q.23)$$

where α is the recombination decay rate. Each decay mechanism has its own rate. Even if the rate is slow, because the source is in a cavity, the wave experiences the complete time change of the permittivity parameter. Otherwise the plasma medium has to be of a large enough width d to experience the complete temporal change of the parameter before it escapes through the boundary to the environment.

Even in a cavity, the amplitudes can diminish due to damping mechanisms if the decay rate is slow and it takes a long time for the magnetoplasma to become free space. The parameter $k_v = v/\omega_{p0}$ is chosen to be small but nonzero to illustrate and get a qualitative feel of the effect of damping.

Since the input wave is a whistler wave with high refractive index, the width of the cavity can be quite small and is given by $d = m\lambda/2$. For the data of Figures Q.4 through Q.8, $\lambda = 12.56$ cm. From the data of Figure Q.9, $\lambda = 0.377$ cm. Thus the large frequency transformation ratio can be obtained using a small ionization source, leading to a compact frequency transformer.

Q.6.4 FDTD Solution for the Time-Varying Medium

Even an approximate (analytical) one-dimensional time-domain solution of the coupled first-order partial differential equations (obtained from Equations Q.14, Q.A1, and Q.A3) for arbitrary $\omega_p^2(z,t)$ is difficult, requiring considerable effort. We have solved then for the unbounded-medium problem for sudden decay or very slow decay [1]. We also succeeded in solving them by using a Green's function and a perturbation technique for a small fall-time [1]. FDTD is indeed a fast and quick method of solving the coupled partial differential equations with time-varying and space-varying coefficients for the problem under study. It may appear that one can use the conventional FDTD for a dielectric medium allowing ω_p to be a function of time in Equation Q.4. It will lead to errors in the numerical solution, since Equation Q.4 is a frequency description of the dielectric function ε_p for the plasma. Since ω_p^2 is itself time-varying, it is appropriate that the constitutive relation relating the current in the plasma and the electric field be described in the time domain. For a time-varying plasma, it should be noted that [1, p. 152]

$$\bar{J}(t) \neq -qN(t)\bar{v}(t), \tag{Q.24a}$$

$$\Delta\bar{J}(t) = -q\Delta N_i\bar{v}_i(t). \tag{Q.24b}$$

In the above equation, ΔN_i is the electron density at t_i and $v_i(t)$ is the velocity at t of these ΔN_i electrons. From Equation Q.24b, we obtain [1] the auxiliary differential equation Q.14.

The solution of Equation Q.14 by 3D FDTD for arbitrary values of ω_p^2 and ω_b by a step-by-step method is given in detail in Ref. [30]. It differs from the work of others [24–29] in two important ways. (1) Instead of using a central-difference approximation for the first derivative and averaging operators to discretize $\bar{J}$ in Equation Q.14, we used as exponential stepping algorithm. It gives an exact solution of Equation Q.14 assuming that v, ω_p^2, and ω_b are constant (their values at the center of the time step) over a time step; and (2) We still maintained a step-by-step leap-frog computation by positioning the sampling point for $\bar{J}$ at the center of the Yee's cube. This algorithm is valid for arbitrary values of v, ω_p^2, and ω_b, in contrast to the algorithm

based on discretizing Equation Q.14 through central-difference approxima-
tion. The latter is valid only for restricted values of ν and ω_b, since it amounts to
approximating the exponential and trigonometric terms in Equations Q.A5
and Q.A6 by the first few terms in their series expansion. Equation Q.A5
is the one-dimensional version of our 3D FDTD [30] for a time-varying
medium. It can handle very well arbitrary specified functions of $\omega_p^2(z,t)$
and $\omega_b(z,t)$.

By coupling Maxwell's equations with equations of decay, we can modify
the FDTD technique where $\omega_p^2(z,t)$ is determined as part of the solution rather
than the assumption made here as specified. The equation of decay comes
from plasma physics and depends on the experimental setup.

Q.6.5   Space-Varying Plasma Profile

The gaseous plasma in the cavity will not be uniform in space, particularly
near the cavity walls. To illustrate the effect of a nonuniform spatial profile,
the spatial density profile is assumed notionally to be

$$\omega_p^2(z,t) = \omega_{p0}^2, \quad t < 0, \tag{Q.25a}$$

$$\omega_p^2(z,t) = \omega_{p0}^2, \quad t > 0, \quad \frac{(d-d_1)}{2} < z < \frac{(d+d_1)}{2}, \tag{Q.25b}$$

$$\omega_p^2(z,t) = \omega_{p0}^2 f(t), \quad t > 0, \quad 0 < z < \frac{d-d_1}{2}, \frac{d+d_1}{2} < z < d. \tag{Q.25c}$$

The plasma density in the interior of the cavity is unchanged but the plasma
density near the walls decays exponentially. As $t \to \infty$, the plasma density
profile is given by $\omega_{p0}^2 g(z)$, where $g(z)$ is a step profile shown in Figure Q.10a.
The chosen spatial profile is somewhat artificial and is constructed to represent the quicker loss of plasma near the walls in the off-state of the pulsed
operation of the ionizing source. The plasma in the center decays only a little
and has considerable strength in the off-state. The plasma in the entire cavity
is quickly rebuilt in the on-state of the ionizing source. Figure 10b shows the
FFT of the output signals when the plasma decays nonuniformly (in space) in
the cavity. The other parameters of Figure Q.10 are the same as those of Figure
Q.4, except that $d_1/d = 0.3$ and E_x is sampled at $z = d/4$ and H_y is sampled at
$z \approx d/4$. The sampling point $z = d/4$ is in free space after the plasma decayed
in the space to zero density. While the dominant mode is still the upshifted
mode with an upshift ratio of approximately 15, there are significant additional modes at much higher frequencies. These arise because the decay
into a nonuniform plasma in space excites higher-order, frequency-shifted
harmonics in the cavity [30,32,35]. The chosen spatial profile is somewhat
artificial and is constructed to illustrate the generation of significant harmonics. Similar qualitative results are obtained for other spatial profiles but are
not discussed here.

Q.6.6   New Plasma Sources

It appears that the existing conventional pulsed plasma sources need significant modifications to provide the basic experimental framework to implement the frequency transformer concepts discussed in this paper. Recent developments in the research on plasma sources allow us to suggest an alternative black box description to affect the frequency transformation. The first alternative is close to the ideal black box, except that the source wave is the whistler wave in a helicon plasma source, which was originally developed for plasma processing of computer chips. At present, it is under intense development for various applications [36–39]. It can produce typically a high-density plasma of $10^{13}/cm^3$ using a special helical twist antenna wrapped around a Pyrex tube. The RF source is about 30 MHz and the background magnetic field is several hundred Gauss. By suitable modifications, one can develop the system to implement the concept of a frequency transformer.

Semiconductor plasmas with e–h pairs form a dense neutral plasma [40–43]. Semiconductor plasmas, as experimental systems, have several advantages over the gaseous ones [42].

Q.7   Concluding Remarks

The principles of frequency transformation of a whistler wave by switching off a magnetoplasma, established earlier [21] by considering an unbounded idealized medium, were confirmed by considering a bounded medium in a cavity. After considering the effects of the following parameters: the rate of collapse, spatial plasma profile, collision frequency as well as the switching angle, we conclude that the effect of conversion of a whistler wave to an electric-field-intensified higher-frequency wave (with an order of magnitude upshift of the frequency) by switching off the ionization source can be achieved in a cavity. These results, based on notional values, are important in guiding the development of a frequency transformer system. Effective use of the new plasma sources being developed will result in a compact and cost-effective device even for a frequency transformer with high frequency-upshift ratio.

A dual problem of switching off the source of the background quasi-static magnetic field can be solved on similar lines. In this case, the magnetoplasma medium is converted into an isotropic plasma, thus achieving again a big change in ε_p. Such a problem was solved in Ref. [44] and it was shown that the whistler wave would be converted into a wiggler (zero frequency, space-varying) magnetic field. A wiggler magnetic field is used in an FEL [45]. When the switching off takes place in a *cavity*, it can be shown that the direction of the wiggler magnetic field is controlled by the switching angle. The results of this investigation will be published soon.

Q.A Appendix

The cavity is sampled for the electric field, current density, and magnetic field as depicted in Figure Q.A1. The vertical axis represents variation in time, while the horizontal axis represents variation in space. The grid is chosen so that E_x and E_y are sampled at the cavity edges. This is done to take advantage of the boundary condition that these fields are always zero at the PEC plates of the cavity.

From the Maxwell equation

$$\nabla \times \mathbf{H} = \varepsilon_0 \frac{\partial \mathbf{E}}{\partial t} + \mathbf{J} \tag{Q.A1}$$

and using central-difference finite difference approximation for the derivatives, one obtains

$$(E_x)_k^{n+1} = (E_x)_k^n - \frac{\Delta t}{\varepsilon_0 \Delta z}\left((H_y)_{k+(1/2)}^{n+(1/2)} - (H_y)_{k-(1/2)}^{n+(1/2)}\right) - \frac{\Delta t}{\varepsilon_0}(J_x)_k^{n+(1/2)}, \tag{Q.A2a}$$

$$(E_y)_k^{n+1} = (E_y)_k^n + \frac{\Delta t}{\varepsilon_0 \Delta z}\left((H_x)_{k+(1/2)}^{n+(1/2)} - (H_x)_{k-(1/2)}^{n+(1/2)}\right) - \frac{\Delta t}{\varepsilon_0}(J_y)_k^{n+(1/2)}. \tag{Q.A2b}$$

From the Maxwell equation

$$\nabla \times \mathbf{E} = -\mu_0 \frac{\partial \mathbf{H}}{\partial t}, \tag{Q.A3}$$

Time n					
$\frac{3}{2}$	$J_x J_y$	$H_x H_y$	$J_x J_y$	$H_x H_y$	$J_x J_y$
1	$E_x E_y$		$E_x E_y$		$E_x E_y$
$\frac{1}{2}$	$J_x J_y$	$H_x H_y$	$J_x J_y$	$H_x H_y$	$J_x J_y$
0	$E_x E_y$		$E_x E_y$		$E_x E_y$
$-\frac{1}{2}$	$J_x J_y$	$H_x H_y$	$J_x J_y$	$H_x H_y$	$J_x J_y$
	0	$\frac{1}{2}$	1	$\frac{3}{2}$	2

Space $\longrightarrow$ k

FIGURE Q.A1 Sampling points in the grid for FDTD computation of various fields.

and central-difference finite difference approximations, we obtain

$$(H_x)_{k+(1/2)}^{n+(1/2)} = (H_x)_{k+(1/2)}^{n-(1/2)} + \frac{\Delta t}{\mu_0 \Delta z}\left[(E_y)_{k+1}^n - (E_y)_k^n\right], \tag{Q.A4a}$$

$$(H_y)_{k+(1/2)}^{n+(1/2)} = (H_y)_{k+(1/2)}^{n-(1/2)} - \frac{\Delta t}{\mu_0 \Delta z}\left[(E_x)_{k+1}^n - (E_x)_k^n\right]. \tag{Q.A4b}$$

A technique similar to exponential time-stepping [22] algorithm is developed to discretize Equation Q.14 and the details of its derivation are given in Refs. [30–32]. However, we give the following results:

$$\begin{bmatrix} (J_x)_k^{n+(1/2)} \\ (J_y)_k^{n+(1/2)} \end{bmatrix} = e^{-(\nu\Delta t)} \begin{bmatrix} \cos[(\omega_b)_k^n \Delta t] & -\sin[(\omega_b)_k^n \Delta t] \\ \sin[(\omega_b)_k^n \Delta t] & \cos[(\omega_b)_k^n \Delta t] \end{bmatrix} \begin{bmatrix} (J_x)_k^{n-(1/2)} \\ (J_y)_k^{n-(1/2)} \end{bmatrix}$$

$$+ \frac{\varepsilon_0(\omega_p^2)_k^n e^{-(\nu\Delta t)}}{\nu^2 + (\omega_b^2)_k^n} \begin{bmatrix} k_{11} & k_{12} \\ k_{21} & k_{22} \end{bmatrix} \begin{bmatrix} (E_x)_k^n \\ (E_y)_k^n \end{bmatrix}, \tag{Q.A5}$$

where

$$\left\{ \begin{array}{l} k_{11} = \nu\left(e^{\nu\Delta t} - \cos[(\omega_b)_k^n \Delta t]\right) + (\omega_b)_k^n \sin[(\omega_b)_k^n \Delta t], \\ k_{12} = -e^{\nu\Delta t}(\omega_b)_k^n + (\omega_b)_k^n \cos[(\omega_b)_k^n \Delta t] + \nu \sin[(\omega_b)_k^n \Delta t], \\ k_{21} = e^{\nu\Delta t}(\omega_b)_k^n - (\omega_b)_k^n \cos[(\omega_b)_k^n \Delta t] - \nu \sin[(\omega_b)_k^n \Delta t], \\ k_{22} = \nu\left(e^{\nu\Delta t} - \cos[(\omega_b)_k^n \Delta t]\right) + (\omega_b)_k^n \sin[(\omega_b)_k^n \Delta t]. \end{array} \right\} \tag{Q.A6}$$

The steady-state solutions given in Section Q.2 are used to determine the data for the field values for the first two rows in the FDTD grid of Figure Q.A1. For chosen profiles for ω_p^2 and ω_b, the values of $(\omega_b)_k^n$ and $(\omega_p)_k^n$ for the FDTD grid are set. In this article, ω_b is taken to be a constant.

Finally, Equations Q.A2 through Q.A6 are used to determine the field values as time increases. Equations Q.A2 through Q.A6 are the leap-frog technique [22,23] of FDTD, which permit step-by-step computation of the fields as the index n of the time step increases.

References

1. Kalluri, D. K., *Electromagnetics of Complex Media*, CRC Pres LLC, Boca Raton, FL, 1999.
2. Auld, B. A., Collins, J. H., and Zapp, H. R., Signal processing in a nonperiodically time-varying magnetoelastic medium, *Proc. IEEE*, 56, 258–272, 1968.
3. Jiang, C. L., Wave propagation and dipole radiation in a suddenly created plasma, *IEEE Trans. Ant. Prop.*, 23, 83–90, 1975.
4. Felsen, B. L. and Whitman, G. M., Wave propagation in time-varying media, *IEEE Trans. Ant. Prop.*, 18, 242–253, 1970.
5. Fante, R. L., Transmission of electromagnetic waves into time-varying media, *IEEE Trans. Ant. Prop.*, 19, 417–424, 1971.

6. Joshi, C. J., Clayton, C. E., Marsh, K., Hopkins, D. B., Sessler, A., and Whittum, D., Demonstration of the frequency upshifting of microwave radiation by rapid plasma creation, *IEEE Trans. Plasma Sci.*, 18, 814–818, 1990.

7. Kuo, S. P., Frequency up-conversion of microwave pulse in a rapidly growing plasma, *Phys. Rev. Lett.*, 65, 1000–1003, 1990.

8. Kuo, S. P. and Ren, A., Experimental study of wave propagation through a rapidly created plasma, *IEEE Trans. Plasma Sci.*, 21, 53–56, 1993.

9. Rader, M., Dyer, F., Matas, A., and Alexeff, I., Plasma-induced frequency shifts in microwave beams, *IEEE Int. Conf. Plasma Sci.*, Oakland, CA, p. 171, 1990.

10. Yablonovitch, E., Spectral broadening in the light transmitted through a rapidly growing plasma, *Phys. Rev. Lett.*, 31, 877–879, 1973.

11. Savage, Jr., R. L., Joshi, C. J., and Mori, W. B., Frequency up-conversion of electromagnetic radiation upon transmission into an ionization front, *Phys. Rev. Lett.*, 68, 946–949, 1992.

12. Lampe, M. and Walker, J. H., Interaction of electromagnetic waves with a moving ionization front, *Phys. Fluids*, 21, 42–54, 1978.

13. Banos, Jr., A., Mori, W. B., and Dawson, J. M., Computation of the electric and magnetic fields induced in a plasma created by ionization lasting a finite interval of time, *IEEE Trans. Plasma Sci.*, 21, 57–69, 1993.

14. Wilks, S. C., Dawson, J. M., and Mori, W. B., Frequency up-conversion of electromagnetic radiation with use of an overdense plasma, *Phys. Rev. Lett.*, 61, 337–340, 1988.

15. Mori, W. B., (Ed), Special issue on generation of coherent radiation using plasmas, *IEEE Trans. Plasma Sci.*, 21(2), 1993.

16. Kalluri, D. K., Effect of switching a magnetoplasma medium on a travelling wave: Longitudinal propagation, *IEEE Trans. Ant. Prop.*, 37, 1638–1642, 1989.

17. Kuo, S. P., Bivolaru, D., Orlick, L., Alexeff, I., and Kalluri, D. K., A transmission line filled with fast switched periodic plasma as a wideband frequency transformer, *IEEE Trans. Plasma Sci.*, 29, 365–370, 2001.

18. Kalluri, D. K., Khalifeh, A. F., and Eker, S., Frequency and polarization transformer: Transverse modes: I zero rise time, *IEEE Trans.. Ant. Prop.*, 55, 1789–1796, 2007.

19. Khalifeh, A. F., Eker, S., and Kalluri, D. K., Frequency and polarization transformer: Transverse modes: II finite rise time, *IEEE Trans. Ant. Prop.*, 55, 2761–2766, 2007.

20. Booker, H. G., *Cold Plasma Waves*, Kluwer, Hingham, MA, 1984.

21. Kalluri, D. K., Frequency upshifting with power intensification of a whistler wave by a collapsing plasma medium, *J. Appl. Phys.*, 79, 3895–3899, 1996.

22. Taflove, A., *Computational Electrodynamics: The Finite-Difference Time-Domain Method*, Artech House Inc., Norwood, MA, 1995; Taflove, A. and Hagness, S., *Computational Electrodynamics: The Finite-Difference Time-Domain Method*, 2nd edition, Artech House Inc., Norwood, MA, 2000.

23. Yee, K., Numerical solution of initial boundary value problems involving Maxwell equations in isotropic media, *IEEE Trans. Ant. Prop.*, 14(3), 302–307, 1966.

24. Young, J. L., Propagation in linear dispersive media: Finite difference time domain methodologies, *IEEE Trans. Ant. Prop.*, 43, 422–426, 1995.

25. Thevenot, M., Berenger, J. P., Monediere, T., and Jecko, F., A FDTD scheme for the computation of VLF-LF propagation in the anisotropic earth-ionosphere waveguide, *Ann. Telecommun.*, 54, 297–310, 1999.

26. Luebbers, R. J., Hunsgerger, F., and Kunz, K. S., A frequency-dependent finite difference time domain formulation for transient propagation in a plasma, *IEEE Trans. Ant. Prop.*, 39, 29–34, 1991.

27. Hu, W. and Cummer, S. A., A FDTD model for low and high altitude lightning-generated EM fields, *IEEE Trans. Ant. Prop.*, 54, 1513–1522, 2006.

28. Ma, Z., Croskey, C. L., and Hale, L.C., The electrodynamic responses of the atmosphere and ionosphere to the lightning discharge, *J. Atoms. Terr. Phys.*, 60, 845–861, 1998.

29. Chevalier, M. W., Chevalier, T. W., and Inan, U. S., A PML utilizing k-vector information as applied to the whistler mode in a magnetized plasma, *IEEE Trans. Ant. Prop.*, 54(8), 2424–2429, 2006.

30. Lee, J. H. and Kalluri, D. K., Three-dimensional FDTD simulation of electromagnetic wave transformation in a dynamic inhomogeneous magnetized plasma, *IEEE Trans. Ant. Prop.*, 47, 1146–1151, 1999.

31. Ehsan, M. M. and Kalluri, D. K., Plasma induced wiggler magnetic field in a cavity II-FDTD method for a switched lossy plasma, *Int. J. Infrared Millim. Waves*, 24, 1655–1676, 2003.

32. Kalluri, D. K. and Ehsan, M. M., Frequency and polarization transformer: Longitudinal modes, *Int. J. Infrared Millim. Waves*, 25, 327–353, 2004.

33. Ehsan, M. M. and Kalluri, D. K., Plasma induced wiggler magnetic field in a cavity, *Int. J. Infrared Millim. Waves*, 24, 1215–1234, 2003.

34. Chen, F. F., *Introduction to Plasma Physics and Controlled Fusion*, 2nd edition, Plenum Press, New York, 1988.

35. Mendonca, J. T. and Oliveria e Silva, L., Mode coupling theory of flash ionization in a cavity, *IEEE Trans. Plasma Sci.*, 24(2), 147–151, 1996.

36. Es'Kin, V. A. and Kudrin, A. V., Excitation of whistler modes by a loop antenna in helical discharge plasmas, 28th ICPIG, Prague, Czech Republic, pp. 2043–2046, July 15–20, 2007.

37. Fridman, A. and Kennedy, L., *Plasma Physics and Engineering*, Taylor & Francis, New York, 2004.

38. Boswell, R. W. and Chen, F. F., Helicons-the early years, *IEEE Trans. Plasma Sci.*, 25, 1229–1244, 1997.

39. Chen, F. F. and Boswell, R. W., Helicons—the past decade, *IEEE Trans. Plasma Sci.*, 25, 1245–1257, 1997.

40. Bongiovanni, G. and Staehli, J. L., Density dependence of electron–hole plasma lifetime in semiconductor quantum wells, *Phys. Rev. B*, 46(15), 9861–9864, 1992.

41. Avitzour, Y., Geltner, I., and Suckewer, S., Laser pulse frequency shifting by ionization and recombination fronts in semiconductor plasma, *J. Phy. B*, 38, 779–787, 2005.

42. Berezhiani, V. I., Mahajan, S. M., and Miklaszewski, R., Frequency up-conversion and trapping of ultrashort laser pulses in semiconductor plasmas, *Phys. Rev. A*, 59(1), 859–864, 1999.

43. Jho, Y. D., Xiaoming Wang, Kono, J. Reitze, Z. H., Wei, X., Belvanin, A. A., Kocharovsky, V. V., and Solomon, G. S., Cooperative recombination of a quantized high-density electron–hole plasma in semiconductor quantum wells, *Phys. Rev. Lett.*, 96, 237401.1–237401.4, 2006.

44. Kalluri, D. K., Conversion of a whistler wave into a controllable helical wiggler magnetic field, *J. Appl. Phys.*, 79, 6770–6774, 1996.

45. Granastein, V. L. and Alexeff, I., *High-Power Microwave Sources*, Artech House, Boston, MA, 1987.

Part IV

Experiments

13

Experiments

Igor Alexeff

Department of Electrical and Computer Engineering
University of Tennessee Knoxville, TN

So far, several highly definitive experiments on frequency shifting have been done. Several experiments were done by Doctor Mark Rader in his thesis work [1] under the supervision of Professor Igor Alexeff. Another experiment was done by Spencer Kuo [2]. Some works were done by Mori [3] et al. and Joshi [4] et al. Surprisingly, it appears that the experimental work had not been carried forward, and no recent publications have appeared. However, the earliest experimental evidence of the principle of frequency shifting by dynamic plasma was given by Yablonovitch in 1973 [5], who observed the spectral broadening of the light by the plasma created by the light.

Mark Rader's work was done in two different modes. In the first mode, a plasma jet was fired into a microwave beam, and the reflected signal was upshifted in frequency. In this work, the plasma entered into the interaction space as a jet with a distinct front, but there was obviously a change in the propagating medium. In the second experiment, the plasma was generated simultaneously in a long plasma tube containing microwaves. And since the wavelength of the waves was fixed while the phase velocity in the medium was increased, the frequency was upshifted. In the case of Spencer Kuo, the microwaves were reflected and upshifted from plasma layers generated by microwaves in a low-pressure gas in a bell jar.

13.1 Mark Rader: 1

Considering Mark Rader's work in more detail, the first experiment involved the use of a Bostick gun. The general schematic of the experiment is shown in Figure 13.1.

Microwaves from a high-frequency, steady-state transmitter are injected through a Lucite window into an evacuated glass tube approximately 1 m long. The microwaves reflected from a rapidly moving plasma jet from a

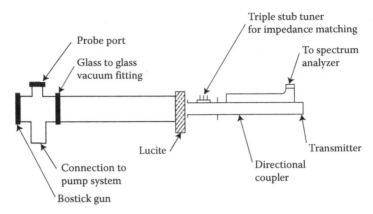

FIGURE 13.1 Experimental setup for frequency shifting using a plasma jet.

Bostick gun are collected in the original transmitting waveguide, are redirected by a directional coupler, and are sent to a spectrum analyzer. The plasma jet from the Bostick gun reflects the microwaves like a moving mirror and upshifts the microwave signal. The frequency upshift can be computed to first order as $v = (1 + (2V/c))v_0$, where v is the upshifted frequency, v_0 is the original frequency, V is the velocity of the plasma jet, and c is the speed of light.

To provide the high-speed plasma jet we used a Bostick gun. The Bostick gun is an extremely simple device and is shown schematically in Figure 13.2.

A high-voltage, high-current pulse arcs over the epoxy surface, creating a dense plasma. The high current in the plasma generates an intense azimuthal magnetic field that interacts with the radial current flow to accelerate the plasma to high velocities. An image converter photograph of the expanding plasma jet is shown in Figure 13.3.

The velocity of the plasma jet was measured by using two microwave probes placed along the glass tube, as shown in Figure 13.4. Numerous measurements gave a jet velocity of 8.06×10^4 m/s.

The experimental results for the frequency shift observed are shown in Figure 13.5. The frequency is clearly upshifted by about 2 MHz by the plasma jet. The data were taken by a Hewlett-Packard 8555 spectrum analyzer with a range from 100 MHz to 18 GHz.

These data can be compared with the 2.6 MHz frequency shift predicted from the measured plasma jet velocity.

An unexpected effect was the frequency downshift that was also observed (as seen in Figure 13.5). This was attributed to the decay of the plasma column. This was confirmed by repeating the experiment with gas present in the plasma column. In this case, the frequency upshift is present, but the downshift disappears. This is shown in Figure 13.6.

To summarize, the frequency upshift predicted by theory was observed. The application of this work was possibly to screen satellites from probing

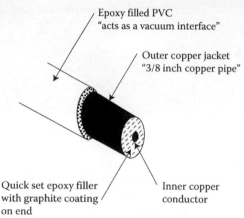

Epoxy filled PVC
"acts as a vacuum interface"

Outer copper jacket
"3/8 inch copper pipe"

Quick set epoxy filler
with graphite coating
on end

Inner copper
conductor

FIGURE 13.2 Schematic of a Bostick plasma gun.

RADAR signals by shifting the reflected signals out of the receiver pass band. Conceptually, this is shown in Figure 13.7. The expected lifetime of a Bostick gun pulsed repeatedly to maintain a continuous radial plasma flow is about 6 months. It takes very little solid material to produce a dense plasma.

13.2 Mark Rader: 2

In this second series of experiments, the plasma density was increased uniformly along the tube length. This experimental work corresponds very closely with the theoretical work of Professor Kalluri. The apparatus is shown schematically in Figure 13.8.

Here, microwaves are injected from the microwave system into a glass pipe, are reflected from the opposite end of the glass pipe, re-enter the microwave

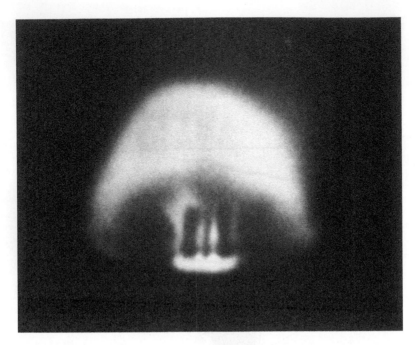

FIGURE 13.3 Image converter photograph of parallel electrode gun firing.

system, are extracted by a directional coupler, and are sent to a frequency analyzer. An intense pulsed electrical discharge along the glass column produces a plasma from the low-pressure residual gas in the system. The glass pipe is coated with aluminum foil to contain the microwaves.

The energy storage system was a 0.5-microfarad capacitor rated at 25 kV. It was a low-inductance model designed to give a very high rise time. There

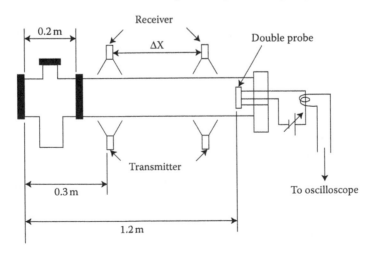

FIGURE 13.4 Schematic of the vacuum system with associated diagnostics.

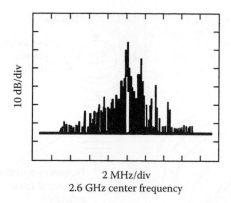

FIGURE 13.5 Spectral shift of a Bostick gun with 2.6 GHz transmitter frequency.

was no switch between the capacitor and the device. The reason for this was to reduce the electrical noise from the switch. Instead the tube was allowed to break down at its natural standoff voltage. This voltage is a nonlinear function of the device length and the background pressure and it cannot be calculated, but depending on gas pressure, it ranges from 3 to 10 kV.

The results for the 1-m-long tube are shown in Figure 13.9. The peaks at the far left and right of the spectrum are artifacts of the system, because a switch was used in the initial test. The frequency was upshifted by approximately 8 MHz.

We next built a new system to try and increase the frequency upshift by increasing the radiation interaction time. The new system was 3 m long and was made from a PVC water pipe covered with aluminum foil. The results of this experiment are shown in Figure 13.10. In this case, the frequency upshift corresponds to 50 MHz.

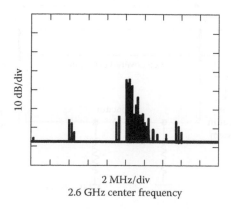

FIGURE 13.6 High-pressure spectral shift with no downshift.

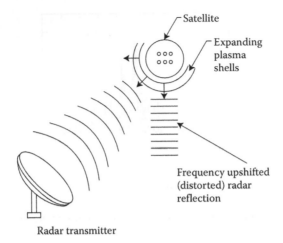

FIGURE 13.7   Satellite stealth system—basic concept of plasma screening.

Computing the expected frequency shift is complicated in that the plasma is not generated instantaneously, but rises during the experiment. Assuming that the wavelength is unchanged due to geometrical considerations, the frequency is upshifted by the increasing phase velocity in the plasma:

$$\omega_1 = k \frac{c}{\sqrt{1 - (\omega_p^2/\omega^2)}}.$$

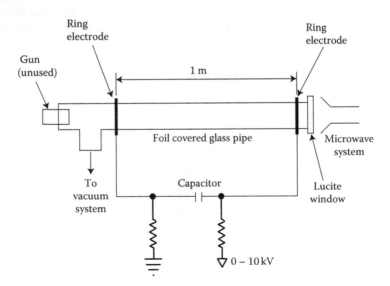

FIGURE 13.8   Schematic of frequency shifting by phase velocity modulation.

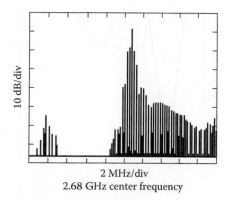

2 MHz/div
2.68 GHz center frequency

FIGURE 13.9 Frequency shift using the 1 m glass system.

Since $k = C/\omega$, we obtain

$$\omega_1 = \frac{\omega}{\sqrt{1 - (\omega_p^2/\omega_1^2)}}.$$

Solving for ω_1, we obtain $\omega_1^2 = \omega_p^2 + \omega^2$. We can compute the change of angular frequency with time as $(\delta\omega/\delta t t) = (\omega_2 - \omega_1)/T$, where T is the transit time for the device. Applying a little calculus, we obtain $\Delta\omega = \left(1/2\sqrt{\omega_p^2 + \omega^2}\right)(d\omega_p^2/dt)T$, which at short times is approximately $\Delta\omega = (1/2\omega)(d\omega_p^2/dt)T$, where $\Delta\omega$ is the observed frequency shift produced during the transit time T. Thus, the critical measurement is to compare the observed frequency shift during the wave transit time with the time derivative of the square of the plasma density, which corresponds to the plasma electron density.

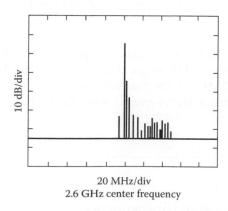

20 MHz/div
2.6 GHz center frequency

FIGURE 13.10 Frequency shift using the 3 m PVC system at 20 MHz/div.

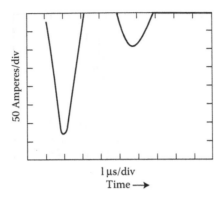

FIGURE 13.11   Time-based discharge current signal on the 3 m PVC system.

To obtain the rise time for the plasma frequency, two sets of measurements were made. First, the peak plasma density was measured by a Langmuir probe technique. Second, the time resolved discharge current was measured using a fast current transformer in the discharge current feed. This second measurement yields the plasma time rate of formation, assuming that the time integral of the absolute value of the discharge current is proportional to the plasma density. (A correction to the original thesis material is made at this point.) The current as a function of time is shown in Figure 13.11.

For the long tube, we can obtain the following values, using frequency in place of angular frequency. (The factors of 2π cancel.)

Transmitter frequency—2.6 GHz

Peak plasma frequency as obtained from Langmuir probe data—10.9 GHz

Transit time for the microwaves—20 ns

Time for plasma formation during oscillating and decaying discharge currents—7 μs

Inserting these values into the above equation, we obtain a theoretical frequency upshift of 65 MHz, in good agreement with the observed value of approximately 50 MHz. Note that no frequency downshift is observed, showing that the plasma decays much more slowly than it is formed.

13.3 Spencer Kuo

Spencer Kuo and coworkers have performed a very ingenious experiment in which they produced a 1 MHz frequency upshift in a 3.27 GHz microwave signal. By splitting the pulsed microwave signal into two beams intersecting at right angles, they created layers of enhanced microwave radiation inside a plastic chamber filled with rarefied air. These enhanced electric field regions created isolated slabs of plasma that grew in density and upshifted the incident microwave radiation.

13.4 Mori and Joshi

In these sophisticated experiments, photoionization by means of a powerful pulsed laser was used to create the plasma. The medium being ionized was an organic gas, TAMI, that has an extremely low photoionization threshold. The initial microwave signal was 33.3 GHz, and the resulting frequency upshift was 1.5 GHz.

References

1. Rader, M. and Alexeff, I., Microwave frequency shifting using photon acceleration, *J. Infrared Millim. Terahertz Waves*, 12(7), 683–702, 1991.
2. Kuo, S. P., Frequency up-conversion of microwave pulse in a rapidly growing plasma, *Phys. Rev. Lett.*, 65(8), 1000, 1990.
3. Wilks, S. C, Dawson, J. M., and Mori, W. B., Frequency up-conversion of electromagnetic radiation with use of an overdense plasma, *Phys. Rev. Lett.*, 61, 337, 1988.
4. Joshi, C. J., Clayton, C. E., Marsh, K., Hopkins, D. B., Sessler, A., and Whittum, D., Demonstration of the frequency upshifting of microwave radiation by rapid plasma creation, *IEEE Trans. Plasma Sci.*, 18(5), 814, 1990.
5. Yablonovitch, E., Spectral broadening in the light transmitted through a rapidly growing plasma, *Phys. Rev. Lett.*, 31, 877, 1973.

14

Problems

P1.1 For a lossy plasma, the frequency-domain relative permittivity is complex and is given by

$$\varepsilon_p(\omega) = 1 - \frac{\omega_p^2}{\omega(\omega - j\nu)}, \tag{P1.1}$$

where ν is the collision frequency (rad/s). The above relation is called Drude dispersion relation, and in the limit $\nu \to 0$, it reduces to Equation 1.50.

Show that the basic field equations for a cold isotropic lossy plasma are given by Equations 1.1, 1.2, and

$$\frac{d\mathbf{J}}{dt} + \nu\mathbf{J} = \varepsilon_0\omega_p^2\mathbf{E}. \tag{P1.2}$$

Consider a plasma of plasma frequency $f_p = \omega_p/2\pi = 30\,\text{GHz}$ and a collision frequency $\nu = 2 \times 10^{10}$ rad/s. Plot the real and imaginary parts of the dielectric constant in the range of 0–90 GHz.

P1.2 Water is an example of materials whose frequency-domain relative permittivity exhibits Debye dispersion:

$$\varepsilon_r(\omega) = \varepsilon_\infty + \frac{\varepsilon_s - \varepsilon_\infty}{1 + j\omega t_0}, \tag{P1.3}$$

where ε_∞ is the infinite frequency relative permittivity, ε_s is the static relative permittivity at zero frequency, and t_0 is the relaxation time.

Show that the basic field equations in this medium may be described by

$$\nabla \times \mathbf{E} = -\mu_0\frac{\partial\mathbf{H}}{\partial t}, \tag{P1.4}$$

$$\nabla \times \mathbf{H} = \frac{\partial\mathbf{D}}{\partial t}, \tag{P1.5}$$

$$t_0\frac{d\mathbf{D}}{dt} + \mathbf{D} = \varepsilon_s\varepsilon_0\mathbf{E} + t_0\varepsilon_\infty\varepsilon_0\frac{d\mathbf{E}}{dt}, \tag{P1.6}$$

and the one-dimensional field Equations 1.20 through 1.22 are modified as

$$\frac{\partial E}{\partial z} = -\mu_0 \frac{\partial H}{\partial t}, \tag{P1.7}$$

$$-\frac{\partial H}{\partial z} = \frac{\partial D}{\partial t}, \tag{P1.8}$$

$$t_0 \frac{dD}{dt} + D = \varepsilon_s \varepsilon_0 E + t_0 \varepsilon_\infty \varepsilon_0 \frac{dE}{dt}. \tag{P1.9}$$

Plot the real and imaginary parts of complex relative permittivity of water in the frequency domain 0–80 GHz. Assume $\varepsilon_s = 81$, $\varepsilon_\infty = 1.8$, and $t_0 = 9.4 \times 10^{-12}$ s.

P1.3 A second-order Lorentz dispersive material has a relative permittivity

$$\varepsilon_r(\omega) = \varepsilon_\infty + \frac{(\varepsilon_s - \varepsilon_\infty)\omega_R^2}{\omega_R^2 + 2j\omega\delta - \omega^2}, \tag{P1.10}$$

where ω_R is the resonant frequency and δ is the damping constant.

a. Show that the basic field equations in this medium are given by

$$\nabla \times \mathbf{E} = -\mu_0 \frac{\partial \mathbf{H}}{\partial t}, \tag{P1.11}$$

$$\nabla \times \mathbf{H} = \frac{\partial \mathbf{D}}{\partial t}, \tag{P1.12}$$

$$\omega_R^2 \mathbf{D} + 2\delta \frac{d\mathbf{D}}{dt} + \frac{d^2\mathbf{D}}{dt^2} = \omega_R^2 \varepsilon_s \varepsilon_0 \mathbf{E} + 2\delta \varepsilon_\infty \varepsilon_0 \frac{d\mathbf{E}}{dt} + \varepsilon_\infty \varepsilon_0 \frac{d^2\mathbf{E}}{dt^2}, \tag{P1.13}$$

and the one-dimensional field Equations 1.20 through 1.22 are modified as

$$\frac{\partial E}{\partial z} = -\mu_0 \frac{\partial D}{\partial t}, \tag{P1.14}$$

$$-\frac{\partial H}{\partial z} = \frac{\partial D}{\partial t}, \tag{P1.15}$$

$$\omega_R^2 D + 2\delta \frac{dD}{dt} + \mu_0 \frac{d^2D}{dt^2} = \omega_R^2 \varepsilon_s \varepsilon_0 E + 2\delta \varepsilon_\infty \varepsilon_0 \frac{dE}{dt} + \varepsilon_\infty \varepsilon_0 \frac{d^2E}{dt^2}. \tag{P1.16}$$

Plot the real and imaginary parts of $\varepsilon_r(\omega)$ using the parameters $\varepsilon_s = 2.25$, $\varepsilon_\infty = 1$, $\omega_R = 4 \times 10^{16}$ rad/s, and $\delta = 0.28 \times 10^{16}$ rad/s. This Lorentz medium has a resonance in the optical range.

P2.1 Let $\varepsilon_p(z, \omega)$ in Equation 2.15 be given by

$$\varepsilon_p(z, \omega) = \varepsilon_1, \quad -\infty < z < 0, \tag{P2.1}$$

$$\varepsilon_p(z, \omega) = \varepsilon_1 + (\varepsilon_2 - \varepsilon_1)\frac{z}{L}, \quad 0 < z < L, \tag{P2.2}$$

$$\varepsilon_p(z, \omega) = \varepsilon_2, \quad L < z < \infty. \tag{P2.3}$$

Show that the solution of Equation 2.15 in the two regions given below may be written as

$$E = Ie^{-jk_1 z} + Re^{+jk_1 z}, \quad -\infty < z < 0, \tag{P2.4}$$

$$E = Te^{-jk_2 z} + Ae^{jk_2 z}, \quad L < z < \infty, \tag{P2.5}$$

where I, R, T, and A are constants and $k_1 = k_0\sqrt{\varepsilon_1}$, $k_2 = k_0\sqrt{\varepsilon_2}$. Give physical reasons for setting $A = 0$. Note that if $I = 1$, R and T give the reflection and transmission coefficients. Assume that the waves are propagating along the z-axis.

P2.2 In this problem we investigate the nature of the solution of Equation 2.15 in the domain $0 < z < L$.

By introducing a new variable $\xi = k_0^2 \varepsilon_P(z, \omega)$, show that Equation 2.15 may be transformed to

$$\frac{d^2 E}{d\xi^2} + \frac{\xi E}{\beta^2} = 0, \tag{P2.6}$$

where

$$\beta = k_0^2 \frac{\varepsilon_2 - \varepsilon_1}{L}. \tag{P2.7}$$

Note that the solution of Equation P2.6 may be written in terms of Airy function Ai and Bi

$$E = c_1 Ai[-\beta^{-2/3}\xi] + c_2 Bi[-\beta^{2/3}\xi], \quad 0 < z < L. \tag{P2.8}$$

P2.3 Determine the reflection coefficient R and plot the power reflection coefficient $|R|^2$ versus L. Take $\varepsilon_1 = (4/3)^2$ and $\varepsilon_2 = 1$. These data approximate the water–air interface at optical frequencies. Assume that the dielectric function is a linear profile given in Equations P2.1 through P2.3.

P2.4 Find an analytical expression for $|R|^2$ if $L = 0$, in P2.3.

P2.5 Plot $|R|^2$ versus θ_i for the data of P2.4 if θ_i is the angle of incidence.

P2.6 Let $z < 0$ be free space and $z > 0$ be a dispersive medium. For the data given in P1.1, find ρ and τ. Find the absorptance defined by $A = 1 - \rho - \tau$.

Justify the definition. Plot A versus f, ρ versus f, and τ versus f in the frequency band $0 < f < 90$ GHz. Assume normal incidence.

P2.7 Let $z > 0$ be water and $z < 0$ be free space. Using the data of P1.2, plot A versus f, ρ versus f, and τ versus f. Assume normal incidence.

P2.8 Let $z > 0$ be the second-order Lorentz dispersive medium and $z < 0$ be free space. Using the data of P1.3, plot A versus ω, ρ versus ω, and τ versus ω in the frequency band $2 \times 10^{16} < \omega < 6 \times 10^{16}$. Comment on the values of ρ, τ, and A around the frequency $\omega \approx 4 \times 10^{16}$/s. Repeat the above calculations for $\delta = 1.4 \times 10^{16}$ s. What do you infer from the two sets of results? Assume normal incidence.

P2.9 Discuss the possibility of having a TE surface wave mode at a metal–vacuum interface.

P2.10 Light of wavelength $0.633\,\mu$m is incident from air on to an air–aluminum interface. Take the dielectric constant of aluminum at $0.633\,\mu$m as $\varepsilon_r = -60.56 - j24.86$.

Plot the reflectivity (ρ) versus θ_i of an s wave where θ_i is the angle of incidence.

P2.11 The half-space $z > 0$ is free space and the half-space $z < 0$ is metal of plasma frequency $\omega_p = 8.2 \times 10^{15}$ rad/s. Find the reflectivity of a p wave in free space incident on the metal at an angle of incidence of $45°$, when

$$\omega = 8.2 \times 10^{15} \text{ rad/s},$$

$$\omega = \sqrt{2}(1.1)8.2 \times 10^{15} \text{ rad/s}.$$

P2.12 The interface $z = 0$ is as described in P2.11. Investigate the propagation of a surface plasmon (TM) on this interface at $\omega = 4 \times 10^{15}$ rad/s

a. Determine the phase velocity of the surface wave on the interface.
b. Determine the distance, in meters, in which the z-component of the electric field in free space reaches $0.3679(= 1/e)$.
c. Determine the distance, in meters, in which the z-component of the electric field in metal reaches $0.3679(= 1/e)$ of its value at the interface.

P2.13 The interface $z = 0$ is as described in P2.11. Investigate the propagation of a surface plasmon (TM) on this interface at $\omega = 9.84 \times 10^{15}$ rad/s

a. Determine the phase velocity of the surface wave on the interface.
b. Determine the distance, in meters, in which the z-component of the electric field in free space reaches 0.3679 $(= 1/e)$ of its value at the interface.

c. Determine the distance, in meters, in which the z-component of the electric field in metal reaches 0.3679 (= $1/e$) of its value at the interface.

P2.14 Derive the dispersion relation and draw the ω–β diagram of an unbounded periodic medium consisting of alternating layers of dielectric layers. Assume the following values for the parameters of the layers in a unit cell

$$\text{First layer}: \varepsilon_1 = 2\varepsilon_0, \quad \mu_1 = \mu_0, \quad L_1 = L_0 = 3\,\text{cm}$$
$$\text{Second layer}: \varepsilon_2 = 3\varepsilon_0, \quad \mu_2 = \mu_0, \quad L_2 = 2L_0$$

Determine the location of the center of the first stop band. Also, determine the bandwidth of this stop band.

P3.1

a. Equations 3.25 through 3.27 specify the initial conditions for the suddenly created plasma. Convert these initial conditions to the initial conditions on H, $\dot{H}$, and $\ddot{H}$.

b. Solve Equation 3.13 with ω_p^2 being a constant and show that Equation 3.15 will be the solution of the higher-order differential equation in H. Determine ω_m and H_m in Equation 3.15.

P3.2 Let ω_1 be the frequency of a propagating wave in a plasma medium of plasma frequency ω_{p1}. At $t = 0$, the electron density suddenly drops. Let $\omega_p = \omega_{p2}$ for $t > 0$, where $\omega_{p2} < \omega_{p1}$. Find the new frequencies and fields of all the modes generated by the switching action.

P3.3 Let the plasma frequency be a step profile in space, that is,

$$\omega_p^2(z, t) = \omega_{p1}^2, \quad z < 0,$$
$$\omega_p^2(z, t) = \omega_{p2}^2, \quad z > 0.$$

a. Find the Green's function for this problem of spatially-unlike media.

b. Develop a perturbation technique for this problem on the lines of Section 3.5 where $\omega_p^2(z)$ is a fast profile with a scale length L.

c. Consider a linear profile with a slope of $\left(\omega_{p2}^2 - \omega_{p1}^2 \right)/L$. Determine the reflection coefficient for this profile to the order of L^2, that is, you may neglect terms in L^3 and higher power of L. Assume that the source wave has a frequency ω_1 and wave number k_1.

P3.4 Solve the problem specified in P3.3 using the exact solution on the lines of Section 3.9 and express your solution in terms of Airy functions. Use any mathematical software to plot the power reflection coefficient ρ versus the scale length L. Compare this result with the result obtained in P3.3c.

P4.1

a. Show that the Laplace transform of the electric field of the reflected field when the suddenly created plasma is lossy with a collision frequency ν is given by

$$\frac{E_{yr}(0,s)}{E_0} = \frac{N_1 N_2}{D_1 D_2},\tag{P4.1}$$

$$N_1(s) = \omega_0^2 A - s^2 \sqrt{A^2 + \omega_p^2},\tag{P4.2}$$

$$N_2(s) = \sqrt{A^2 + \omega_p^2} - A,\tag{P4.3}$$

$$D_1(s) = s^2 + \omega_0^2,\tag{P4.4}$$

$$D_2(s) = s^3 + \nu s^2 + s(s^2 + \omega_p^2) + \nu \omega_0^2,\tag{P4.5}$$

$$A(s) = \sqrt{s(s+\nu)}.\tag{P4.6}$$

b. For the case of a low-loss plasma ($\nu/\omega_p \gg 1$), show that $D_2(s)$ may be factored as

$$D_2(s) = \left[s + \frac{\nu \omega_0^2}{\omega^2} \right] \left[(s + \alpha_1)^2 + \omega^2 \right],\tag{P4.7}$$

where

$$\alpha_1 = \frac{\nu}{2} \frac{\omega_p^2}{\omega^2},\tag{P4.8}$$

$$\omega^2 = \omega_0^2 + \omega_p^2.\tag{P4.9}$$

c. From the results of (b) deduce that the damping time constant t_{pB} of the B wave is

$$t_{pB} = \frac{2\omega^2}{\nu \omega_p^2},\tag{P4.10}$$

and the damping time constant of the wiggler magnetic field is

$$t_w = \frac{\omega^2}{\nu \omega_0^2}.\tag{P4.11}$$

P4.2 Let $f_0 = 10$ GHz. Find N_0 if the frequency of the upshifted wave is 23 GHz. If $v/\omega_p = 0.01$, find the collision frequency. Calculate the damping constants t_p and z_p.

P4.3 For the data given in P4.2, find the value of z_1 in Figure 4.2 if the negative-going B wave attenuates to 50% of the original value at z_1 by the time it reaches the interface.

P5.1 Show that the sum of $R_A(s)$ and $R_B(s)$ given by Equations 5.32 and 5.33 equals $(A_{1R}(s)/E_0)$ given by Equation 4.42.

P5.2 Show that in the limit $d \to \infty$, Equation 5.24 becomes Equation 5.32 and Equation 5.25 becomes Equation 5.33.

P5.3 For the data given in P4.2, find the minimum slab width d, if the strength of the A wave, in the region $z > d$, is negligible (less than 1% of the strength of the incident wave).

P6.1 Derive Equation 6.25.

P6.2 Show that Equations 6.25 and 6.28 are the same.

P6.3 Given $f = 2.8$ GHz and $B_0 = 0.3$ T, find the electron density of the plasma at which the L wave has a cutoff.

P6.4 Faraday rotation of an 8-mm-wavelength microwave beam in a uniform plasma in a 0.1 T magnetic field is measured. The plane of polarization is found to be rotated by $90°$ after traversing 1 m of plasma. What is the electron density in the plasma?

P6.5 Derive Equations 6.52 and 6.54.

P6.6 Derive Equations 6.56, 6.57, and 6.58.

P6.7

a. For a whistler mode, that is, $\omega \ll \omega_p$, $\omega \ll \omega_b$, show that the group velocity is given approximately by

$$v_g \sim \frac{\sqrt{\omega \omega_b}}{\omega_p}. \tag{P6.1}$$

b. Assume that due to lighting, signals in the frequency range of 1–10 kHz are generated. If these signals are guided from one hemisphere to the other hemisphere of the earth along the earth's magnetic field over a path length of 5000 km, calculate the travel time of the signals as a function of frequency. Take the following representative values for the parameters: $f_p = 0.5$ MHz and $f_b = 1.5$ MHz.

P6.8 A right circularly polarized electromagnetic wave of frequency $f_0 = 60$ Hz is propagating in the z-direction in the metal potassium, at a very low temperature of a few degrees Kelvin, in the presence of a z-directed static magnetic flux density field B_0.

Determine (a) the phase velocity and (b) the wavelength of the electromagnetic wave in the metal.

Assume that electron density N_0 of the metal is such that $\omega_p = 10^{16}$ rad/s.

Assume that B_0 value is such that the electron gyrofrequency $\omega_p = 10^{12}$ rad/s.

Assume that the collision frequency v is negligible.

The calculations become simple if approximations are made based on the inequalities mentioned above. Relate the specification of a very low temperature to one of the assumptions given above.

P6.9   A right circularly polarized wave is propagating in the z-direction in a semiconductor in the presence of a z-directed B_0 field. Assuming that $\omega_b \gg \omega_p$ and $v/\omega_b \ll 1$, investigate the wave propagation near cyclotron resonance $\omega = \omega_b$, that is, find the expression for $k(\omega)$, where k is a complex wave number.

Plot $(\operatorname{Im} k(\omega)/\operatorname{Im} k(\omega_b))$ versus ω/ω_b for two values of the parameter, (a) $v/\omega_b = 0.1$ and (b) $v/\omega_b = 0.01$.

Assume that $\omega_b = 10^{12}$ rad/s and $\omega_p = 10^{10}$ rad/s for computation.

Discuss the resonance characteristic of the response indicated by the results of (a) and (b).

Suppose this sample is put in a waveguide and we launch an electromagnetic wave frequency ω (R wave). Then we apply a B_0 field in the z-direction and measure the output of the electromagnetic wave, while the strength of B_0 is varied. For what value of ω will the output be minimal?

P6.10   An electromagnetic wave of frequency ω and wave number k is propagating in a magnetoplasma medium at an angle θ to the direction of the static magnetic field. Determine the dispersion relation and sketch the ω–k diagram. Identify the cutoff frequencies and the resonant frequencies.

P6.11   A p wave in free space of frequency ω_0 is obliquely incident on a magnetoplasma half-space. Le $z < 0$ be free space and $z > 0$ be the magnetoplasma half-space and let the static magnetic field be in the z-direction. Let the angle of incidence with the z-axis be θ_I. Find the power reflection coefficient ρ. Express your answer in terms of θ_I, ω_b, ω_p, and ω_0. Examine the existence of the Brewster angle for this problem.

P6.12   Study the dispersion relation of a surface magnetoplasmon.

P7.1   Show that the wave impedance given by Equation 7.20 is the same as that given by Equation 7.30.

P7.2   The source frequency f_0 in free space is 10 GHz. The frequency upshift ratio of the R1 wave is 1.5. If $\omega_b = \omega_p$ find N_0 and B_0. Find the damping constants t_p and z_p if $v/\omega_p = 0.01$.

P7.3 Repeat the calculation of P7.2 for L1, X1, and X2 waves (see Appendix B for the R and L waves and Appendix C for the X waves).

P7.4 Let $\omega_b = 0$. Repeat the calculation of P7.2 for the O wave.

P7.5 The frequency downshift ratio is given as 0.4. Repeat the calculations of P7.2 for the R3 wave.

P7.6 The frequency downshift ratio is given as 0.7. Repeat the calculations of P7.2 for the X2 wave (see Appendix C for the X wave).

P7.7 A perpendicularly polarized wave (s wave) of frequency ω_0 and wave number k is propagating in free space in the presence of a static magnetic field $\mathbf{B}_0$, which is in the z-direction. The angle between $\mathbf{B}_0$ and the wave vector $\mathbf{k}$ is θ_I. At $t = 0$ an unbounded plasma medium of plasma frequency ω_p is created. Find the frequencies and the fields of the newly created wave modes. Assume that the plasma is lossless.

P8.1 Describe the differential equation for longitudinal propagation of an R wave in an inhomogeneous magnetoplasma with a space-varying profile $\omega_p^2(z)$. The gyrofrequency ω_b is constant.

P8.2 Find the reflection and transmission coefficients if $\omega_p^2(z)$ is a spatial step profile $\tilde{\omega}_p^2(z)$. The step change is from ω_{p1} to ω_{p2} at $z = 0$.

P8.3 Find the Green's function for longitudinal propagation of an R wave in spatially-unlike media. Medium 1 has a plasma frequency ω_{p1} and medium 2 has a plasma frequency ω_{p2}. The gyrofrequency is ω_b.

P8.4 Consider a linear profile for the electron density, that is,

$$\omega_p^2(z) = \begin{cases} 0, & z < -L/2 \\ \dfrac{\omega_p^2}{2} + \dfrac{\omega_p^2}{L}z, & -L/2 < z < L/2 \\ \omega_p^2, & z > L/2 \end{cases} \tag{P8.1}$$

Calculate the power reflection coefficient of a source wave of frequency ω_1 and wave number k_1. Assume that the profile is a fast profile so that terms in L^3 and higher may be neglected.

P9.1 Perform the adiabatic analysis of longitudinal propagation of an R wave in a magnetized plasma with spatially varying electron density profile $\omega_p^2(z)$.

Find $E_m(z)/E_0$ when $E_m(z)$ is the amplitude of the modified source wave and E_0 is the amplitude at $z = 0$.

P9.2 An R wave of frequency ω_0 is propagating in anisotropic plasma of plasma frequency ω_p. At $t = 0$, a slowly varying nonoscillatory magnetic field in the direction of propagation is created, that is,

$$\omega_b = 0, \quad t < 0, \tag{P9.1}$$

$$\omega_b = \omega_b(t), \quad t > 0, \tag{P9.2}$$

where ω_b varies slowly with time. Apply the adiabatic analysis to determine the modification of the frequency and the magnetic field of the source wave.

P10.1   An O wave of frequency ω_0 is propagating in free space. At $t = 0$ a Lorentz medium discussed in Section 10.5 is suddenly created. Determine the frequencies and the fields of the frequency-shifted waves created by the switching action.

P10.2   Consider that a Lorentz medium is created suddenly in the presence of a static magnetic field B_0 in the direction of propagation of an R source wave of frequency ω_0. Determine the frequency and the fields of the frequency-shifted waves created by the switching action.

P10.3   A source wave of frequency 10 GHz is propagating in free space. At $t = 0$, an unbounded periodic medium of plasma layers shown in Figure 2.1 is created. If $L = 1.8$ cm, $d = 0.6$ cm, and $f_{p0} = 12$ GHz, determine the frequencies of the first three modes in the increasing order of frequencies (positive numbers). Identify the downshifted frequency mode.

P10.4   Show that R and L waves are the natural modes of propagation in a chiral medium. Find the wave number and the wave impedance of the (a) R wave and (b) L wave.

P10.5

a. Draw a ω–k diagram for an unbounded periodic medium consisting of alternating layers of chiral media. Assume the following values for the parameters of the layers in a unit cell:

First layer: $\varepsilon_1 = 2\varepsilon_0$, $\mu_1 = \mu_0$, $\xi_{c1} = 10^{-3}$, $L_1 = L_0 = 3$ cm

Second layer: $\varepsilon_2 = 3\varepsilon_0$, $\mu_2 = \mu_0$, $\xi_{c2} = -4\xi_{c1}$, $L_2 = 2L_0$

b. Determine the location of the center of the first stop band. Also, determine the bandwidth of this stop band.

c. The same data as above but $\xi_{c1} = 0$, that is, the layers are not chiral. Determine the first stop band and its width.

Index

Printed and bound by CPI Group (UK) Ltd, Croydon, CR0 4YY

21/10/2024

01777044-0016